Operator Theory
Advances and Applications
Vol. 112

Editor:
I. Gohberg

Asymptotic Theory of Elliptic Boundary Value Problems in Singularly Perturbed Domains

Volume II

Vladimir Maz'ya
Serguei Nazarov
Boris Plamenevskij

Translated from the German by Boris Plamenevskij

Springer Basel AG

Authors:

Vladimir Maz'ya
Department of Mathematics
Linköping University
58183 Linköping
Sweden
e-mail: vlmaz@mai.liu.se

Serguei Nazarov
Laboratory of Mathematical Methods in Mechanics of Solids
Institute of Mathematics and Mechanics
St. Petersburg University
Bibliotechnaya pl. 2
198904 St. Petersburg
Russia
e-mail: serna@snark.ipme.ru

Boris A. Plamenevskij
Department of Mathematical Physics
Faculty of Physics
St. Petersburg State University
Ulyanova 1
Stary Petershof
198904 St. Petersburg
Russia
e-mail: Boris.Plamenevskij@pobox.spbu.ru

Originally published by Akademie Verlag GmbH, Leipzig, Germany, under the title "Asymptotische Theorie Elliptischer Randwertaufgaben in singulär gestörten Gebieten".

2000 Mathematics Subject Classification 35B25, 35J25; 73B27, 35B40, 73C02, 35J40

A CIP catalogue record for this book is available from the
Library of Congress, Washington D.C., USA

Deutsche Bibliothek Cataloging-in-Publication Data
Maz'ja, Vladimir G.:
Asymptotic theory of elliptic boundary value problems in singularly pertubed domains / Vladimir Maz'ya ; Serguei Nazarov ; Boris Plamenevskij. Transl. from the German by Boris Plamenevskij
– Basel ; Boston ; Berlin : Birkhäuser
Einheitssacht.: Asimptotika rešenij elliptičeskich kraevych zadač pri singuljarnych vozmuščenijach oblasti <dt.>
ISBN 978-3-0348-9564-4 ISBN 978-3-0348-8432-7 (eBook)
DOI 10.1007/978-3-0348-8432-7

Vol. 2. – (2000)
(Operator theory ; Vol. 112)
ISBN 978-3-0348-9564-4

Originally published by Birkhäuser Verlag, Basel - Boston - Berlin in 2000
Softcover reprint of the hardcover 1st edition 2000

Printed on acid-free paper produced from chlorine-free pulp. TCF ∞
Cover design: Heinz Hiltbrunner, Basel

ISBN 978-3-0348-9564-4

9 8 7 6 5 4 3 2 1

Contents

Volume I

Part II
General Elliptic Boundary Value Problems in Domains Perturbed Near Isolated Singularities of the Boundary

Volume II

Part VI
Behaviour of Solutions of Boundary Value Problems in Thin Domains

Chapter 14 The Dirichlet Problem in Domains with Thin Ligaments

Chapter 15 Boundary Value Problems of Mathematical Physics in Thin Domains

Preface

This book is devoted to the development and applications of asymptotic methods to boundary value problems for elliptic equations in "singularly" perturbed domains (rounded corners and edges, small holes, small slits, thin ligaments etc.). The boundary value problems will be considered in a domain $\Omega(\varepsilon)$ that depends on a small parameter ε, where the boundary of the domain $\Omega(0)$ is not smooth and contains a number of singular points, contours, or surfaces. A transition from $\Omega(0)$ to $\Omega(\varepsilon)$, results in the fact that isolated points convert into small cavities, contours convert into thin tubes and surfaces into flat holes, or the boundary of the domain near a conical point or an edge becomes smooth, and so on. These perturbations of the domain are said to be singular, in contrast to regular perturbations, when the boundaries of domains $\Omega(0)$ and $\Omega(\varepsilon)$ are closed smooth surfaces. We investigate the behaviour of solution u_ε of the boundary problem, eigenvalues of the correponding operator, and the behaviour of different functionals (like energy, capacity, etc.) as $\varepsilon \to 0$.

The asymptotic theory of boundary value problems in singularly perturbed domains has turned out to be very useful in numerical methods. Dependence of solution on small and large parameters can be taken into consideration, and the originally complicated problem can be decomposed into several simpler subproblems.

Problems considered here emerged from problems in hydrodynamics and aerodynamics, the theory of elasticity, fracture mechanics, electrostatics and others. A substantial body of results has been accumulated on the applications of asymptotic methods to physical problems. This knowledge has been particularly useful in a broad range of engineering problems. Systematic presentations can be found in, among others, the monographs of VAN DYKE [1], COLE [1], NAYFEH [1], [2], and CHEREPANOV [1], [2].

In the present book a general approach to construction of asymptotics of solutions of elliptic boundary value problems in singularly perturbed domains will be developed. Starting from the original problems, certain "limit" problems will be derived that do not depend on ε and whose solutions enter asymptotic expansions as coefficients. For these problems, statements concerning solvability in special classes of functions and asymptotic behaviour of solutions in a neighbourhood of singularities or infinity are required. For this purpose the classical theory of elliptic boundary value problems in domains with a smooth boundary is not sufficient. During the last twenty years, the theory of general elliptic boundary value problems in domains with non-smooth boundaries has been significantly developed (KONDRATYEV [1], MAZ'YA/PLAMENEVSKI [1]–[7] and others). Thus we are able, in this book, to treat in the same manner the problems of asymptotics of solutions in domains with different types of perturbations.

We use here, in essence, a modification of the method of compound asymptotic expansions. A characteristic feature of this method is that all limit problems are solved in the same function space, which does not depend on the corresponding iteration step, so that singularities of their solutions do not become stronger during successive constructions of terms in asymptotic series. For this purpose we systematically apply the so-called "method of redistribution of discrepancies" between different limit problems. This approach for construction of complete asymptotic expansions is, in our opinion, simple and universal.

In this book, a lot of attention is paid to particular problems of mathematical physics. They serve as illustrations of general algorithms, but they are also independent subjects of study. Most of applications emerge from the theory of elasticity (torsion problems, planar and three-dimensional problems for the Lamé system, theory of thin plates, crack and fracture mechanics etc.).

Most of the material presented here is based on results of the authors and has been partly published in scientific journals. The book does not have an essential overlap with other monographs dedicated to the theory of elliptic boundary value problems. The two volumes of the work are divided into parts, and these parts into chapters. The first volume contains parts I–IV, in which boundary value problems with perturbations near isolated singularities of the boundary of the domain are studied. The second volume contains parts V–VII, which deal with other kinds of perturbations (problems with perturbations of the boundary of singular manifolds, problems in thin domains, and problems with rapid oscillations of the boundary of domain or coefficients of differential operators). The first part has an introductory character. In two chapters, examples of the Dirichlet and the Neumann problems for the Laplace operator, basic features of the general method for construction of asymptotics of solutions of problems in domains with singularly perturbed boundaries is explained. First, results concerning the solvability of boundary value problems for the Laplace operator in domains with conical (or corner) points, and concerning the behaviour of solutions near singularities, are presented. These results are used in the second chapter to discuss our main theme for the examples just mentioned, namely expansion of the solutions in an asymptotic series of powers of the small parameter, which represents a perturbation of the boundary.

In the second part, the method of compound expansions, which was described for examples in the second chapter, is developed for application to general boundary value problems. The third chapter contains a survey of results concerning solvability and properties of the solution of general elliptic problems in domains with conical points. The asymptotic method itself is presented in Chapter 4, whereas examples, modifications and consequences are described in Chapter 5. Results and methods of the second part are used throughout the whole book.

The third and fourth parts (Chapter 6–10) deal with expansion of certain functionals over solutions of boundary value problems (stress intensity factors, energy) and eigenvalues in the asymptotic series in the small parameter. These two parts of the book are closely connected with the mechanics of solids.

While in the first volume perturbations of the boundary near isolated singular point are the centre of attention, in Part V, which opens the second volume, analogous problems for perturbations near singular submanifolds are discussed. Chapter 11, with results from the theory of boundary value problems in domains with edges, provides a basis for construction of the asymptotics. Chapter 12 is dedicated to particular equations, and in Chapter 13 the theory for elliptic equations of higher order is developed in application to the Dirichlet problem.

Behaviour of the solutions of boundary value problems in thin domains, dependent on the thickness ε of the domain, is investigated in Part VI. Chapter 14 is dedicated to the Dirichlet problem for the Laplace operator in a domain with a slender ligament (in particular, the condensator problem). Chapter 15 is devoted to a number of particular problems of mathematical physics, and Chapter 16 to general elliptic problems.

The last part contains solutions of four different asymptotic problems, for which averaging of one or another asymptotic structure is the common feature. In Chapter 17 construction of the complete asymptotic expansion of solutions of boundary value problems for second order elliptic equations with periodic, rapidly oscillating (both in the interior of the domain and up to a planar boundary) coefficients will be considered. In the subsequent chapter, asymptotic effects that appear in certain types of rapidly oscillating perturbations of the boundary of the domain will be studied. Chapters 19 and 20 are dedicated to the asymptotics of the solutions of ordinary differential and difference equations. In Chapter 19 equations on a fine periodic grid and in Chapter 20 on a discrete grid are investigated.

The reader will find in the introductory remarks to each chapter more detailed information about the contents of the book.

Acknowledgements
We are grateful to Jan Åslund and Nikolay Kuznetsov for reading the manuscript and valuable comments.

Part V

Boundary Value Problems in Domains Perturbed Near Multidimensional Singularities of the Boundary

Chapter 11

Boundary Value Problems in Domains with Edges on the Boundary

In this chapter we begin to study the asymptotics of solutions to elliptic problems in domains perturbed near multidimensional singularities of the boundary. As such singularities, one takes edges of various dimensions on the boundary of an n-dimensional domain Ω or smooth d-dimensional submanifolds within Ω, $1 \leq d \leq n-2$.

As before, when studying the singularly perturbed boundary value problems, we make use of results on elliptic problems in domains with piecewise smooth boundary. The chapter is devoted to presenting these results. Thus Chapter 11 is related to Chapters 12, 13 in the same manner as Chapters 1, 3 to Chapters 2, 4–10.

Some problems for the Laplace operator in three-dimensional domains are discussed in detail in Section 1–3. We prove solvability theorems in weighted spaces and derive the asymptotic expansions for solutions near singularities (one-dimensional edges and contours within the domain). Section 4 contains the statements of similar results on the Sobolev problems for a scalar strongly elliptic operator of order $2m$.

11.1 The Dirichlet Problem for the Laplace Operator

11.1.1 Statement of the problem

Let Ω be a domain in $\mathbb{R}^3$ with compact closure $\bar{\Omega}$ and boundary $\partial\Omega$, and let M be a simple smooth closed curve in Ω. We assume that $\partial\Omega \setminus M$ is a $\mathbf{C}^\infty$-manifold and in a neighborhood of every point $P \in M$ the set $\bar{\Omega}$ is diffeomorphic to a wedge. Thus, the two tangent planes to $\partial\Omega$ are determined at every point P. Suppose that the angle between them (as viewed from Ω) does not depend on $P \in M$ and is equal to $\vartheta_0 \in (0, 2\pi)$.

Let $\rho(x)$ be the distance from $x \in \Omega$ to M, $\beta \in \mathbb{R}$, and $l = 0, 1$. We introduce the space $\mathbf{V}^l_{2,\beta}(\Omega; M)$ of functions in Ω endowed with the norm

$$\|u; \mathbf{V}^l_{2,\beta}(\Omega; M)\| = \Big(\sum_{|\alpha|\leq l} \|\rho^{\beta-l+|\alpha|} D^\alpha_x v; \mathbf{L}_2(\Omega)\|^2 \Big)^{1/2}. \tag{1}$$

Denote by $\mathbf{V}^{l-1/2}_{2,\beta}(\partial\Omega; M)$ with $l = 1, 2, \dots$ the space of traces on $\partial\Omega \setminus M$ of functions in $\mathbf{V}^l_{2,\beta}(\Omega; M)$, while

$$\|v; \mathbf{V}^{l-1/2}_{2,\beta}(\partial\Omega;\, M)\| = \inf\{\|u; \mathbf{V}^l_{2,\beta}(\Omega;\, M\| : u(x) = v(x) \quad \text{at} \quad \partial\Omega \setminus M\}. \tag{2}$$

We study the question of solvability of the problem

$$-\Delta u(x) = f(x), \quad x \in \Omega; \qquad u(x) = \varphi(x), \quad x \in \partial\Omega \setminus M, \tag{3}$$

in $\mathbf{V}^{l+2}_{2,\beta}(\Omega;\, M)$ for $f \in \mathbf{V}^l_{2,\beta}(\Omega;\, M)$, $\varphi \in \mathbf{V}^{l+3/2}_{2,\beta}(\partial\Omega;\, M)$, $\quad l = 0, 1, \dots$.

One can verify the next assertion in the same way as Proposition 1.3.2.

Proposition 11.1.1. *Let* $\varphi \in \mathbf{V}^{1/2}_{2,0}(\partial\Omega;\, M)$ *and* $f \in \mathbf{V}^{0}_{2,1}(\Omega;\, M)$*. Then there exists a unique (generalized) solution* $u \in \mathbf{V}^{1}_{2,0}(\Omega;\, M)$ *of the problem* (3) *and*

$$\|u;\, \mathbf{V}^{1}_{2,0}(\Omega;\, M)\| \le c\big(\|f;\, \mathbf{V}^{0}_{2,1}(\Omega;\, M)\| + \|\varphi;\, \mathbf{V}^{1/2}_{2,0}(\partial\Omega;\, M)\|\big). \tag{4}$$

Thus, the question of solvability of problem (3) reduces to analyzing the differential properties of the generalized solution for a right-hand side with improved properties.

11.1.2 Model problem in a wedge

As usual, the basic point is the consideration of the model problem (cf. Section 1.2 and 2.3).

Let (y_1, y_2) be Cartesian coordinates and (r, ϑ) the corresponding polar coordinates in $\mathbb{R}^2$ and $K = \{y \in \mathbb{R}^2 \colon 0 < r < \infty, |\vartheta| < \vartheta_o/2\}$. We denote by D the wedge $K \times \mathbb{R} = \{x = (y, z) \colon y \in K,\ z \in \mathbb{R}\}$ with edge Γ. Introduce the space $\mathbf{V}^{l}_{2,\beta}(D; \Gamma)$ of functions in D with norm (1) where Ω, M, ρ are replaced by D, Γ, r. The notation $\mathbf{V}^{l-1/2}_{2,\beta}(\partial D;\, \Gamma)$ has similar meaning (see (2)).

We seek $u \in \mathbf{V}^{l+2}_{2,\beta}(D;\, \Gamma)$ satisfying

$$-\Delta u(x) = f(x), \quad x \in D; \qquad u(x) = \varphi(x), \quad x \in \partial D \setminus \Gamma, \tag{5}$$

with $f \in \mathbf{V}^{l}_{2,\beta}(D;\, \Gamma)$, $\varphi \in \mathbf{V}^{l+3/2}_{2,\beta}(\partial D;\, \Gamma)$. We apply the Fourier transform in z to equations (1.5), denote the dual variable by ξ, and set

$$\begin{aligned} &\eta = |\xi| y, \quad U(\eta, \xi) = \hat{u}(|\xi|^{-1}\eta, \xi), \\ &F(\eta, \xi) = |\xi|^{-2} \hat{f}(|\xi|^{-1}\eta, \xi), \\ &\Phi(\eta, \xi) = \hat{\varphi}(|\xi|^{-1}\eta, \xi), \end{aligned}$$

where $\hat{v}$ is the Fourier transform of v. Then we arrive at the problem

$$-\Delta_\eta U + U = F \quad \text{in} \quad K, \qquad U = \Phi \quad \text{at} \quad \partial K \setminus \{O\}. \tag{6}$$

Let $\mathbf{E}^{l}_{2,\beta}(K), l \in \mathbb{N}_0$ be the space of functions in K with norm

$$\|W;\, \mathbf{E}^{l}_{2,\beta}(K)\| = \Big(\sum_{|\gamma| \le l} \|r^\beta (1 + r^{|\gamma| - l})\, D^\gamma_y W;\, \mathbf{L}_2(K)\|^2\Big)^{1/2}, \tag{7}$$

and $\mathbf{E}^{l-1/2}_{2,\beta}(\partial K), l \in \mathbb{N}$ the corresponding space of traces on $\partial K \setminus \{O\}$.

Lemma 11.1.2. *The norm* $\|w;\, \mathbf{V}^{l}_{2,\beta}(D;\, \Gamma)\|$ *is equivalent to the norm*

$$\Big(\int_{\mathbb{R}} |\xi|^{2(l-\beta-1)} \|W(\cdot, \xi);\, \mathbf{E}^{l}_{2,\beta}(K)\|^2 d\xi\Big)^{1/2}$$

with $W(\eta, \xi) = \hat{w}(|\xi|^{-1}\eta, \xi)$.

Proof. Let us apply the Parseval equality and then pass into the variables η:

$$\sum_{|\alpha|\le l}\int_D |y|^{2(\beta-l+|\alpha|)}|D_x^\alpha v|^2\,dx$$
$$=\sum_{|\alpha|\le l}\int_{\mathbb{R}}|\xi|^2\int_K |y|^{2(\beta-l+|\alpha|)}\,|D_y^{\alpha'}\hat v(y,\xi)|^2\,dy\,d\xi$$
$$=\sum_{|\alpha|\le l}\int_{\mathbb{R}}|\xi|^{2(l-\beta-1)}\sum_{|\alpha|\le l}\int_K |\eta|^{2(\beta-l+|\alpha|)}\,|D_\eta^{\alpha'}W(\eta,\xi)|^2 d\eta d\xi$$
$$=\int_{\mathbb{R}}|\xi|^{2(l-\beta-1)}\sum_{|\alpha'|\le l}\int_K\left[\sum_{\alpha_3=0}^{l-|\alpha'|}|\eta|^{2(|\alpha|-l)}\right]|\eta|^{2\beta}\,|D_\eta^{\alpha'}W(\eta,\xi)|^2 d\eta d\xi.$$

To conclude the proof we note that the expression $\Xi(\eta)$ in the braces satisfies

$$(1+|\eta|^{|\alpha'|-l})^2/2\le\Xi(\eta)\le l\big(1+|\eta|^{|\alpha'|-l}\big)^2. \qquad \square$$

According to Lemma 1.2, the solvability of problem (6) in $\mathbf{E}^{l+2}_{2,\beta}(K)$ and the estimate

$$||U;\,\mathbf{E}^{l+2}_{2,\beta}(K)||\le c\left(||F;\,\mathbf{E}^{l}_{2,\beta}(K)||+||\Phi;\,\mathbf{E}^{l+3/2}_{2,\beta}(\partial K)||\right) \tag{8}$$

imply the solvability of problem (5) in $\mathbf{V}^{l+2}_{2,\beta}(D;\,\Gamma)$ and the estimate

$$||u;\,\mathbf{V}^{l+2}_{2,\beta}(D;\,\Gamma)||\le c\big(||f;\,\mathbf{V}^{l}_{2,\beta}(D;\,\Gamma)||+||\varphi;\,\mathbf{V}^{l+3/2}_{2,\beta}(\partial D;\,\Gamma)||\big). \tag{9}$$

Proposition 11.1.3. *Let l be a nonnegative integer, $F\in\mathbf{E}^{l}_{2,\beta}(K)$, $\Phi\in\mathbf{E}^{l+3/2}_{2,\beta}(\partial K)$. Then the inequality $|\beta-l-1|<\pi/\vartheta_0$ is necessary and sufficient for the problem (6) to be solvable.*

Proof. Without loss of generality we can assume $\Phi=0$. Since for $\beta-l-1>-\pi/\vartheta_0$ the Dirichlet problem for the operator $\Delta-1$ in a bounded domain G with angular point is solvable in $\mathbf{V}^{l+2}_{2,\beta}(G)$ (see Proposition 1.3.18 and Theorem 3.3.1), it suffices to consider (6) with F equal to zero near the vertex of K. Indeed, let F_1 denote a function in $\mathbf{V}^{l+2}_{2,\beta}(G)$, where $\{x\in K\colon |x|<1\}\subset G\subset K$ and $F_1=F$ near the point O. We assume that $U_1\in\mathbf{V}^{l+2}_{2,\beta}(G)$ satisfies

$$-\Delta U_1+U_1=F_1\quad\text{in}\quad G;\qquad U_1=0\quad\text{on}\quad\partial G.$$

The function U_1 is subject to the inequality

$$||U_1;\,\mathbf{V}^{l+2}_{2,\beta}(G)||\le c\,||F;\,\mathbf{V}^{l}_{2,\beta}(G)||. \tag{10}$$

Then for the difference $U-\zeta_1U_1$, where $\zeta_1\in\mathbf{C}_0^\infty(\bar K)$, $\zeta_1=1$ near 0, and $\zeta_1=0$ outside G, we obtain the problem with the right-hand side $F-\zeta_1F_1+[\Delta,\,\zeta_1]\,U_1$, which equals zero in a neighborhood of the vertex of K. In what follows $U-\zeta_1\,U_1$ and $F-\zeta_1F_1+[\Delta,\,\zeta_1]\,U_1$ stand for U and F.

Now we suppose that $F=0$ near O and $\Phi=0$ on ∂K and prove the existence of a unique generalized solution of problem (6) satisfying

$$\int_K(1+|\eta|^2)^\beta\left(|\nabla U(\eta)|^2+|U(\eta)|^2\right)d\eta\le c\int_K(1+|\eta|^2)^\beta|F(\eta)|^2d\eta. \tag{11}$$

Let $\overset{\circ}{\mathbf{H}}{}^1_\beta(K)$ be the space of functions in K that equal zero on ∂K and have the finite norm

$$\|W;\mathbf{H}^1_\beta(K)\| = \Big\{ \int_K \big(1+\tau^2|\eta|^2\big)^\beta \big(|\nabla W(\eta)|^2 + |W(\eta)|^2\big)d\eta \Big\}^{1/2}.$$

We multiply the equation in K in problem (6) by $\overline{W(\eta)}(1+\tau^2|\eta|^2)^\beta$, where $\tau > 0$ and W is an arbitrary element in $\overset{\circ}{\mathbf{H}}{}^1_\beta(K)$. Integrating by parts, we obtain the identity for the generalized solution

$$\begin{aligned}&\int_K \big(1+\tau^2|\eta|^2\big)^\beta \Big\{\nabla U(\eta)\cdot\nabla\overline{W(\eta)} + U(\eta)\overline{W(\eta)}\Big\}d\eta\\ &+2\beta\tau^2\int_K \big(1+\tau^2|\eta|^2\big)^{\beta-1}\overline{W(\eta)}\,\eta\cdot\nabla U(\eta)\,d\eta \\ &= \int_K \big(1+\tau^2|\eta|^2\big)^\beta F(\eta)\,\overline{W(\eta)}\,d\eta.\end{aligned} \tag{12}$$

The right-hand side in (12) is a linear functional on $\overset{\circ}{\mathbf{H}}{}^1_\beta(K)$. By Riesz's theorem on the general form of functionals on a Hilbert space, one can rewrite it as $\langle\tilde F, W\rangle$, where $\tilde F \in \overset{\circ}{\mathbf{H}}{}^1_\beta(K)$ and $\langle .,.\rangle$ is the inner product on $\overset{\circ}{\mathbf{H}}{}^1_\beta(K)$. Since $2\tau|\eta|\,(1+\tau^2|\eta|^2)^{\beta-1} \le (1+\tau^2|\eta|^2)^\beta$, the second term on the left does not exceed $\beta\,\tau\|U;\overset{\circ}{\mathbf{H}}{}^1_\beta(K)\|\,\|W;\overset{\circ}{\mathbf{H}}{}^1_\beta(K)\|$. Therefore, it can be represented in the form $\langle AU,\, W\rangle$, while

$$\|A;\overset{\circ}{\mathbf{H}}{}^1_\beta(K)\to\overset{\circ}{\mathbf{H}}{}^1_\beta(K)\| \le \tau\cdot\text{const}\,. \tag{13}$$

According to (12), the desired solution U satisfies

$$\langle U + AU - \tilde F, W\rangle = 0 \qquad \forall\ \ W\in\overset{\circ}{\mathbf{H}}{}^1_\beta(K).$$

For small τ the operator $1+A$ is invertible by virtue of (13). Hence, the function U, defined by $U=(1+A)^{-1}\tilde F$, is a generalized solution of problem (6). Setting $F=0$, $W=U$ in (1.12) and taking into account (13), we prove the uniqueness of the solution.

Now we show that the generalized solution belongs to $\mathbf{E}^{l+2}_{2,\beta}(K)$. By Proposition 1.3.18 and Theorem 3.3.1, $\zeta U\in\mathbf{E}^{l+2}_{2,\beta}(K)$, where $\zeta\in\mathbf{C}^\infty_0(\bar K)$ and $\zeta(\eta)=1$ for $|\eta|<2R$ with a large R. Also,

$$\|\zeta U;\mathbf{E}^{l+2}_{2,B}(K)\| \le c\|F;\mathbf{E}^l_{2,\beta}\left(\{\eta\in K\colon |\eta|<3R\}\right)\|. \tag{14}$$

Let $\{Q_j\}$ be a locally finite covering of the set $\{\eta\in K\colon |\eta|>R\}$ with squares $Q_j\subset\bar K$ whose sides are equal to 1. We assume that Q_j is an open square if none of its sides belongs to ∂K. Otherwise the side of Q_j located in ∂K is included in Q_j. Let $\{\chi_j\}$ be a partition of unity subject to the covering $\{Q_j\}$, $\psi_j\in\mathbf{C}^\infty_0(Q_j)$, $\psi_j\chi_j=\chi_j$, and $|D^\gamma\chi_j|+|D^\gamma\psi_j|\le c_\gamma$ for all j. Recall the known local estimate

$$\|\chi_j U;\mathbf{W}^{l+2}_2(Q_j)\| \le c\big(\|\psi_j F;\,\mathbf{W}^l_2(Q_j)\| + \|\psi_j U;\mathbf{L}_2(Q_j)\|\big) \tag{15}$$

(see Theorem 3.1.2). We multiply (15) by d_j^β, where d_j is the distance of O to the centre of Q_j. Since $cd_j < |\eta| < Cd_j$ for $\eta \in Q_j$, we arrive at the inequality

$$\|\chi_j U; \mathbf{E}^{l+2}_{2,\beta}(K)\| \le c\,(\|\psi_j F; \mathbf{E}^l_{2,\beta}(K)\| + \|\psi_j |\eta|^\beta U; \mathbf{L}_2(K)\|). \tag{16}$$

Adding (14) and (16) for all j, we obtain

$$\|U;\, \mathbf{E}^{l+2}_{2,\beta}(K)\| \le c\,(\|F; \mathbf{E}^l_{2,\beta}(K)\| + \|\,|\eta|^\beta U; \mathbf{L}_2(\{\eta \in K\colon |\eta| > R\})\|). \tag{17}$$

The estimate (17) implies (8) for $\Phi = 0$ because U is subject to (11).

Thus, the unique solvability of problem (6) in $\mathbf{E}^{l+2}_{2,\beta}(K)$ is proved for $F = 0$ near the point O. The last restriction can be dropped with the help of (10).

The sufficiency of assumptions of the proposition is proved. Let us check the necessity.

If $\beta - l - 1 = \pm\pi/\vartheta_0$, then the operator of the problem (6) ceases to be Fredholm, which can be established similarly to Proposition 1.3.11. In the case $\beta - l - 1 > \pi/\vartheta_0$ the function $Z(\eta) = K_{\pi/\vartheta_0}(|\eta|)\cos(\pi\vartheta/\vartheta_0)$ is a solution to the homogeneous problem (6) in $\mathbf{E}^{l+2}_{2,\beta}(K)$ (see Remark 1.3.17). So there is no uniqueness of solution in the case mentioned. Finally, let $\beta - l - 1 < \pi/\vartheta_0$. We consider (6) with $F = -[\Delta, \zeta]Z$ and $\Phi = 0$. When considering the generalized solution of (6) with a right-hand side F equal to zero near the point O, we did not use the condition $|\beta - l - 1| < \pi/\vartheta_0$. Therefore, if $U \in \mathbf{E}^{l+2}_{2,\beta}(K)$ is a solution to (6), then U is a generalized solution to the same problem. Since the generalized solution is unique, we have $U = \zeta Z$. However, the function ζZ is not in $\mathbf{E}^{l+2}_{2,\beta}(K)$ because of the behavior of MacDonald's function K_{π/ϑ_0} near the origin (see Remark 1.3.17). It follows that for $\beta - l - 1 < \pi/\vartheta_0$ and $F = -[\Delta, \zeta]Z$ the problem (6) has no solution in $\mathbf{E}^{l+2}_{2,\beta}(K)$. □

The verification of the next assertion reduces to applying Lemma 1.2 and Proposition 1.3.

Proposition 11.1.4. *Let l be a nonnegative integer, $f \in \mathbf{V}^l_{2,\beta}(D;\Gamma)$, $\varphi \in \mathbf{V}^{l+3/2}_{2,\beta}(\partial D;\Gamma)$, and $|\beta - l - 1| < \pi/\vartheta_0$. Then there exists a unique solution $u \in \mathbf{V}^{l+2}_{2,\beta}(D;\Gamma)$ of the problem* (5), *and the estimate* (9) *holds.*

Remark 11.1.5. The condition $|\beta - l - 1| < \pi/\vartheta_0$ is not only sufficient for the unique solvability of the problem (5), but necessary as well. One can show (see Section 3 below) that for $\beta - l - 1 > \pi/\vartheta_0$ ($\beta - l - 1 < -\pi/\vartheta_0$) problem (5) has infinite dimensional kernel (cokernel). In the case $\beta - l - 1 = \pm\pi/\vartheta_0$ the range of the operator

$$A\colon \mathbf{V}^{l+2}_{2,\beta}(D;\Gamma) \to \mathbf{V}^l_{2,\beta}(D;\Gamma) \times \mathbf{V}^{l+3/2}_{2,\beta}(\partial D;\Gamma)$$

of the problem is not closed. So for $|\beta - l - 1| \ge \pi/\vartheta_0$ the operator A is not even Fredholm.

Proposition 11.1.6. *Let $f \in \mathbf{V}^l_{2,\beta_1}(D;\Gamma) \cap \mathbf{V}^l_{2,\beta_2}(D;\Gamma)$ and $\varphi \in \mathbf{V}^{l+3/2}_{2,\beta_1}(\partial D;\Gamma) \cap \mathbf{V}^{l+3/2}_{2,\beta_2}(\partial D;\Gamma)$, while the numbers β_1 and β_2 are subject to $|\beta_j - l - 1| < \pi/\vartheta_0$, $j = 1, 2$. Then the solution $u \in \mathbf{V}^{l+2}_{2,\beta_1}(D;\Gamma)$ of problem* (5) *belongs to $\mathbf{V}^{l+2}_{2,\beta_2}(D;\Gamma)$ as well.*

Proof. To be specific, we assume $\beta_1 < \beta_2$. Let u_j denote the unique solution of problem (5) in $\mathbf{V}^{l+2}_{2,\beta_j}(D;\Gamma)$, $j = 1,2$. We have to prove the equality $u_1 = u_2$. By Lemma 1.2, the function $\eta \mapsto U_j(\eta,\xi) = \hat{u}_j\left(|\xi|^{-1}\eta,\xi\right)$ is in $\mathbf{E}^{l+2}_{2,\beta_j}(K)$ and satisfies (6) with

$$F \in \mathbf{E}^{l}_{2,\beta_1}(K) \cap \mathbf{E}^{l}_{2,\beta_2}(K), \quad \Phi \in \mathbf{E}^{l+3/2}_{2,\beta_1}(\partial K) \cap \mathbf{E}^{l+3/2}_{2,\beta_2}(\partial K).$$

Since $\beta_1 > \beta_2$, from the definition of the norm (7) it follows that $(1-\zeta)U_1 \in \mathbf{E}^{l+2}_{2,\beta_2}(K)$ with the same cut-off function ζ as in Proposition 11.1.3. According to Proposition 1.2.2 (see also Theorem 3.3.1), $\zeta U_1 \in \mathbf{E}^{l+2}_{2,\beta_2}(K)$. Hence, $U_1 \in \mathbf{E}^{l+2}_{2,\beta_2}(K)$. In virtue of the uniqueness of a solution of problem (6) (Proposition 11.1.3), $U_1 = U_2$. Therefore, $u_1 = u_2$. □

11.1.3 Boundary value problem in Ω

Propositions 11.1.4 and 11.1.6 on properties of solutions of the model problem (6) in a wedge enable us to prove the unique solvability of problem (1.3) in the domain Ω.

Theorem 11.1.7. *Let l be a nonnegative integer, $f \in \mathbf{V}^{l}_{2,\beta}(\Omega;M)$, $\varphi \in \mathbf{V}^{l+3/2}_{2,\beta}(\partial\Omega;M)$, and $|\beta - l - 1| < \pi/\vartheta_0$. Then there exists a unique solution $u \in \mathbf{V}^{l+2}_{2,\beta}(\Omega;M)$ of problem* (3), *and*

$$\|u;\mathbf{V}^{l+2}_{2,\beta}(\Omega;M)\| \leq c\left(\|f;\mathbf{V}^{l}_{2,\beta}(\Omega;M)\| + \|\varphi;\mathbf{V}^{l+3/2}_{2,\beta}(\partial\Omega;M)\|\right). \tag{18}$$

Proof. We first suppose that $\beta \leq l+1$. Then the right-hand side f,φ of problem (3) satisfies the assumptions of Proposition 11.1.1 and, therefore, there exists a unique generalized solution $u \in \mathbf{V}^{1}_{2,0}(\Omega;M)$. We will show that $u \in \mathbf{V}^{l+2}_{2,\beta}(\Omega;M)$.

Let P be an arbitrary point in the edge M and χ a smooth function with support in a small neighborhood U of P equal to 1 near P. Let also $\varkappa$ be a diffeomorphism sending $\bar{\Omega} \cap U$ into $\bar{D} \cap B_1$. Then

$$\begin{aligned} &-\Delta(\chi u \circ \varkappa) = \chi f \circ \varkappa + \tilde{L}(u \circ \varkappa) =: f_1 \quad \text{in} \quad D,\\ &\chi u \circ \varkappa = \chi f \circ \varkappa =: \varphi_1 \quad \text{on} \quad \partial D \setminus \Gamma, \end{aligned} \tag{19}$$

where $\tilde{L}$ is a first order differential operator with smooth coefficients. It is clear that $f_1 \in \mathbf{V}^{0}_{2,\beta_1}(D;\Gamma)$ and $\varphi_1 \in \mathbf{V}^{3/2}_{2,\beta_1}(\partial D;\Gamma)$ with $\beta_1 = \max\{\beta - l, 0\}$.

According to Proposition 11.1.6, $\chi u \circ \varkappa \in \mathbf{V}^{2}_{2,\beta_1}(D;\Gamma)$. In view of the embedding $\mathbf{V}^{s+1}_{\gamma+1}(D;\Gamma) \subset \mathbf{V}^{s}_{2,\gamma}(D;\Gamma)$, we find that $f_1 \in \mathbf{V}^{l_2}_{2,\beta_2}(D;\Gamma)$ and $\varphi_1 \in \mathbf{V}^{l_2+3/2}_{2,\beta_2}(\partial D;\Gamma)$ with $l_2 = \min\{l,1\}$ and $\beta_2 = \max\{\beta_1 - 1, \beta - l\} + l_2$.

Now Proposition 11.1.6 provides the inclusion $\chi u \circ \varkappa \in \mathbf{V}^{l+2}_{2,\beta_2}(D;\Gamma)$. Improving step by step the properties of χu, we obtain $\chi u \circ \varkappa \in \mathbf{V}^{l+2}_{2,\beta}(D;\Gamma)$.

The traditional local estimates (see (1.3.2)) yield smoothness of the generalized solution far from the edge M.

Thus, the theorem is proved under the supplementary restriction $\beta \leq l+1$. Now let $\beta \in (l+1, l+1+\pi/\vartheta_0)$. The assumption of Proposition 11.1.1 is not fulfilled and one cannot use the generalized solution.

Let us construct a function $v \in \mathbf{V}^{l+2}_{2,\beta}(\Omega;M)$ subject to the inequality (18) and inclusions $f + \Delta v \in \mathbf{V}^{l}_{2,l+1}(\Omega;M)$ and $\varphi - v \in \mathbf{V}^{l+3/2}_{2,l+1}(\partial\Omega;M)$.

Let $\{U_j\}$ be a fine covering of a neighborhood of M, $\{\zeta_j\}$ a partition of unity subject to the covering, and $\varkappa_j$ a diffeomorphism of $U \cap \bar{\Omega}$ onto $B_1 \cap \bar{D}$. We denote by $w_j \in \mathbf{V}_{2,\beta}^{l+2}(D;\Gamma)$ the solution of problem (5) with right-hand side $\zeta_j f \circ \varkappa_j$ and $\zeta_j \varphi \circ \varkappa_j$, which exists by virtue of Proposition 11.1.4. Set $w = \sum \chi_j w_j$, where $\chi_j \in \mathbf{C}_0^\infty(U_j)$ and $\chi_j \zeta_j = \zeta_j$. We have

$$-\Delta w = \sum \zeta_j f + \sum \tilde{L}_j w_j,$$

$\tilde{L}_j$ being first order operators with smooth coefficients. Since $\sum \zeta_j = 1$ near the edge, $f_1 = \Delta w + f \in \mathbf{V}_{2,\beta-1}^{l}(\Omega; M)$ and $\varphi_1 = \varphi - w \in \mathbf{V}_{2,\beta-1}^{l+3/2}(\partial\Omega; M)$. As the function v, one can take the function w in the case $\beta - 1 \leq l + 1$.

If $\beta - 1 > l + 1$ the procedure has to be repeated for the problem (3) with new right-hand side f_1, φ_1. So for $\beta - 2 \leq l + 1$ the desired function v has been constructed. In the case $\beta - 2 > l + 1$ we take such a step once again, etc.

Denote by $Z \in \mathbf{V}_{2,0}^1(\Omega; M)$ the generalized solution to problem (1.3) with right-hand side $f + \Delta v \in \mathbf{V}_{2,l+1}^{l}(\Omega; M)$, $\varphi - v \in \mathbf{V}_{2,l+1}^{l+3/2}(\partial\Omega; M)$. According to the first part of the proof, $Z \in \mathbf{V}_{2,l+1}^{l+2}(\Omega; M) \subset \mathbf{V}_{2,\beta}^{l+2}(\Omega; M)$. Therefore, the function $u = v + Z \in \mathbf{V}_{2,\beta}^{l+2}(\Omega; M)$ satisfies problem (3) and admits estimate (18).

The existence of a solution has been established. To prove uniqueness it suffices to check that a solution $u_0 \in \mathbf{V}_{2,\beta}^{l+2}(\Omega; M)$ of the homogeneous problem (3) is in $\mathbf{V}_{2,0}^1(\Omega; M)$. As in the first part of the proof, it can be obtained by applying Proposition 11.1.6 to the problem (19) with $f = 0$, $\varphi = 0$. □

11.2 The Neumann Problem for the Laplace Operator

In this Section we prove the solvability of the Neumann problem in weighted classes in a three-dimensional domain Ω with edge M on the boundary. Besides, in Subsect.2.4 we consider a problem where the role of edge is given to a smooth contour located in Ω.

11.2.1 Statement of the problem

Let Ω be the same domain as in Section 1. We consider the problem

$$-\Delta u(x) = f(x), \quad x \in \Omega; \quad (\partial/\partial\nu)\, u(x) = \psi(x), \quad x \in \partial\Omega \setminus M, \tag{1}$$

ν being the outward normal to $\partial\Omega \setminus M$. The next well-known assertion is a consequence of Poincaré's inequality and Riesz's theorem on the general form of functionals on a Hilbert space.

Proposition 11.2.1. *Let $f \in \mathbf{L}_2(\Omega)$, $\psi \in \mathbf{L}_2(\partial\Omega)$, and let*

$$\int_\Omega f(x)dx + \int_{\partial\Omega} \psi(x) ds_x = 0. \tag{2}$$

Then there exists a generalized solution $u \in \mathbf{W}_2^1(\Omega)$ to problem (1).

The solution is determined up to an additive constant. For the solution subject to

$$\int_\Omega u(x)dx = 0 \tag{3}$$

there is the estimate

$$\|u;\mathbf{W}_2^1(\Omega)\| \le c\left(\|f;\mathbf{L}_2(\Omega)\| + \|\psi;\mathbf{L}_2(\partial\Omega)\|\right). \tag{4}$$

11.2.2 Model problems

As in the preceding section, the basic point consists in studying the model problem in a wedge D:

$$-\Delta u(x) = f(x), \quad x\in D; \quad (\partial/\partial\nu)u(x) = \psi(x), \quad x\in\partial D\setminus\Gamma. \tag{5}$$

By means of the Fourier transform, problem (5) reduces to the problem in the cone K

$$-\Delta_\eta U + U = F \quad \text{in} \quad K; \qquad (\partial/\partial\nu)U = \Psi \quad \text{on} \quad \partial K\setminus\{O\}, \tag{6}$$

where $\eta = |\xi|\, y$, $U(\eta,\xi) = \hat{u}\,(|\xi|^{-1}\,\eta,\,\xi)$, $F(\eta,\xi) = |\xi|^{-2}\,\hat{f}\,(|\xi|^{-1}\,\eta,\xi)$, $\Psi(\eta,\xi) = |\xi|^{-1}\,\hat{\psi}(|\xi|^{-1}\,\eta,\xi)$ (cf. Section 1.2).

Let $\beta\in\mathbb{R}$ and

$$I(F,\Psi) := \int_K (1+|\eta|^2)^\beta\,|F(\eta)|^2\,d\eta + \int_{\partial K}(1+|\eta|^2)^\beta\,|\Psi(\eta)|^2\,d\eta < \infty. \tag{7}$$

We denote by $\tilde{\mathbf{H}}_\beta^1(K)$ the space of functions in K with norm

$$\Big(\int_K (1+\tau^2\,|\eta|^2)^\beta\,(|\nabla U(\eta)|^2 + |U(\eta)|^2)d\eta\Big)^{1/2},$$

where $\tau>0$. A function $U\in\tilde{\mathbf{H}}_\beta^1(K)$ is called a generalized solution to problem (6) if it satisfies the integral identity

$$\begin{aligned}
&\int_K (1+\tau^2\,|\eta|^2)^\beta\left\{\nabla U(\eta)\cdot\nabla\overline{W(\eta)} + U(\eta)\overline{W(\eta)}\right\}d\eta \\
&+ 2\beta\tau^2\int_K (1+\tau^2\,|\eta|^2)^{\beta-1}\nabla U(\eta)\cdot\eta\,\overline{W(\eta)}d\eta \\
&= \int_K (1+\tau^2\,|\eta|^2)^\beta\,F(\eta)\,\overline{W(\eta)}d\eta + \int_{\partial K}(1+\tau^2\,|\eta|^2)^\beta\,\Psi(\eta)\,\overline{W(\eta)}ds_\eta
\end{aligned}\tag{8}$$

with arbitrary W in $\tilde{\mathbf{H}}_\beta^1(K)$. The right-hand side in (8) can be treated as a linear functional on $\tilde{\mathbf{H}}_\beta^1(K)$ and rewritten in the form $\langle\tilde{F},W\rangle$, where $\langle,\rangle$ is the inner product on $\tilde{\mathbf{H}}_\beta^1(K)$, and $\tilde{F}\in\tilde{\mathbf{H}}_\beta^1(K)$. Since $2\tau|\eta|\,(1+\tau^2\,|\eta|^2)^{\beta-1}\le(1+\tau^2|\eta|^2)$, the second term on the left in (8) can be written as $\langle AU,W\rangle$, while A is an operator on $\tilde{\mathbf{H}}_\beta^1(K)$ with norm $O(\tau)$. Therefore, (6) takes the form

$$\langle U + AU - \tilde{F}, W\rangle = 0 \quad \forall \quad W\in\tilde{\mathbf{H}}_\beta^1(K).$$

For small τ the operator $1+A$ is invertible. The function $U = (1+A)^{-1}\,\tilde{F}$ is a generalized solution to problem (6).

Thus, we established the following assertion.

Lemma 11.2.2. *Suppose that $\beta \in \mathbb{R}$ and F and Ψ are subject to* (7). *Then there exists a unique generalized solution $U \in \tilde{\mathbf{H}}^1_\beta(K)$ to the problem* (6), *and*

$$\int_K (1+|\eta|^2)^\beta \left(|\nabla U(\eta)|^2 + |U(\eta)|^2\right) d\eta \le c\, I(F,\Psi),$$

$I(F,\Psi)$ being defined in (7).

In contrast to the Dirichlet problem (6), uniquely solvable in $\mathbf{E}^{l+2}_{2,\beta}(K)$ (Proposition 1.3), the Neumann problem does not possess this property. The reason is that for a smooth right-hand side and $\beta < l+1$ the generalized solution U does not fit in $\mathbf{E}^{l+2}_{2,\beta}(K)$ because $U(\eta) = O(1)$ near the point $\eta = 0$ (see, for example, Proposition 1.4.6). Therefore, in the case $\beta < l+1$ problem (6) ceases to be solvable. On the other hand, for $\beta > l+1$ the homogeneous problem (6) has the nontrivial solution $Z \in \mathbf{E}^{l+2}_{2,\beta}(K)$, $Z(\eta) = \log|\eta| + O(1)$ as $|\eta| \to 0$ (Proposition 1.4.6). The number $\beta = l+1$ is forbidden (cf. Proposition 1.3.11).

Let us prove the solvability of problem (6) in the space $\tilde{\mathbf{E}}^{l+2}_{2,\beta}(K)$ with norm

$$\begin{aligned} \|U;\, \tilde{\mathbf{E}}^{l+2}_{2,\beta}(K)\| = \Big(&\| |\eta|^\beta (1+|\eta|^{-l-1})\, U;\, \mathbf{L}_2(K)\|^2 \\ &+ \sum_{0<|\gamma|\le l+2} \| |\eta|^\beta (1+|\eta|^{|\gamma|-l-2})\, D^\gamma_\eta U;\, \mathbf{L}_2(K)\|^2 \Big)^{1/2}. \end{aligned} \tag{9}$$

One has to choose β so that the generalized solution U belongs to $\tilde{\mathbf{E}}^{l+2}_{2,\beta}(K)$ while Z does not. Since $U(\eta) = O(1)$ and $D^\gamma_\eta U(\eta) = O(|\eta|^{\pi/\vartheta_0 - |\gamma|})$ for $\eta \to 0$ and $|\gamma| > 0$, the inclusion $U \in \tilde{\mathbf{E}}^{l+2}_{2,\beta}(K)$ is provided by the condition $\beta > l + \max\{0, 1 - \pi/\vartheta_0\}$. By virtue of $D^\gamma_\eta Z(\eta) = O(|\eta|^{-|\gamma|})$ as $\eta \to 0$ and $|\gamma| \ge 1$, the requirement $Z \notin \tilde{\mathbf{E}}^{l+2}_{2,\beta}(K)$ is fulfilled with $\beta < l+1$.

Proposition 11.2.3. *Let $\beta \in (l + \max\{0, 1-\pi/\vartheta_0\}, l+1)$, where l is a nonnegative integer. Let also $F \in \mathbf{E}^l_{2,\beta}(K)$ and $\Psi \in \mathbf{E}^{l+1/2}_{2,\beta}(\partial K)$. There exists a unique solution $U \in \tilde{\mathbf{E}}^{l+2}_{2,\beta}(K)$ to problem* (6), *and*

$$\|U;\, \tilde{\mathbf{E}}^{l+2}_{2,\beta}(K)\| \le c\left(\|F;\, \mathbf{E}^l_{2,\beta}(K)\| + \|\Psi;\, \mathbf{E}^{l+1/2}_{2,\beta}(\partial K)\|\right). \tag{10}$$

Proof. Assume that G is a bounded domain in $\mathbb{R}^2$ such that $B_1 \cap K \subset G \subset K$, $B_1 \subset \mathbb{R}^2$ being the open unit disk. By Proposition 1.4.6, there exists a solution v of the problem

$$-\Delta v(\eta) = \zeta(\eta)\, F(\eta),\ \eta \in G; \quad (\partial/\partial\nu) v(\eta) = \zeta(\eta)\, \Psi(x), \quad \eta \in \partial G, \tag{11}$$

where $\zeta \in \mathbf{C}^\infty_0(B_1)$ while $\zeta = 1$ near $\eta = 0$. This solution is determined up to an additive constant c and admits the representation

$$v(\eta) = c_1 \log|\eta| + c + v_1(\eta), \quad v_1 \in \mathbf{V}^{l+2}_{2,\beta}(G)$$

with $c_1 = c_1(\zeta F, \zeta\Psi)$. We take $c = 0$. Then by Proposition 1.4.6,

$$|c_1| + \|v_1;\, \mathbf{V}^{l+2}_{2,\beta}(G)\| \le c\left(\|F;\, \mathbf{E}^l_{2,\beta}(K)\| + \|\Psi;\, \mathbf{E}^{l+1/2}_{2,\beta}(\partial K)\|\right). \tag{12}$$

Let $U_1 = \zeta v_1$ and let U_2 be a solution to the problem

$$\begin{aligned} -\Delta U_2(\eta) + U_2(\eta) &= F(\eta) + \Delta U_1(\eta) - U_1(\eta), \quad \eta \in K; \\ (\partial/\partial\nu) U_2(\eta) &= \Psi(\eta) - (\partial/\partial\nu) U_1(\eta), \quad \eta \in \partial K. \end{aligned} \tag{13}$$

It is clear that $F + \Delta U_1 = 0$ in K near the point O and $\Psi - \partial U_1/\partial\nu = 0$ in $\partial K \setminus \{O\}$ near O. According to (12),

$$\begin{aligned}&||U_1; \mathbf{E}^{l+2}_{2,\beta}(K)|| + ||\tilde{F}; \mathbf{E}^{l}_{2,\beta}(K)|| + ||\Phi; \mathbf{E}^{l+1/2}_{2,\beta}(\partial K)||\\ &\leq c(||F; \mathbf{E}^{l}_{2,\beta}(K)|| + ||\Psi; \mathbf{E}^{l+1/2}_{2,\beta}(\partial K)||),\end{aligned} \tag{14}$$

where $\tilde{F}$ and Φ stand for the right-hand side of problem (13). Besides, $U_1 \in \mathbf{V}^{l+2}_{2,\beta}(G) \subset \mathbf{V}^{l}_{2,\beta-2}(G)$ and, consequently,

$$I(\tilde{F}, \Phi) \leq c(||F;\ \mathbf{E}^{l}_{2,\beta}(K)||^2 + ||\Psi;\ \mathbf{E}^{l+1/2}_{2,\beta}(\partial K)||^2), \tag{15}$$

(see (7)). By Lemma 2.2, there exists a generalized solution $U_2 \in \tilde{\mathbf{H}}^1_\beta(K)$ to problem (13). We show that $U_2 \in \tilde{\mathbf{E}}^{l+2}_{2,\beta}(K)$.

Let χ_j, ψ_j be the same as in the proof of Proposition 1.3. Repeating with obvious changes the derivation of (16), we arrive at

$$\begin{aligned}||\chi_j U_2;\ \mathbf{E}^{l+2}_{2,\beta}(K)|| &\leq c(||\psi_j \tilde{F};\ \mathbf{E}^{l}_{2,\beta}(K)|| + ||\psi_j \Phi;\ \mathbf{E}^{l+1/2}_{2,\beta}(K)||\\ &+ ||\psi_j |\eta|^\beta\, U_2;\ \mathbf{L}_2(K)||).\end{aligned} \tag{16}$$

Since $\tilde{F}$ and Φ satisfy (14), from Proposition 1.2.5 it follows that

$$U_2(\eta) = c_2 + v_2(\eta) \tag{17}$$

with a constant c_2 depending on $\tilde{F}$ and Φ. Also,

$$|c_2| + ||v_2;\ \mathbf{V}^{l+2}_{2,\beta}(G)|| \leq c(||\tilde{F};\ \mathbf{E}^{l}_{2,\beta}(K)|| + ||\Phi;\ \mathbf{E}^{l+1/2}_{2,\beta}(\partial K)||). \tag{18}$$

Taking the definition of the norm (4) into account, from (17) and (18) we directly deduce the estimate

$$||\zeta U_2;\ \mathbf{E}^{l+2}_{2,\beta}(K)|| \leq c(||\tilde{F};\ \mathbf{E}^{l}_{2,\beta}(K)|| + ||\Phi;\ E^{l+1/2}_{2,\beta}(\partial K)||).$$

We add the last inequality and estimates (16) and, using (15) and Lemma 2.2, find that

$$||U_2;\ \tilde{\mathbf{E}}^{l+2}_{2,\beta}(K)|| \leq c(||F;\ \mathbf{E}^{l}_{2,\beta}(K)|| + ||\Psi;\ \mathbf{E}^{l+1/2}_{2,\beta}(\partial K)||). \tag{19}$$

Combining the inclusion $\mathbf{E}^{l+2}_{2,\beta}(K) \subset \tilde{E}^{l+2}_{2,\beta}(K)$, (14) and (19), we obtain that the function $U = U_1 + U_2$ satisfies (6) by virtue of (13) and admits estimate (10).

Thus, the existence of the solution $U \in \tilde{\mathbf{E}}^{l+2}_{2,\beta}(K)$ has been established. In view of $\tilde{\mathbf{E}}^{l+2}_{2,\beta}(K) \subset \tilde{\mathbf{H}}^1_\beta(K)$, the uniqueness follows from Lemma 2.2. □

Now we check that the operator A of problem (6) implements a continuous mapping

$$A : \tilde{\mathbf{E}}^{l+2}_{2,\beta}(K) \to \mathbf{E}^{l}_{2,\beta}(K) \times \mathbf{E}^{l+1/2}_{2,\beta}(\partial K) \tag{20}$$

for all $\beta \in \mathbb{R}$ and $l = 0, 1, \dots$. From the definition of the norm in $\mathbf{E}^{s}_{2,\beta}(K)$ (see (11.1)) it follows directly that

$$||\Delta U;\ \mathbf{E}^{l}_{2,\beta}(K)|| + ||\partial U/\partial\nu;\ \mathbf{E}^{l+1/2}_{2,\beta}(\partial K)|| \leq c||U;\ \tilde{\mathbf{E}}^{l+2}_{2,\beta}(K)||. \tag{21}$$

Besides, by virtue of (9),

$$||U;\ \mathbf{E}^{l}_{2,\beta}(K)|| \leq ||U;\ \tilde{\mathbf{E}}^{l+2}_{2,\beta}(K)||.$$

Together with (21), it gives the continuity of the mapping (2.20). Now Proposition 11.2.3 leads to the following assertion.

Proposition 11.2.4. *Let $\beta \in (l+\max\{0, 1-\pi/\vartheta_0\}, l+1)$, where l is a nonnegative integer. Then the mapping* (20) *is an isomorphism.*

We turn to problem (5) in a wedge. The replacement of $\mathbf{E}^{l+2}_{2,\beta}(K)$ by $\tilde{\mathbf{E}}^{l+2}_{2,\beta}(K)$ requires the corresponding replacement of $\mathbf{V}^{l+2}_{2,\beta}(D;\Gamma)$. Denote by $\tilde{\mathbf{V}}^{l+2}_{2,\beta}(D;\Gamma)$ the space with norm

$$\Big(\int_D |y|^{2(\beta-l-1)}(1+|y|)^{-2}\,|w(x)|^2 dx + \sum_{1\le|\alpha|\le l+2} \int_D |y|^{2(\beta-l-2+|\alpha|)}|D^\alpha_x\, w(x)|^2\, dx\Big)^{1/2}. \tag{22}$$

Lemma 1.2 describes the relation between the norms in $\mathbf{E}^{l+2}_{2,\beta}(K)$ and $\mathbf{V}^{l+2}_{2,\beta}(D;\Gamma)$. It turns out that the lemma is not valid for $\tilde{\mathbf{E}}^{l+2}_{2,\beta}(K)$ and $\tilde{\mathbf{V}}^{l+2}_{2,\beta}(D;\Gamma)$.

Lemma 11.2.5. *i) The following norms are equivalent:*

$$\begin{aligned}\Big(\int_{\mathbb{R}}\int_K &|y|^{2(\beta-l-1)}\,\big(|\xi|^{-1}+|y|\big)^{-2}\,|\hat{w}(y,\xi)|^2\,dy\,d\xi\\ &+\sum_{1\le|\alpha|\le l+2}\int_{\mathbb{R}}|\xi|^{2\alpha_3}\int_K |y|^{2(\beta-l-2+|\alpha|)}|D^{\alpha'}_y\,\hat{w}(y,\xi)|^2\,dyd\xi\Big)^{1/2}\end{aligned} \tag{23}$$

and

$$\Big(\int_{\mathbb{R}} |\xi|^{2(l-\beta+1)}\,\|W(\cdot,\xi);\,\tilde{\mathbf{E}}^{l+2}_{2,\beta}(K)\|^2\,d\xi\Big)^{1/2}; \tag{24}$$

here $W(\eta,\xi)=\hat{w}(|\xi|^{-1}\eta;\xi)$. After replacing $(|\xi|^{-1}+|y|)^{-2}$ by $(1+|y|)^{-2}$, the norm (23) *coincides with* (22).

ii) For functions w with supports in $\{x\in\bar{D}\colon |x_3|<\text{const}\}$ the norms (22) *and* (24) *are equivalent.*

Proof. i) After the change of variables $y\mapsto\eta=|\xi|y$ the norm (23) takes the form

$$\begin{aligned}&\Big(\int_{\mathbb{R}}|\xi|^{2(l-\beta+1)}\int_K|\eta|^{2(\beta-l-1)}(1+|\eta|)^{-2}\,|W(\eta,\xi)|^2\,d\eta d\xi\\ &+\int_{\mathbb{R}}|\xi|^{2(l-\beta+1)}\sum_{1\le|\alpha|\le l+2}\int_K|\eta|^{2(\beta-l-2+|\alpha|)}|D^{\alpha'}_\eta\,W(\eta,\xi)|^2\,d\eta d\xi\Big)^{1/2}\\ &=\Big\{\int_{\mathbb{R}}|\xi|^{2(l-\beta+1)}\int_K|\eta|^{2\beta}\Big(\Big[|\eta|^{-2(l+1)}(1+|\eta|)^{-2}+\sum_{\alpha_3=1}^{l+2}|\eta|^{2(\alpha_3-l-2)}\Big]\,|W(\eta,\xi)|^2\\ &+\sum_{1\le|\alpha'|\le l+2}\Big[\sum_{\alpha_3=0}^{l+2-|\alpha'|}|\eta|^{2(|\alpha|-l-2)}\Big]|D^{\alpha'}_\eta\,W(\eta,\xi)|^2\Big)\,d\eta d\xi\Big\}^{1/2}.\end{aligned} \tag{25}$$

The expressions in square brackets are equivalent to $(1+|\eta|^{-(l+1)})^2$ and $(1+|\eta|^{|\alpha|-l-2)})^2$ respectively. Therefore, the right-hand side in (25) is equivalent to the norm (24).

ii) It suffices to show that

$$\Big(\int_{\mathbb{R}}\int_K|y|^{2(\beta-l-1)}\big(|\xi|^{-1}+|y|\big)^{-2}|\hat{w}(y,\xi)|^2\,dyd\xi\Big)^{1/2} \tag{26}$$

is majorized by the norm (22) multiplied by a constant and

$$\begin{aligned}&\Big(\int_D |y|^{2(\beta-l-1)}\big(1+|y|\big)^{-2}\,|w(x)|^2\,dx\Big)^{1/2}\\ &=\Big(\int_{\mathbb{R}}\int_K |y|^{2(\beta-l-1)}\big(1+|y|\big)^{-2}\,|\hat{w}(y,\xi)|^2\,dyd\xi\Big)^{1/2}\end{aligned} \tag{27}$$

is majorized by the norm (23). Since $(|\xi|^{-1}+|y|)^{-2}\le|\xi|^2$, the expression (26) squared does not exceed the term

$$\int_D |y|^{2(\beta-l-1)}|(\partial/\partial x_3)\,w(x)|^2\,dx \tag{28}$$

in (22).

On the other hand, in view of the one-variable Friedrichs inequality,

$$\int_D |y|^{2(\beta-l-1)}|(\partial/\partial x_3)w(x)|^2dx\ge c\int_D |y|^{2(\beta-l-1)}|w(x)|^2dx,$$

i.e.,

$$\int_{\mathbb{R}}\int_K |y|^{2(\beta-l-1)}|\xi|^2|\hat{w}(y,\xi)|^2\,dy\,d\xi\ge c\int_{\mathbb{R}}\int_K |y|^{2(\beta-l-1)}\,|\hat{w}(y,\xi)|^2\,dy\,d\xi.$$

Since $(1+|y|)^{-2}\le 1$, the expression (27) squared does not exceed the integral (28) in the norm (23). □

11.2.3 Solvability of the Neumann problem

Now we are in a position to prove the solvability of problem (1) in a domain Ω. Denote by $\tilde{V}^{l+2}_{2,\beta}(\Omega;M)$ the space of functions in Ω with norm

$$\begin{aligned}&||u;\tilde{\mathbf{V}}^{l+2}_{2,\beta}(\Omega;M)||\\ &=\Big(||\rho^{\beta-l-1}u;\mathbf{L}_2(\Omega)||^2+\sum_{1\le|\alpha|\le l+2}||\rho^{\beta-l-2+|\alpha|}\,D^\alpha\,u;\mathbf{L}_2(\Omega)||^2\Big)^{1/2},\end{aligned} \tag{29}$$

where ρ is the distance of a point x to M.

Theorem 11.2.6. *Let l be a nonnegative integer and $l+\max\{0,1-\pi/\vartheta_0\}<\beta<l+1$. We suppose that $f\in\mathbf{V}^l_{2,\beta}(\Omega;M)$ and $\varphi\in\mathbf{V}^{l+1/2}_{2,\beta}(\partial\Omega;M)$ while (2) is valid. Then there exists a unique solution of the problem (1) subject to (3). Besides,*

$$||\,u;\,\tilde{\mathbf{V}}^{l+2}_{2,\beta}(\Omega;M)||\le c(||\,f;\,\mathbf{V}^l_{2,\beta}(\Omega;M)||+||\,\psi;\,\mathbf{V}^{l+1/2}_{2,\beta}(\partial\Omega;M)||).$$

Proof. The fact that the integrals in (2) and (3) take finite values follows from the inclusions given in the statement of the theorem. □

When proving Proposition 11.1.6 and Theorem 11.1.7, we applied Proposition 1.4 and Lemma 1.2 only to functions with compact supports. Now it enables us to use literally the same scheme replacing Lemma 1.2 with Lemma 2.4.

11.2.4 The problem in a domain with a contour excluded

Let M be a smooth simple closed contour situated within a three-dimensional domain Ω with smooth boundary. We consider the problem

$$-\Delta u(x) = f(x),\ x \in \Omega \setminus M; \quad u(x) = \varphi(x),\ x \in \partial\Omega, \tag{30}$$

where $f \in \mathbf{V}^l_{2,\beta}(\Omega; M)$ and $\varphi \in \mathbf{W}^{l+3/2}_2(\partial\Omega)$. We seek a solution u in $\tilde{\mathbf{V}}^{l+2}_{2,\beta}(\Omega; M)$. The norm in this space is defined by (29). The basic model problem for problem (30) is the following:

$$-\Delta u(x) = f(x), \quad x \in \mathbb{R}^3 \setminus \Gamma,$$

with $f \in \mathbf{V}^l_{2,\beta}(\mathbb{R}^3; \Gamma)$ and $\Gamma = \{x\colon x_1 = x_2 = 0\}$, $u \in \tilde{\mathbf{V}}^{l+2}_{2,\beta}(\mathbb{R}^3; \Gamma)$. The investigation of the model problem is similar to the reasoning in Section 2.2. The only distinction is that instead of Propositions 1.4.3 and 1.4.5 on the Neumann problem in a plane domain with angular point one should use Proposition 1.5.1 on the problem in a domain with a point deleted. Therefore, the condition $l + \max\{0, 1 - \pi\vartheta_0^{-1}\} < \beta < l + 1$ is replaced by $l < \beta < l + 1$. The next assertion is similar to Theorem 11.2.6.

Theorem 11.2.7. *Let l be a nonnegative integer and $\beta \in (l, l+1)$. Let also $f \in \mathbf{V}^l_{2,\beta}(\Omega; M)$, $\varphi \in \mathbf{W}^{l+3/2}_2(\partial\Omega)$. Then there exists a unique solution $u \in \tilde{\mathbf{V}}^{l+3/2}_{2,\beta}(\Omega; M)$ to problem* (30), *and*

$$\|u;\ \tilde{\mathbf{V}}^{l+2}_{2,\beta}(\Omega; M)\| \le c(\|f;\ \mathbf{V}^l_{2,\beta}(\Omega; M)\| + \|\varphi;\ \mathbf{W}^{l+3/2}_2(\partial\Omega)\|).$$

Remark 11.2.8. The evident analog of Proposition 11.2.3 is true. So, under the assumptions of Theorem 2.5, the operator A of problem (30) implements an isomorphism

$$A : \tilde{\mathbf{V}}^{l+2}_{2,\beta}(\Omega; M) \to \mathbf{V}^l_{2,\beta}(\Omega; M) \times \mathbf{W}^{l+3/2}_2(\partial\Omega).$$

11.3 The Asymptotics Near an Edge of Solutions to Boundary Value Problems for the Laplace Operator

In this section we derive the asymptotic formulas for solutions to the boundary value problems considered in the two preceding sections. The proofs for all of the problems are similar. We discuss only the Dirichlet problem in detail and give the corresponding formulations for the Neumann problem and for the problem in a domain with excluded contour.

11.3.1 Estimates for the derivatives of solutions along an edge

Let Ω be the same three-dimensional domain with edge M as in Section 1.1. When studying the asymptotics, we may consider solutions only in a small neighborhood of the edge. So, without loss of generality, we assume that the set Ω is situated in a tubular neighborhood of M and represented as $G \times M$, where G is a bounded plane domain with angular point O and $\partial G \setminus O$ is a smooth curve. We introduce coordinates (y, z) in Ω, where $y = (y_1, y_2)$ are Cartesian coordinates in the plane normal to M at the point $z \in M$. In view of the smoothness of M, the equality

$$(\partial/\partial z)\Delta = \Delta(\partial/\partial z) + \tilde{L}(x, \partial/\partial x), \tag{1}$$

holds, $\tilde{L}$ being a second order operator with coefficients smooth in $\bar{\Omega}$.

In what follows we suppose that the right-hand side of problem (3) smoothly depends on z. More precisely, we assume that

$$(\partial/\partial z)^k f \in \mathbf{V}^l_{2,\beta}(\Omega; M), \quad (\partial/\partial z)^k \varphi \in \mathbf{V}^{l+3/2}_{2,\beta}(\partial\Omega; M), \qquad k = 0, \dots, N+1, \tag{2}$$

with $|\beta - l - 1| < \pi/\vartheta_0$ while l and some nonnegative integers.

Differentiating equations (3) along the edge and using (1), we obtain that $u^{(1)} = \partial u/\partial z$ satisfies the boundary value problem

$$\begin{aligned} -\Delta u^{(1)}(x) &= (\partial/\partial z) f(x) + \tilde{L}(x, \partial/\partial x) u(x), \quad x \in \Omega; \\ u^{(1)}(x) &= (\partial/\partial z)\varphi(x), \quad x \in \partial\Omega \setminus M. \end{aligned} \tag{3}$$

By virtue of Theorem 11.1.7, $u \in \mathbf{V}^{l+2}_{2,\beta}(\Omega; M)$ and hence $\tilde{L}u \in \mathbf{V}^l_{2,\beta}(\Omega; M)$. Now from (2) it follows that the right-hand sides F and Φ of the equations (3) are subject to

$$\begin{aligned} &\| F; \mathbf{V}^l_{2,\beta}(\Omega; M) \| + \|\Phi; \mathbf{V}^{l+3/2}_{2,\beta}(\partial\Omega; M)\| \\ &\le c \sum_{k=0}^{1} \Big(\|(\partial/\partial z)^k f; \mathbf{V}^l_{2,\beta}(\Omega; M) \| + \|(\partial/\partial z)^k \varphi; \mathbf{V}^{l+3/2}_{2,\beta}(\partial\Omega; M) \| \Big). \end{aligned}$$

We apply Theorem 11.1.7 to problem (3) and obtain that $\partial u/\partial z \in \mathbf{V}^{l+2}_{2,\beta}(\Omega; M)$ and

$$\begin{aligned} &\|(\partial/\partial z)^n u; ; \mathbf{V}^{l+2}_{2,\beta}(\Omega; M)\| \\ &\le c \sum_{k=0}^{1} \Big(\| (\partial/\partial z)^k f; \mathbf{V}^l_{2,\beta}(\Omega; M)\| + \|(\partial/\partial z)^k \varphi; \mathbf{V}^{l+3/2}_{2,\beta}(\partial\Omega; M)\| \Big) \end{aligned} \tag{4}$$

for $n = 1$. The same reasoning for $\partial u/\partial z, \dots, \partial^{n-1}u/\partial z^{n-1}$ leads to the following assertion.

Lemma 11.3.1. *Let* (2) *hold with* $|\beta - l - 1/ < \pi/\vartheta_0$. *Then the solution* u *of problem* (3) *satisfies the inequality* (4) *for* $n = 0, 1, \dots, N+1$.

11.3.2 The asymptotics of solutions to the Dirichlet problem

Lemma 11.3.1 provides the possibility to differentiate in z without deterioration of the weight exponent. It enables us to reduce the derivation of the asymptotics of solutions of the problem to that of the Dirichlet problem in the domain G.

We rewrite (3), 11.1, in the form

$$\begin{aligned} -\Delta_y u(y,z) &= f(y,z) + \tilde{M}(y, z, \partial/\partial y, \partial/\partial z) u(y,z), \quad y \in G; \\ u(y,z) &= \varphi(y,z), \quad y \in \partial G \setminus \{0\}, \end{aligned} \tag{5}$$

where $\tilde{M}$ is a differential operator with smooth coefficients containing only $\partial/\partial y$, $\partial/\partial z$, $\partial^2/\partial y \partial z$, $\partial^2/\partial z^2$ and a multiplication operator.

Suppose that

$$(\partial/\partial z)^k f \in \mathbf{V}^l_{2,\gamma}(\Omega; M), \quad (\partial/\partial z)^k \varphi \in \mathbf{V}^{l+3/2}_{2,\gamma}(\partial\Omega; M) \tag{6}$$

for $k = 0, \dots, N+1$ and $\gamma < \beta$. Using Lemma 11.3.1 with $N \ge 1$, we obtain that the solution $u \in \mathbf{V}^{l+2}_{2,\beta}(\Omega; M)$ to problem (3), 11.1, satisfies

$$\tilde{M} u \in \mathbf{V}^{l+1}_{2,\beta}(\Omega; M) \subset \mathbf{V}^l_{2,\beta-1}(\Omega; M).$$

So the right-hand sides f_1 and φ in (5) are subject to the inclusions

$$f_1 \in \mathbf{V}^l_{2,\gamma_1}(\Omega; M), \qquad \varphi \in \mathbf{V}^{l+3/2}_{2,\gamma_1}(\partial\Omega; M) \tag{7}$$

with $\gamma_1 = \max\{\gamma, \beta - 1\}$. It follows that

$$f_1(\cdot, z) \in \mathbf{V}^l_{2,\gamma_1}(G), \qquad \varphi(\cdot, z) \in \mathbf{V}^{l+3/2}_{2,\gamma_1}(\partial G)$$

for almost all $z \in M$.

Now we are in a position to apply Proposition 1.3.7 on the asymptotics of solutions to the Dirichlet problem in a plane domain with corner point. By the proposition,

$$u(y, z) = c_1(z)\, r^{\pi/\vartheta_0} \sin(\pi\vartheta/\vartheta_0) + v_1(y, z), \tag{8}$$

(r, ϑ) being polar coordinates in the plane of y, $r = |y|$, $\vartheta \in [0, \vartheta_0]$, and $v_1(\cdot, z) \in \mathbf{V}^{l+2}_{2,\gamma_1-\delta}(G)$ for almost all z and small $\delta > 0$. If $\gamma_1 \geq l + 1 - \pi/\vartheta_0$, then $c_1(z) = 0$ in (8). In the case $\gamma_1 < l + 1 - \pi/\vartheta_0$ we have $c_1 \in \mathbf{L}_2(M)$ by virtue of the estimate

$$\int_M |c_1(z)|^2\, dz \leq c \int_M \left(\|f_1(\cdot, z);\, \mathbf{V}^l_{2,\gamma_1}(G)\|^2 + \|\varphi(\cdot, z);\, \mathbf{V}^{l+3/2}_{2,\gamma_1}(\partial G)\|^2 \right) dz$$

$$\leq c\left(\|f;\, \mathbf{V}^l_{2,\gamma_1}(\Omega; M)\|^2 + \|\varphi;\, \mathbf{V}^{l+3/2}_{2,\gamma}(\partial\Omega; M)\|^2\right)$$

(the first inequality follows from Corollary 1.3.9).

According to Lemma 3.1, $\partial u/\partial z \in \mathbf{V}^{l+2}_{2,\beta}(\Omega; M)$. Hence, $\partial u/\partial z \in \mathbf{V}^{l+2}_{2,\beta}(G)$ for almost all $z \in M$. Repeating the above reasoning to $\partial u/\partial z$, we obtain

$$(\partial/\partial z)u(y, z) = d(z)\, r^{\pi/\vartheta_0} \sin(\pi\vartheta/\vartheta_0) + w(y, z), \tag{9}$$

where $w(\cdot, z) \in \mathbf{V}^{l+2}_{2,\gamma_1-\delta}(G)$ for almost all $z \in M$ and $d \in \mathbf{L}_2(M)$. The terms on the right in (8) belong to different function classes. So the possibility to differentiate the sum enables us to differentiate term by term. Therefore, (9) takes the form

$$(\partial/\partial z)\, u(y, z) = (d/dz)c_1(z)\, r^{\pi/\vartheta_0} \sin(\pi\vartheta/\vartheta_0) + (\partial/\partial z)\, v_1(y, z).$$

If the inclusions (6) are fulfilled for $k = 1, 2, \ldots$, then we may argue as above to see that the functions c and $z \mapsto v(\cdot, z)$ are infinite differentiable.

Let f and φ admit the asymptotic expansions

$$f(y, z) = \sum_{j=0}^{J} f_j(\vartheta, \log r, z)\, r^{\mu_j - 2} + \tilde{f}_J(y, z), \tag{10}$$

$$\varphi(y, z) = \sum_{j=0}^{J} \varphi_j(\vartheta, \log r, z)\, r^{\mu_j} + \tilde{\varphi}_J(y, z), \tag{11}$$

where $\{\mu_j\}$ is an increasing positive sequence, $\lim \mu_j = \infty$, f_j and φ_j are polynomial in $\log r$ with coefficients smoothly depending on ϑ and z. The remainders are subject to the inclusions

$$(\partial/\partial z)^k)\tilde{f}_J(\cdot, z) \in \mathbf{V}^l_{2,l+1-\mu_J}(G), \quad (\partial/\partial z)^k \tilde{\varphi}_J(\cdot, z) \in \mathbf{V}^{l+3/2}_{2,l+1-\mu_J}(\partial G). \tag{12}$$

In order to obtain the asymptotic expansions for solutions to the problem (3), 11.1, one should combine the above derivation of asymptotic formulas of type (8) with consecutive computation of the terms in the asymptotic series for the solution to the plane Dirichlet problem having a special right-hand side (see Section 1.3.5).

Proposition 11.3.2. *Let the right-hand sides of equations* (3), 11.1, *satisfy* (12) *and* (13). *Then the solution u to problem* (3), 11.1, *admits the representation*

$$u(y,z) = \sum_{j=0}^{J} u_j(\vartheta, \log r, z)\, r^{\lambda_j} + \tilde{u}_J(y,z)$$

with any nonnegative integer J and polynomials u_j in $\log r$ whose coefficients smoothly depend on ϑ and z. The remainder $\tilde{u}_J$ is subject to

$$(\partial^k \tilde{u}_J/\partial z^k)(\cdot, z) \in V^{l+2}_{2,l+1-\lambda_J}(G)$$

and the sequence $\{\lambda_j\}$ consists of numbers of the form

$$\mu_j + h \quad \text{or} \quad k\pi\vartheta_0^{-1} + h, \tag{13}$$

where $j, h = 0, 1, \dots$, $k = 1, 2, \dots$.

11.3.3 The asymptotics of solutions to the Neumann problem

One can check the next assertion in the same way as Proposition 11.3.2.

Proposition 11.3.3. *Assume that f satisfies* (10) *and* (12) *and ψ admits the expansion*

$$\psi(y,z) = \sum_{j=0}^{J} \psi_j(\vartheta, \log r, z)\, r^{\mu_j - 1} + \tilde{\phi}_J(y,z),$$

while $(\partial^k \tilde{\psi}_J/\partial z^k)(\cdot, z) \in \mathbf{V}^{l+1/2}_{2,l+1-\mu_J}(\partial G)$ and the coefficients ψ_j possess the same properties as φ_j in (11). *If u is a solution to problem* (1) *then*

$$u(y,z) = u_0(z) + \sum_{j=1}^{J} u_j(\vartheta, \log r, z)\, r^{\lambda_j} + \tilde{u}_J(y,z). \tag{14}$$

Here $u_0 \in \mathbf{C}^\infty(M)$, the functions u_j and $\tilde{u}_J$ are such as in Proposition 11.3.2, *and the sequence $\{\lambda_j\}_{j=0}^{\infty}$ consists of numbers of the form* (13) *with $j, h, k = 0, 1, \dots$.*

11.3.4 The asymptotics of solutions to the boundary value problem in a domain with excluded contour

Proposition 11.3.4. *Let* (10) *and* (12) *hold for f in a small neighborhood of the contour M while $\{\mu_j\}_{j=0}^{\infty}$ is a sequence of positive integers. If u is a solution to the problem* (30), *then* (14) *holds while the sequence $\{\lambda_j\}_{j=1}^{\infty}$ consists of positive integers.*

The next assertion will be used in Chapter 12.

Corollary 11.3.5. *For any $\gamma \in \mathbf{C}^\infty(M)$ there exists a unique solution U_γ to the homogeneous problem* (8) *such that*

$$U_\gamma(y,z) = (2\pi)^{-1}\gamma(z)\log(1/r) + U^{(0)}(z) + + \sum_{j=1}^{J} U^{(j)}(\vartheta, \log r, z)\, r^j + \tilde{U}_J(y,z). \tag{15}$$

The notation is similar to that in (14).

Proof. Let ζ be a smooth nonnegative function that is 1 near M and 0 far from the contour. By Theorem 11.2.6, there exists a solution $u \in \tilde{\mathbf{V}}^{l+2}_{2,l+1/2}(\Omega; M)$ of problem (30) with $f(x) = (2\pi)^{-1}[\Delta, \zeta]\,(\gamma(z)\log|y|^{-1})$ and $\varphi = 0$. According to Proposition 11.3.4, the solution admits the representation (14). So $u(x) + \zeta(x)\,(2\pi)^{-1}\,\gamma(z)\,\log|y|^{-1}$ satisfies the homogeneous problem (8) and is represented by (15). The same theorem provides the uniqueness of the solution.

Remark 11.3.6. It is clear that $U_\gamma \in \mathbf{V}^{l+2}_{2,\beta}(\Omega; M)$ for $\beta > l+1$. Therefore, the kernel of the operator

$$A : \mathbf{V}^{l+2}_{2,\beta}(\Omega; M) \to \mathbf{V}^{l}_{2,\beta}(\Omega; M) \times \mathbf{W}^{l+3/2}_{2}(\partial\Omega)$$

is infinite dimensional for $\beta > l+1$. In the case $\beta \in (l-1, l+1)$ the cokernel turns out to be infinite dimensional, which follows from the relations $\zeta u_0 \notin \mathbf{V}^{l+2}_{2,\beta}(\Omega; M)$ and $\Delta(\zeta u_0) \in \mathbf{V}^{l}_{2,\beta}(\Omega; M)$ with $u_0 \in \mathbf{C}^\infty(M)$. One can show that the cokernel remains infinite dimensional as β decreases. The role of $\zeta(x)\,u_0(z)$ is successively played by the terms $\zeta(x)\,r^k\,\sin(k\vartheta)$ in the asymptotic series of the solution. Thus the operator A is not Fredholm for any β. □

11.4 Sobolev Problems

We state without proof some theorems on the solvability of the Sobolev problems in weighted classes and give asymptotic formulas for solutions. These assertions generalize the results obtained in Section 2 and 3 for the Laplace operator and will be of use in Chapter 13.

11.4.1 Statement of the problem

Let Ω_0 be a domain in $\mathbb{R}^n$ with smooth $(n-1)$-dimensional boundary $\partial\Omega_0$ and M a smooth d-dimensional submanifold in $\mathbb{R}^n$ without boundary while $M \subset \Omega_0$ and $0 < d < n-1$. We denote by $\tilde{P}$ a strongly elliptic operator of order $2m$ with smooth coefficients in $\bar{\Omega}_0$.

In the domain $\Omega = \Omega_0 \setminus M$, we consider the boundary value problem

$$\begin{aligned} &\tilde{P}(x, D_x)\,u(x) = f(x), \quad x \in \Omega; \\ &u(x) = \varphi_0(x), \ldots, (\partial/\partial\nu)^{m-1}\,u(x) = \varphi_{m-1}(x), \quad x \in \partial\Omega_0. \end{aligned} \tag{1}$$

11.4.2 Solvability of the problem

Introduce the space $\mathbf{V}^{l}_{2,\beta}(\Omega; M)$ with norm

$$||v; \mathbf{V}^{l}_{2,\beta}(\Omega; \Gamma)|| = \Big(\int_\Omega \sum_{|\alpha|=0}^{l} \rho(x)^{2(\beta-l+|\alpha|)}|D^\alpha u(x)|^2\,dx\Big)^{1/2},$$

where $\rho(x) = \mathrm{dist}(x, M)$ (cf. (1),11.1).

Theorem 11.4.1. *i) Let $n - d > 2m$ and $l + 2m - (n-d)/2 < \beta < l + (n-d)/2$. Then the operator*

$$A : \mathbf{V}^{l+2m}_{2,\beta}(\Omega; M) \to \mathbf{V}^{l}_{2,\beta}(\Omega; M) \times \prod_{j=0}^{m-1} \mathbf{W}^{l+2m-j-1/2}_{2}(\partial\Omega_0) \tag{2}$$

of problem (1) *is an isomorphism.*

ii) Let $n-d<2m$, $n-d=2k+1\,(k=1,2,\dots)$ and $l+m-1/2<\beta<l+m+1/2$. Then (2) is an isomorphism.

Remark 11.4.2. If β does not satisfy the assumptions of Theorem 11.4.1, then the corresponding operator A is not Fredholm.

Theorem 11.4.3. *Let $n-d=2k$ and $k\le m$. Then for $\beta\in(l+m-1,\,l+m)$ the operator*

$$A:\mathbf{V}^{l+2m}_{2,\beta}(\Omega;M)\to\mathbf{V}^{l}_{2,\beta}(\Omega;M)\times\prod_{j=0}^{m-1}\mathbf{W}_2^{l+2m-j-1/2}(\partial\Omega_0) \tag{3}$$

of the problem (1) is an isomorphism. Here $\mathbf{V}^{l+2m}_{2,\beta}(\Omega;M)$ is the space with norm

$$\|u;\tilde{\mathbf{V}}^{l+2m}_{2,\beta}(\Omega;M)\|=\Big(\|u;\mathbf{V}^{m-1}_{2,\beta-l-m+2}(\Omega;M)\|^2+\sum_{|\alpha|=m}\|D^\alpha u;\mathbf{V}^{l+m}_{2,\beta}(\Omega;M)\|^2\Big)^{1/2}.$$

(cf. (22), 11.2).

Remark 11.4.4. In the case $n-d=2k$, $k\le m$ the mapping (2) is not Fredholm for any $\beta\in\mathbb{R}$. If β does not satisfy the assumptions of Theorem 11.4.3, the mapping (3) is not Fredholm.

11.4.3 Asymptotics of a solution

Let $s\in M$ and let y be a point in the $(n-d)$-dimensional space normal to M at s. The pair (s,y) can be taken as local coordinates in a neighborhood U of the manifold M. We denote by χ a function in $\mathbf{C}_0^\infty(U)$ that equals 1 near M.

Theorem 11.4.5. *Let $\varphi\in\mathbf{C}^\infty(\partial\Omega_0)$, $j=0,\dots,m-1$ and let*

$$f(x)=\chi(x)\,\rho^{\mu-2m}\sum_{j=0}^{N}\rho^j f_j(s,\psi,\log\rho)+f^{(N)}(x), \tag{4}$$

where N is a positive integer, (ρ,ψ) are spherical coordinates in $\mathbb{R}^{n-d}$, and f_j are polynomials in $\log\rho$ whose coefficients are smooth functions in $s\in M$ and $\psi\in S^{n-d-1}$ while

$$D_x^\alpha f^{(N)}(x)=o(\rho^{\mu-2m+N-|\alpha|})\quad\text{in}\quad U.$$

Then the following assertions are valid:

i) If $n-d>2m$ and μ is an integer in $(2m-n+d,0]$, then

$$u(x)=\chi(x)\,\rho^{\mu}\sum_{j=0}^{N}\rho^j u_j(s,\psi,\log\rho)+u^{(N)}(x), \tag{5}$$

where u_j is a unique solution to problem (1), u_j being of the same type as f_j, and

$$D_x^\alpha u^{(N)}(x)=o(\rho^{\mu+N-|\alpha|})\quad\text{in}\quad U.$$

ii) If $n-d=2k+1<2m$ and $\mu=m-k$, then (5) holds, where u is a unique solution to the problem (1).

iii) If $n - d = 2k \leq 2m$ *and* $\mu = m - k + 1$, *then*

$$u(x) = \chi(x)\,(p_{m-k}\,(s,y) + \rho^{m-k+1}\sum_{j=0}^{N}\rho^j u_j\,(s,\psi,\log\rho) = u^{(N)}(x),$$

where p_{m-k} *are homogeneous polynomials of degree* $m - k$ *in* y *having coefficients in* $\mathbf{C}^\infty(M)$, *and the remaining notation has the same sense as in* (5).

Chapter 12

Asymptotics of Solutions to Classical Boundary Value Problems in a Domain with Thin Cavities

In the present chapter we describe the asymptotics of solutions some of boundary value problems in two- or three-dimensional domains perturbed near a closed curve or a segment. In Section 12.1 and 12.2 we consider the Neumann and Dirichlet problems for the Laplace operator in the exterior of a thin closed tube. The first boundary value problem for the Lamé system in the same geometric situation is studied in Section 12.3. Then we return to the Laplace operator and discuss the Dirichlet problem in a domain perturbed near an interval; Section 12.4 is devoted to the two-dimensional case and Section 12.5 to the three-dimensional one. In Section 12.6 we derive the asymptotics of a solution to the nonlinear equation $\Delta u + u^2 = f$ in a domain with boundary smoothed near an edge.

12.1 The Asymptotics of a Solution of the Neumann Problem in the Exterior of a Thin Tube

The asymptotics will be obtained by iterative processes similar to those in 2.2. It is intuitively clear that the algorithm reduces to solving two limiting problems: the Poisson equation on the whole space and the exterior two-dimensional Neumann problem. In comparison with the case of perturbation near an isolated singularity, the new fact here is the dependence of the second limiting problem on a parameter. The tube shrinks to a contour and the parameter runs over the contour. Both of the limiting problems are uniquely solvable. Owing to this, the algorithm turns out to be relatively simple.

12.1.1 Statement of the problem

Let M be the unit circle $\{x = (x_1, x_2, x_3) \in \mathbb{R}^3 : x_1^2 + x_2^2 = 1,\ x^3 = 0\}$ in $\mathbb{R}^3$ and $\varphi \in [0, 2\pi)$ the natural parameter. In the half-plane Π_φ containing the axis x_3 and a point $\varphi \in M$ we introduce Cartesian coordinates $y = (y_1, y_2)$ with origin at φ (see Fig. 12.1).

Denote by ω a domain in $\mathbb{R}^2$ and assume that its boundary $\partial\omega$ is a simple closed piece-wise smooth contour (Fig. 12.2). We suppose that ω contains the origin O and introduce the set $\omega_\varepsilon = \{y \in \mathbb{R}^2 : \varepsilon^{-1} y \in \omega\}$ with small positive ε . Finally, let

$$\Omega_\varepsilon = \{x \in \mathbb{R}^3 : \varphi \in M,\ y \in \omega_\varepsilon\}. \tag{1}$$

We consider the boundary value problem

$$-\Delta u(\varepsilon, x) = f(x), \quad x \in \mathbb{R}^3 \setminus \bar{\Omega}_\varepsilon; \tag{2}$$

$$(\partial/\partial n) u(\varepsilon, x) = g(\varphi, \varepsilon^{-1} y), \quad x \in \partial\Omega_\varepsilon, \tag{3}$$

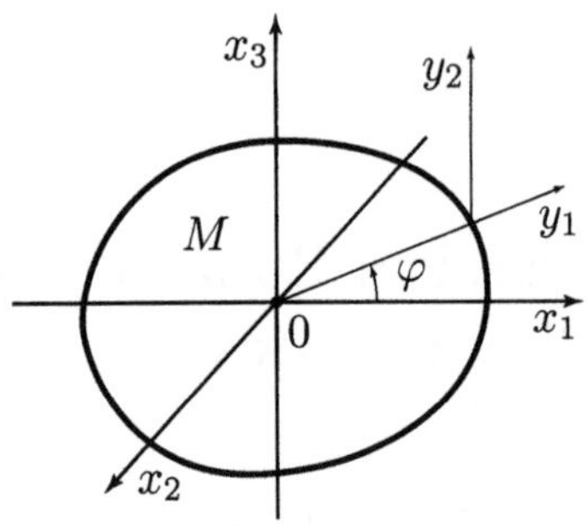

Fig. 12.1

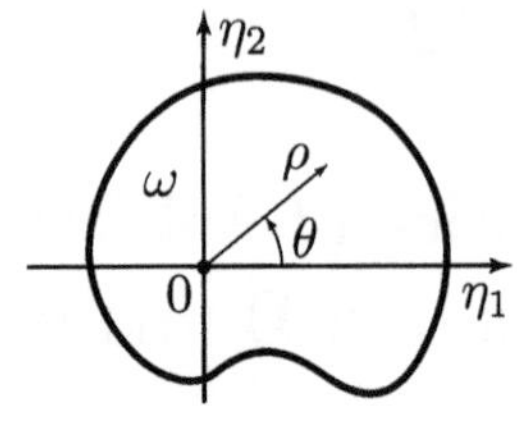

Fig. 12.2

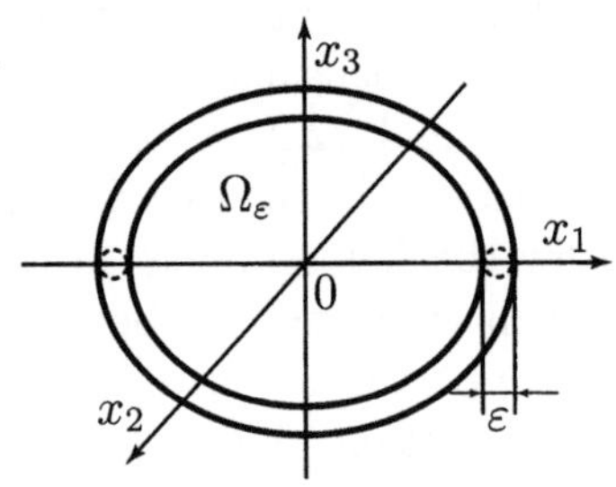

Fig. 12.3

with the supplementary requirement

$$u(\varepsilon, x) = O(|x|^{-1}), \quad |x| \to +\infty. \tag{4}$$

Here $f \in \mathbf{C}_0^\infty(\mathbb{R}^3)$, $g \in \mathbf{C}^\infty(M \times \partial\omega)$, and n is the outward normal on $\partial\Omega_\varepsilon$. It is known that there exists a unique solution to the exterior Neumann problem having the finite Dirichlet integral (or subject to (4)).

We associate to (2), (3) two limiting problems: the equation

$$-\Delta v^{(0)}(x) = f(x), \quad x \in \mathbb{R}^3 \setminus M, \tag{5}$$

and the Neumann problem

$$-\Delta_\eta W(\eta) = F(\eta), \quad \eta \in \mathbb{R}^2 \setminus \bar{\omega}; \quad (\partial/\partial\nu)W(\eta) = G(\eta), \quad \eta \in \partial\omega. \tag{6}$$

12.1.2 Principal terms of asymptotics

We construct two principal terms of the asymptotic expansion in ε of a solution to the problem (2), (3). The solution of equation (5) which vanishes at infinity is given by

$$v^{(0)}(x) = (4\pi)^{-1} \int_{\mathbb{R}^3} f(\xi)|x-\xi|^{-1} d\xi. \tag{7}$$

The potential (7) satisfies (2) and gives (3) the discrepancy

$$g(\varphi, y/\varepsilon) - (\partial/\partial n)v^{(0)}(x) = g(\varphi, y/\varepsilon) - (\partial/\partial n)\big(y \cdot \nabla_y v^{(0)}(\varphi, 0)\big) + O(\varepsilon). \tag{8}$$

We make it up with the help of $\varepsilon w^{(0)}(\varphi, \varepsilon^{-1}y)$. Let us write the operator $-\Delta_x$ near M in the coordinates (φ, y_1, y_2),

$$-\Delta_x = -\Delta_y - (1+y_2)^{-1}(\partial/\partial y_2) - (1+y_2)^{-2}(\partial/\partial\varphi)^2. \tag{9}$$

After the change of variables $y \to \eta = \varepsilon^{-1}y$, the section ω_ε of Ω_ε by the plane Π_φ coincides with the domain ω independent of ε and

$$-\Delta_x = -\varepsilon^{-2}\Delta_\eta - \varepsilon^{-1}\tilde{L}(\varepsilon\eta_2, \partial/\partial\eta_2, \partial/\partial\varphi), \tag{10}$$

where $\tilde{L} = -(1+\varepsilon\eta_2)^{-1}(\partial/\partial\eta_2) - \varepsilon(1+\varepsilon\eta_2)^{-2}(\partial/\partial\varphi)^2$.

Let χ be a cut-off function,

$$\chi \in \mathbf{C}^\infty(\mathbb{R}); \qquad \chi(t) = 1, \quad |t| < 1/3; \quad \chi(t) = 0, \quad |t| \geq 2/3. \tag{11}$$

We apply the operator $-\Delta$ to the sum $v^{(0)}(x) + \varepsilon\chi(|y|)\, w^{(0)}(\varphi, \varepsilon^{-1}y)$ in $\mathbb{R}^3 \setminus \bar{\Omega}_\varepsilon$ and calculate its normal derivative on $\partial\Omega_\varepsilon$. The discrepancy in (2) will be small only if the function $\eta \to w^{(0)}(\varphi, \eta)$ is harmonic in $\mathbb{R}^2 \setminus \bar{\omega}$ (see (10)). Taking (8) into account, we arrive at the problem

$$\begin{aligned} &-\Delta_\eta w^{(0)}(\varphi, \eta) = 0, \qquad \eta \in \mathbb{R}^2 \setminus \bar{\omega}; \\ &(\partial/\partial\nu) w^{(0)}(\varphi, \eta) = g(\varphi, \eta) - (\partial/\partial\nu)\big(\eta \cdot \nabla_y v^{(0)}(\varphi, 0)\big), \quad \eta \in \partial\omega, \end{aligned} \tag{12}$$

where $\eta \cdot \nabla_y v^{(0)}(\varphi, 0) = \eta_1 (\partial/\partial y_1) v^{(0)}|_{y=0} + \eta_2 (\partial/\partial y_2) v^{(0)}|_{y=0}$ is a smooth function in φ. A solution of problem (12) is determined up to an additive constant and increases as $\log|\eta|$ at infinity (see Proposition 1.5.4). We choose the solution which satisfies

$$w^{(0)}(\varphi, \eta) = C(\varphi) \log|\eta| + O(|\eta|^{-1}), \quad |\eta| \to \infty. \tag{13}$$

Green's formula gives an expression for $C(\varphi)$. Indeed, the function $w^{(0)}$ being harmonic,

$$\begin{aligned} -\int_{\partial\omega} (\partial/\partial\nu)\, w^{(0)}(\varphi, \eta)\, dl_\eta &= R \int_0^{2\pi} (\partial/\partial\rho) w^{(0)}(\varphi, \eta)\, d\vartheta \\ &= \int_0^{2\pi} R\big(C(\varphi)/R + O(R^{-2})\big)\, d\vartheta = 2\pi C(\varphi) + O(R^{-1}). \end{aligned}$$

Letting $R \to \infty$ shows that

$$C(\varphi) = -(2\pi)^{-1} \int_{\partial\omega} \{g(\varphi, \eta) - (\partial/\partial\nu)\, (\eta \cdot \nabla_y v^{(0)}(\varphi, 0))\}\, dl_\eta.$$

Since

$$-\int_{\partial\omega} (\partial/\partial\nu)(\eta \cdot \Delta_y v^{(0)}(\varphi, 0))\, dl_\eta = \int_\omega \sum_{j=1}^{2} (\partial/\partial y_j) v^{(0)}|_{y=0} \Delta_\eta\, \eta_j\, d\eta = 0,$$

we obtain

$$C(\varphi) = -(2\pi)^{-1} \int_{\partial\omega} g(\varphi, \eta)\, dl_\eta.$$

The difference $R^{(0)}(\varepsilon, x) = u(\varepsilon, x) - v^{(0)}(x) - \varepsilon\chi(|y|) w^{(0)}(\varphi, \varepsilon^{-1}y)$ satisfies a boundary value problem of the form (2), (3) with right-hand side $O(\varepsilon(|y|^{-1} + |\log\varepsilon|))$ on $\mathbb{R}^3 \setminus \bar{\Omega}_\varepsilon$ and $O(\varepsilon)$ on $\partial\omega_\varepsilon$. From the standard energy estimate it follows that $R^{(0)}$ is small. The principal term of asymptotics is thus constructed.

The function $\varepsilon\chi(|y|)w^{(0)}(\varphi, \varepsilon^{-1}y)$ compensates the main term of discrepancy of $v^{(0)}$ in (3). However, there remains the discrepancy in (2). According to (10), the Laplace operator admits the asymptotic expansion

$$-\Delta_x \sim -\varepsilon^{-2}\Delta_\eta - \varepsilon^{-2}\sum_{k=1}^{\infty}\varepsilon^k L_k(\eta_2, \partial/\partial\eta_2, \partial/\partial\varphi) \tag{14}$$

with

$$L_k(\eta_2, \partial/\partial\eta_2, \partial/\partial\varphi) = (-\eta_2)^{k-1}\,\partial/\partial\eta_2 + (k-1)(-\eta_2)^{k-2}\,\partial^2/\partial\varphi^2.$$

Therefore, the mentioned discrepancy takes the form

$$-\varepsilon[\Delta_x, \chi(r)]\,w^{(0)}(\varphi,\eta) - \chi(r)\sum_{k=1}^{\infty}\varepsilon^{k-1}L_k(\eta_2, \partial/\partial\eta_2, \partial/\partial\varphi)w^{(0)}(\varphi,\eta), \tag{15}$$

where $[P, Q] = PQ - QP$, $r = |y|$, and χ is the cut-off function (11). Since $w^{(0)}$ is a harmonic function in $\mathbb{R}^3 \setminus \bar{\omega}$, we have the convergent series

$$w^{(0)}(\varphi,\eta) = C(\varphi)\log\rho = \sum_{k=1}^{\infty}\left(A_k(\varphi)\sin k\vartheta + B_k(\varphi)\cos k\vartheta\right)\rho^{-k} \tag{16}$$

outside the disc $B_d = \{\eta \in \mathbb{R}^2 : |\eta| = \rho < d\} \subset \bar{\omega}$. Using (16), we rewrite the first term in (15) in the form

$$\begin{aligned} &-[\Delta_x, \chi(r)]\left\{C(\varphi)\log r - C(\varphi)\log\varepsilon\right\} \\ &-\sum_{k=1}^{\infty}\varepsilon^{k+1}[\Delta_x, \chi(r)]\left\{(A_k(\varphi)\sin k\vartheta + B_k(\varphi)\cos k\vartheta)\,r^{-k}\right\}. \end{aligned} \tag{17}$$

The coefficients of the operator $[\Delta_x, \chi(r)]$ vanish for $r \leq 1/3$ and for $r \geq 2/3$, because they contain derivatives of χ. So every term in (17) is a smooth function on $\mathbb{R}^3$. Therefore, the terms of series (17) can be compensated with the help of solutions of the first limiting problem. All the remaining terms in (15) are equal to $\chi(r)$ multiplied by functions depending on φ and η. Hence, it is convenient to take them into account by means of boundary layer functions. However, there is the following difficulty. The coefficient Φ_k by ε^k, $k = 1, 2, \dots$, is equal to $O(|\eta|^{k-2})$ as $|\eta| \to \infty$, and the boundary layer function, satisfying

$$\Delta_\eta W(\varphi,\eta) = \Phi_k(\varphi,\eta), \quad \eta \in \mathbb{R}^2 \setminus \bar{\omega},$$

increases at infinity as $O(|\eta|^k)$, which is inadmissible. To circumvent the difficulty we use the simplest version of the process of redistribution of discrepancies; such a process was not used at all in Chapters 2 and 4. Namely, we separate the terms whose orders exceed that of $|\eta|^{-2}|\log|\eta||^k$, rewrite them into the coordinates x and compensate them when we solve the first limit problem, i.e., an equation of the form (5). More precisely, taking

$$\begin{aligned} L_k(\eta_2, \partial/\partial\eta_2, \partial/\partial\varphi) &= \varepsilon^{2-k}L_k(y_2, \partial/\partial y_2, \partial/\partial\varphi), \\ L_k(\eta_2, \partial/\partial\eta_2, \partial/\partial\varphi)\,\rho^s &= O(\rho^{s+k-2}) \end{aligned}$$

and (16) into account, we rewrite the series in (15) :

$$\chi(r)\sum_{k=1}^{\infty}\varepsilon^{k-1}L_k(\eta_2,\partial/\partial\eta_2,\partial/\partial\varphi)\Big\{C(\varphi)\log\rho+\sum_{j=1}^{\infty}\rho^{-j}(A_j(\varphi)\sin j\vartheta+B_j(\varphi)\cos j\vartheta)\Big\}$$

$$=\chi(r)\sum_{k=1}^{\infty}\varepsilon^{k-1}L_k(\eta_2,\partial/\partial\eta_2,\partial/\partial\varphi)\sum_{j=k}^{\infty}\rho^{-j}(A_j(\varphi)\sin j\vartheta+B_j(\varphi)\cos j\vartheta)$$

$$+\chi(r)\sum_{k=1}^{\infty}\varepsilon L_k(y_2,\partial/\partial y_2,\partial/\partial\varphi)$$

$$\times\Big\{C(\varphi)(\log r-\log\varepsilon)+\sum_{j=1}^{\infty}\varepsilon^j r^{-j}(A_j(\varphi)\sin j\vartheta+B_j(\varphi)\cos j\vartheta)\Big\}. \tag{18}$$

Let us consider the coefficients of the first series on the right in (18)

$$\chi(r)\sum_{k=1}^{\infty}\varepsilon^{k-1}L_k(\eta_2,\partial/\partial\eta_2,\partial/\partial\varphi)$$
$$\times\Big\{w^{(0)}(\varphi,\eta)-C(\varphi)\log\rho-\sum_{j=1}^{k-1}\rho^{-j}(A_j(\varphi)\sin j\vartheta+B_j(\varphi)\cos j\vartheta)\Big\}. \tag{19}$$

They are $O(|\eta|^{-2})$ as $|\eta|\to\infty$ and may be taken as right-hand sides for the second limit problem (6). To assign a sense to the last series in (18) we interchange the order of summations

$$\varepsilon\chi(r)\sum_{k=1}^{\infty}L_k(y_2,\partial/\partial y_2,\partial/\partial\varphi)\{C(\varphi)(\log r-\log\varepsilon)\}$$

$$+\chi(r)\sum_{j=1}^{\infty}\varepsilon^{j+1}\sum_{k=j+1}^{\infty}L_k(y_2,\partial/\partial y_2,\partial/\partial\varphi)\{r^{-j}(A_j(\varphi)\sin j\vartheta+B_j(\varphi)\cos j\vartheta)\}$$
$$=\varepsilon\chi(r)(\Delta_x-\Delta_y)\{C(\varphi)(\log r-log\varepsilon)\}$$

$$+\chi(r)\sum_{j=1}^{\infty}\varepsilon^{j+1}\Big(\Delta_x-\Delta_y-\sum_{k=1}^{j}L_k(y_2,\partial/\partial y_2,\partial/\partial\varphi)\Big)$$
$$\times\{r^{-j}(A_j(\varphi)\sin j\vartheta+B_j(\varphi)\cos j\vartheta)\}. \tag{20}$$

Every term on the right in (20) is $O(r^{-1}|\log r|)$ as $r=|y|\to 0$. From (7) it follows that the solution of equation (5) with such a right-hand side is continuous. So the coefficients in (20) may be taken as right-hand sides for the first limit problem.

Thus, the redistribution of discrepancies is complete for the first step of the algorithm. We return now to deriving the asymptotics of the solution to the problem (2), (3). We seek $v^{(1)}$ and $w^{(1)}$ that give the second approximation

$$v^{(0)}(x)+\varepsilon\big(\chi(r)w^{(0)}(\varphi,\varepsilon^{-1}y)+v^{(1)}(x,\log\varepsilon)\big)+\varepsilon^2\chi(r)w^{(1)}(\varphi,\varepsilon^{-1}y,\log\varepsilon)$$

to the solution u of the original problem. By virtue of (20), the function $v^{(1)}$ linearly depends on $\log\varepsilon$ and satisfies

$$-\Delta_x v^{(1)}(x,\log\varepsilon)=\{[\Delta_x,\chi(r)]+\chi(r)(\Delta_x-\Delta_y)\}\{C(\varphi)(\log r-\log\varepsilon)\}. \tag{21}$$

We check formally that $v^{(1)}$ admits the asymptotic expansion

$$v^{(1)}(x, \log\varepsilon) = C^{(1)}(\varphi, \log\varepsilon) + r v_1^{(1)}(\varphi, \vartheta, \log r) + O(|\log\varepsilon| r^2 |\log r|^2). \tag{22}$$

One can use the results in Section 1.3 and 1.5 to justify the reasoning in what follows. In (22) $v_1^{(1)}$ smoothly depends on ϑ and φ and linearly on $\log r$. To find this function we substitute (22) in (21) and, using the expansion (14) for the Laplace operator, separate the terms of order $O(r^{-1})$ as $r \to 0$. Then

$$-r^{-2}\big((r\partial/\partial r)^2 + (\partial/\partial\vartheta)^2\big)\big(r v_1^{(1)}(\varphi, \vartheta, \log r)\big) = C(\varphi)\,(\partial/\partial y_2)\,\log r. \tag{23}$$

We split $v_1^{(1)}$ into a sum

$$\varkappa(\varphi)\log r\,\sin\vartheta + \Psi(\varphi, \vartheta). \tag{24}$$

Substituting (24) in (23), we obtain that $\varkappa$ and Ψ satisfy the differential equation on the circle

$$(\partial/\partial\vartheta)^2\Psi(\varphi, \vartheta) + \Psi(\varphi, \vartheta) + 2\varkappa(\varphi)\,\sin\vartheta = -C(\varphi)\sin\vartheta.$$

Therefore,

$$\varkappa(\varphi) = -C(\varphi)/2 = (4\pi)^{-1}\int_{\partial\omega} g(\varphi, \eta)\,dl_\eta$$

and

$$\Psi(\varphi, \vartheta) = V_{11}^{(1)}(\varphi)\,\sin\vartheta + V_{12}^{(1)}(\varphi)\,\cos\vartheta,$$

while $V_{1,q}^{(1)}$ are arbitrary functions of φ. So the solution of equation (21) takes the form

$$v^{(1)}(x, \log\varepsilon) = C^{(1)}(\varphi, \log\varepsilon) + \chi(r)(4\pi)^{-1}y_2\log r\int_{\partial\omega} g(\varphi, \eta)\,dl_\eta + \tilde V^{(1)}(x, \log\varepsilon), \tag{25}$$

where $C^{(1)} \in \mathbf{C}^\infty(M)$ and $\tilde V^{(1)} \in \mathbf{C}^1(\mathbb{R}^3)$.

Now we state the problem to determine the boundary layer $w^{(1)}$. Using (19) and the discrepancy of $v^{(0)} + \varepsilon\chi w^{(0)} + \varepsilon v^{(1)}$, we obtain the following boundary value problem:

$$-\Delta_\eta w^{(1)}(\eta, \log\varepsilon) = (\partial/\partial\eta_2)\{w^{(0)})\varphi, \eta) - C(\varphi)\log\rho\}, \qquad \eta \in \mathbb{R}^2 \setminus \bar\omega; \tag{26}$$

$$\begin{aligned}(\partial/\partial\nu)w^{(1)}(\eta) = -(\partial/\partial\nu)\{&\eta\cdot\nabla_{2,y}v^{(0)}(\varphi, 0)\cdot\eta + \eta\cdot\nabla_y\tilde V^{(1)}(\varphi, 0, \log\varepsilon)\\ &+ (4\pi)^{-1}\eta_2\int_{\partial\omega} g(\varphi, \eta)\,dl_\eta\,(\log\rho + \log\varepsilon)\}, \quad \eta \in \partial\omega.\end{aligned} \tag{27}$$

Here we used (14), (25), and the notation

$$\eta\cdot\nabla_{2,y}v^{(0)}(\varphi, 0)\cdot\eta = 2^{-1}\sum_{p,q=1}^{2}\eta_p\eta_q(\partial/\partial y_p)(\partial/\partial y_q)v^{(0)}\big|_{y=0}.$$

Let us turn to the asymptotic expansion of the solution to problem (26). By virtue of (16), one can, far from $\bar\omega$, write the right-hand side of (26) in the form

$$\sum_{j=1}^{\infty} j\rho^{-j-1}\big(A_j(\varphi)\cos(j+1)\vartheta - B_j(\varphi)\sin(j+1)\vartheta\big). \tag{28}$$

The series

$$4^{-1}\sum_{k=0}^{\infty}\rho^{-k}\big(A_{k+1}(\varphi)\cos(k+2)\vartheta - B_{k+1}(\varphi)\sin(k+2)\vartheta\big) \tag{29}$$

is evidently a special solution of the Poisson equation with right-hand side (28). Hence, outside $B_d \supset \bar{\omega}$ the solution of problem (26), (27) is the sum of the series (29) and an analytic function that is represented by

$$C^{(1)}(\varphi,\log\varepsilon)\,\log\rho + \sum_{j=1}^{\infty}\rho^{-j}\{A_j^{(1)}(\varphi,\log\varepsilon)\,\sin j\vartheta + B_j^{(1)}(\varphi,\log\varepsilon)\cos j\vartheta\}. \tag{30}$$

We describe dependence of $C^{(1)}$ on $\log\varepsilon$ in more detail. By Green's formula

$$\begin{aligned}
&-\int_{\partial\omega}\cos(\eta_2,\nu)\tilde{H}(\varphi,\eta)\,dl_\eta - R\int_0^{2\pi}\sin\vartheta\,\tilde{H}(\varphi,\eta)|_{|\eta|=R}\,d\vartheta\\
&= \int_{B_R\setminus\omega}\Delta_\eta\, w^{(1)}(\varphi,\eta)\,d\eta\\
&= \int_{\partial\omega} h(\varphi,\eta,\log\varepsilon)\,dl_\eta + R\int_0^{2\pi}(\partial/\partial\rho)\,w^{(1)}(\varphi,\eta,\log\varepsilon)|_{|\eta|=R}\,d\vartheta,
\end{aligned}$$

where $\tilde{H}(\varphi,\eta) = w^{(0)}(\varphi,\eta) - C(\varphi)\,\log\rho$ while $h(\varphi,\eta,\log\varepsilon)$ is the right-hand of (27). Now we use the asymptotics (16) of $w^{(0)}$ and the fact that $w^{(1)}$ is equal to the sum of series (29) and (30). Then

$$\begin{aligned}
&-\int_{\partial\omega}\big(\cos(\eta_2,\nu)\tilde{H}(\varphi,\eta) + h(\varphi,\eta,\log\varepsilon)\big)\,dl_\eta\\
&= R\int_0^{2\pi}\Big\{R^{-1}C^{(1)}(\varphi,\log\varepsilon) + \sin\vartheta\,\{A_1(\varphi)\,\sin\vartheta + B_1(\varphi)\cos\vartheta)\,\}R^{-1} + O(R^{-2})\Big\}\,d\vartheta\\
&= 2\pi C^{(1)}(\varphi,\log\varepsilon) + \pi A_1(\varphi) + O(R^{-1}).
\end{aligned} \tag{31}$$

Also,

$$\begin{aligned}
\int_{\partial\omega} h(\varphi,\eta,\log\varepsilon)\,dl_\eta &= \int_{\omega}\Delta_\eta\,\eta\cdot\nabla_{2,y}v^{(0)}(\varphi,0)\cdot\eta + \eta\cdot\nabla_y\,\tilde{V}^{(1)}(\varphi,0,\log\varepsilon)\\
&\quad + (4\pi)^{-1}\,\eta_2\Big(\log\rho + \log\varepsilon\Big)\int_{\partial\omega} g(\varphi,\eta)dl_\eta\Big)\,d\eta\\
&= \int_{\omega}\Big(\Delta_y\,v^{(0)}|_{y=0} + \rho^{-2}\eta_2(2\pi)^{-1}\int_{\partial\omega} g(\varphi,\eta)dl_\eta\Big)\,d\eta.
\end{aligned}$$

Letting $R \to \infty$ in (31), we obtain

$$C^{(1)}(\varphi, \log\varepsilon) = -A_1(\varphi)/2 - (2\pi)^{-1} \int\limits_{\partial\omega} \cos(\eta_2, \nu)\{w^{(0)}(\varphi, \eta) - C(\varphi)\log\rho\}\, dl_\eta$$
$$- (2\pi)^{-1}\mathrm{mes}_2\,\omega\Delta_y v^{(0)}(\varphi, 0) - (2\pi)^{-2} \int\limits_{\partial\omega} g(\varphi,\eta) dl_\eta \int\limits_{\omega} \rho^{-1}\sin\vartheta\, d\eta, \tag{32}$$

where $\mathrm{mes}_2\,\omega$ denotes the measure of ω. In particular, from (32) it follows that $C^{(1)}$ in (30) does not depend on $\log\varepsilon$.

12.1.3 Asymptotic series

In 12.1.2 we found two first terms of the asymptotics of the solution to problem (2), (3). We are going to get the asymptotic expansion

$$u(\varepsilon, x) \sim \sum_{k=0}^{\infty} \varepsilon^k \left(v^{(k)}(x, \log\varepsilon) + \varepsilon\chi(r) w^{(k)}(\varphi, \varepsilon^{-1}y, \log\varepsilon)\right) \tag{33}$$

for the solution. Here $v^{(0)}$ and $v^{(1)}$ are the above solutions to (5) and (21), $w^{(0)}$ and $w^{(1)}$ satisfy the boundary value problem (21) and (26), (27), respectively; $v^{(k)}$ and $w^{(k)}$ with $k \geq 2$ are similar solutions to the first and second limit problems whose right-hand sides will be defined later on; finally, $v^{(k)}$ and $w^{(k)}$ are polynomials in $\log\varepsilon$.

To find the coefficients in (33) we need the asymptotic expansions of solutions to the limit problems as $\rho \to \infty$ or $r \to 0$. The next two assertions are contained in Proposition 1.5.5 and 11.2.6 as particular cases.

Lemma 12.1.1. *Let $G \in \mathbf{C}^\infty(\partial\omega)$, $F \in \mathbf{C}^\infty(\mathbb{R}^2 \setminus \omega)$, and*

$$F(\eta) = \sum_{j=0}^{J} \rho^{-j-2}\, F_j(\vartheta, \log\rho) + F^{(J)}(\eta), \tag{34}$$

where $F_j(\vartheta, t)$ is a polynomial in t with coefficients smooth on the circle S^1 and (ρ, ϑ) are polar coordinates. Suppose that

$$|(\partial/\partial\eta_1)^p\,(\partial/\partial\eta_2)^q\, F^{(J)}(\eta)| \leq c_{pq}^{(J)} |\eta|^{-J-3-p-q+\delta}, \quad \delta > 0.$$

Then there exists a solution W to the problem (6) that is dominated by a power of $\log|\eta|$ as $|\eta| \to \infty$. The solution is determined up to an additive constant and admits the representation

$$W(\eta) = \sum_{j=0}^{\infty} \rho^{-j}\, W_j(\vartheta, \log\rho) + \tilde{W}_J(\eta) \tag{35}$$

for any integer J. Here $W_j(\vartheta, t)$ is a polynomial in t with coefficients smooth on S^1, $\deg W_q(\vartheta, t) \leq 1 + \deg F_q(\vartheta, t)$, $q \in \mathbb{N}$, $\deg W_0(\vartheta, t) \leq 2 + \deg F_0(\vartheta, t)$, and

$$|(\partial/\partial\eta_1)^p\,(\partial/\partial\eta_2)^q\, W_J(\eta)| \leq C_{pq}^{(J)} |\eta|^{-1-J-p-q+\delta}, \quad \delta > 0.$$

Lemma 12.1.2. *Let the right-hand side $f \in \mathbf{C}^\infty(\mathbb{R}^3 \setminus M)$ in (5) have a compact support in $\mathbb{R}^3$ and admit the representation*

$$f(x) = \chi(r) \sum_{j=0}^{J} r^{j-1} f_j(\vartheta, \log r, \varphi) + f^{(J)}(x), \tag{36}$$

where $f_j(\vartheta, t, \varphi)$ is a polynomial in t with coefficients smoothly depending on $\vartheta \in S^1$ and $\varphi \in M$ while χ is the cut-off function (11). Suppose that

$$|(\partial/\partial x_1)^p (\partial/\partial x_2)^q (\partial/\partial x_3)^n f^{(J)}(x)| \leq c^{(J)}_{pqn} r^{J-p-q-n-\delta}, \quad \delta > 0.$$

Then for the solution $v^{(0)}$ of equation (5) which satisfies (4) there is the asymptotic formula

$$v^{(0)}(x) = \chi(r)\Big(c(\varphi) + \sum_{j=1}^{J} r^j v^{(0)}_j(\vartheta, \log r, \varphi)\Big) + \tilde{v}^{(0)}_J(x). \tag{37}$$

Here J is an arbitrary integer, $c \in \mathbf{C}^\infty(M)$, and $v^{(0)}_j(\vartheta, t, \varphi)$ are polynomials in t whose coefficients smoothly depend on $\vartheta \in S^1$ and $\varphi \in M$. Finally,

$$|(\partial/\partial x_1)^p (\partial/\partial x_2)^q (\partial/\partial x_3)^n \tilde{v}^{(0)}_J(x)| \leq C^{(J)}_{pqn} r^{J+1-p-q-n-\delta}, \quad \delta > 0.$$

We assume that $v^{(k)}$ and $w^{(k)}$ in (33) admit representations of the form (37) and (35), respectively, and denote by $v^{(k)}_j(\varphi, \vartheta, \log r, \log\varepsilon)$ and $w^{(k)}_j(\varphi, \vartheta, \log\rho, \log\varepsilon)$ the coefficients by r^j and ρ^{-j} in the representations. Note that for $w^{(0)}_j$ and $w^{(1)}_j$ the same role is played by (16), (29), and (30). The function $v^{(0)}$ defined by (7) is smooth on $\mathbb{R}^3$ and, consequently, has the form (37). Two principal terms in the asymptotics of $v^{(1)}$ as $r \to 0$ are given in (25).

Substitute (33) in (2) and (3). Using (37) for $v^{(k)}$, we pass to the coordinates (φ, η) in (3) and equate the coefficients by the same degrees of ε. We thus obtain

$$\begin{aligned}(\partial/\partial\nu) w^{(k)}(\varphi, \eta, \log\varepsilon) &= \delta_{k0}\, g(\varphi, \eta) \\ &- (\partial/\partial\nu) \sum_{j=1}^{k+1} \rho^j v^{(k-j+1)}_j(\varphi, \vartheta, \log\rho + \log\varepsilon, \log\varepsilon), \quad \eta \in \partial\omega.\end{aligned} \tag{38}$$

Now we consider (2), and represent every expression

$$\varepsilon^k \Delta_x\big(\chi(r)\, w^{(k)}(\varphi, \varepsilon^{-1}y, \log\varepsilon)\big)$$

as the sum of

$$\varepsilon^k \chi(r) \Delta_y w^{(k)}(\varphi, \varepsilon^{-1}y, \log\varepsilon)$$

and the series similar to (17), (19), and (20). Equating the coefficients by ε^k written in the same coordinates, we obtain the equations to determine $v^{(k)}$ and $w^{(k)}$

$$\begin{aligned}-\Delta_x v^{(k)}(x, \log\varepsilon) &= \delta_{k0}\, f(x) + [\Delta_x, \chi(r)] \sum_{j=0}^{k-1} r^{-j}\, w^{(k-1-j)}(\varphi, \vartheta, \log r - \log\varepsilon, \log\varepsilon) \\ &+ \sum_{j=0}^{k-1} \Big(\Delta_x - \Delta_y - \sum_{q=1}^{j} L_k(y_2, \partial/\partial y_2, \partial/\partial\varphi)\Big) \\ &\times (r^{-j} w^{(k-1-j)}(\varphi, \vartheta, \log r - \log\varepsilon, \log\varepsilon)), \quad x \in \mathbb{R}^3;\end{aligned} \tag{39}$$

$$-\Delta_\eta w^{(k)}(\varphi,\eta,\log\varepsilon) = \sum_{j=1}^{k} L_j(\eta_2,\partial/\partial\eta_2,\partial/\partial\varphi)\Big(w^{(k-j)}(\varphi,\eta,\log\varepsilon) - \sum_{q=0}^{j-1} w_q^{(k-j)}(\varphi,\vartheta,\log\rho,\log\varepsilon)\Big), \quad \eta\in\mathbb{R}^2\setminus\bar\omega. \tag{40}$$

We have shown how to determine $v^{(k)}$ and $w^{(k)}$ in (33). When solving (39), one has to know $w^{(0)},\dots,w^{(k-1)}$. The functions $w^{(0)},\dots,w^{(k-1)}$ and $v^{(0)},\dots,v^{(k)}$ must be known in (40) and (38). Asymptotic expansions of the form (37) and (35) for the solutions of problems (39) and (40),(38) can easily be derived by induction from Lemma 12.1.1 and 12.1.2. Notice that

$$v_j^{(q)}(\varphi,\vartheta,\log\rho+\log\varepsilon,\log\varepsilon) \qquad \text{and} \qquad w_j^{(p)}(\varphi,\vartheta,\log r-\log\varepsilon,\log\varepsilon)$$

are polynomials in $\log\varepsilon$. Before proving the asymptotics (33), we state two immediate corollaries of the one-dimensional Hardy inequalities (see Lemma 1.4.2).

Lemma 12.1.3. *Let $\tilde K\subset\mathbb{R}^3$ be the cone in S^2 cutting out a domain $\tilde S$ on the unit sphere S^1. (The case $\tilde K=\mathbb{R}^3$, $\tilde S=S^2$ is not excluded.) Then*

$$\int_{\tilde K} |\nabla_x u(x)|^2 dx \ge 4^{-1}\int_{\tilde K} |u(x)|^2\,|x|^{-2}\,dx.$$

Proof. Passing to spherical coordinates (R,τ) and using Lemma 1.4.2, we obtain

$$\int_{\tilde K} |\nabla_x u(x)|^2 dx \ge \int_0^\infty\int_{\tilde S} |(\partial/\partial R)u(x)|^2\,R^2\,dRd\tau$$
$$\ge 4^{-1}\int_0^\infty\int_{\tilde S} |u(x)|^2\,dRd\tau = 4^{-1}\int_{\tilde K} |u(x)|^2|x|^{-2}dx.$$

□

Lemma 12.1.4. *Let $\tilde T_d$ be the torus $\{x\in\mathbb{R}^3\colon \varphi\in M,\ |y|<d\}$. Then*

$$\int_{\tilde T_{1/3}\setminus\Omega_\varepsilon} |\log|y||^{-2}\,|u(x)|^2\,|y|^{-2}dx \le 4^{-1}\int_{\tilde T_{2/3}\setminus\Omega_\varepsilon} |\nabla_x u(x)|^2 dx + \int_{\tilde T_{2/3}\setminus\tilde T_{1/3}} |u(x)|^2 dx.$$

Proof. According to (8), 1.4,

$$\int_{\tilde T_{1/3}\setminus\Omega_\varepsilon} |\log|y||^{-2}\,|u(x)|^2\,|y|^{-2}dx \le \int_{\tilde T_{1/3}\setminus\Omega_\varepsilon} |\log|y||^{-2}\,|\chi(r)u(x)|^2\,|y|^{-2}dx$$
$$\le 4^{-1}\int_{\tilde T_{2/3}\setminus\Omega_\varepsilon} |(\partial/\partial r)\,(\chi(r)\,u(x))|^2 dx$$
$$\le 4^{-1}\int_{\tilde T_{2/3}\setminus\Omega_\varepsilon} |\nabla_x u(x)|^2 dx + C_\chi\int_{\tilde T_{2/3}\setminus\tilde T_{1/3}} |u(x)|^2 dx,$$

where χ is the cut-off function (11) and $C_\chi=\max\{|\chi'(t)|\colon t\in[1/3,\,2/3]|\}$. □

Now we turn to the proof of the main result of the section. Let us consider the partial sum of series (33):

$$u^N(\varepsilon, x) = v^{(0)}(x) + \sum_{k=0}^{N-1} \varepsilon^k \left(v^{(k+1)}(x, \log\varepsilon) + \chi(r)\, w^{(k)}(\varphi, \varepsilon^{-1}y, \log\varepsilon)\right). \quad (41)$$

Theorem 12.1.5. *The solution $u(\varepsilon, x)$ of problem (2), (3) admits the estimate*

$$\int_{\mathbb{R}^3\setminus\Omega_\varepsilon} \left(|\nabla_x u(\varepsilon,x) - \nabla_x u^{(N)}(\varepsilon,x)|^2 + d(x)\,|u(\varepsilon,x) - u^{(N)}(\varepsilon,x)|^2\right) dx \le C_N\, \varepsilon^{2(N+1)-\delta}, \quad (42)$$

where $d(x) = \min\{1, |x|^{-2}, |\log|y||^{-2}, |y|^{-2}\}$, C_N is a constant independent of ε, and δ is an arbitrary positive number.

Proof. The difference $\tilde{R}^{(N)} = u - u^{(N)}$ satisfies the following boundary value problem:

$$-\Delta_x \tilde{R}^{(N)}(\varepsilon, x) = \tilde{F}^{(N)}(\varepsilon, x), \qquad x \in \mathbb{R}^3 \setminus \bar{\Omega}_\varepsilon; \quad (43)$$

$$(\partial/\partial n) R^{(N)}(\varepsilon, x), = \tilde{G}^{(N)}(\varepsilon, x), \qquad x \in \partial\Omega_\varepsilon. \quad (44)$$

We can estimate the right-hand side in (43) and in (44) with the help of (38) and (40). Omitting simple and tedious calculations, we have

$$|\tilde{F}^{(N)}(\varepsilon, x)| \le c_{1N} \varepsilon^{N+1} |\log\varepsilon|^{T_{1N}} |\log r|^{T_{2N}} r^{-2}, \quad (45)$$

$$|\tilde{G}^{(N)}(\varepsilon, x)| \le c_{2N} \varepsilon^{N} |\log\varepsilon|^{T_{3N}}, \quad (46)$$

where c_{jN} are some constants and T_{jN} are nonnegative integers. Notice that $\tilde{F}^{(N)}(\varepsilon, x) = 0$ for $x \notin \tilde{T}_{2/3}$.

We multiply (43) by $\tilde{R}^{(N)}$ and integrate by parts using (44). Then

$$\int_{\mathbb{R}^3\setminus\Omega_\varepsilon} |\nabla_x \tilde{R}^{(N)}(\varepsilon, x)|^2 dx = \int_{\tilde{T}_{2/3}\setminus\Omega_\varepsilon} \tilde{F}^{(N)}(\varepsilon, x)\, \tilde{R}^{(N)}(\varepsilon, x) dx + \int_{\partial\Omega_\varepsilon} \tilde{G}^{(N)}(\varepsilon, x)\, \tilde{R}^{(N)}(\varepsilon, x) ds. \quad (47)$$

From Lemma 12.1.3 and 12.1.4 it follows that

$$2^{-1} \int_{\mathbb{R}^3\setminus\Omega_\varepsilon} |\nabla_x \tilde{R}^{(N)}(\varepsilon, x)|^2\, dx \le c_1 \int_{\mathbb{R}^3\setminus\Omega_\varepsilon} d(x)\, |\tilde{R}^{(N)}(\varepsilon, x)|^2\, dx. \quad (48)$$

In view of (45),

$$\left|\int_{\tilde{T}_{2/3}\setminus\Omega_\varepsilon} \tilde{F}^{(N)}(\varepsilon, x) \tilde{R}^{(N)}(\varepsilon, x)\, dx\right| \le \delta_1 \int_{\tilde{T}_{2/3}\setminus\Omega_\varepsilon} d(x) |\tilde{R}^{(N)}(\varepsilon, x)|^2 dx + (4\delta_1)^{-1} \int_{\tilde{T}_{2/3}\setminus\Omega_\varepsilon} d(x)^{-1} |\tilde{F}^{(N)}(\varepsilon, x)|^2 dx; \quad (49)$$

$$\int_{\tilde{T}_{2/3}} d(x)^{-1}|\tilde{F}^{(N)}(\varepsilon,x)|^2 dx \le c_2\,\varepsilon^{2(N+1)}|\log\varepsilon|^{2T_{1N}} \int_{\varepsilon}^{1} |\log r|^{2T_{2N}+2}\, r^{-1} dr$$
$$\le c_3 \varepsilon^{2(N+1)}|\log\varepsilon|^{2T_{1N}+2T_{2N}+3}.$$

We estimate the second integral on the right in (47) by means of (46):

$$\begin{aligned} &\Big|\int_{\partial\Omega_\varepsilon} \tilde{G}^{(N)}(\varepsilon,x)\tilde{R}^{(N)}(\varepsilon,x)ds\Big| \\ &\le (\varepsilon|\log\varepsilon|^2)^{-1}\delta_2 \int_{\partial\Omega_\varepsilon} |\tilde{R}^{(N)}(\varepsilon,x)|^2 ds + (4\delta_2)^{-1}\,\varepsilon|\log\varepsilon|^2 \int_{\partial\Omega_\varepsilon} |\tilde{G}^{(N)}(\varepsilon,x)|^2 ds \\ &\le (\varepsilon|\log\varepsilon|^2)^{-1}\delta_2 \int_{\partial\Omega_\varepsilon} |\tilde{R}^{(N)}(\varepsilon,x)|^2 ds + c_4\varepsilon^{2(N+1)}|\log\varepsilon|^{2T_{3N}+2}. \end{aligned} \tag{50}$$

Further,

$$\begin{aligned} (\varepsilon|\log\varepsilon|^2)^{-1} \int_{\partial\Omega_\varepsilon} |\tilde{R}^{(N)}(\varepsilon,x)|^2 ds &\le |\log\varepsilon|^{-2}c_5 \int_M\int_{\partial\omega} |\tilde{R}^{(N)}(\varepsilon,x)|^2 dl\, d\varphi \\ &\le |\log\varepsilon|^{-2}c_6 \int_M\int_{\partial\omega} (|\tilde{R}^{(N)}(\varepsilon,x)|^2 + |\nabla_\eta \tilde{R}^{(N)}(\varepsilon,x)|^2) d\eta\, d\varphi \\ &\le |\log\varepsilon|^{-2}c_7 \int_{\tilde{T}_{R\varepsilon}\setminus\Omega_\varepsilon} (|\nabla_y \tilde{R}^{(N)}(\varepsilon,x)|^2 + \varepsilon^2|\tilde{R}^{(N)}(\varepsilon,x)|^2)dx \\ &\le c_8 \int_{\tilde{T}_{2/3}\setminus\Omega_\varepsilon} (|\nabla_x \tilde{R}^{(N)}(\varepsilon,x)|^2 + d(x)|\tilde{R}^{(N)}(\varepsilon,x)|^2)dx. \end{aligned} \tag{51}$$

Combining (48)–(51) and taking small positive $\delta_1 > 0$ and $\delta_2 > 0$ such that $\delta_2 c_8 < 1/4$ and $\delta_1 + \delta_2 c_8 < c_1/2$, we see that the left-hand side in (42) does not exceed

$$C^{(N)}\varepsilon^{2(N+1)}\,|\log\varepsilon|^{2\max\{T_{3N}-1/2,\,T_{1N}+T_{2N}\}+3}.$$

It remains to apply the evident inequality $|\log\varepsilon| \le C(\delta)\varepsilon^{-\delta}$, which holds for any $\delta > 0$. □

Corollary 12.1.6. *Let $v^{(0)}$ and $v^{(1)}$ be the solution to* (5) *and* (21), *respectively, and $w^{(0)}$ the solution of problem* (12). *Then*

$$\int_{\mathbb{R}^3\setminus\Omega_\varepsilon} (|\nabla_x u(\varepsilon,x) - \nabla_x u^{(1)}(\varepsilon,x)|^2 + d(x)\,|u(\varepsilon,x) - u^{(1)}(\varepsilon,x)|^2)\,dx \le C\varepsilon^4\,|\log\varepsilon|^2 \tag{52}$$

with $u^{(1)}(\varepsilon,x) = v^{(0)}(x) + \varepsilon\big(v^{(1)}(x,\log\varepsilon) + \chi(r)w^{(0)}(\varphi,\varepsilon^{-1}y)\big)$.

As was said at the end of Section 12.1.2, $v^{(2)}(x,\log\varepsilon)$ linearly depends on $\log\varepsilon$ and the asymptotics of $w^{(1)}$ as $\rho\to\infty$ is given by the sum of series (28) and (33). Therefore,

$$\int_{\mathbb{R}^3\setminus\Omega_\varepsilon} \big|\nabla_x\{\varepsilon^2(v^{(2)}(x,\log\varepsilon) + \chi(r)w^{(2)}(\varphi,\varepsilon^{-1}y,\log\varepsilon))\}\big|^2 dx \le C_1\,\varepsilon^4|\log\varepsilon|^2. \tag{53}$$

Combining (42) for $N = 2$ and (53), we obtain (52).

12.2 The Asymptotics of Solutions to the Dirichlet Problem in the Exterior of a Thin Tube

We consider the Poisson equation in the same domain as in the preceding section. However, now we set the Dirichlet condition on the boundary of the tube Ω_ε which leads to an essentially more complicated asymptotic expansion for the solution.

Let us try to repeat the reasoning given in Section 12.1. As the principal term of asymptotics far from the tube, we take the solution of the Poisson equation (5), 12.1. Then the discrepancy in the boundary condition is $O(1)$ and one has to compensate for it by means of a boundary layer $w(\varphi, \varepsilon^{-1}y)$. The function $\eta \to w(\varphi,\eta)$ must be harmonic in $\mathbb{R}^2 \setminus \bar{\omega}$ with prescribed values on $\partial\omega$. There exists a unique bounded function with such properties. In Section 12.1 the boundary layer function has the form $\varepsilon w(\varphi, \varepsilon^{-1}y)$ and gives the discrepancy $O(\varepsilon|\log\varepsilon|)$ far from Ω_ε, but here the corresponding residual is only bounded and, therefore, inadmissible.

We would like to solve the exterior Dirichlet problem in the class of functions vanishing at infinity. As the principal approximation outside a neighborhood of Ω_ε, we take the solution to the Poisson equation in $\mathbb{R}^3$ with right-hand side $f(x) + \delta(y)\,\gamma(\varphi)$. Here δ is Dirac's measure while the function γ on the circle M is as yet unknown. It could be determined by the requirement that the boundary layer function decays. This leads to an integral equation for γ, which is not solvable for all small ε. In Section 12.2.2 we show how one can overcome the difficulty.

12.2.1 Statement of the problem and the boundary layer

We make use of the same notation M, ω, and Ω_ε as in Section 12.1.1. Consider the Dirichlet problem

$$-\Delta u(\varepsilon,x) = f(x), \qquad x \in \mathbb{R}^3 \setminus \bar{\Omega}_\varepsilon; \tag{1}$$

$$u(\varepsilon,x) = g(\varphi, \varepsilon^{-1}y), \qquad x \in \partial\Omega_\varepsilon, \tag{2}$$

where $f \in \mathbf{C}_0^\infty(\mathbb{R}^3)$ and $g \in \mathbf{C}^\infty(M \times \partial\omega)$. It is known that there exists a unique solution U subject to

$$U(\varepsilon,x) = O(|x|^{-1}), \qquad |x| \to \infty. \tag{3}$$

Notice that the solution is smooth in $\mathbb{R}^3 \setminus \Omega_\varepsilon$. We seek its asymptotics as $\varepsilon \to 0$.

As in Section 12.1.2, the boundary layer $w(\varphi, \varepsilon^{-1}y)$ is described by means of the following Dirichlet problem

$$-\Delta_\eta w(\eta) = F(\eta), \quad \eta \in \mathbb{R}^2 \setminus \bar{\omega}; \quad w(\eta) = G(\eta), \quad \eta \in \partial\omega. \tag{4}$$

We list the properties of problem (4) that will be of use in what follows. Let

$$\begin{gathered} G \in \mathbf{C}^\infty(\partial\omega), \qquad F \in \mathbf{C}^\infty(\mathbb{R}^2 \setminus \omega), \\ |(\partial/\partial\eta_1)^p\,(\partial/\partial\eta_2)^q\,F(\eta)| \le c_{pq}\,|\eta|^{-3-p-q}, \quad p,q \in \mathbb{N}_0, \end{gathered} \tag{5}$$

where $p,q = 0,1\dots$ and c_{pq} are constant. The bounded solution $w \in \mathbf{C}^\infty(\mathbb{R}^2 \setminus \omega)$, admits the representation

$$w(\eta) = C + O(|\eta|^{-1}), \qquad |\eta| \to \infty, \tag{6}$$

and one may differentiate it.

Let us express the constant C in terms of the right-hand sides F and G in (4). There exists a harmonic function E in $\mathbb{R}^2 \setminus \bar{\omega}$ vanishing on $\partial\omega$ and satisfying

$$E(\eta) = (2\pi)^{-1}\left(\log|\eta| + \Xi\right) + O(|\eta|^{-1}), \quad |\eta| \to \infty, \tag{7}$$

with a constant Ξ. Indeed, since $\log|\eta|$ is harmonic in $\mathbb{R}^2 \setminus \{O\}$, the function E is the sum of $(2\pi)^{-1}\log|\eta|$ and the bounded solution of the problem

$$\Delta e(\eta) = 0, \quad \eta \in \mathbb{R}^2 \setminus \bar{\omega}; \qquad e(\eta) = -(2\pi)^{-1}\log|\eta|, \quad \eta \in \partial\omega. \tag{8}$$

Using (5) and Green's formula, we have

$$\begin{aligned}
\int\limits_{\mathbb{R}^2\setminus\omega} F(\eta)E(\eta)\,d\eta &= -\lim_{R\to\infty} \int\limits_{B_R\setminus\omega} E(\eta)\Delta w(\eta) d\eta \\
&= -\int\limits_{\partial\omega} \big(E(\eta)(\partial/\partial\nu)w(\eta) - w(\eta)(\partial/\partial\nu)E(\eta)\big)\,dl \\
&\quad - \lim_{R\to\infty}\Big\{ \int\limits_{B_R\setminus\omega} w(\eta)\Delta E(\eta) d\eta \\
&\quad + \int\limits_{\partial B_R} \big(E(\eta)(\partial/\partial\rho)w(\eta) - w(\eta)(\partial/\partial\rho)E(\eta)\big)\,dl\Big\} \\
&= \int\limits_{\partial\omega} G(\eta)(\partial/\partial\nu)E(\eta) dl + \lim_{R\to\infty} \int\limits_{\partial B_R} \big(w(\eta)(\partial/\partial\rho)E(\eta) \\
&\quad - E(\eta)(\partial/\partial\rho)w(\eta)\big)\,dl,
\end{aligned} \tag{9}$$

where $\rho = |\eta|$ and $B_R = \{\eta \in \mathbb{R}^2 \colon |\eta| \le R\}$. From (6) and (7) it follows that the last limit in (9) takes the form

$$\lim_{R\to\infty} \int\limits_0^{2\pi} R((2\pi R)^{-1}C + O(R^{-2}\log R))d\vartheta = C.$$

Therefore,

$$C = \int\limits_{\mathbb{R}^2\setminus\omega} F(\eta)E(\eta)\,d\eta - \int\limits_{\partial\omega} G(\eta)(\partial/\partial\nu)E(\eta) dl.$$

We denote the right-hand side by $E(F, G)$.

Remark 12.2.1. From (7) and what was said before (8) it follows that $E(0, -(2\pi)^{-1} \times \log|\eta|) = (2\pi)^{-1}\Xi$. It is also clear that $E(0, 1) = 1$.

12.2.2 The integral equation on M

The solution (7), 12.1 of equation (5), 12.1, gives the discrepancy in (2) with principal term

$$G(\varphi, \eta) = g(\varphi, \eta) - v(\varphi, 0). \tag{10}$$

Generally, the solution $w(\varphi, \cdot)$ to the problem (4) with $F = 0$ and G defined by (10) does not decay (see (6)). Therefore, to get a decaying boundary layer one has to expand the class of admissible solutions to equation (5), 12.1. According to Corollary 11.3.5, the solution to the equation $-\Delta_x V(\gamma; x) = \gamma(\varphi)\delta, \quad x \in \mathbb{R}^3$ takes the form

$$V(\gamma; x) = (4\pi)^{-1} \int\limits_M \gamma(\tau)\,|x - (\tau, 0)|^{-1}\,dl_\tau, \tag{11}$$

$|x - (\tau, 0)|$ being the distance from x to $\tau \in M$.

We study the asymptotics of integral (11) as $|y| \to 0$. Let x be a point in the half-plane $\Pi\varphi$. If $|y| \to 0$, then

$$4\pi V(\gamma; x)$$
$$= \gamma(\varphi) \int_M |x - (\tau, 0)|^{-1} dl_\tau + \int_M |(\varphi, 0) - (\tau, 0)|^{-1} (\gamma(\tau) - \gamma(\varphi))\, dl_\tau + O(|y|)$$
$$= \gamma(\varphi) \int_M |x - (\tau, 0)|^{-1} dl_\tau + 2^{-1} \int_0^{2\pi} |\sin((\tau - \varphi)/2)|^{-1} (\gamma(\tau) - \gamma(\varphi))\, d\tau + O(|y|). \tag{12}$$

Denote by $\tilde{K}\gamma(\varphi)$ the second integral in (12). Taking the equality

$$|x - (\tau, 0)| = \left(x_3^2 + 1 + R(x)^2 - 2R(x)\cos(\tau - \varphi)\right)^{1/2}$$

with $R(x) = (x_1^2 + x_2^2)^{1/2}$ into account, we rewrite the first integral in the form

$$\int_M |x - (\tau, 0)|^{-1}\, dl_\tau = \int_0^{2\pi} \left(x_3^2 + 1 + R^2 - 2R\cos(\tau - \varphi)\right)^{-1/2} d\tau$$
$$= 4\left(d(x)/R(x)\right)^{1/2} K(d(x))$$

(see GRADSHTEYN and RYZHIK [1], 3.674.1). Here K is the complete elliptic integral of the first kind, $d(x) \in (0, 1)$, and

$$d(x)^{-1/2} - d(x)^{1/2} = \{(x_3^2 + (R(x) - 1)^2)/R(x)\}^{1/2} = |y| R(x)^{-1/2}.$$

Using the relations $R(x) = 1 + O(|y|)$ and $d(x) = 1 - |y| + O(|y|^2)$ and the asymptotic formulas for the elliptic integrals (see GRADSHTEYN and RYZHIK [1]), we obtain

$$(4\pi)^{-1} \int_M |x - (\tau, 0)|^{-1} dl_\tau = -(2\pi)^{-1} \log(|y|/8) + O(|y|), \quad |y| \to 0. \tag{13}$$

According to (10), (12), and (13), the discrepancy of $v + V(\gamma; \cdot)$ takes the form

$$G(\varphi, \eta) = g(\varphi, \eta) - v(\varphi, \eta) + (2\pi)^{-1}\left(\gamma(\varphi) \log(|\eta|\varepsilon/8) - 2^{-1}\tilde{K}\gamma(\varphi)\right) + O(\varepsilon). \tag{14}$$

The condition $E(0, G(\varphi, \eta)) = 0$, which provides the decay of the boundary layer, enables us to find γ. By virtue of Remark 12.2.1, one can rewrite the condition:

$$(-\log\varepsilon - \Xi + 3\log 2)\, \gamma(\varphi) + 2^{-1}\tilde{K}\gamma(\varphi) = 2\pi\Big(E(0, g(\varphi, \eta)) - v(\varphi, 0)\Big). \tag{15}$$

The corresponding boundary layer is given by

$$w(\varphi, \eta) = W(\varphi, \eta) - E(0, g(\varphi, \eta)) - \gamma(\varphi)\left(e(\eta) - (2\pi)^{-1}\Xi\right), \tag{16}$$

where e is the solution to problem (8) while W satisfies

$$-\Delta W(\varphi, \eta) = 0, \quad \eta \in \mathbb{R}^2 \setminus \bar{\omega}; \qquad W(\varphi, \eta) = g(\varphi, \eta), \quad \eta \in \partial\omega,$$

and admits the representations

$$W(\varphi, \eta) = E\big(0, g(\varphi, \eta)\big) + O(|\eta|^{-1}), \quad |\eta| \to \infty, \tag{17}$$

$\varphi \in M$ being a parameter. From (7) and (17) it follows that the boundary layer (16) is decaying as $|\eta| \to \infty$.

Now we turn to the integral equation (15).

Lemma 12.2.2. *The following equalities*

$$\tilde{K}h_{\pm k}(\varphi) = -2\lambda_k h_{\pm k}(\varphi), \quad k \in \mathbb{N}_0, \tag{18}$$

hold with $h_p(\varphi) = \exp(ip\varphi)$ *and*

$$\lambda_k = 2\sum_{j=0}^{k-1}(2j+1)^{-1}.$$

Proof. We have

$$\begin{aligned}\tilde{K}h_{\pm k}(\varphi) &= 2^{-1}\int_0^{2\pi} |\sin\Big((\tau-\varphi)/2\Big)|^{-1}\Big(h_{\pm k}(\tau) - h_{\pm k}(\varphi)\Big)\,d\tau \\ &= 2^{-1}\exp(\pm ik\varphi)\int_0^{2\pi} |\sin\Big((\tau-\varphi)/2\Big)|^{-1}\Big(\exp(\pm ik(\tau-\varphi)\Big) - 1\Big)d\tau \\ &= 2^{-1}h_{\pm k}\int_0^{2\pi}\Big(\sin(\vartheta/2)\Big)^{-1}\Big(\exp(\pm ik\vartheta) - 1\Big)\,d\vartheta.\end{aligned}$$

Since $(\sin(\vartheta/2))^{-1}(\exp(\pm ik\vartheta) - 1) = \pm 2i\sum_{j=0}^{k-1}\exp\big(\pm i(j+1/2)\vartheta\big)$,

$$2^{-1}\int_0^{2\pi}\Big(\sin(\vartheta/2)\Big)^{-1}\Big(\exp(\pm ik\vartheta) - 1\Big)d\vartheta = -4\sum_{j=0}^{k-1}(2j+1)^{-1} = -2\lambda_k. \qquad \square$$

Lemma 12.2.3. *The following relation*

$$\lambda_k = \log k + 2\log 2 + c + o(1), \quad k \to \infty, \tag{19}$$

is valid with the Euler constant c.

Proof. It is known that

$$S(n) := \sum_{j=1}^{n} 1/j = \log n + c + o(1), \quad n \to \infty.$$

In view of the equality

$$\lambda_k = 2S(2k) - S(k),$$

we obtain (19). $\square$

Let a_p be the Fourier coefficients of the right-hand side in (15),

$$2\pi\Big(E(0, g(\varphi,\eta)) - v(\varphi,0)\Big) = \sum_{p=-\infty}^{\infty} a_p \exp(ip\varphi). \tag{20}$$

Then from Lemma 12.2.2 it follows that

$$\gamma(\varphi) = -\sum_{p=-\infty}^{\infty}\big(\log\varepsilon + \Xi - 3\log 2 + \lambda_{|p|}\big)^{-1} a_p \exp(ip\varphi) \tag{21}$$

can be regarded as a (formal) solution to the equation (15). The formula (21) shows that the solution does not exist for all small $\varepsilon > 0$. Indeed, the denominator by $\exp(ik\varphi)$ vanishes at $\varepsilon = \varepsilon_k = 8\exp(-(\lambda_k + \Xi))$. In virtue of (19), the sequence $\{\varepsilon_k\}$

tends to zero. However, if the sum in (20) is finite, i.e., $2\big(E(0,g(\varphi,\eta))-v(\varphi,0)\big)$ is a trigonometric polynomial of degree p, then the function γ is defined to be a trigonometric polynomial for $\varepsilon<\varepsilon_p$.

For our purposes we have to keep a discrepancy in (10) no greater than $O(\varepsilon)$. So we may restrict our consideration to an approximative solution to (15).

To construct the desired γ for an arbitrary right-hand side (20) of the equation (15) we recall that the Fourier coefficients of $E(0,g(\varphi,\eta))-v(\varphi,0)\in\mathbf{C}^\infty(M)$ rapidly decrease, namely,

$$|a_p|\le c_n(|p|+1)^{-n},\quad p\in\mathbb{N}_0, \tag{22}$$

with any $n\in\mathbb{N}$.

Lemma 12.2.4. *Let*

$$\gamma(\varphi)=-\sum_{|p|<\varepsilon^{-1/2}}(\log\varepsilon+\Xi-3\log 2+\lambda_{|p|})^{-1}a_p\exp(ip\varphi). \tag{23}$$

Then

$$\max_{\varphi\in[0,2\pi)}\left|(-\log\varepsilon-\Xi+3\log 2)\gamma(\varphi)+2^{-1}\tilde K\gamma(\varphi)-2\pi\Big(E(0,g(\varphi,\eta))-v(\varphi,0)\Big)\right| \le C^{(q)}\varepsilon^q, \tag{24}$$

with any integer $q\in\mathbb{N}$, $\varepsilon\in(0,1)$, *and* $C^{(q)}=const.$

Proof. From (23) it follows that

$$(-\log\varepsilon-\Xi+3\log 2)\gamma(\varphi)+2^{-1}\tilde K\gamma(\varphi)-2\pi\Big(E(0,g(\varphi,\eta))-v(\varphi,0)\Big) =-\sum_{|p|>\varepsilon^{-1/2}}a_p\exp(ip\varphi).$$

Using (22), we find that

$$\left|\sum_{|p|>\varepsilon^{-1/2}}a_p\exp(ip\varphi)\right|\le 2C^{(2q+2)}\sum_{s>\varepsilon^{-1/2}}(1+s)^{-2(q+1)}$$
$$\le 2C^{(2q+2)}(1+\varepsilon^{-1/2})^{-2q}\sum_{s=0}^{\infty}(1+s)^{-2}\le c_q\varepsilon^q.$$

□

The function (23) does not satisfy the equation (15). However, the corresponding discrepancy in the equation is small (see (24)). Therefore, the function can be used to construct the asymptotics of the solution to problem (1), (2).

12.2.3 Justification of the asymptotics

Taking into account what was said at the end of Section 12.2.2, we represent the principal term of the asymptotics of the solution $u(\varepsilon,\cdot)$ to the problem (1), (2) as the sum

$$U(\varepsilon,x)=v(x)+V(\gamma;x)+w(\varphi,\varepsilon^{-1}x)\,\chi(|y|), \tag{25}$$

where v, V, and w are defined by (7), 12.1, (11), and (16), respectively. The role of γ is played by the sum (23), and χ is the cut-off function (11), 12.1.

The following assertion justifies the asymptotic expansion (25) for the solution u to a problem (1)–(3).

Theorem 12.2.5. *The inequality*

$$\max_{x\in\mathbb{R}^3\setminus\Omega_\varepsilon} |u(\varepsilon,x)-U(\varepsilon,x)| \le C\varepsilon|\log\varepsilon| \tag{26}$$

holds with a constant C independent of ε.

Proof. Combining (14)–(16) and (24), we obtain

$$Z(\varepsilon,x)=O(\varepsilon), \quad x\in\partial\Omega_\varepsilon, \tag{27}$$

where $Z=u-U$. Since V is a harmonic function and v satisfies (5), 12.1, the formula (9), 12.1, implies

$$\Delta Z(\varepsilon,x)=\Delta\big(w(\varphi,\varepsilon^{-1}y)\,\chi(|y|)\big)=[\Delta,\chi(|y|)]\,w(\varphi,\varepsilon^{-1}y)$$
$$+\chi(|y|)\left((1+y_2)^{-1}(\partial/\partial y_2)+(1+y_2)^{-2}(\partial/\partial\varphi)^2\right)w(\varphi,\varepsilon^{-1}y).$$

Therefore,

$$\operatorname{supp}\ \Delta Z\subset\{x\in\mathbb{R}^3\colon |y|\le 2/3\} \tag{28}$$

and

$$|\Delta Z(\varepsilon,x)|\le C_1(\varepsilon+\varepsilon^2|y|^{-2}+\varepsilon|y|^{-1})\le C_2\,\varepsilon\,|y|^{-1}, \quad x\in\mathbb{R}^3\setminus\Omega_\varepsilon.$$

We represent Z as the sum of solutions to the problems

$$-\Delta Z_1(\varepsilon,x)=0, \quad x\in\mathbb{R}^3\setminus\bar\Omega_\varepsilon; \qquad Z_1(\varepsilon,x)=Z(\varepsilon,x), \quad x\in\partial\Omega_\varepsilon;$$
$$\Delta Z_2(\varepsilon,x)=-\Delta Z(\varepsilon,x), \quad x\in\mathbb{R}^3\setminus\bar\Omega_\varepsilon; \qquad Z_2(\varepsilon,x)=0 \quad x\in\partial\Omega_\varepsilon.$$

From (27) and the maximum principle it follows that

$$\max_{x\in\mathbb{R}^3\setminus\Omega_\varepsilon}|Z_1(\varepsilon,x)|\le\max_{x\in\partial\Omega_\varepsilon}|Z(\varepsilon,x)|\le C_3\,\varepsilon.$$

Further, from (28) and the estimate

$$|\tilde G(x,\xi)|\le C_4\,|x-\xi|^{-1}, \quad x\in\mathbb{R}^3\setminus(\Omega_\varepsilon\cup\{\xi\}),$$

for Green's function $\tilde G$ of the Dirichlet problem in $\mathbb{R}^3\setminus\bar\Omega_\varepsilon$ we derive the inequality

$$|Z_2(\varepsilon,\xi)|=\Big|\int_{\mathbb{R}^3\setminus\Omega_\varepsilon}\tilde G(x,\xi)\Delta Z(\varepsilon,x)\,dx\Big|\le C_5\,\varepsilon\int_M\int_{B_{3/2}}|x-\xi|^{-1}\,|y|^{-1}\,d\varphi\,dy. \tag{29}$$

Let (φ,y) and (τ,z) be the coordinates of x and ξ in the coordinate system that was introduced at the beginning of Section 12.1.1. If $|z|\ge1$, then from (29) it follows that $|Z_2(\varepsilon,\xi)|\le C_6\,\varepsilon$. In the case $|z|<1$ the last integral in (29) does not exceed

$$C_7\varepsilon\int_M\int_{B_{3/2}}(|y-z|+|\varphi-\tau|)^{-1}\,|y|^{-1}\,d\varphi\,dy\le C_8\varepsilon\int_{B_{3/2}}(1+|\log|y-z||)\,|y|^{-1}\,dy.$$

The integral on the right can be represented as the sum I_1+I_2 of the integrals over the sets $B_{3/2}\cap\{y\colon |y-z|\le|z|/2\}$ and $B_{3/2}\setminus\{y\colon|y-z|\le|z|/2\}$. We have

$$I_1\le\int_{B_{|z|/2}}(1+|\log|\zeta||)\,|z+\zeta|^{-1}\,d\zeta\le C_9\int_0^{|z|/2}(1+|\log r|)\,(|z|-r)^{-1}r\,dr\le C_{10},$$

$$I_2\le C_{11}|\log|z||\int_{B_{3/2}}|y|^{-1}\,dy\le C_{12}|\log|z||.$$

Since the origin of the coordinate $y = (y_1, y_2)$ belongs to ω_ε, the inequality $|\log|z|| \le C_{13}|\log\varepsilon|$ holds for $z \in B_2 \setminus \Omega_\varepsilon$. Using the above estimates, we arrive at (26). □

Thus, we obtained the asymptotics with remainder $O(\varepsilon|\log\varepsilon|)$. As well as for the Neumann problem (see Section 12.1.3) one could construct the asymptotic series. Here we do not deal with it because the corresponding procedure is described in Chapter 13 for arbitrary order equations.

The next section contains the application of the asymptotics obtained in Section 12.2.2 to a problem in electrostatics.

12.2.4 Asymptotics of the capacity of a thin toroidal domain

Using (25), we find the asymptotics of the electrostatic capacity of the set Ω_ε,

$$\operatorname{cap}\Omega_\varepsilon = (4\pi)^{-1} \int\limits_{\mathbb{R}^3\setminus\Omega_\varepsilon} |\nabla_x u(\varepsilon, x)|^2\, dx \tag{30}$$

(see Polya and Szegö [1] and Landkof [1]). Here u is the solution to the problem (1)–(3) with $f = 0$ and $g = 1$. In view of Green's formula,

$$\begin{aligned}\operatorname{cap}\Omega_\varepsilon &= (4\pi)^{-1} \int\limits_{\partial\Omega_\varepsilon} u(\varepsilon, x)(\partial/\partial n)u(\varepsilon, x)\, ds \\ &= (4\pi)^{-1} \int\limits_{\partial\Omega_\varepsilon} (\partial/\partial n)\, u(\varepsilon, x)\, ds = (4\pi)^{-1} \int\limits_{B_R\setminus\Omega_\varepsilon} \Delta u(\varepsilon, x)\, dx \\ &+ (4\pi)^{-1} \int\limits_{\partial B_R} (\partial/\partial r)\, u(\varepsilon, x)\, ds = (4\pi)^{-1} \int\limits_{\partial B_R} (\partial/\partial r)\, u(\varepsilon, x)\, ds,\end{aligned} \tag{31}$$

where $r = |x|$, and $B_R = \{x \in \mathbb{R}^3 \colon |x| < R\}$ and $B_R \supset \Omega_\varepsilon$, i.e., $R > 1$.

Theorem 12.2.6. *Let $c_{\log}(\omega) = \exp(\Xi)$ be the logarithmic capacity* [1] *of ω and Ξ the constant in* (7). *Then*

$$\operatorname{cap}\Omega_\varepsilon = 2\Big\{\log\left(8/\varepsilon c_{\log}(\omega)\right)\Big\}^{-1} + O(\varepsilon\log(1/\varepsilon)). \tag{32}$$

Proof. It is clear that in the case under consideration we have $v = 0$,

$$\gamma(\varphi) = -(\log\varepsilon + \Xi - 3\log 2)^{-1} =: \Gamma(\varepsilon);$$
$$\omega(\varphi, \eta) = -\Gamma(\varepsilon)\left(e(\eta) - (2\pi)^{-1}\,\Xi\right)$$

(see (7), 12.1, (15), and (16)). Because $\chi(|y|) = 0$ for $|y| > 2/3$, the equality

$$U(\varepsilon, x) = \Big(4\pi(\log(8/\varepsilon) - \Xi)\Big)^{-1} \int\limits_M |x - \xi|^{-1}\, dl_\xi$$

holds outside the ball B_2. Since the functions U and u are harmonic, from (26) and the representation

$$(\partial/\partial x_j)\,(u - U)\,(\varepsilon, x) = (2\pi)^{-1} \int\limits_{|\xi - x| = 1} (\xi_j - x_j)\,\big(u(\varepsilon, \xi) - U(\varepsilon, \xi)\big)\, ds_\xi$$

[1] The logarithmic capacity for a number of domains in $\mathbb{R}^2$ is tabulated in Polya and Szegö [1], p. 215 f., and in Landkof [1], p. 330. There are various estimates for $c_{\log}(\omega)$ in Polya and Szegö [1]

it follows that

$$|\nabla_x (u-U)(\varepsilon,x) \le C\varepsilon|\log\varepsilon|, \qquad x\in\mathbb{R}^3\setminus B_3.$$

Using (31), we obtain

$$\begin{aligned}\operatorname{cap}\Omega_\varepsilon &= (4\pi)^{-1}\int_{\partial B_4}(\partial/\partial r)U(\varepsilon,x)\,ds + O(\varepsilon|\log\varepsilon|)\\ &=\big(\log(8/\varepsilon)-\Xi\big)^{-1}\int_M\int_{\partial B_4}(4\pi)^{-1}\big((\partial/\partial n_x)(1/|x-\xi|)\big)\,dx\,dl_\xi + O(\varepsilon|\log\varepsilon|)\\ &=\big(\log(8/\varepsilon)-\Xi\big)^{-1}\int_M dl_\xi + O(\varepsilon|\log\varepsilon|) = 2\pi\big(\log(8/\varepsilon)-\Xi\big)^{-1} + O(\varepsilon|\log\varepsilon|).\end{aligned}$$

□

12.3 Stress and Strain State of the Space with a Thin Toroidal Inclusion

In this section we generalize the reasoning presented in Section 12.2 to describe the asymptotic behavior of solutions to the Dirichlet problem for the Lamé system. We consider an elastic half-space with a rigid inclusion having the form of a thin torus.

12.3.1 Statement of the problem

Let M be the circle $\{x\in\mathbb{R}^3\colon x_1^2+x_2^2=R^2;\, x_3=0\}$ with natural parameter $\varphi\in[0,2\pi)$. In the planes normal to M we introduce Cartesian $y=(y_1,y_2)$ and polar (r,ϑ) coordinate systems. Denote by $\Omega_\varepsilon=\{x\in\mathbb{R}^3\colon \varphi\in[0,2\pi),\, y\in\omega_\varepsilon\}$ the thin toroidal body with section $\omega_\varepsilon=\{y\in\mathbb{R}^2\colon \varepsilon^{-1}y\in\omega\}$ (Fig. 12.3). Here ω is a fixed two-dimensional simply connected domain containing the origin (Fig. 12.2). We suppose that the ratio ε/R is small; let $R=1$ and so ε is a small dimensionless parameter.

In $\mathbb{R}^3\setminus\Omega_\varepsilon$ we consider the Lamé system

$$\mu\Delta u(\varepsilon,x)+(\lambda+\mu)\operatorname{grad}\operatorname{div}u(\varepsilon,x)-f(x)=0,\quad x\in\mathbb{R}^3\setminus\bar\Omega_\varepsilon, \tag{1}$$

with conditions

$$u(\varepsilon,x)=\psi(\varepsilon^{-1}y,\varphi)\quad\text{on } x\in\partial\Omega_\varepsilon, \tag{2}$$

and

$$|u(\varepsilon,x)|\to 0\quad\text{as } |x|\to\infty. \tag{3}$$

Here $u(\varepsilon,x)$ is the displacement vector, λ and μ are the Lamé coefficients, $\mathbf{C}^\infty(\mathbb{R}^3)\ni f$ volume forces, and $\mathbf{C}^\infty(\partial\omega_\varepsilon\times M)\ni\psi$ is a displacement given at the boundary.

12.3.2 Preliminaries

Let v satisfy (1) in the whole space $\mathbb{R}^3$. It is known that

$$v(x)=\int_{\mathbb{R}^3}\Gamma(x-\xi)\,f(\xi)\,d\xi,\quad \Gamma(\eta)=\big(\Gamma_{jk}(\eta)\big)_{j,k=1}^3, \tag{4}$$

Γ being the Kelvin-Somigliana tensor,

$$\Gamma_{jk}(\eta)=\big(8\pi\mu(\lambda+2\mu)\,|\eta|\big)^{-1}\big((\lambda+\mu)\eta_j\,\eta_k\,|\eta|^{-2}+(\lambda+3\mu)\delta_{jk}\big). \tag{5}$$

It follows that under the condition $|f(x)| = O(|x|^{\delta-2})$ with $\delta > 0$ as $|x| \to \infty$, there exists the solution v vanishing at infinity.

We introduce the matrix differential operator $L(\partial/\partial y)$ with entries

$$\begin{aligned} L_{jk}(\partial/\partial y) &= \delta_{jk}\mu\Delta_y + (\lambda+\mu)\,(\partial/\partial y_j)(\partial/\partial y_k), \quad j,k=1,2, \\ L_{3j}(\partial/\partial y) &= L_{j3}(\partial/\partial y) = 0, \quad j=1,2, \quad L_{33}(\partial/\partial y) = \mu\Delta_y \end{aligned} \tag{6}$$

and consider the boundary value problem

$$L(\partial/\partial y)w(y) = 0, \qquad \text{in } \mathbb{R}^2 \setminus \bar{\omega}; \qquad w(y) = g(y) \quad \text{on } \partial\omega. \tag{7}$$

This problem decomposes into the following two: the two-dimensional problem of the theory of elasticity (the first two lines) and the problem of antiplane deformation (the third line). We formulate a few known assertions about solutions. They are specifications of the results in Chapter 3.

Problem (7) is uniquely solvable in the class of bounded functions and its solution w satisfies

$$w(y) = C + O(|y|^{-1}) \quad \text{as } |y| \to \infty \tag{8}$$

with a constant vector $C = (C_1, C_2, C_3)$ independent of g (cf. Proposition 1.5.4).

There exist exactly three linearly independent solutions W^j to the homogeneous problem (7) in the class of functions majorized by a certain power of $\log|y|$ at infinity. These solutions admit the representations

$$W_3^3(y) - (2\pi)^{-1}\log(1/|y|) = O(1), \quad W_k^j(y) - \gamma_{jk}(y) = O(1), \quad j,k=1,2, \tag{9}$$

where γ_{jk} are the elements of two-dimensional Somigliana's tensor:

$$\gamma_{jk}(y) = \big(4\pi\mu(\lambda+2\mu)\big)^{-1}(\lambda+3\mu)\big(\delta_{jk}\log(1/|y|) + (\lambda+3\mu)^{-1}(\lambda+\mu)|y|^{-2}\,y_j\,y_k\big).$$

The remaining elements $W_3^j(y)$ and W_j^3 $(j=1,2)$ are zero.

The constants C_j in the asymptotics (8) of the solution w of problem (7) can be expressed in terms of g with the help of (12), 3.3, taking the form

$$\begin{aligned} C_3 &= -\mu\int_{\partial\omega} g_3(s)\,(\partial/\partial n)\,W_3^3\,(s)\,ds, \\ C_j &= -\int_{\partial\omega}(g_1,g_2)(s)\cdot T_n(W_1^j,W_2^j)\,(s)\,ds, \quad j=1,2, \end{aligned} \tag{10}$$

where $T_n(W_1^j, W_2^j)$ is the force that is generated by the displacement (W_1^j, W_2^j) and acts on the area element with normal n.

From (8) it follows that the solution of problem (7) vanishes at infinity if $(g_1,g_2,0)$ and $(0,0,g_3)$ are orthogonal to $(T_{ny_1},T_{ny_2},0)\,(W_1^j,W_2^j)$ and $(0,0,\partial W_3^3/\partial n)$, respectively, on $\partial\omega$.

We consider the displacement field b given by

$$b(x) = \int_M \Gamma(x-\eta_\tau)\,\beta(\tau)\,d\tau \tag{11}$$

with $\eta_\tau = (\cos\tau, \sin\tau, 0)$ and a smooth vector-valued function β. Using (5), one can obtain the following asymptotic formula:

$$\begin{aligned} b(x) &= \int_M \Gamma(x-\eta_\tau)\,d\tau\,\beta(\varphi) + \int_M \Gamma(x-\eta_\tau)\,(\beta(\tau)-\beta(\varphi)d\tau \\ &= \big(4\pi\mu(\lambda+2\mu)\big)^{-1}\Big\{(\lambda+3\mu)\,\log(8/r)\mathbf{1}\,\beta(\varphi) + (\lambda+\mu)\,I(\varphi,\vartheta)\,\beta(\varphi) \\ &\quad + \int_0^{2\pi} K(\tau,\varphi)\,\big(\beta(\tau)-\beta(\varphi)\big)\,d\tau\Big\} + O(r) \quad as \quad r\to 0, \end{aligned} \tag{12}$$

where

$$K(\tau,\varphi) = \\ = k\begin{pmatrix} (3\lambda+7\mu)-(\lambda+\mu)\cos(\tau+\varphi) & -(\lambda+\mu)\sin(\tau+\varphi) & 0 \\ -(\lambda+\mu)\sin(\tau+\varphi) & (3\lambda+7\mu)+(\lambda+\mu)\cos(\tau+\varphi) & 0 \\ 0 & 0 & \lambda+3\mu \end{pmatrix}$$

with $k = \{8\,|\sin[(\tau-\varphi)/2]|\}^{-1}$ and

$$I(\varphi,\vartheta) = \begin{pmatrix} \cos^2\varphi(1+\cos^2\vartheta) & \sin\varphi\cos\varphi(1+\cos^2\vartheta) & \cos\varphi\sin\vartheta\cos\vartheta \\ \cos\varphi\sin\varphi(1+\cos^2\vartheta) & \sin^2\varphi(1+\cos^2\vartheta) & \sin\varphi\sin\vartheta\cos\vartheta \\ \cos\varphi\sin\varphi\cos\vartheta & \sin\varphi\sin\vartheta\cos\vartheta & \sin^2\vartheta \end{pmatrix}.$$

Note that $K(\tau,\varphi)$ is the trace of the tensor (5) on M. The first and second terms in the braces in (12) are the asymptotics of the mean value of $\Gamma(x-\eta_\tau)$ on M. Denote by $\mathbf{K}\beta(\varphi)$ the last integral in the braces in (12). We state some properties of the matrix I and operator $\mathbf{K}$ that will be of use in what follows. Let

$$\Pi(\varphi) = \begin{pmatrix} \cos\varphi & \sin\varphi & 0 \\ -\sin\varphi & \cos\varphi & 0 \\ 0 & 0 & 1 \end{pmatrix}, \qquad \Pi^{-1}(\varphi) = \begin{pmatrix} \cos\varphi & -\sin\varphi & 0 \\ \sin\varphi & \cos\varphi & 0 \\ 0 & 0 & 1 \end{pmatrix}.$$

Lemma 12.3.1. *The following equalities are valid:*

$$\begin{aligned} S(\vartheta):\ &= \Pi(\varphi)\,I(\varphi,\vartheta)\,\Pi^{-1}(\varphi) = \begin{pmatrix} 1+\cos^2\vartheta & 0 & \sin\vartheta\cos\vartheta \\ 0 & 0 & 0 \\ \sin\vartheta\cos\vartheta & 0 & \sin^2\vartheta \end{pmatrix}, \\ Q\big(\alpha\exp(iq\varphi)\big):\ &= \Pi(\varphi)\,K\,\Pi^{-1}(\varphi)\,[\alpha\exp(iq\varphi)] = k^{(q)}\alpha\exp(iq\varphi), \end{aligned} \tag{13}$$

where $a = (\alpha_1, \alpha_2, \alpha_3)$ *is any vector,*

$$k^{(q)}/2 = \\ \begin{pmatrix} -(\lambda+3\mu)a_1^{(q)}+(\lambda+\mu)a_2^{(q)} & i(\lambda+\mu)a_3^{(q)} & 0 \\ i(\lambda+\mu)a_3^{(q)} & -(\lambda+3\mu)a_1^{(q)}-(\lambda+\mu)(a_1^{(q)}+a_2^{(q)}) & 0 \\ 0 & 0 & -(\lambda+3\mu)a_1^{(q)} \end{pmatrix},$$

$$a_1^{(q)} = 1^{-1}+3^{-1}+\ldots+(2|q|-1)^{-1}, \quad a_2^{(q)} = 2q^2(1-4q^2)^{-1}, \quad a_3^{(q)} = q(1-4q^2)^{-1}.$$

Lemma 12.3.2. *Let the right-hand side in* (7) *have the form*

$$\left(4\pi\mu(\lambda+2\mu)\right)^{-1}\left((\lambda+3\mu)\log(8\varepsilon^{-1}|y|^{-1})\,B(\varphi)+(\lambda+\mu)S(\vartheta)\,B(\varphi)+Q\,B(\varphi)\right)$$

with $B(\varphi)=\Pi(\varphi)\,\beta(\varphi)$. *Denote by* $C(\beta)$ *the constant vector in the representation* (8) *of the solution to the problem. Then*

$$C(\beta)=\left(4\pi\mu(\lambda+2\mu)\right)^{-1}\left((\lambda+3\mu)\log(8\varepsilon^{-1})\,B(\varphi)+DB(\varphi)+QB(\varphi)\right), \tag{14}$$

where D *is the matrix with entries*

$$d_{lk}=\int_{\partial\omega}\{(\lambda+3\mu)\,\delta_{k1}\log(1/\rho)+(\lambda+\mu)\,S_{1k}(\vartheta),$$

$$(\lambda+3\mu)\,\delta_{k2}\log(1/\rho)+(\lambda+\mu)\,S_{2k}(\vartheta)\}\,T_n\,(W_1^l,\,W_2^l)(s)\,ds,\quad l=1,2,\quad k=1,2,3,$$

$$d_{3j}=\int_{\partial\omega}\left((\lambda+3\mu)\,\delta_{j3}\log(1/\rho)+(\lambda+\mu)\,S_{3j}(\vartheta)\right)(\partial/\partial n)\,W_3^3(s)\,ds,\quad j=1,2,3.$$

The proof of Lemma 12.1.3 is a direct calculation. When deriving (13), we used the representation

$$(\sin(\varphi/2))^{-1}\,(\exp(iq\varphi)-1)=2i\sum_{j=0}^{|k|-1}\exp(i(2j+1)\,\varphi/2)$$

and tables of integrals in GRADSHTEYN and RYZHIK [1]. The relation (14) follows from (10) and the evident equality $C=g^0$ for the solution of problem (7) with $g=g^0=\text{const}$. □

Remark 12.3.3. We introduce more notation. Let $\Lambda^{(j)}(y)$ be the solution of problem (7) with $g(y)=(\delta_{j1},\delta_{j2},\delta_{j3})\log(1/\rho)$ and $\Sigma^{(j)}(y)$ the solution with $g(y)=(S_{1j},S_{2j},S_{3j})(\vartheta)$. For these solutions (8) takes the form

$$\Lambda^{(j)}(y)=\Lambda^{(0,j)}(y)+l_j,\quad \Sigma^{(j)}(y)=\Sigma^{(0,j)}(y)+s_j,\quad j=1,2,3,$$

while the functions $\Lambda^{(0,j)}(y)$ and $\Sigma^{(0,j)}(y)$ vanish at infinity, and l_j and s_j are constant vectors.

Introduce the matrices

$$\Lambda^{(0)}(y)=(\Lambda^{(0,1)},\,\Lambda^{(0,2)},\,\Lambda^{(0,3)})\,(y),\quad \Sigma^{(0)}(y)=(\Sigma^{(0,1)},\,\Sigma^{(0,2)},\,\Sigma^{(0,3)})\,(y),$$
$$l=(l_1,l_2,l_3),\qquad\qquad s=(s_1,s_2,s_3).$$

Then the matrix D in (14) can be written in the form

$$D=(\lambda+3\mu)\,l+(\lambda+\mu)\,s \tag{15}$$

with $l_{k3}=l_{3k}=0$, $s_{j2}=0$, $k=1,2$, $j=1,2,3$.

12.3.3 Asymptotics of the solution

The vector-valued function v defined by (4) satisfies (1) and (3). In (2) the function gives the discrepancy $\psi(\varepsilon^{-1}y,\varphi)-v(x(y,\varphi))=\psi(\varepsilon^{-1}y,\varphi)-v(x(0,\varphi))+O(\varepsilon)$. According to (2), the boundary layer compensating the discrepancy has to satisfy the problem of type (7) which depends on the parameter $\varphi\in[0,2\pi)$ and can be obtained from the Lamé system by means of the change of variables $x\to(\eta,\varphi)=(\varepsilon^{-1}y,\varphi)$, multiplication by ε^2, and letting $\varepsilon=0$, cf. Section 12.1.2. As this takes place, every section of the set $\mathbb{R}^3\setminus\Omega_\varepsilon$ by the plane $\{\varphi=\varphi_0\}$ turns to the exterior of a fixed two-dimensional domain ω. However, the solution of problem (7) does not always play the role of the boundary layer because, generally, it does not vanish at infinity.

As in Section 12.2.2, we introduce the supplementary displacement field (11). From (12) it follows that the principal term of the discrepancy of $v+b$ in (2) can be written in the form

$$\begin{aligned}&\psi(\varepsilon^{-1}y,\varphi)-v(x(0,\varphi))-\left(4\pi\mu(\lambda+2\mu)\right)^{-1}\left((\lambda+3\mu)\log(8/\rho\varepsilon)\mathbf{1}\,\beta(\varphi)\right.\\&\left.+(\lambda+\mu)\,I(\varphi,\vartheta)\,\beta(\varphi)+\mathbf{K}\beta(\varphi)\right)+O(\varepsilon).\end{aligned}\tag{16}$$

We find β using the decay condition at infinity of the boundary layer that compensates the discrepancy (16). Let us turn the coordinate system by $-\varphi$ in the plane $\{z=0\}$. Then, by virtue of Lemmas 12.3.1 and 12.3.2, the condition takes the form

$$\left((\lambda+3\mu)\log(8/\varepsilon)\mathbf{1}+D+Q\right)B(\varphi)=4\pi\mu(\lambda+2\mu)\,\Pi(\varphi)\left(\Psi(\varphi)-v(x(0,\varphi))\right).\tag{17}$$

Here $\Psi(\varphi)$ is the vector C in (8) for the solution of the problem

$$\begin{aligned}&h(\eta,\varphi)=h^{\circ}(\eta,\varphi)+\Psi(\varphi),\quad h^{\circ}(\eta,\varphi)=O(|\eta|^{-1})\quad as\quad |\eta|\to\infty,\\&L(\partial/\partial\eta)\,h(\eta,\varphi)=0\quad in\quad \mathbb{R}^2\setminus\bar{\omega};\quad h(\eta,\varphi)=\psi(\eta,\varphi)\quad on\quad \partial\omega.\end{aligned}$$

We seek the principal term of the asymptotics of u in the form

$$u(\varepsilon,x)\sim v(x)+b(x)+\chi(z)\,H(\varepsilon^{-1}y,\varphi)=u^{\varepsilon}(\varepsilon,x),\tag{18}$$

where v is the solution (4) to the equation (1) in $\mathbb{R}^3$, b is the displacement field corresponding to a density β, which is as yet unknown, and H is the decreasing solution of problem (7):

$$\begin{aligned}&H(\eta,\varphi)=h^{\circ}(\eta,\varphi)\\&\quad\cdot\left(4\pi\mu(\lambda+2\mu)\right)\Pi^{-1}(\varphi)\left(\lambda+3\mu)\Lambda^{(0)}(y)+(\lambda+\mu)\,\Sigma^{(0)}(y)\right)\Pi(\varphi)\,\beta(\varphi)\end{aligned}\tag{19}$$

with cut-off function χ defined in (11), 12.1, (see Remark 12.3.3 and (17)). Thus, it remains to find β by means of the integral equation (17). We expand the right-hand side in the Fourier series

$$G(\varphi)=4\pi\mu(\lambda+2\mu)\,\Pi(\varphi)\left(\Psi(\varphi)-v(x(0,\varphi))\right)=\sum_{q=-\infty}^{\infty}G^{(q)}\exp(iq\varphi).\tag{20}$$

Since $G\in\mathbf{C}^{\infty}(M)$, we have

$$\left|G^{(q)}\right|\le c_N\,|q|^{-N}\tag{21}$$

for any positive integer N. The vector-valued function $B(\varphi)$ can be expanded in the series

$$B(\varphi)=\sum_{q=-\infty}^{\infty}B^{(q)}\exp(iq\varphi).\tag{22}$$

Substituting (20), (22) in (17) and using Lemma 12.3.1, we obtain

$$(\lambda+3\mu)\log(8/\varepsilon)B^{(q)}+DB^{(q)}+k^{(q)}\,B^{(q)}=G^{(q)},\tag{23}$$

where $q=0,\pm1,\dots$, while D and $k^{(q)}$ are the matrices (13) and (15).

Remark 12.3.4. The matrix

$$-(\lambda+3\mu)^{-1}k^{(q)}/a_1^{(q)}$$

tends to the identity one as $q\to\infty$. Since $a_1^{(q)}=\log(2|q|^{1/2})+c+o(1)$ with the Euler constant c as $q\to\infty$, the eigenvalues of $k^{(q)}$ and, consequently, of $D+k^{(q)}$ are $O(-\log|q|)$. Therefore, system (17) is not solvable for all small $\varepsilon>0$.

As in Section 12.2.2, we find the asymptotic solution satisfying (17) up to $O(\varepsilon^N)$ for any N. Namely,

$$B^\varepsilon(\varphi)=\sum_{|q|\le\varepsilon^{-1/2}} B^{(q)}\exp(iq\varphi),$$

where

$$B^{(q)}=\left((\lambda+3\mu)\log(8/\varepsilon)\,\mathbf{1}+D+k^{(q)}\right)^{-1}G^{(q)}. \tag{24}$$

Note that by virtue of Remark 12.3.4 the matrix $(\lambda+3\mu)\log(8/\varepsilon)\,\mathbf{1}+D+k^{(q)}$ is invertible for small ε and $|q|\le\varepsilon^{-1/2}$.

We have

$$(\lambda+3\mu)\log(8/\varepsilon)\,B^\varepsilon(\varphi)+DB^\varepsilon(\varphi)+QB^\varepsilon(\varphi)=\sum_{|q|\le\varepsilon^{-1/2}} G^{(q)}\exp(iq\varphi).$$

Therefore, the discrepancy of $B^\varepsilon(\varphi)$ in (17) is equal to $\sum G^{(q)}\exp(iq\varphi)$. According to (21), for $|q|>\varepsilon^{-1/2}$ and any N it admits the estimate

$$\sum_{|q|>\varepsilon^{-1/2}} G^{(q)}\,|\exp(iq\varphi)|\le c_N\sum_{|q|>\varepsilon^{-1/2}}|q|^{-N}\,|\exp(iq\varphi)|\le C_N\,\varepsilon^{N/2-1},$$

i.e., it tends to zero more rapidly than any power of ε.

The difference $U=u-u^\varepsilon$ of the solution to problem (1)–(3) and the vector (18) satisfies

$$\mu\Delta U(\varepsilon,x)+(\lambda+\mu)\,\mathrm{grad}\ \mathrm{div}\,U(\varepsilon,x)-f^\varepsilon(\varepsilon,x)=0\quad\text{in}\quad\mathbb{R}^3\setminus\bar\Omega_\varepsilon;$$
$$U(\varepsilon,x)=\psi^\varepsilon(\varepsilon,y,\varphi)\quad\text{on}\quad\partial\Omega_\varepsilon;\quad |U(\varepsilon,x)|\to0\quad\text{as}\quad|x|\to\infty.$$

It is straightforward to show that $f^\varepsilon(\varepsilon,x)=0$ for $r>2/3$ and

$$|(\partial/\partial y_1)^p\,(\partial/\partial y_2)^q\,(\partial/\partial\varphi)^n\,\psi^\varepsilon(\varepsilon,y,\varphi)|=O(\varepsilon^{1-p-q}),$$
$$|(\partial/\partial y_1)^p\,(\partial/\partial y_2)^q\,(\partial/\partial\varphi)^n\,f^\varepsilon(\varepsilon,y,\varphi)|=O(\varepsilon|y|^{2-p-q}),$$

where p,q,n are arbitrary nonnegative integers. In the same way as in Section 12.1.3, we derive

$$E(u-u^\varepsilon,\,u-u^\varepsilon)\le\mathrm{const}\,\varepsilon^2\,|\log\varepsilon|^3, \tag{25}$$

$E(z,z)$ being the elastic energy,

$$E(z^{(1)},\,z^{(2)})=2^{-1}\sum_{j,k=1}^{3}\int_{\Omega_\varepsilon}\sigma_{jk}\,(z^{(1)};x)\,e_{jk}\,(z^{(2)};x)\,dx, \tag{26}$$

where σ_{jk} and e_{jk} are the Cartesian components of the stress and strain tensors,

$$e_{jk}(z)=2^{-1}\,(\partial z_k/\partial x_j+\partial z_j/\partial x_k),$$

$$\sigma_{jk}(z)=2\mu\,e_{jk}(z)+\lambda\delta_{jk}\sum_{p=1}^{3}e_{pp}(z).$$

12.4 Asymptotics of Solutions to the Dirichlet Problem in a Plane Domain with a Thin Cavity

So far in the chapter we dealt with boundary value problems in the exterior of a thin neighborhood of a manifold without boundary. There are new phenomena in the case when the manifold has boundary. We show this in Section 12.4 and 12.5, where we consider the Dirichlet problem for the Poisson equation in domains with thin cavities.

In this section we derive asymptotics of solutions to the problem in a plane domain having a thin cavity that shrinks into an interval as $\varepsilon \to 0$. Similar problems were discussed in Section 12.5.5. Here, in contrast to Section 12.5.5, we do not refer to the general results of Chapter 4.

12.4.1 Statement of the problem

Let G_0 be a bounded domain in $\mathbb{R}^2$ containing the interval $M = \{x \in \mathbb{R}^2 \colon x_2 = 0, |x_1| \le 1\}$. We introduce the thin set $\omega_\varepsilon = \{x \in \mathbb{R}^2 \colon |x_1| \le 1, |x_2| \le \varepsilon h(x_1)\}$, where h is a smooth positive function on $(-1, 1)$ such that $h(\pm 1) = 0$. Suppose that

$$h(x_1) = (1 - x_1)^{\gamma_+} (1 + x_1)^{\gamma_-} H(x_1), \tag{1}$$

where $\gamma_\pm$ are positive numbers and H is a smooth positive function on $[-1, 1]$. We put $G = G_0 \setminus M$ and $\Omega(\varepsilon) = G \setminus \omega_\varepsilon$ and consider the Dirichlet problem

$$-\Delta u(\varepsilon, x) = f(x), \quad x \in \Omega(\varepsilon); \qquad u(\varepsilon, x) = 0, \quad x \in \partial\Omega(\varepsilon), \tag{2}$$

with $f \in \mathbf{C}^\infty(\bar{G}_0)$. The purpose is to construct the asymptotics of the bounded solution U to problem (2) as $\varepsilon \to 0$.

12.4.2 A cavity whose shores encounter each other at zero angles

We first suppose that $\gamma_\pm \ge 1$ in (1) (see Fig. 12.4). The asymptotic expansion of u we seek in the form

$$u(\varepsilon, x) \sim \sum_{k=0}^{\infty} \varepsilon^k u_k(x). \tag{3}$$

Substituting (3) the Poisson equation, we conclude that

$$-\Delta u_k(x) = \delta_{0k} f(x), \quad x \in G, \tag{4}$$

δ_{jk} being the Kronecker delta. Similarly, using the Dirichlet condition $u(\varepsilon, x) = 0$ on the exterior contour ∂G_0, we have

$$u_k(x) = 0, \quad x \in \partial G_0. \tag{5}$$

Now we find the trace of the series (3) on the boundary of the cavity. Expand $u_k(x)$ in Taylor's series in x_2. Then

$$u(\varepsilon, x_1, \pm\varepsilon h(x)) \sim \sum_{p=0}^{\infty} \varepsilon^p \sum_{q=0}^{p} (q!)^{-1} (\pm h(x_1))^q (\partial/\partial x_2)^q u_{p-q}(x_1, \pm 0), \quad |x_1| < 1.$$

Taking account of the homogeneous Dirichlet condition on $\partial\omega_\varepsilon$, we find the boundary conditions for u_k on the shores $M_\pm$ of M:

$$u_k(x_1, \pm 0) = -\sum_{q=1}^{k} (q!)^{-1} (\pm h(x_1))^q (\partial/\partial x_2)^q u_{k-q}(x_1, \pm 0), \quad x_1 \in [-1, 1]. \tag{6}$$

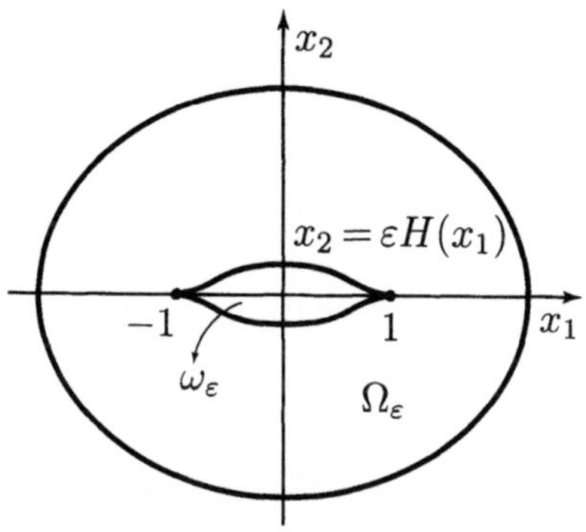

Fig. 12.4

It remains to check that problems (4)–(6) are solvable. The next assertion follows from Theorems 3.3.1 and 3.3.2 for Hölder weighted spaces. (Propositions 1.3.1 and 1.3.14 give the corresponding assertion on the solvability of the Dirichlet problem in $\mathbf{V}^2_{2,\beta}(G)$.)

Proposition 12.4.1. *i) Let the right-hand sides of the equations*

$$\begin{aligned} &-\Delta v(x) = g(x), \quad x \in G;\\ &v(x) = \varphi_\pm(x_1), \quad x \in M_\pm; \qquad v(x) = \varphi(x), \quad x \in \partial G_0, \end{aligned} \tag{7}$$

be subject to the conditions: $\varphi \in \mathbf{C}^\infty(\partial G_0)$;

$$\begin{aligned} &|D^\alpha_x\, g(x)| \le c_\alpha\, d(x)^{\sigma-2-|\alpha|}, \quad x \in G;\\ &|D^{\alpha_1}_x\, \varphi_\pm(x_1)| \le c_\alpha\, (1-|x_1|)^{\sigma-\alpha_1}, \quad x_1 \in [-1,1], \end{aligned} \tag{8}$$

with $\sigma \in [0,1/2)$, $d(x) = dist\{x; (\pm 1, 0)\}$. *Then there exists a unique bounded solution* v *of problem* (7) *and*

$$|D^\alpha_x\, v(x)| \le c_{\alpha\sigma}\, d(x)^\sigma, \quad x \in G.$$

ii) In addition to the above conditions we suppose that in neighborhoods $U_\pm$ *of the endpoints of* M

$$\begin{aligned} &g(x) = \sum_{n=0}^N r_\pm^{\sigma_n - 2}\, g^{(\pm,n)}\,(\vartheta_\pm, \log r_\pm) + \tilde g_N^{(\pm)}(x),\\ &D^\alpha_x\, \tilde g_N^{(\pm)}(x) = o(r_\pm^{\sigma_N - |\alpha| - 2}),\\ &\varphi_\pm(x_1) = \sum_{n=0}^N r_\pm^{\sigma_n}\, \varphi_\pm^{(\pm,n)}(\log r_\pm) + \tilde\varphi_\pm^{(\pm,N)}(x_1),\\ &D^{\alpha_1}_x\, \tilde\varphi_\pm^{(\pm,N)}\,(x_1) = o(r_\pm^{\sigma_N - \alpha_1}), \end{aligned} \tag{9}$$

where $\{\sigma_n\}_{n=0}^\infty$ *is an increasing sequence of positive numbers,* $\sigma_n \to \infty$ *as* $n \to \infty$; N *is an arbitrary positive integer;* $(r_\pm, \vartheta_\pm)$ *are the polar coordinate systems with origins* $(\pm 1, 0)$, $\vartheta_\pm \in (0, 2\pi)$; $g^{(\pm,n)}$ *and* $\varphi^{(\pm,n)}$ *are polynomials in* $\log r_\pm$; *the functions* $g^{(\pm,n)}$, $v^{(\pm,n)}$ *smoothly depend on* $\vartheta_\pm \in [0, 2\pi]$.

Then the bounded solution of problem (7) *admits the following asymptotic representation*

$$v(x) = \sum_{n=0}^{N} r_{\pm}^{\lambda_n} v^{(\pm,n)} (\vartheta_{\pm}, \log r_{\pm}) + \tilde{v}_N^{(\pm)}(x), \quad x \in U^{\pm} \cap G, \tag{10}$$

where the increasing sequence $\{\lambda_n\}$ *consists of numbers* σ_p *and* $(p+1)/2, (p = 0, 1, \dots)$, $D_x^\alpha \tilde{v}_N^{(\pm)}(x) = o(r_{\pm}^{\lambda_N - |\alpha|})$, *and the remaining notation is identical to that in* (9).

We return to the problems (4)–(6). We will show by induction that there exist solutions u_k satisfying

$$|D_x^\alpha u_k(x)| \le c_{k\alpha} d(x)^{\sigma - |\alpha|}, \tag{11}$$

where α is an arbitrary multiindex and σ is any number in $(0, 1/2)$.

For $k = 0$ the right-hand side in (6) equals zero. Hence, by Proposition 12.4.1, the estimates (11) follow from the inclusion $f \in \mathbf{C}^\infty(\bar{G})$. Now assume that (4.11) is established for $k = 0, 1, \dots, K-1$ and check it for $k = K$. By virtue of Proposition 12.4.1, i), it suffices to show that the right-hand sides $\varphi_{\pm}$ in (6) admit the estimate (8). From the induction hypothesis and from (11) we obtain that

$$\varphi_{\pm} = O\Big(\sum_{q=1}^{K} r_{\pm}^{q\gamma_{\pm}} \cdot r_{\pm}^{\sigma - q}\Big) = O(r_{\pm}^{\sigma + q(\gamma_{\pm} - 1)}).$$

Since $\gamma_{\pm} \ge 1$, this implies the desired estimate.

The existence of bounded solutions to problems (4)–(6) is proven. Using the maximum principle (cf. Section 12.1.1–12.1.4), we derive the following assertion.

Proposition 12.4.2. *Let* $\gamma_{\pm} \ge 1$ *in* (1). *Then for the solution to problem* (2) *the asymptotic formula*

$$\max_{x \in \bar{\Omega}(\varepsilon)} \Big| u(\varepsilon, x) - \sum_{k=0}^{J} \varepsilon^k u_k(x) \Big| \le \text{const } \varepsilon^{J+1}$$

holds, where J *is an arbitrary nonnegative integer and* u_k *are the bounded solutions to the recurrent sequence of problems* (4)–(6).

12.4.3 A thin cavity with smooth boundary

If γ_+ (or γ_-) in (1) is less than 1, from (6) it follows that the coefficients of the series (3) do not satisfy (11). For example, the solution u_1 to the problem

$$-\Delta u_1(x) = 0, \quad x \in G;$$
$$u_1(x) = 0, \quad x \in \partial G_0; \qquad u_1(x) = \pm h(x_1)\, (\partial/\partial x_2)\, u_0(x_1, \pm 0), \quad x \in M_{\pm},$$

admits the following representation near the point $(1, 0)$:

$$u_1(x) = 2^{\gamma_- - 1} H(1)\, r_+^{\gamma_+ - 1/2} C_+ \{ \cos((\gamma_+ - 1/2)\, \vartheta_+) - \Big(1 + \cos((2\gamma_+ - 1)\, \pi)\Big)$$
$$\times \Big(\sin((2\gamma_+ - 1)\pi) \Big)^{-1} \sin((\gamma_+ - 1/2)\vartheta_+) \} + o(r^{\gamma_+ - 1/2}); \tag{12}$$

here (r_+, ϑ_+) are the polar coordinates with origin $(1, 0)$, and C_+ is the coefficient in the asymptotics

$$u_0(x) = C_+ r_+^{1/2} \sin(\vartheta_+/2) + O(r_+) \quad \text{as} \quad r_+ \to 0, \tag{13}$$

of the solution to problem (4)–(6) for $k = 0$. (Formulas (12) and (13) can be obtained from Propositions 1.3.7 and 1.3.14.) If $\gamma_+ \in (0, 1/2)$, then εu_1 cannot be taken as the second term in the asymptotics of the solution to problem (2) because this function is not bounded near the point $(1, 0)$.

The singularities in the principal terms of the asymptotic series (3) necessitate a modification of the algorithm described in the preceding section. As will be shown, in contrast to the case $\gamma_\pm \geq 1$ the asymptotics of the solution to problem (1) contains not only terms written in the coordinates x but those of the boundary layer type as well. Namely, there appear boundary layers near the tip $(1, 0)$ of M for $\gamma_+ \in (0, 1)$ and near the tip $(-1, 0)$ for $\gamma_- \in (0, 1)$. The boundary layers have the form $w^\pm(\varepsilon^{\alpha_\pm}(\pm x_1 - 1, x^2))$ with $\alpha_\pm = -(1 - \gamma_\pm)^{-1}$.

We are now ready to construct the asymptotics. For simplicity we assume that $\gamma_\pm = 1/2$ in (1) and

$$h(x_1) = a(1 - |x_1|)^{1/2} \quad \text{for} \quad 1 - |x_1| < 1/2. \tag{14}$$

Thus, near the tips of M the variation of the domain consists in changing from the interval to the parabolic arcs (see Fig. 12.5), i.e., the cavity ω_ε is close to a slender ellipse. The first restriction simplifies the set of exponents of ε, and the second one allows us to circumvent the difficulties connected with the redistribution of discrepancies.

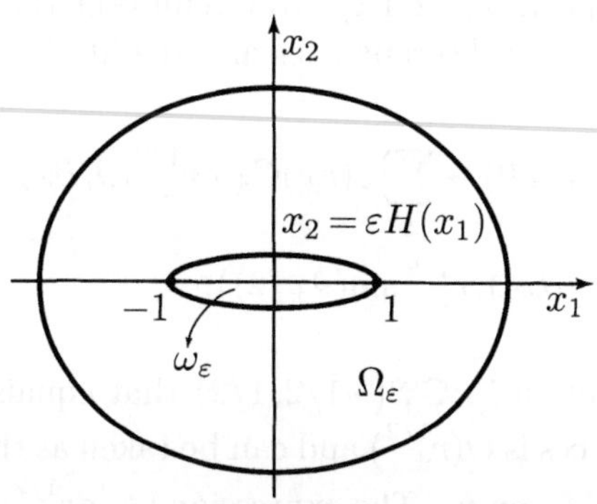

Fig. 12.5

To describe the boundary layer we introduce the coordinates $\xi^\pm = \varepsilon^{-2}(|x_1| - 1, x_2)$ in a neighborhood of the point $(\pm 1, 0)$. According to (14), the inequality $|x_2| < \varepsilon h(x_1)$ defininig the slender domain ω_ε takes the form $|\xi_2| < a_\pm \xi_1^{1/2}$ (we write the variables without indices "$\pm$"). Thus, we have the Dirichlet problem for the Laplace operator in the exterior of the parabola $\partial\Pi_\pm$ to find solutions of the boundary layer type; we put $\Pi_\pm = \{\xi \in \mathbb{R}^2 : \xi_1 > 0, \quad |\xi_2| < a_\pm \xi_1^{1/2}\}$. The next assertion follows from the results of Chapters 1 and 3.

Proposition 12.4.3. *Let the right-hand side ψ of the problem*

$$\Delta_\xi w(\xi) = 0, \quad \xi \in \mathbb{R}^2 \setminus \overline{\Pi}_\pm; \qquad w(\xi) = \psi(\xi), \quad \xi \in \partial\Pi_\pm, \tag{15}$$

admit the representations

$$\begin{gathered} \psi(\xi_1, \pm a\xi_1^{1/2}) = \sum_{j=0}^{J} |\xi|^{-j/2}\, \psi_j^\pm(\log|\xi|) + \tilde{\psi}_\pm^{(J)}(\xi_1), \\ (\partial/\partial\xi_1)^k\, \psi_\pm^{(J)}(\xi_1) = o(\xi_1^{-k-J/2}) \end{gathered} \tag{16}$$

as $\xi_1 \to +\infty$; *here* k, J *are arbitrary nonnegative integers and* $\psi_j^{\pm}$ *are polynomials. Then there exists a unique solution* w *of problem* (15) *that increases no more rapidly than a power of* $\log|\xi|$. *The function* w *has the asymptotics*

$$\begin{gathered} w(\xi) = \sum_{j=0}^{J} |\xi|^{-j/2}\, W_j(\vartheta, \log|\xi|) + \tilde{w}^{(J)}(\xi), \\ D_\xi^\alpha\, w^{(J)}(\xi) = o(|\xi|_1^{-|\alpha|-J/2}) \quad \text{as} \quad |\xi| \to \infty, \end{gathered} \tag{17}$$

where w_j *is a polynomial in* $\log|\xi|$ *with coefficients in* $\mathbf{C}^\infty[0, 2\pi]$, α *being an arbitrary multiindex.*

We turn to the asymptotics of the solution of problem (2) with right-hand side $f \in \mathbf{C}^\infty(\bar{G})$. Since the domain Ω_ε converts to G as $\varepsilon \to 0$, one can take the bounded solution u_0 of problem (4)–(6) as the principal term of the asymptotics. By Proposition 1.3.7, (13) is valid. If the next term is taken in the form $\varepsilon u_1(x)$, then the right-hand side $\varphi_1^{\pm}$ in the boundary condition

$$u_1(x_1, \pm 0) = \pm h(x_1)\,(\partial/\partial x_2)\, u_0(x_1, \pm 0) \tag{18}$$

(see (6) for $k = 1$) turns out to be only bounded. So the solution u_1 of problem (4)–(6) for $k = 1$ does not vanish at the endpoints of M and $\partial u_1/\partial x_2$ is $O(r_\pm^{-1/2})$ as $x \to (\pm 1, 0)$ (see Proposition 12.4.1). To circumvent the difficulty we apply the procedure of redistribution of discrepancies and rewrite the right-hand side of (18) in the form

$$\begin{gathered} \pm\Big\{\{h(x_1)\,(\partial/\partial x_2)\, u_0(x_1, \pm 0) - \sum_{\pm} \zeta(r_\pm) C_\pm\, a r_\pm^{1/2} (\partial/\partial x_2)(r_\pm^{1/2} \sin(\vartheta_\pm/2))\Big\} \\ \pm \sum_{\pm} \zeta(r_\pm) C_\pm\, a r_\pm^{1/2} (\partial/\partial x_2)(r_\pm^{1/2} \sin(\vartheta_\pm/2)), \end{gathered} \tag{19}$$

where ζ is a cut-off function in $\mathbf{C}_0^\infty(-1/2, 1/2)$ that equals 1 on $[-1/4, 1/4]$. The first summand in the braces is $O(r_\pm^{1/2})$ and can be taken as the right-hand side of the boundary condition for $M_\pm$ on u_1. The expression $C_\pm\, a r_\pm^{1/2}(\partial/\partial x_2)(r_\pm^{1/2} \sin(\vartheta_\pm/2))$ takes the form $C_\pm\, a\rho_\pm^{1/2}(\partial/\partial \xi_2^\pm)(\rho_\pm^{1/2} \sin(\vartheta_\pm/2))$ in the coordinates $\xi^\pm$ and coincides with the principal term of the asymptotics of the trace of $C_\pm\, \rho_\pm^{1/2} \sin(\vartheta_\pm/2)$ on the arc $\Gamma_\pm$ of the parabola $\{\xi^\pm \in \partial\Pi_\pm : \xi_1^\pm, \xi_2^\pm > 0\}$ as $\rho_\pm = |\xi^\pm| \to \infty$. Therefore, the last terms in (19) are compensated by the sum $\zeta(r_+) w_1^+(\xi^+) + \zeta(r_-) w_1^-(\xi^-)$, where $w_1^\pm$ are solutions to the problems

$$\Delta w_1^\pm(\xi) = 0, \quad \xi \in \mathbb{R}^2 \setminus \overline{\Pi_\pm}; \qquad w_1^\pm(\xi) = -C_\pm \rho^{1/2} \sin(\vartheta/2), \quad \xi \in \partial\Pi_\pm. \tag{20}$$

Remark 12.4.4. Since $\sin\vartheta = a r^{-1/2}(\cos\vartheta)^{1/2}$ on $\Gamma_\pm$, the trace of $\sin(\vartheta/2)$ on $\Gamma_\pm$ is an analytic function in $\rho^{-1/2}$ for $\rho = |\xi| > 1$. This trace is $O(1)$ as $\rho \to \infty$. So the right-hand side of the boundary condition in (20) admits the expansion (16), and (17) holds for the solution to problem (20).

We are now ready to construct the asymptotic series for the solution to problem (2) with right-hand side $f \in \mathbf{C}^\infty(\overline{G}_0)$. We seek the series in the form

$$v_0(x) + \sum_{k=1}^{\infty} \varepsilon^k \big(v_k(x, \log\varepsilon) + \sum_{\pm} \zeta(r_\pm)\, w_k^\pm(\varepsilon^{-2}(|x_1| - 1),\, \varepsilon^{-2} x_2, \log\varepsilon)\big), \tag{21}$$

where v_0 is the solution to the Dirichlet problem

$$-\Delta v_0(x) = f(x), \quad x \in G; \qquad v_0(x) = 0, \quad x \in \partial G.$$

We list some properties of v_k and $w_k^{\pm}$. These functions polynomially depend on $\log\varepsilon$. The coefficients of the polynomials $t \to w_k^{\pm}(\xi, t)$ admit the expansions (17). We denote by $W_{k,j}^{\pm}(\vartheta, \log\rho, \log\varepsilon)$ the factors by $|\xi|^{-j/2}$ in the expansion of $w_k^{\pm}(\xi, \log\varepsilon)$. For v_k there are the following representations near the endpoints of M:

$$\begin{aligned} v_k(x, \log\varepsilon) &= \sum_{j=1}^{J} r_{\pm}^{j/2}\, V_{kj}^{(\pm)}(\vartheta_{\pm}, \log r_{\pm}, \log\varepsilon) + \tilde{v}_{kj}^{(\pm)}(x, \log\varepsilon), \\ D_x^{\alpha}\, \tilde{v}_{kj}^{(\pm)}(x, \log\varepsilon) &= O(r_{\pm}^{(J+1)/2-|\alpha|}\, |\log r_{\pm}|^{Q_{jk}}\, |\log\varepsilon|^{P_{jk}}), \end{aligned} \tag{22}$$

where Q_{jk} and P_{jk} are nonnegative integers (cf. (10)). Finally, outside the corner points the function $x \to v_k(x, \log\varepsilon)$ is smooth up to the boundary.

Lemma 12.4.5. *Let $t \to V(\vartheta, t)$ be a polynomial with coefficients in $\mathbf{C}^{\infty}[0, 2\pi]$ and j an integer. Then for the trace X of the function $Y\colon \xi \to \rho^{j/2}\, V(\vartheta, \log\rho)$ on $\partial\Pi_{\pm}$ we have*

$$X(\xi) = \sum_{p=0}^{j-1} (p!)^{-1}\, (a\rho^{1/2})^p\, (\partial/\partial\xi_2)^p\, Y(\xi_1, \pm 0) + \Psi(\varepsilon), \quad \pm\xi_2 > 0, \tag{23}$$

where Ψ admits the representation (16).

In view of Remark 12.4.4 the proof is a straightforward computation.

We write expansions of the form (23) for the terms $r_{\pm}^{j/2}\, V_{kj}^{(\pm)}(\vartheta_{\pm}, \log r_{\pm}, \log\varepsilon)$ in (22) and denote the corresponding remainders by $\Psi_{kj}^{(\pm)}(\xi^{\pm})$.

Now we find the boundary value problems for the coefficients of the series (21). Substitute (21) in the Poisson equation and factor out the cut-off functions from the Laplace operator. In the corresponding commutators we change from the variables $r_{\pm}$ to $\rho_{\pm}$ and, using (17), expand the result in powers of ε:

$$\begin{aligned} &\sum_{k=0}^{\infty} \varepsilon^k \Delta_x(\zeta(r_{\pm}) w_k^{(\pm)}(\xi^{\pm}, \log\varepsilon)) - \zeta(r_{\pm}) \sum_{k=0}^{\infty} \varepsilon^{k-4} \Delta_{\xi} w_k^{(\pm)}(\xi^{\pm}, \log\varepsilon) \\ &+ \sum_{p,k=0}^{\infty} \varepsilon^{k+p}\, [\Delta_x, \zeta(r_{\pm})]\, (r_{\pm}^{-p/2}\, W_{kp}^{(\pm)}(\vartheta_{\pm}, \log r_{\pm} - 2\log\varepsilon, \log\varepsilon)). \end{aligned} \tag{24}$$

The coefficients of the first series on the right must vanish, which leads to the equations for $w_k^{\pm}$:

$$-\Delta_{\xi} w_k^{\pm}(\xi, \log\varepsilon) = 0, \quad \xi \in \mathbb{R}^2 \setminus \overline{\Pi_{\pm}}. \tag{25}$$

One should take account of the coefficients of the second series in the right-hand sides of the equations for v_k:

$$\begin{aligned} &-\Delta_x v_k(x, \log\varepsilon) = \delta_{k0}\, f(x) \\ &+ \sum_{\pm} \sum_{j=0}^{k} [\Delta_x, \zeta(r_{\pm})]\, (r_{\pm}^{-p/2}\, W_{k-j,j}^{(\pm)}(\vartheta_{\pm}, \log r_{\pm} - 2\log\varepsilon, \log\varepsilon)), \quad x \in G. \end{aligned} \tag{26}$$

Note that the supports of the commutators in (26) are separated from the corner points $(\pm 1, 0)$.

Substituting (21) in the homogeneous Dirichlet condition on ∂G_0, we find that

$$v_k(x, \log\varepsilon) = 0, \quad x \in \partial G_0. \tag{27}$$

From (22) and Lemma 12.4.5 it follows that

$$\begin{aligned} v_k(x_1, \pm\varepsilon h(x_1)) &= v_k(x_1, \pm 0) \\ &+ \sum_{p=0}^{\infty} \varepsilon^p \, (p!)^{-1} (\pm h(x_1))^p \Big(Z(x_1) \, (\partial/\partial x_2)^p \, v_k(x_1, \pm 0) + \sum_{\pm} \zeta_\pm(r_\pm) \\ &\times (\partial/\partial x_2)^p \, \tilde{v}_{kp}^{(\pm)}(x_1, \pm 0) \Big) + \sum_{q=1}^{\infty} \varepsilon^q \sum_{\pm} \zeta_\pm(r_\pm) \Psi_{kp}^{(\pm)}(\xi^\pm) \quad \text{on} \quad \partial\omega_\varepsilon, \end{aligned} \tag{28}$$

where $Z(x_1) = 1 - \sum_\pm \zeta(r_\pm)$. We relate the factors by ε^p in (28) to the first limit problem in the domain G and those by ε^q to the second problem in the exterior of the parabola. Then v_k satisfies the following boundary conditions on M:

$$\begin{aligned} v_k(x_1, \pm 0) &= -\sum_{j=1}^{k} (j!)^{-1} \big(\pm h(x_1) \big)^j \\ &\times \Big(Z(x_1) (\partial/\partial x_2)^j v_{k-j}(x_1, \pm 0) + \sum_{\pm} \zeta_\pm(r_\pm) (\partial/\partial x_2) v_{k-j,j}^{(\pm)}(x_1, \pm 0) \Big), \\ & |x_1| < 1. \end{aligned} \tag{29}$$

According to (22), the right-hand side of (29) admits the expansion (9) with $\sigma_n = (n+1)/2$.

The boundary condition for $w_k^\pm$ has the form

$$w_k^\pm(\xi) = -\sum_{j=1}^{k} \Psi_{k-j,j}^\pm(\xi), \qquad \xi \in \partial\Pi_\pm. \tag{30}$$

By Lemma 12.4.5, the representation (16) holds for every summand on the right in (30).

Taking the mentioned properties of the right-hand sides of (26), (29), and (30) into account, we derive from Propositions 12.4.1 and 12.4.3 that (22) and (17) are true for the solutions v_k and $w_k^\pm$ to problems (27), (29) and (25), (30), respectively. Observe that $w_0^\pm = 0$; first we define $w_k^\pm$ as solutions to problems (25) and (30), whereupon v_k can be obtained from (26), (27), and (29).

As before, the asymptotic expansion is justified by the maximum principle.

Proposition 12.4.6. *Let N be a nonnegative integer and $U_N(\varepsilon, x)$ the partial sum of the series (21). Then the solution u to problem (2) is subject to the estimate*

$$\max_{x \in \Omega(\varepsilon)} |u(\varepsilon, x) - U_N(\varepsilon, x)| \le c_\delta \, \varepsilon^{N+1-\delta},$$

where δ is an arbitrary positive number.

12.4.4 A remark on application of the results of Chapter 4

It was established in Section 5.5 that by means of a change of variables the problem (1) in a plane domain with slender cavity can be reduced to a singularly perturbed problem in the domain G whose operator is an ε-admissible in the sense of Section 4.1.2. We formulate an assertion that follows from Theorem 4.4.3 and enables us to estimate the derivatives of the difference of the solution to problem (2) and the asymptotic approximation constructed above.

Let us recall the norm in $\mathbf{W}_2^l(\Omega(\varepsilon))$ depending on ε that was introduced in Section 4.1.6:

$$\left\{ \sum_{|\alpha|\le l} \|[d(x)+\varepsilon^2]^{\beta-l+|\alpha|} D_x^\alpha u; \mathbf{L}_2(\Omega(\varepsilon))\|^2 \right\}^{1/2}. \tag{31}$$

In contrast to Section 4.1 and 5.1 here we have the weight factor $r+\varepsilon^2$ instead of $r+\varepsilon$, which is related to the fact that we used the dilatation coefficient ε^{-2} in passing from the coordinates x to $\xi^{\pm}$. Denote by $\mathbf{V}_{2,\beta}^l(\Omega(\varepsilon))$ the space $\mathbf{W}_2^l(\Omega(\varepsilon))$ with norm (31) and by $\mathbf{V}_{2,\beta}^{l-1/2}(\partial\Omega(\varepsilon))$ the corresponding space of traces on $\partial\Omega(\varepsilon)$.

Proposition 12.4.7. *Let $\gamma_\pm = 1/2$ in (1) and let (14) be valid. If l is a nonnegative integer and $\beta \in (l+1/2, l+3/2)$, then the norm of the operator of problem (2):*

$$\mathbf{V}_{2,\beta}^{l+2}(\Omega(\varepsilon)) \to \mathbf{V}_{2,\beta}^l(\Omega(\varepsilon)) \times \mathbf{V}_{2,\beta}^{l+3/2}(\partial\Omega(\varepsilon))$$

and the norm of the inverse operator are bounded by constants independent of ε.

12.5 Asymptotics of Solutions to the Dirichlet Problem in a Three-dimensional Domain with a Thin Cavity

As in Section 12.2, to construct the principal term $v(x)$ of the asymptotics outside a neighborhood of the cavity Q_ε one should distribute sources with certain density $\gamma(x_3)$ along the limit interval $M = \{x\colon |x_3| \le 1,\, x_1 = x_2 = 0\}$. Similarly to Section 12.2, the decay condition for the boundary layer leads to the integral equation with the operator

$$\gamma(z) \to (T\gamma)(z) = \int_{-1}^{1} (\gamma(z) - \gamma(\xi))\, |z-\xi|^{-1}\, d\xi,$$

which is generally not solvable. However, to construct the asymptotics it suffices to satisfy this equation up to $O(\varepsilon)$. Such an approximate solution γ will be found in Section 12.5.4.

Besides the boundary layer $w(x_3, \varepsilon^{-1}y)$ in the planes orthogonal to M with $|x_3| < 1$ (see Section 12.5.3), there arise two boundary layers for the endpoints of M. These layers are constructed in Section 12.4.5 and 12.4.6. An estimate on the remainder in the asymptotic formula is given in Section 12.4.7.

As application, we obtain the asymptotics of the harmonic capacity $\operatorname{cap} Q_\varepsilon$ (see Section 12.4.8).

In Section 12.4.9 we consider the case where Q_ε is a rotation ellipsoid (w is the unit disc). In particular, for the problem

$$-\Delta u = 0 \quad \text{in} \quad \mathbb{R}^3 \setminus \bar{Q}_\varepsilon, \quad u = \varphi \quad \text{on} \quad \partial Q_\varepsilon$$

with $\varphi \in \mathbf{C}^\infty(\mathbb{R}^3)$ the general asymptotic formula in Section 12.4.7 can be simplified and reduced to the form

$$4\pi u(x,\varepsilon) = \int_{-1}^{1} (r^2 + (x_3 - z)^2)^{-1/2}\,\gamma(z)dz$$
$$-\sum_{\pm} \gamma(\pm 1)\,\chi(r^2 + (x_3 \pm 1)^2) \int_{1-x_3^2-\varepsilon}^{1-x_3^2} (4r^2 + z^2)^{-1/2} dz + O(\varepsilon|\log\varepsilon|).$$

Here $r^2 = x_1^2 + x_2^2$, $\chi \in \mathbf{C}_0^\infty(-1,1)$, $\chi(0) = 1$ and

$$\gamma(z) = -\sum_{n\varepsilon^{1/2}\le 1} \left(\log(\varepsilon/2) + \sum_{k=1}^{n} k^{-1}\right)^{-1} (2n+1)P_n(z) \int_{-1}^{1} \varphi(0,\zeta)\, P_n(\zeta)\, d\zeta,$$

P_n being the Legendre polynomial of degree n.

12.5.1 Statement of the problem

Let $\Omega \subset \mathbb{R}^3$ be a subdomain in M with compact closure $\bar{\Omega}$ and smooth boundary, containing $M = \{x \in \mathbb{R}^3 \colon y = (x_1, x_2) = 0,\ |x_3| \le 1\}$. Denote by ω the domain $\{y \in \mathbb{R}^2 \colon |y| < \varkappa(\vartheta)\}$, where ϑ is a polar angle and $\varkappa$ is a smooth positive function on the unit circle. We introduce the set $Q_\varepsilon = \{x \in \mathbb{R}^3 \colon \varepsilon^{-1}(1-x_3^2)^{-1/2}\, y \in \omega,\ |x_3| < 1\}$ with a small ε such that $\bar{Q}_\varepsilon \subset \Omega$. Consider the problem

$$-\Delta u(x) = f(x), \quad x \in \Omega \setminus \bar{Q}_\varepsilon, \tag{1}$$
$$u(x) = 0, \quad x \in \partial\Omega, \tag{2}$$
$$u(x) = \varphi(x), \quad x \in \partial Q_\varepsilon, \tag{3}$$

where $f, \varphi \in \mathbf{C}^\infty(\bar{\Omega})$. It is known that there exists a unique solution $u \in \mathbf{W}_2^1(\Omega \setminus \bar{Q}_\varepsilon)$.

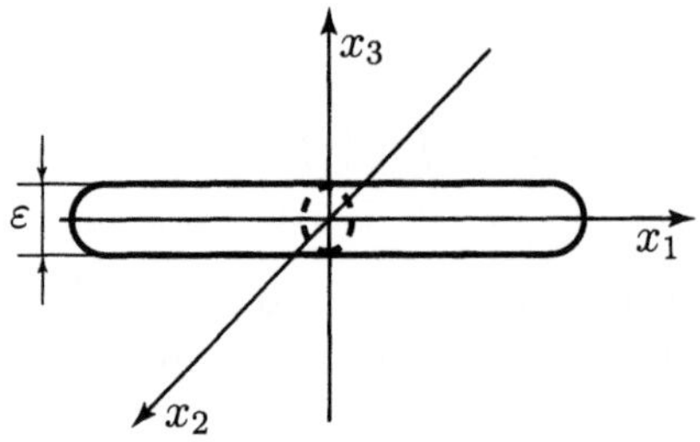

Fig. 12.6

12.5.2 Asymptotics of the solution in the exterior of $\bar{\mathbf{Q}}_\varepsilon$

Denote by v the solution of the boundary value problem

$$-\Delta v(x) = f(x) + \gamma(x_3)\,\delta(M), \quad x \in \Omega; \qquad v(x) = 0, \quad x \in \partial\Omega. \tag{4}$$

Here δ is the Dirac measure and γ is a function in $\mathbf{C}_\infty[-1,1)$ that has to be determined. The solution to problem (4) will be taken as approximation for the solution to the original problem outside a neighborhood of $\bar{Q}_\varepsilon$. It is clear that

$$v(x) = v_0(x) + \int_{-1}^{1} G\big(x, (0, \xi_3)\big)\, \gamma(\xi_3)\, d\xi_3, \tag{5}$$

where $G(x,\xi) = (4\pi|x-\xi|)^{-1} + \Gamma(x,\xi)$ is Green's function for the domain Ω,

$$v_0 = \int_\Omega G(x,\xi)\, f(\xi)\, d\xi.$$

We seek the asymptotics of v as $y \to 0$, $|x_3| \le 1-\sigma$, and $\sigma > 0$. We have

$$\begin{aligned} v(x) &= v_0(x) + (4\pi)^{-1}\gamma(x_3) \int_{-1}^{1} |x-(0,\xi_3)|^{-1}\, d\xi_3 \\ &+ (4\pi)^{-1} \int_{-1}^{1} \big(\gamma(\xi_3)-\gamma(x_3)\big)|x-(0,\xi_3)|^{-1}\, d\xi_3 + \int_{-1}^{1} \gamma(\xi_3)\,\Gamma(x,(0,\xi_3)\, d\xi_3. \end{aligned} \tag{6}$$

Since

$$\begin{aligned} L(x) &= \int_{-1}^{1} |x-(0,\xi_3)|^{-1}\, d\xi_3 \\ &= \log\Big(\big\{(|y|^2+(1+x_3)^2)^{1/2}-1-x_3\big\}^{-1}\big\{(|y|^2+(1-x_3)^2)^{1/2}+1-x_3\big\}\Big), \end{aligned} \tag{7}$$

we obtain that

$$v(x) = v_0(0,x_3) + V(x) + O\big(|y_2|^2(1-x_3^2)^{-2}\big), \tag{8}$$

where

$$\begin{aligned} V(x) &= -(2\pi)^{-1}\log|y|\,\gamma(x_3) + (4\pi)^{-1}\log\big(4(1-x_3^2)\big)\,\gamma(x_3) \\ &+ \int_{-1}^{1} \gamma(\xi_3)\,\Gamma\big((0,x_3),(0,\xi_3)\big)d\xi_3 + (4\pi)^{-1}\int_{-1}^{1} \big(\gamma(\xi_3)-\gamma(x_3)\big)|x_3-\xi_3|^{-1}\, d\xi_3. \end{aligned} \tag{9}$$

12.5.3 The two-dimensional boundary layer

The function v satisfies (1) and (2), but there remains a discrepancy in the boundary condition (3). To compensate for the discrepancy we construct a boundary layer in a neighborhood of the segment. Making the change of variables $\eta = \varepsilon^{-1}(1-x_3^2)^{-1/2}y$, $\zeta = x_3$, we obtain

$$\Delta_x = \varepsilon^{-2}(1-\zeta^2)^{-1}\Delta_\eta + \big(\partial/\partial\zeta + (1-\zeta^2)^{-1}\zeta\eta\cdot\partial/\partial\eta\big)^2. \tag{10}$$

Far from the endpoints of the segment the principal part of (5) with respect to ε is the operator $\varepsilon^{-2}(1-\zeta^2)^{-1}\Delta_\eta$. Therefore outside neighborhoods of the points $(0,0,\pm1)$ it is natural to define the boundary layer as the solution to the Dirichlet problem for the Laplace operator in $\mathbb{R}^2 \setminus \bar{\omega}$, while the problem depends on the parameter $\zeta \in (-1,1)$. The principal part of the discrepancy of v in (3) written in the coordinates (η,ζ) has the form

$$\begin{aligned} &-(2\pi)^{-1}\log(\varepsilon|\eta|/2)\,\gamma(\zeta) + (4\pi)^{-1}\int_{-1}^{1}(\gamma(z)-\gamma(\zeta))|z-\zeta|^{-1}\,dz \\ &+\int_{-1}^{1}\gamma(z)\,\Gamma((0,\zeta),\,(0,z))\,dz + v_0(0,\zeta)-\varphi(0,\zeta). \end{aligned} \tag{11}$$

As in 12.2.1, we denote by E the harmonic function in $\mathbb{R}^2\setminus\bar{\omega}$ determined by

$$\begin{aligned} &E(\eta) = (2\pi)^{-1}(\log|\eta|^{-1}+\Xi) = o(1), \quad |\eta|\to\infty; \\ &E(\eta)=0 \quad \text{for} \quad \eta\in\partial\omega; \quad \Xi = \text{const}. \end{aligned} \tag{12}$$

Set

$$e(\eta) = E(\eta) - (2\pi)^{-1}(\log|\eta|-\Xi) = O(|\eta|^{-1}). \tag{13}$$

The harmonic function e vanishes at infinity and satisfies the boundary condition

$$e(\eta) = (2\pi)^{-1}(\log|\eta|-\Xi), \quad \eta\in\partial\omega. \tag{14}$$

The constant Ξ is defined by

$$\Xi = -\int_{\partial\omega}\log|\eta|(\partial E/\partial\nu)\,ds_\eta, \tag{15}$$

where ν is the normal to $\partial\omega$ at the point η directed to ω. We take the function

$$w(\eta,\zeta) = \gamma(\zeta)\,e(\eta) \tag{16}$$

as boundary layer outside a neighborhood of the endpoints of the segment M.

12.5.4 Construction of the function γ

The principal part of the discrepancy of $v+w$ in (3) has the form

$$\begin{aligned} &-(2\pi)^{-1}(\log\varepsilon/2)+\Xi)\gamma(\zeta) + (4\pi)^{-1}\int_{-1}^{1}(\gamma(z)-\gamma(\zeta))|z-\zeta|^{-1}\,dz \\ &+\int_{-1}^{1}\gamma(z)\,\Gamma\big((0,\zeta),\,(0,z)\big)\,dz + v_0(0,\zeta)-\varphi(0,\zeta) \end{aligned} \tag{17}$$

(see (11) and (14)). We are going to choose γ so that the discrepancy is small.

Let

$$\begin{aligned} &(T\gamma)(\zeta) = -2^{-1}\int_{-1}^{1}(\gamma(z)-\gamma(\zeta))|z-\zeta|^{-1}\,dz, \\ &(K\gamma)(\zeta) = -2\pi\int_{-1}^{1}\gamma(z)\,\Gamma\big((0,\zeta),\,(0,z)\big)\,dz. \end{aligned} \tag{18}$$

The quadratic form $\langle T\gamma, \gamma\rangle$ of the operator T has the form

$$4^{-1}\int_{-1}^{1}\int_{-1}^{1}\big(\gamma(z)-\gamma(\zeta)\big)^2|z-\zeta|^{-1}\,dz\,d\zeta.$$

We denote by H the completion of $\mathbf{C}^\infty([-1,1])$ with respect to the norm

$$\big(\langle T\gamma,\gamma\rangle + \|\gamma;\mathbf{L}_2(-1,1)\|^2\big)^{1/2}.$$

It is easy to verify that the space H is compactly embedded in $\mathbf{L}_2(-1,1)$. Therefore the operator T with domain H has discrete spectrum. (The eigenvalues and eigenfunctions of T are explicitly described in Lemma 12.5.1 below.) Since K is an integral operator with symmetric kernel, the operator $T+K$ has a sequence of real eigenvalues λ_k such that $\lambda_k \to \infty$. If the discrepancy (17) vanishes, we obtain the equation

$$\big(\log(\varepsilon/2)+\Xi\big)\gamma + T\gamma + K\gamma = 2\pi(v_0-\varphi)\,(0,\cdot). \tag{19}$$

From the aforementioned properties of $T+K$ it follows that such an equation is not solvable for all right-hand sides if $\varepsilon = 2\exp(-\lambda_k-\Xi)$. However, in what follows it will suffice to find a function γ satisfying (19) up to $O(\varepsilon)$. To this end we need the following assertion.

Lemma 12.5.1. *Let P_n be the Legendre polynomial, $n=0,1,\dots$. Then*

$$(TP_n)\,(\zeta) = P_n(\zeta)\sum_{k=1}^{n}k^{-1}.$$

Proof. The argument is by induction on n. For $n=0$ and $n=1$ the assertion is evident. From the formula $(n+1)P_{n+1}(\zeta) = (2n+1)\,\zeta P_n(\zeta) - n\,P_{n-1}(\zeta)$ we obtain

$$\begin{aligned}-2(n+1)(TP_{n+1}(\zeta) = (2n+1)\int_{-1}^{1}\operatorname{sgn}(z-\zeta)P_n(z)\,dz&\\ -2(2n+1)\,\mu_n\,\zeta\,P_n(\zeta)+2n\mu_{n-1}\,P_{n-1}(\zeta)&,\end{aligned}$$

where $\mu_0=0$ and

$$\mu_n=\sum_{k=1}^{n}k^{-1}.$$

Since

$$n\int_{-1}^{1}\operatorname{sgn}(z-\zeta)P_n(z)\,dz = -2P_{n+1}(\zeta)+2\zeta P_n(\zeta),$$

we have

$$\begin{aligned}&-2(n+1)(T\,P_{n+1})(\zeta)\\ &=2(2n+1)(n^{-1}-\mu_n)\,\zeta\,P_n(\zeta) - 2(2n+1)n^{-1}P_{n+1}(\zeta)+2n\,\mu_{n-1}\,P_{n-1}(\zeta)\\ &=-2[(n+1)\mu_{n-1}-n^{-1}(2n+1)]\,P_{n+1}(\zeta)\\ &=-2(n+1)\mu_{n+1}\,P_{n+1}(\zeta).\end{aligned}$$

□

Thus, the operator T has the complete orthonormal system of eigenfunctions $[(2n+1)/2)^{1/2}]\, P_n$, $n = 0, 1, \dots$ in $\mathbf{L}_2([-1,1])$ and the sequence of eigenvalues μ_n. Hence

$$T\gamma = \sum_{n=0}^{\infty} (n+1/2)\, \mu_n \langle P_n, \gamma\rangle\, P_n.$$

In the generally nonsolvable equation (19) we replace T by its finite-dimensional approximation

$$T_\varepsilon = \sum_{n\varepsilon^{1/2}\le 1} (n+1/2)\, \mu_n \langle P_n, \cdot\rangle\, P_n, \tag{20}$$

and show that the new equation

$$\bigl(\log(\varepsilon/2) + \Xi\bigr)\gamma + T_\varepsilon\gamma + K\gamma = 2\pi(v_0 - \varphi)\,(0,\cdot) \tag{21}$$

is solvable. Moreover, the discrepancy (17) which corresponds to its solution γ is $O(\varepsilon)$. The equation

$$\{(\log(\varepsilon/2) + \Xi)\mathbf{1} + T_\varepsilon\}\, b = g$$

has the solution

$$\begin{aligned} b(\zeta) = &\sum_{n\varepsilon^{1/2}\le 1} (n+1/2)\bigl(\log(\varepsilon/2) + \Xi + \mu_n\bigr)^{-1} \langle g, P_n\rangle, P_n(\zeta) \\ &+ \sum_{n\varepsilon^{1/2}>1} (n+1/2)(\log\varepsilon/2) + \Xi)^{-1}\langle g, P_n\rangle. \end{aligned}$$

It is clear that

$$\|b; \mathbf{L}_2(-1,1)\| \le c|\log\varepsilon|^{-1}\, \|g;\, \mathbf{L}_2(-1,1)\|. \tag{22}$$

We denote by $\tilde{\mathbf{H}}^k$ the completion of $\mathbf{C}^\infty[-1,1]$ with respect to the norm

$$\|u; \tilde{\mathbf{H}}^k\| = \|u; \mathbf{W}_2^{k-1}\,(-1,1)\| + \|(1-z^2)d^k u/dz^k; \mathbf{L}_2(-1,1)\|.$$

One can check that the norm in $\tilde{\mathbf{H}}^k$ is equivalent to the norm

$$\|D^{k/2}\, u; \mathbf{L}_2(-1,1)\| + \|u; \mathbf{L}_2(-1,1)\|$$

with $D = -(d/dz)(1-z^2)(d/dz)$. Since the operators D and $\{(\log(\varepsilon/2)+\Xi)\mathbf{1}+T_\varepsilon\}^{-1}$ commute, from (22) it follows that

$$\|\{\bigl(\log(\varepsilon/2) + \Xi\bigr)\mathbf{1} + T_\varepsilon\};\, \tilde{\mathbf{H}}^k \to \tilde{\mathbf{H}}^k\| \le c|\log\varepsilon|^{-1}.$$

The operator K is bounded on $\tilde{\mathbf{H}}^k$. Therefore the equation (21) is uniquely solvable in $\tilde{\mathbf{H}}^k$ for any right-hand side in $\tilde{\mathbf{H}}^k$ and

$$\|R_\varepsilon;\, \tilde{\mathbf{H}}^k\| \to \tilde{\mathbf{H}}^k\| \le c|\log\varepsilon|^{-1}, \tag{23}$$

where $R_\varepsilon = \{(\log\varepsilon/2) + \Xi)\mathbf{1} + T_\varepsilon\ + K\}^{-1}$.

Now we estimate $(T_\varepsilon - T)\gamma$, i.e., the discrepancy (17). We have

$$(T_\varepsilon - T)\gamma = -\sum_{n\varepsilon^{1/2}>1} (n+1/2)\mu_n \langle P_n, \gamma\rangle\, P_n.$$

Hence

$$\|(T_\varepsilon - T)\gamma; \tilde{\mathbf{H}}^k\|^2 \le c \sum_{n\varepsilon^{1/2}>1} (n+1/2)\mu_n^2 \big(\langle P_n, D^{k/2}\gamma\rangle^2 + \langle P_n,\gamma\rangle^2\big)$$

$$\le c \sum_{n\varepsilon^{1/2}>1} (n^2+n)^{-2m}(n+1/2)\mu_n^2 \big(\langle P_n, D^{(k+m)/2}\gamma\rangle^2$$

$$+ \langle P_n, D^{m+2}\gamma\rangle^2\big).$$

It follows that

$$\|(T_\varepsilon - T)\gamma; \tilde{\mathbf{H}}^k\| \le c\varepsilon^m |\log\varepsilon|\, \|\gamma; \tilde{\mathbf{H}}^{k+m}\|$$

with a constant c not depending on ε. Since

$$\gamma = 2\pi R_\varepsilon\big((v_0 - \varphi)(0,\cdot)\big), \tag{24}$$

the estimate (23) implies

$$\|(T_\varepsilon - T)\gamma; \tilde{\mathbf{H}}^k\| \le c\varepsilon^m \|v_0(0,\zeta) - \varphi(0,\zeta); \tilde{\mathbf{H}}^k\|. \tag{25}$$

Thus, the function γ defined by (24) gives the discrepancy $O(\varepsilon)$ in (19). Obviously, instead of γ one can take any other function with the same property. In Section 12.4.9 the role of γ will be played by the solution to (21) with the right-hand side

$$\sum_{n\varepsilon^{1/2}\le 1} c_n P_n(\zeta),$$

where c_n are the coefficients of the expansion of $2\pi\big(v_0(0,\zeta) - \varphi(0,\zeta)\big)$ in Legendre polynomials.

12.5.5 Boundary layer near the endpoints of segment M (the first term)

As a result of the preceding discussion we have the following approximation for the solution to problem (1)–(3):

$$v_0(x) + \int_{-1}^{1} G(x,(0,\xi_3))\,\gamma(\xi_3)\,d\xi_3 + \begin{cases} 0, & |x_3| \ge 1, \\ \gamma(x_3)\, e(\varepsilon^{-1}(1-x_3^2)^{-1/2}\, y), & |x_3| < 1. \end{cases} \tag{26}$$

Here v_0, γ, and e are given by (11), (24), and (12), respectively. Though the function (26) may have discontinuities at the planes $\{x: x_3 = \pm 1\}$, it contributes the small discrepancy in (1)–(3) at the remaining points of $\Omega \setminus \bar{Q}_\varepsilon$. In this section we compensate for the discrepancy of the function (26) near the endpoints $p_\pm = (0,0,\pm 1)$ of M. It suffices to consider p_+. In a small neighborhood of the point we introduce coordinates $\xi = (\xi_1, \xi_2, \xi_3)$ with $\xi' = (\xi_1, \xi_2) = \varepsilon^{-2} y$, $\xi_3 = \varepsilon^{-2}(1 - x_3^2)/2$. In these coordinates the set Q_ε is the "paraboloid" $\Pi = \{\xi \in \mathbb{R}^3 : \xi'/\sqrt{2\xi_3} \in \omega\}$ and the Laplace operator takes the form

$$\Delta_x = \varepsilon^{-4}\{\Delta_{\xi'} + ((1 - 2\varepsilon^2\,\xi_3)^{1/2}\,\partial/\partial\xi_3)^2\}. \tag{27}$$

The principal part of (27) coincides with $\varepsilon^{-4}\Delta_\xi$.

Up to $O(\varepsilon)$ the discrepancy of the function (26) in (3) is equal to

$$(4\pi)^{-1}\gamma(1)\Big\{\log\Big\{\Big(\big(|y|^2+(1+x_3)^2\big)^{1/2}-1-x_3\Big)^{-1}\Big(\big(|y|^2+(1-x_3)^2\big)^{1/2}+1-x_3\Big)\Big\}$$
$$+\log|y|^2 - \log\big(4(1-x_3^2)\big)\Big\}$$

(see (7), (9), and (17)). We rewrite the last expression in the coordinates ξ, separating the main term

$$(4\pi)^{-1}\gamma(1)\log\{((|\xi'|^2/\xi_3^2+1)^{1/2}+1)/2\}+O(\varepsilon). \tag{28}$$

Let us consider the problem

$$\Delta_\varepsilon h(\xi)=0, \qquad \xi\in\mathbb{R}^3\setminus\bar{\Pi}, \tag{29}$$

$$h(\xi)=h_0(\xi), \qquad \xi\in\partial\Pi, \tag{30}$$

$$h(\xi)=o(1), \qquad |\xi|\to\infty, \tag{31}$$

where

$$h_0(\xi)=(4\pi)^{-1}\log\{((|\xi'|^2+1)^{1/2}+1)/2\}.$$

Lemma 12.5.2. *There exists a unique solution to problem* (29)–(31) *in the space* $\mathbf{V}^1_{2,0}(\mathbb{R}^3\setminus\Pi)$ *(see* 3.2.3 *and* 1.3.7*). This solution satisfies the inequalities*

$$|\nabla_k h(\xi)|\le c_k\,|\xi|^{-k}, \qquad \xi\in\mathbb{R}^3\setminus\Pi,\quad |\xi|<\text{const},\qquad k\in\mathbb{N}, \tag{32}$$

$$\begin{aligned}&|\nabla_l h(\xi)|\le c_l\,|\xi|^{-1-l}, && \xi\in\mathbb{R}^3\setminus\Pi,\quad |\xi|>\text{const},\quad |\xi'|>\xi_3,\\ &|\nabla_l h(\xi)|\le c_l\,|\xi'|^{-l}\xi_3^{-1}, && \xi\in\mathbb{R}^3\setminus\Pi,\quad |\xi|>\text{const},\quad |\xi'|<\xi_3,\quad l\in\mathbb{N}_0.\end{aligned} \tag{33}$$

Proof. In order to establish the solvability of problem (29)–(31) it suffices to show that $h_0|_{\partial\Pi}\in\mathbf{V}^{1/2}_{2,0}(\partial\Pi)$. Let

$$h_1(\xi)=-(4\pi)^{-1}\log\Big(((4\varkappa(\vartheta)^4/|\xi'|^2+(2\xi_3-|\xi'|^2)/\varkappa(\vartheta)^2+1)^{1/2}+1)/2\Big)\\ \times\chi(|\xi'|/|\xi|),$$

where χ is a function in $\mathbf{C}^\infty(\mathbb{R})$ such that

$$\chi(t)=0,\quad\text{for}\quad t<1/3,\qquad \chi(t)=1,\quad\text{for}\quad t>2/3. \tag{34}$$

It is clear that $h_0-h_1=0$ on $\partial\Pi$ near the point $\xi=0$. The following estimate is valid:

$$|D^\beta_\xi h_1(\xi)|\le c_\beta\,|\xi|^{-|\beta|},\qquad \xi\in\mathbb{R}^3\setminus\Pi,\quad |\xi|<\text{const},\quad |\beta|\ge 1.$$

Therefore, $h_1\in\mathbf{W}^1_{2,\mathrm{loc}}(\mathbb{R}^3\setminus\Pi)$ and consequently $h_0\in\mathbf{W}^{1/2}_{2,\mathrm{loc}}(\partial\Pi)$.

If $|\xi|\to\infty$ and $\xi\in\partial\Pi$ then

$$h_0(\xi)=-(8\pi)^{-1}\varkappa(\vartheta)^2\,\xi_3^{-1}+O(\xi_3^{-2}).$$

Therefore, the function $\xi\to h_2(\xi)=h_0(\xi)\,\chi(\xi_3)\chi(\xi_3/|\xi'|)$ belongs to $\mathbf{V}^1_{2,0}(\mathbb{R}^3\setminus\Pi)$. Since $h_0(\xi)=h_2(\xi)$ and $\xi_3>\text{const}>0$, we have $h_0\in\mathbf{V}^{1/2}_{2,0}(\partial\Pi)$. Thus, the unique solvability of problem (29)–(31) is proven. □

Now we obtain the estimate (33). Let G be Green's function of the Dirichlet problem for the Laplace operator in $\mathbb{R}^3\setminus\Pi$. Since $\Delta h_2=O(|\xi|^{-3})$ and $0\le G(p,q)\le c|p-q|^{-1}$, we have $|h(\xi)-h_2(\xi)|\le c|\xi|^{-1}$. Hence $|h(\xi)|\le c|\xi|^{-1}$.

Let $\xi\in\mathbb{R}^3\setminus\Pi$, $|\xi|>2$, and let $\overline{B_R(\xi)}$ be the closure of the ball with center ξ and radius R, where $R=\sigma|\xi'|$ for $|\xi'|<\xi_3$ and $R=\sigma|\xi|$ for $|\xi'|\ge\xi_3$, σ being a small positive constant. We next make use of the known estimate

$$|\nabla_l h(\xi)|\le c\Big(\sum_{j=0}^{l}R^{j-l}\,\|\nabla_j\,h_2|_{\partial\Pi};\mathbf{C}(\overline{B_R(\xi)}\cap\partial\Pi)\|+R^{-1}\|h;\mathbf{C}(\overline{B_R(\xi)}\setminus\Pi)\|\Big) \tag{35}$$

(see Theorem 3.1.4). Since $\nabla_j h_2(\xi) = O(\xi_3^{-j-1})$ on $\partial\Pi$ and $h(\xi) = O(|\xi|^{-1})$ in $\mathbb{R}^3 \setminus \Pi$, from (35) it follows that $\nabla_l h(\xi) = O(|\xi'|^{-l}/\xi_3)$ for $|\xi'| \leq \xi_3$ and $\nabla_l h(\xi) = O(|\xi|^{-l-1})$ for $|\xi'| > \xi_3$.

It remains to verify (32). Near $\xi = 0$ the function $h - h_1$ satisfies the boundary value problem

$$\begin{aligned} &\Delta(h - h_1)(\xi) = F(\xi/|\xi|)\,|\xi|^{-2} + O(|\xi|^{-1}), \quad \xi \notin \bar{\Pi}, \\ &(h - h_1)(\xi) = 0, \quad \xi \in \partial\Pi, \end{aligned} \tag{36}$$

where F is a smooth function on the unit sphere S^2. Therefore (see Lemma 1.3.13 and Proposition 1.5.3),

$$(h - h_1)(\xi) = \Phi(\xi/|\xi|) + O(|\xi|), \quad \xi \to 0. \tag{37}$$

One may differentiate the asymptotic representations (36) and (37). The estimate (33) results from (37) and the form of h_1.

Adding up (26) and the expression

$$\Big(\gamma(1)\,\chi(x_3) + \gamma(-1)\,\chi(-x_3)\Big)\, h(\varepsilon^{-2}\, y,\, \varepsilon^{-2}(1 - x_3^2)/2),$$

we obtain the approximation for the solution to the original problem satisfying the boundary condition on ∂Q_ε up to $O(\varepsilon)$ (see (28)).

12.5.6 Boundary layer near the endpoints of the segment M (the second term)

It remains to eliminate the discontinuities at the planes $\{x \colon x_3 = \pm 1\}$ of the approximate solution constructed in Section 12.5.4 and 12.5.5. To this end we match the boundary layer (16) with a solution that is smooth near the planes. We consider the problem

$$\Delta H(\xi) = 0, \quad \xi \in \mathbb{R}^3 \setminus \bar{\Pi}, \tag{38}$$

$$H(\xi) = e(\xi'(2\xi_3)^{-1/2}), \quad \xi \in \partial\Pi, \tag{39}$$

where e is the function defined by (13).

Lemma 12.5.3. *There exists at most one bounded solution to problem* (38), (39).

Proof. Let $G(p,q)$ be Green' function of the Dirichlet problem for the operator $-\Delta$ in $\mathbb{R}^3 \setminus \bar{\Pi}$. We set $\chi_R(\xi) = \chi(|\xi|/R)$ and denote by H_0 a bounded solution to the homogeneous problem (38), (39). Obviously,

$$H_0(p) = \int\limits_{\mathbb{R}^3 \setminus \Pi} H_0(q)[\Delta, \chi_R]\, G(p,q)\, dq.$$

Therefore,

$$\begin{aligned} |H_0(p)| &\leq c\Big\{R^{1/2}\Big(\int\limits_{\Xi_1} |\nabla G(p,q)|^2\, dq\Big)^{1/2} + R^{-2}\int\limits_{\Xi} G(p,q) dq\Big\} \\ &\leq cR^{-1/2}\Big(\int\limits_{\Xi_2} G(p,q)^2\, dq\Big)^{1/2}, \end{aligned}$$

where $\Xi_1 = \{\xi \in \mathbb{R}^3 \setminus \Pi \colon 1/3 \leq |\xi|/R \leq 2/3\}$ and $\Xi_2 = \{\xi \in \mathbb{R}^3 \setminus \Pi \colon 1/4 \leq |\xi|/R \leq 1\}$. Since $G(p,q) = o(|q|^{-1})$ for a fixed p and $q \to \infty$ (see, e.g., the estimate (32) in MAZ'YA [2]), we have $|H_0(p)| = o(1)$ as $R \to \infty$. □

Lemma 12.5.4. *There exists a solution e_1 to the problem*

$$\Delta_\eta e_1(\eta) = -2^{-1}(\eta\cdot\nabla_\eta + 3)(\eta\cdot\nabla_\eta) e(\eta), \quad \eta\in\mathbb{R}^2\setminus\bar\omega, \tag{40}$$

$$e_1(\eta)=0, \quad \eta\in\partial\omega; \qquad e_1(\eta) = O(|\eta|\log|\eta|), \quad |\eta|\to\infty. \tag{41}$$

Proof. Since

$$e(\eta)\sim\sum_{k=1}^{\infty}|\eta|^{-k}(a_k\sin k\vartheta + b_k\cos k\vartheta), \quad |\eta|\to\infty,$$

the right-hand side of (40) has the asymptotics

$$\sum_{k=1}^{\infty}|\eta|^{-k}2^{-1}k(k-3)(a_k\sin k\vartheta + b_k\cos k\vartheta), \quad |\eta|\to\infty.$$

Hence there exists a solution e_1 to problem (40), (41) of the form

$$2^{-1}|\eta|\log|\eta|(a_1\sin\vartheta + b_1\cos\vartheta) + 4^{-1}(a_2\sin 2\vartheta + b_2\cos 2\vartheta) + d_0 + o(1). \qquad \square$$

Lemma 12.5.5. *There exists a bounded solution $H(\xi)$ to problem* (38), (39). *The following estimates are valid:*

$$|(\partial/\partial\xi_3)^l H(\xi)| \le c_l\,|\xi_3|^{1/2-l}\,|\xi'|^{-1}\,|\log|\xi'||, \quad \xi\in K_1\setminus\Pi, \quad |\xi|>\text{const}, \tag{42}$$

$$|(\partial/\partial\xi_3)^l H(\xi)| \le c_l\,|\xi|^{-1/2-l}\,|\log|\xi||, \quad \xi\in\mathbb{R}^3\setminus K_1, \quad |\xi|>\text{const}, \tag{43}$$

$$D_\xi^\beta H(\xi) = O(|\xi|^{-|\beta|}), \quad \xi\in\mathbb{R}^3\setminus\Pi, \quad |\xi|<\text{const}, \tag{44}$$

where β is an arbitrary multiindex, $l = 0,1,2,\dots$, and $K_\alpha = \{\xi\in\mathbb{R}^3\colon |\xi'|\le \alpha\xi_3\}, \alpha>0$.

Proof. We set

$$\begin{aligned} E(\xi) &= e(\xi'(2\xi_3)^{-1/2}),\\ E_1(\xi) &= e_1(\xi'(2\xi_3)^{-1/2}),\\ X(\xi) &= \chi(x_3)\chi(3\xi_3(2|\xi'|^{-1}), \end{aligned} \tag{45}$$

where χ is the function (34). Since $\Delta_\eta e(\eta)=0$ and e_1 satisfies (40), we have

$$\begin{aligned} &\Delta_\xi((E+\xi_3^{-1}E_1)X)\\ &= X((\partial/\partial\xi_3)^2(\xi_3^{-1}E_1) + (\partial/\partial\xi_3)^2 E + \xi_3^{-1}\Delta_{\xi'}E_1) - [\Delta_{\xi'}, X](E+\xi_3^{-1}E_1)\\ &= X(\partial/\partial\xi_3)^2(\xi_3^{-1}E_1) - [\Delta_\xi, X](E+\xi_3^{-1}E_1). \end{aligned} \tag{46}$$

The support of the first term on the right in (46) is located in $K_2\setminus\Pi$ while the support of the second one in $K_2\setminus K_1$. Note also that

$$\begin{aligned} |D_\eta^\beta e(\eta)| &\le c|\eta|^{-1-|\beta|}, & \eta&\in\mathbb{R}^2\setminus\omega,\\ |D_\xi^\gamma X(\xi)| &\le c|\xi|^{-|\gamma|}, & \xi&\in\mathbb{R}^3,\\ |D_\xi^\sigma e_1(\eta)| &\le c|\eta|^{1-\sigma}|\log|\eta||, & \eta&\in\mathbb{R}^2\setminus\bar\omega. \end{aligned}$$

Therefore,

$$\Delta_\xi(X(E+\xi_3^{-1}E_1)) = O(|\xi'|\xi^{-7/2}|\log|\xi'||) = O(|\xi|^{-5/2}|\log|\xi||) \tag{47}$$

and these relations can be differentiated. We consider the problems

$$\Delta_\xi H_1(\xi) = 0 \quad\text{in}\quad \mathbb{R}^3\setminus\bar\Pi; \qquad H_1(\xi) = E(1-X)(\xi) \quad\text{on}\quad \partial\Pi, \tag{48}$$

$$\Delta_\xi H_2(\xi) = \Delta_\xi(X(E+\xi_3^{-1}E_1)) \quad\text{in}\quad \mathbb{R}^3\setminus\bar\Pi; \qquad H_2(\xi)=0 \quad\text{on}\quad \partial\Pi. \tag{49}$$

The trace of E on $\partial\Pi$ coincides with the trace of a smooth function ν in ϑ. It is clear that $\nu(\vartheta)\chi(|\xi'|/|\xi|) - E = O(|\xi|)$ on $\partial\Pi$. Since $\nu(\vartheta)\chi(|\xi'|/|\xi|) \in \mathbf{W}^1_{2,\mathrm{loc}}(\mathbb{R}^3 \setminus \Pi)$, we have $E \in \mathbf{W}^{1/2}_{2,\mathrm{loc}}(\partial\Pi)$. Note that the support of $E(1-X)$ is compact. Hence there exists a unique bounded solution H_1 to problem (48). Such a solution admits the estimate $H_1(\xi) = O(|\xi|^{-1})$, $|\xi| > \mathrm{const}$. Now, taking account of (35) written for H_1, we obtain

$$|\nabla_l H_1(\xi)| \le c_l\, |\xi|^{-1-l}, \quad \xi \in \mathbb{R}^3 \setminus K_1, \quad |\xi| > \mathrm{const},$$
$$|\nabla_l H_1(\xi)| \le c_l\, |\xi'|^{-1-l}, \quad \xi \in \mathbb{R}^3 \setminus \Pi, \quad |\xi| > \mathrm{const}, \qquad l \in \mathbb{N}_0.$$

The reasoning in the derivation of (37) applied to the function $H_1 - \nu(\vartheta)\chi(|\xi'|/|\xi|)$ leads to the inequality

$$|D^\beta_\xi H_1(\xi)| \le c|\xi|^{-|\beta|}, \quad \xi \in \mathbb{R}^3 \setminus \Pi, \qquad |\xi| < \mathrm{const}. \tag{50}$$

From (47) and from the inequalities $0 \le G(p,q) \le c|p-q|^{-1}$ it follows that problem (49) is solvable and its solution H_2 admits the estimate

$$H_2(\xi) = O(|\xi|^{-1/2}|\log|\xi||), \ |\xi| > \mathrm{const}.$$

Because the function H_2 is harmonic near the point $\xi = 0$ and $H_2|_{\partial\Pi} = 0$, the estimate (50) on H_2 is evident.

Let ξ and $B_R(\xi)$ be the same as in (35). We make use of the known estimate

$$|\nabla_l H_2(\xi)| \le c\Big(\sum_{j=0}^{l-1} R^{j-l+2}\, \|\nabla_j\, \Delta H_2; \mathbf{C}(\overline{B_R(\xi)} \setminus \Pi)\| + R^{-l}\|H_2; \mathbf{C}(\overline{B_R(\xi)} \setminus \Pi)\|\Big) \tag{51}$$

(see Theorem 3.1.4). In view of $H_2(\xi) = O(|\xi|^{-1/2}|\log|\xi||)$ and $|\nabla_j \Delta H_2(\xi)| = O(|\xi|^{-j-5/2}|\log|\xi||)$, we have

$$\nabla_l H_2(\xi) = O(|\xi'|^{-l}\xi_3^{-1/2}|\log|\xi||) \quad \text{on} \quad K_1 \setminus \Pi \tag{52}$$

and $\nabla_l H_2(\xi) = O(|\xi|^{-l-1/2}|\log|\xi||)$ on $\mathbb{R}^3 \setminus K_1$. One can check by differentiation that the function $X(E + \xi_3^{-1}E_1)$ satisfies (42) and (43).

Thus, we have proved the existence of a bounded solution $H = (E+\xi_3^{-1}E_1)X + H_1 + H_2$ to problem (38), (39) and the inequalities (42) and (43). □

12.5.7 Justification of the asymptotics

As an asymptotic solution to problem (1)–(3) we take

$$\begin{aligned} u_0(x,\varepsilon) &= v(x) + \big(\gamma(x_3) - \gamma(1)\chi(x_3) - \gamma(-1)\big)\chi(-x_3)\chi\big((1-x_3^2)/\varepsilon\big) \\ &\times e\big((1-x_3^2)^{-1/2}y/\varepsilon\big) + \big(\gamma(1)\chi(x_3) + \gamma(-1)\chi(-x_3)\big) \\ &\times h\big(\varepsilon^{-2}y, \varepsilon^{-2}(1-x_3^2)/2\big) + H\big(\varepsilon^{-2}y, \varepsilon^{-2}(1-x_3^2/2\big), \end{aligned} \tag{53}$$

where v, γ, χ, e are defined by (5), (24), (34), and (13) while h and H satisfy problem (29)–(31) and problem (38), (39), respectively.

Denote by F_0, Ψ_0, Φ_0 the discrepancies of the function (53) in (1)–(3). The function v vanishes on $\partial\Omega$. Besides, from (19), (32), and (41) it follows that

$$e\big((1-x_3^2)^{-1/2}y/\varepsilon\big) = O(\varepsilon), \quad x \in \partial\Omega, \quad x_3 \in (-1,1),$$
$$h\big(\varepsilon^{-2}y, \varepsilon^{-2}(1-x_3^2)/2\big) = O(\varepsilon^2), \quad x \in \partial\Omega, \quad x_3 \notin (-1/3, 1/3),$$
$$H\big(\varepsilon^{-2}y, \varepsilon^{-2}(1-x_3^2)/2\big) = O(\varepsilon), \quad x \in \partial\Omega, \quad x_3 \notin (-1/3, 1/3).$$

Thus,

$$|\Psi_0(x,\varepsilon)| \le c\varepsilon, \qquad x \in \partial\Omega. \tag{54}$$

Taking (11), (22), (30), and (39) into account, we obtain

$$\begin{aligned}\Phi_0(x,\varepsilon) = &\Big\{\varphi(0,x_3) - \varphi(x)\Big\} + \Big\{v(x) - (4\pi)^{-1}L(x) - \big(v - (4\pi)^{-1}L\big)(0,x_3)\Big\}\\ &+ (4\pi)^{-1}\Big\{\big(\gamma(x_3) - \gamma(1)\chi(x_3) - \gamma(-1)\chi(-x_3)\big)\\ &\times \Big(L(x) - \log\big(4(1-x_3^2)|y|^{-2}\big) - \log\Big(\big((|\xi'|^2/\xi_3^2 + 1)^{1/2} + 1\big)/2\Big)\Big)\Big\}\\ &+ \Big\{\big(\gamma(x_3) - \gamma(1)\chi(x_3) - \gamma(-1)\chi(-x_3)\big)\Big(1 - \chi\big((1-x_3^2)/\varepsilon\big)\Big)\\ &\times e\big((1-x_3^2)^{-1/2}y/\varepsilon\big)\Big\} + \Big\{(T_\varepsilon - T)\gamma(x_3)\Big\},\end{aligned} \tag{55}$$

where L is the function (7), T and T_ε are operators (18) and (20). Let $I_1, I_2, \dots, I_5$, signify the expressions in braces on the right-hand side of (55). We see that $I_1 + I_2 = O(\varepsilon)$ because φ is a smooth function in $\bar\Omega$ and $v - (4\pi)^{-1}L$ satisfies the Lipschitz condition. From (8) and (28) it follows that $I_3 = O(\varepsilon)$. The support of I_4 is located in the ε neighborhood of $p_\pm$, hence $I_4 = O(\varepsilon)$. The estimate $I_5 = O(\varepsilon)$ immediately follows from (25). As a result, we have

$$|\Psi_0(x,\varepsilon)| \le c\varepsilon, \quad x \in \partial Q_\varepsilon. \tag{56}$$

The function v satisfies $-\Delta v = f$ so $F_0(x,\varepsilon) = \Delta(u_0 - v)$. We rewrite the last expression in the form

$$\begin{aligned}F_0(x,\varepsilon) = &M_1(x_3,\eta,\partial/\partial x_3,\partial/\partial\eta)\Big\{\big(\gamma(x_3) - \Gamma(x_3)\big)\,\chi\big((1-x_3^2)/\varepsilon\big)\,e(\eta)\Big\}\\ &+ \big(\varepsilon^{-4}\Delta_\xi + M_2(\varepsilon,x_3,\partial/\partial\xi_3)\big)\Big\{\Gamma(x_3)\big(h(\xi) + H(\xi)\big)\Big\},\end{aligned} \tag{57}$$

where $\eta = (1-x_3^2)^{-1/2}\,y/\varepsilon$, $\xi' = \varepsilon^{-2}y$, $\xi_3 = (1-x_3^2)/2\varepsilon$, $\Gamma(x_3) = \gamma(1)\chi(x_3) + \gamma(-1)\chi(-x_3)$, $M_1 = (\partial/\partial x_3 + (1-x_3^2)^{-1}\,x_3\eta\cdot\nabla_\eta)^2$, $M_2 = (1-x_3^2)(\partial/\partial\xi_3)^2 + \varepsilon^2 x_3^{-1}(\partial/\partial\xi_3)$. To estimate the first term $F_0^{(1)}$ on the right in (47) we make use of the following properties of $\gamma - \Gamma$, χ, e :

$$\begin{aligned}&|(d/dx_3)^k(\gamma(x_3) - \Gamma(x_3)| \le c_k, && k \in \mathbb{N}; && (\gamma - \Gamma)(\pm 1) = 0;\\ &|(d/dx_3)^k\chi((1-x_3^2)/\varepsilon)| \le c_k\varepsilon^{-k}, && k \in \mathbb{N}; && \operatorname{supp}\chi' \subset [1/3,\,2/3];\\ &|(\eta\cdot\nabla_\eta)^j\,e(\eta)| \le c_j|\eta|^{-1}, && \eta \in \mathbb{R}^2\setminus\omega, && j \in \mathbb{N}_0.\end{aligned}$$

This immediately implies

$$|F_0^{(1)}(x,\varepsilon)| \le c\varepsilon(1-x_3^2)^{-1/2}\,|y|^{-1}, \quad x \in \bar\Omega\setminus Q_\varepsilon, \quad |x_3| \le 1. \tag{58}$$

We represent the second term on the right in (57) as the sum $F_0^{(2)} + F_0^{(3)}$ with $F_0^{(2)} = \varepsilon^{-4}\Delta_\xi(\Gamma(h+H))$ and $F_0^{(3)} = M_2(\Gamma(h+H))$. The functions h and H are harmonic outside the "paraboloid" Π so

$$F_0^{(2)} = \varepsilon^{-4}[(\partial/\partial\xi_3)^2, \Gamma](h+H).$$

In order to estimate $F_0^{(2)}$ we need the following properties of h and H (see (33), (42), and (43)). If $|x_3| \in [1/3\,, 2/3]$, then

$$|h(\xi)| \leq c|\xi|^{-1}; \quad |(\partial/\partial\xi_3)h(\xi)| \leq c|\xi'|^{-1}\xi_3^{-1};$$

$$|H(\xi)| \leq c\xi_3^{1/2}|\xi'|^{-1}\,|\log|\xi||; \quad |(\partial/\partial\xi_3)H(\xi)| \leq c\xi_3^{-1/2}|\xi'|^{-1}|\log|\xi||.$$

Since $\Gamma \in \mathbf{C}^2[-1,1]$ and $\operatorname{supp}\Gamma' \subset \{x\colon 1 \leq 3|x_3| \leq 2\}$, we have

$$|F_0^{(2)}(x,\varepsilon)| \leq c\varepsilon|y|^{-1}\,|\log\varepsilon|, \quad x \in \bar{\Omega} \setminus Q_\varepsilon. \tag{59}$$

To estimate $F_0^{(3)}$ we apply the inequalities

$$|H(\xi)| \leq c, \quad |h(\xi)| \leq c|\log|\xi|| \quad \text{for} \quad |\xi| < 1/2, \quad |h(\xi)| \leq c \quad \text{for} \quad |\xi| > 1/2;$$

$$\begin{aligned}|(\partial/\partial\xi_3)h(\xi)| + |(\partial/\partial\xi_3)H(\xi)| + |\xi|^{-1}\big(|(\partial/\partial\xi_3)^2 h(\xi)| \\ + |(\partial/\partial\xi_3)^2 H(\xi)|\big) \leq c|\xi|^{-1}\log(|\xi|+2),\end{aligned}$$

which follow from (32), (33), (42)–(44). Then

$$|F_0^{(3)}(x,\varepsilon)| \leq c\varepsilon\big(|y|^2 + (1-x_3^2)^2\big)^{-1/2}. \tag{60}$$

Taking account of

$$|u(s,\varepsilon) - u_0(s,\varepsilon)| \leq \int\limits_{\Omega\setminus Q_\varepsilon} G(s,x)|F_0(x,\varepsilon)|dx + \sup_{x\in\partial\Omega}|\Psi_0(x,\varepsilon| + \sup_{x\in\partial Q_\varepsilon}|\Phi_0(x,\varepsilon|)$$

and combining (54), (56), (58)–(60), we obtain

$$\begin{aligned}|u(s,\varepsilon) - u_0(s,\varepsilon)| &\leq c\varepsilon\Big\{1 + \int\limits_{\Omega\cap\{|x_3|\leq 1\}} (|x-s|\,|y|\,(1-x_3^2)^{1/2})^{-1}dx \\ &+ |\log\varepsilon| \int\limits_\Omega (|x-s|\,|y|)^{-1}\,dx + \int\limits_\Omega (|x-s|(|y| + |1-x_3^2|))^{-1}dx\Big\} \\ &= O(\varepsilon|\log\varepsilon|).\end{aligned} \tag{61}$$

Thus, we have proved the following assertion.

Theorem 12.5.6. *The solution u to problem (1)–(3) can be written in the form $u(x,\varepsilon) = u_0(x,\varepsilon) + U(x,\varepsilon)$, where u_0 is defined by (53) and the remainder U admits the estimate $|U(x,\varepsilon)| \leq c\varepsilon|\log\varepsilon|$ with a constant c independent of ε.*

Remark 12.5.7. The requirement for the domain Ω to be bounded was essential for (61). One can drop this restriction making the formula (53) for u_0 more complicated. Namely, it suffices to multiply $W = (\gamma - \Gamma)\chi e + \Gamma(h+H)$ in (53) by $1 - \chi(|x|/4)$. Then the right-hand side in (1) takes the form

$$f + (1-\chi(|x|/4))F - [\Delta, \chi(|x|/4)]W.$$

The function $(1-\chi)F$ can be estimated in the same manner as for bounded domains. From the inclusion $\operatorname{supp}([\Delta,\chi]W) \subset \{x \in \bar{\Omega}\colon 4/3 \leq |x| \leq 8/3\}$ and from (13), (33), (42), and (43) we derive

$$|[\Delta, \chi(|x|/4)]W(x,\varepsilon)| \leq c\varepsilon|\log\varepsilon|.$$

Having the new definition of u_0, we obtain Theorem 5.6 for an arbitrary domain Ω, i.e.,

$$\begin{aligned} u_0(x,\varepsilon) = v(x) + (1-\chi(|x|/4))\Big\{ &\big(\gamma(x_3) - \gamma(1)\chi(x_3) - \gamma(-1)\chi(-x_3)\big) \\ &\times \chi\big((1-x_3^2)/\varepsilon\big) e\big((1-x_3^2)^{-1/2} y/\varepsilon\big) + \big(\gamma(1)\chi(x_3) + \gamma(-1)\chi(-x_3)\big) \\ &\times \big(h(\varepsilon^{-2}y, \varepsilon^{-2}(1-x_3^2)/2) + H\big(\varepsilon^{-2}\, y, \varepsilon^{-2}(1-x_3^2/2)\big)\big)\Big\}. \end{aligned} \tag{62}$$

Remark 12.5.8. The terms on the right-hand side of (53) turn out to be small in a subdomain of Ω (see the beginning of the section). Therefore, (53) can be rewritten in the following equivalent form

$$\begin{aligned} &u(x) \sim v(x) \quad \text{for} \quad x \notin U_{\delta_1}(M), \\ &u(x) \sim v(x) + \gamma(x_3) e((1-x_3^2)^{-1/2} y/\varepsilon) \\ &\qquad \text{for} \quad x \in U_{2\delta_1}(M), \quad x \notin U_{\delta_2}((0,1)), \\ &u(x) \sim v(x) + \gamma(x_3) e((1-x_3^2)^{-1/2} y/\varepsilon) + \gamma(\pm 1) H(\varepsilon^{-2} y, \varepsilon^{-2}(1-x_3^2)/2) \\ &\qquad \text{for} \quad x \in U_{2\delta_2}((0,\pm 1)), \quad 1 - x_3^2 > \delta_3 \varepsilon, \\ &u(x) \sim v(x) + \gamma(\pm 1) H(\varepsilon^{-2} y, \varepsilon^{-2}(1-x_3^2)/2) \\ &\qquad \text{for} \quad x \in U_{2\delta_2}((0,\pm 1)) \setminus U_{\delta_4 \varepsilon}((0,\pm 1)), \quad 1 - x_3^2 < 2\delta_3 \varepsilon; \\ &u(x) \sim v(x) + \gamma(\pm 1)(H + h)(\varepsilon^{-2} y,\, \varepsilon^{-2}(1-x_3^2)/2) \\ &\qquad \text{for} \quad x \in U_{\delta_4 \varepsilon}((0,\pm 1)). \end{aligned}$$

Here $a \sim b$ means that $a - b = O(\varepsilon|\log \varepsilon|)$ and δ_j is a positive constant.

12.5.8 Asymptotics of capacity of the "ellipsoid" Q_ε

We shall derive from (53) the asymptotics of capacity of the set

$$\operatorname{cap} Q_\varepsilon = (4\pi)^{-1} \int\limits_{\partial Q_\varepsilon} (\partial/\partial n)\, u\, d\sigma$$

(cf. Section 12.2.4). Here $d\sigma$ is an area element, n is the inward directed normal to ∂Q_ε, and u is the solution to problem (1)–(3) with $\Omega = \mathbb{R}^3, f = 0, \varphi = 1$.

Theorem 12.5.9. *It holds*

$$\operatorname{cap} Q_\varepsilon = 1/\log\Big\{2\big(\varepsilon c_{\log}(\omega)\big)^{-1}\Big\} + O(\varepsilon|\log\varepsilon|),$$

where $c_{\log}(\omega)$ *is the logarithmic capacity of* ω, *i.e.*

$$c_{\log}(\omega) = \exp \Xi,$$

Ξ *being the constant in* (15).

Proof. The functions u and v are harmonic and $u - v = O(\varepsilon|\log\varepsilon|)$ for $x \in B_2(0)$. Therefore, $\nabla(u - v) = O(\varepsilon|\log\varepsilon|)$ outside the same ball. In view of

$$\operatorname{cap} Q_\varepsilon = (4\pi)^{-1} \int\limits_{\partial B_2(0)} (\partial/\partial n) u d\sigma,$$

it follows that

$$\operatorname{cap} Q_\varepsilon = (4\pi)^{-1} \int\limits_{\partial B_3(0)} (\partial/\partial n) v d\sigma + O(\varepsilon |\log \varepsilon|).$$

Since for the space $\mathbb{R}^3$ the function $G(x,\xi)$ coincides with $(4\pi|x-\xi|)^{-1}$, equation (19) takes the form

$$\big(\log(\varepsilon/2) + \Xi\big)\gamma + T\gamma = -2\pi.$$

Obviously, $\gamma = -2\pi/\log(\varepsilon/2) + \Xi$ satisfies this equation. Hence,

$$v(x) = -2^{-1}\big(\log(\varepsilon/2) + \Xi\big)^{-1} \int\limits_{-1}^{1} |x - (0,\xi_3)|^{-1}\, d\xi$$

and, consequently,

$$(4\pi)^{-1} \int\limits_{\partial B_3(0)} (\partial/\partial n) v d\sigma + (\log(\varepsilon/2) - \Xi)^{-1}. \qquad \square$$

12.5.9 The case of a rotation ellipsoid

We suppose that the domain $\omega \subset \mathbb{R}^2$ introduced in 12.5.1 coincides with the unit disc $\{y \in \mathbb{R}^2 \colon |y| < 1\}$, i.e., $\varkappa = 1$. Then Q_ε is the rotation ellipsoid with semiaxes $\varepsilon, 1$ and focuses $(0,0,\pm(1-\varepsilon^2)^{1/2})$. We show that in this case the formula (53) can be simplified.

A function that is harmonic outside the unit disk and satisfies (12) is defined by

$$g(\eta) = -(4\pi)^{-1} \log|\eta|.$$

Therefore, $e(\eta) = 0$ and $H(\xi) = 0$ (see (14), (38), and (39)). The equation for γ takes the form

$$\log(\varepsilon/2)\gamma(\zeta) + (T\gamma)(\zeta) + (K\gamma)(\zeta) = 2\pi\big(v_0(0,\zeta) - \varphi(0,\zeta)\big) \tag{63}$$

(the operators T and K are defined in 12.5.1 and v_0 is the function (6)). Now we find a solution to problem (29)–(31). The function

$$\xi \to \log\big((|\xi'|^2 + \xi_3^2)^{1/2} + \xi_3\big)$$

is harmonic so the function

$$\xi \to h(\xi) = -(4\pi)^{-1} \log\Big\{ \Big((|\xi'|^2 + (\xi_3 - 1/2)^2)^{1/2} + \xi_3 - 1/2 \Big)^{-1} \times \big((|\xi'|^2 + \xi_3^2)^{1/2} + \xi_3\big) \Big\},$$

too, satisfies the equation $\Delta h = 0$. In our case Π is the paraboloid $\{\xi \in \mathbb{R}^3 \colon |\xi'| < (2\xi_3)^{1/2}\}$. Hence, on $\partial\Pi$

$$\begin{aligned} h(\xi) &= -(4\pi)^{-1} \log\Big\{ \Big((2\xi_3^2 + (\xi_3 - 1/2)^2)^{1/2} + \xi_3 - 1/2 \Big)^{-1} \big((|\xi'|^2 + \xi_3^2)^{1/2} + \xi_3\big) \Big\} \\ &= -(4\pi)^{-1} \log\Big\{ \big((|\xi'|^2 + \xi_3^2)^{1/2} + \xi_3\big)/2\xi_3 \Big\}, \end{aligned}$$

and, consequently, h is subject to the boundary condition (30). The relation (31) is evident.

Thus, for the solution u to problem (1)–(3) we have the asymptotic formula

$$\begin{aligned} u(x) = &\int_\Omega G(x,s)f(s)ds + \int_{-1}^{1} G(x,(0,s_3))\gamma(s_3)ds_3 \\ &- (4\pi)^{-1}\Big(\gamma(1)\chi\big(|y|^2+(x_3-1)^2\big) + \gamma(-1)\chi\big(|y|^2+(x_3+1)^2\big)\Big) \\ &\times \log\Big\{\Big(\big(4|y|^2+(1-x_3^2-\varepsilon^2)^2\big)^{1/2}+1-x_3^2-\varepsilon^2\Big)^{-1} \\ &\times \Big(\big(4|y|^2+(1-x_3^2)^2\big)^{1/2}+1-x_3^2\Big)\Big\} + O(\varepsilon|\log\varepsilon|), \end{aligned} \tag{64}$$

where $x = (y, x_3)$; G is Green's function in the domain Ω; γ is any function that gives the discrepancy $O(\varepsilon|\log\varepsilon|)$ in (63) (the construction of such a function was described in 12.5.4); χ is the function (34). Note that (64) holds for unbounded domains Ω, too, because the cut-off functions in (64) are compactly supported (cf. Remark 12.5.7).

If $\Omega = \mathbb{R}^3$, the formula (64) becomes all the more explicit. In this case equation (19) (or (63)) changes to

$$\log(\varepsilon/2)\,\gamma(\zeta) + (T\gamma)\,(\zeta) = 2\pi(v_0(0,\zeta) - \varphi(0,\zeta)) \tag{65}$$

with

$$v_0(x) = (4\pi)^{-1}\int_{\mathbb{R}^3} f(\xi)\,|x-\xi|^{-1}\,d\xi.$$

The asymptotic solution to equation (65) constructed in 12.5.4 has the form

$$\gamma(\zeta) = \sum_{n\varepsilon^{1/2}\le 1} 2\pi\Big(\log(\varepsilon/2) + \sum_{k=1}^{n} k^{-1}\Big)^{-1} c_n P_n(\zeta), \tag{66}$$

where P_n is the Legendre polynomial of degree n and c_n are the coefficients of expansion of the function $\zeta \to v_0(0,\zeta) - \varphi(0,\zeta)$ in Legendre polynomials.

If $v_0(0,\zeta) - \varphi(0,\zeta)$ is a polynomial of degree N, the function (66),too, turns out to be a polynomial of degree N and satisfies (65),

$$\gamma(\zeta) = \sum_{n=0}^{N} 2\pi\Big(\log(\varepsilon/2) + \sum_{k=1}^{n} k^{-1}\Big)^{-1} c_n P_n(\zeta).$$

12.6 Smoothing the Boundary Near an Edge

The algorithms described in the preceding sections of this chapter apply to other cases, too, of variation of the boundary of a domain near singular submanifolds. In the section this is shown for a three-dimensional domain with boundary perturbed near an edge (see Fig. 12.7). In order to illustrate the possibilities of the algorithm for nonlinear problems we consider here the Dirichlet problem for the equation $\Delta u(x) + \gamma u^2(x) = f(x)$ with a constant γ (in particular, if $\gamma = 0$ we deal with the Poisson equation).

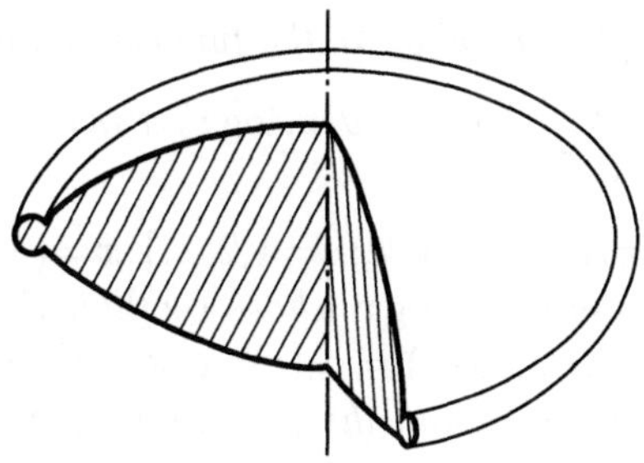

Fig. 12.7

12.6.1 The domain

Let Ω be a domain in $\mathbb{R}^3$ with compact closure $\bar{\Omega}$ and boundary $\partial\Omega$ and let M be a smooth closed curve on $\partial\Omega$. We suppose that $\partial\Omega \setminus M$ is a manifold of class $\mathbf{C}^\infty$ and in a neighborhood of any point $P \in M$ the set Ω is diffeomorphic to a wedge. Thus, two tangent planes to $\partial\Omega$ are determined at every point P. We assume that the angle between them (directed into Ω) does not depend on P and is equal to $\alpha, 0 < \alpha < 2\delta$.

Let $P \in M$ and $\Pi(P)$ be the normal plane to M at P. By z we denote a local coordinate of P on M. Introduce curvilinear coordinates (y_1, y_2) on $\Pi(P)$ smoothly depending on z such that the two arcs located in the intersection $\partial\Omega \cap \Pi(P)$ with the common endpoint P are given by $y_2 = 0$ and $y_2 = y_1 \operatorname{tg} \alpha$. It is clear that (y_1, y_2, z) are coordinates in a small d-neighborhood of the curve M. Let K_α designate the corner on the plane Ω between the tangents to $\partial\Omega \cap \Pi(P)$ at P (directed into Ω). We introduce a family of domains $\omega(z), z \in M$, possessing the following properties. The domain $\omega(z)$ has compact closure and its complement to K_α has smooth boundary. Suppose that $K_\alpha \setminus \overline{\omega(z)}$ smoothly depends on $z \in M$ (in the sense that there exists a diffeomorphism of a neighborhood of $K_\alpha \setminus \overline{\omega(z)}$ that smoothly depends on z and sends the domain to a standard one). Set

$$\begin{aligned} \omega_\varepsilon(z) &= \{(y_1, y_2) \colon \varepsilon^{-1}(y_1, y_2) \in \omega(z)\}, \\ \omega_\varepsilon &= \{x \in \mathbb{R}^3 \colon z \in M, (y_1, y_2) \in \omega_\varepsilon(z)\}, \\ \Omega_\varepsilon &= \Omega \setminus \overline{\omega_\varepsilon}. \end{aligned}$$

As in 5.7, we consider the problem

$$\begin{aligned} &\Delta u(x, \varepsilon) + \gamma u(x, \varepsilon)^2 = f(x), \quad x \in \Omega_\varepsilon; \\ &u(x, \varepsilon) = \varphi(x), \quad x \in \partial\Omega_\varepsilon, \end{aligned} \tag{1}$$

where γ is a constant, $f \in \mathbf{C}_0^\infty(\bar{\Omega} \setminus M)$, and $\varphi \in \mathbf{C}_0^\infty(\partial\Omega \setminus M)$. The functions f and φ have small norms in $\mathbf{C}(\bar{\Omega})$ and $\mathbf{C}(\partial\Omega)$, respectively. It is clear that there exists a unique solution to problem (1).

As principal term v_0 of the asymptotics of the solution u it is natural to take the small solution to the problem

$$\begin{aligned} &\Delta v_0(x) + \gamma v_0(x)^2 = f(x), \quad x \in \Omega; \\ &v_0(x) = \varphi(x), \quad x \in \partial\Omega. \end{aligned} \tag{2}$$

The next assertion follows from Proposition 11.3.2.

Lemma 12.6.1. *In a d-neighborhood of M the function v_0 admits the representation*

$$v_0(x) = \sum_{\beta \in X_0, \beta < N} r^\beta P_\beta(\vartheta, z, \log r) + o(r^N), \quad r \to 0,$$

where $X_0 = \{\beta \in \mathbb{R}^1_+; \beta = k\pi/\alpha + m - 1, k, m = 1, 2, \ldots\}, x = (y, z)$, (r, ϑ) *are polar coordinates of* y; P_β *is a polynomial in* $\log r$ *whose coefficients are smooth functions in* $\vartheta \in [0, \alpha]$ *and* $z \in M$; $P_{\pi/\alpha}(\vartheta, z, \log r) = c_0(z)\sin(\pi\vartheta/\alpha)$; *finally,* N *is an arbitrary positive number. (If* π/α *is irrational, the functions do not depend on* $\log r$.*)*

12.6.2 The principal term of asymptotics

In order to construct a function $w_{\pi/\alpha}$ compensating the discrepancy in the boundary condition on $\partial\omega_\varepsilon \setminus \partial\Omega$ we rewrite the equation in the coordinates (y_1, y_2, z):

$$\Delta_y u + \gamma P(y, z, \partial/\partial y, \partial/\partial z)u + u^2 = f.$$

Here P is a second order differential operator whose coefficients by

$$\partial^2/\partial y_1^2,\ \partial^2/\partial y_1 \partial y_2,\ \partial^2/\partial y_2^2$$

are $O(|y|)$. After the change of variables $y \to \xi = \varepsilon^{-1} y$ we represent the operator as the formal series

$$P(\varepsilon\xi, z, \varepsilon^{-1}\partial/\partial\xi, \partial/\partial z) \sim \sum_{j=0}^{\infty} (\varepsilon|\xi|)^{j-1} P_j(\xi|\xi|^{-1}, z, |\xi|\partial/\partial\xi, \partial/\partial z). \qquad (3)$$

Denote by $w_{\pi/\alpha}$ the solution of the problem

$$\Delta_\xi w_{\pi/\alpha} = 0 \quad \text{in} \quad K_\alpha \setminus \overline{\omega(z)};$$
$$w_{\pi/\alpha} = -c_0(z)\rho^{\pi/\alpha}\sin(\pi\vartheta/\alpha) \quad \text{on} \quad \partial(K_\alpha \setminus \overline{\omega(z)}).$$

One can verify that $w_{\pi/\alpha}$ smoothly depends on $z \in M$.

We write the solution u in the form

$$u(x, \varepsilon) = v_0(x) + \varepsilon^{\pi/\alpha} w_{\pi/\alpha}(\varepsilon^{-1} y, z)\eta(|y|) + R_0(x, \varepsilon), \qquad (4)$$

where $\eta \in \mathbf{C}^\infty(\mathbb{R}^1)$,

$$\eta(t) = 1 \quad \text{for} \quad t \le d/4; \quad \eta(t) = 0 \quad \text{for} \quad t \ge d/2.$$

Substituting (4) into (1) we arrive at the following problem for R_0:

$$\Delta R_0 + 2\gamma(v_0 + \varepsilon^{\pi/\alpha} w_{\pi/\alpha}\eta) R_0 + \gamma R_0^2$$
$$= \varepsilon^{\pi/\alpha}\left(\gamma(\varepsilon^{\pi/\alpha} w_{\pi/\alpha}\eta + v_0) w_{\pi/\alpha}\eta + \gamma P\{w_{\pi/\alpha}\eta\} + [\Delta_y, \eta] w_{\pi/\alpha}\right) \quad \text{in} \quad \Omega_\varepsilon; \ (5)$$
$$R_0 = -v_0 + c_0(z) r^{\pi/\alpha}\sin(\pi\vartheta/\alpha) \quad \text{on} \quad \partial\omega_\varepsilon \cap \partial\Omega_\varepsilon;$$
$$R_0 = 0 \quad \text{on} \quad \partial\Omega \cap \partial\Omega_\varepsilon. \qquad (6)$$

The right-hand side F in (5) admits the estimate

$$F(x, \varepsilon) = O(\varepsilon^{\pi/\alpha} r^{-\pi/\alpha - 1}), \quad x \in \Omega_\varepsilon.$$

Furthermore,

$$R_0(x, \varepsilon) = O(\varepsilon^{\pi/\alpha + \min\{1, \pi/\alpha\}}), \quad x \in \partial\Omega_\varepsilon.$$

Since

$$\|u; \mathbf{C}(\bar\Omega_\varepsilon)\| \le c(\|r^{2-\sigma}\Delta u; \mathbf{C}(\bar\Omega_\varepsilon)\| + \|u; \mathbf{C}(\partial\Omega_\varepsilon)\|)$$

with $\sigma > 0$, we obtain

$$\|R_0; C(\bar\Omega_\varepsilon)\| \le c_\sigma \varepsilon^{\pi/\alpha + \min\{1-\sigma, \pi/\alpha\}}. \qquad (7)$$

12.6.3 The complete asymptotic expansion

As in 5.7.3, we represent the solution of problem (1) in the form

$$\begin{aligned} u(x,\varepsilon) = \sum_{l=0}^{[N\alpha/\pi]} \varepsilon^{l\pi/\alpha}\Big(\sum_{k+l\pi/\alpha\le N, k\in\mathbb{N}_0} \varepsilon^k \tilde{V}_l^{(k)}(x,\log\varepsilon) \\ + \varepsilon^{\pi/\alpha}\eta(|y|) \sum_{k+(l+1)\pi/\alpha\le N, k\in\mathbb{N}_0} \varepsilon^k \tilde{W}_l^{(k)}(y,z,\log\varepsilon)\Big) + R_N(x,\varepsilon), \end{aligned} \tag{8}$$

where $\tilde{V}_0^{(0)} = v_0$, $\tilde{W}_0^{(0)} = w_{\pi/\alpha}$, and $\tilde{V}_l^{(k)}(x,t)$, $\tilde{W}_l^{(k)}(y,z,t)$ are polynomials in t (if the number π/α is irrational, these coefficients do not depend on t). As in the similar situation in 5.7.3, we rewrite the series (8) in the form

$$u(x,\varepsilon) = v_0(x) + \sum_{\gamma\in Z,\gamma\le N} \varepsilon^\gamma\big((v_\gamma(x,\log\varepsilon) + \eta(|y|)\, w_\gamma(\varepsilon^{-1}y, z, \log\varepsilon)\big) + R_N(x,\varepsilon) \tag{9}$$

with $Z = \{\beta\colon \beta = k\pi/\alpha + m, k\in\mathbb{N}_0, m\in\mathbb{N}_0\}$ (cf. (61), 5.7). Introduce the sets X_γ and Y_γ for $\gamma\in Z$ by

$$X_\gamma = \{\beta\colon \beta = l\pi/\alpha + m, \quad l\in\mathbb{Z}, \quad m\in\mathbb{N}_0, \quad \gamma\ge -l\pi/\alpha\}, \tag{10}$$

$$Y_\gamma = \{\beta\colon \beta = l\pi/\alpha - m, \quad l\in\mathbb{N}, \quad m\in\mathbb{N}_0, \quad m<\gamma/2\}. \tag{11}$$

(Note that we obtained (10) and (11) replacing m by $m/2$ in (62), 5.7 and (63), 5.7.) The functions v_γ and w_γ admit the expansions

$$v_\gamma(x,\log\varepsilon) = \sum_{\tau\in X_\gamma,\tau<T} r^\tau v_\gamma^{(\tau)}(\vartheta,\log r,\log\varepsilon,z) + V_\gamma^{(T)}(x,\log\varepsilon),\ T\in X_\gamma, \tag{12}$$

$$w_\gamma(\xi,z,\log\varepsilon) = \sum_{\varkappa\in Y_\gamma,\varkappa<K} \rho^{-\varkappa} w_\gamma^{(\varkappa)}(\vartheta,\log\rho,\log\varepsilon,z) + W_\gamma^{(K)}(\xi,z,\log\varepsilon),\ K\in Y_\gamma, \tag{13}$$

where $r = |y|$ and $\rho = \varepsilon^{-1}|y|$ while $v_\gamma^{(\tau)}(\vartheta,l_1,l_2,z)$ and $w_\gamma^{(\varkappa)}(\vartheta,l_1,l_2,z)$ are polynomials in l_1 and l_2 whose coefficients smoothly depend on $\vartheta\in[0,\alpha]$ and $z\in M$. For the remainders the inequalities

$$|V_\gamma^{(T)}(x,\log\varepsilon)| \le c_\sigma\, r^T(\varepsilon r)^{-\sigma}, \quad |W_\gamma^{(K)}(\xi,z,\log\varepsilon)| \le c_\sigma\, \rho^{-K}(\rho/\varepsilon)^\sigma$$

hold with any $\sigma>0$.

We substitute (9) into the equation $\Delta u + u^2 = f$ in Ω_ε. As in 5.7.3, one has to distribute the terms with the same degree of ε among the right-hand sides of the limit problems in Ω and $K_\alpha\setminus\overline{\omega(z)}$. The terms $\varepsilon^{\gamma+\varkappa}\gamma v_\gamma v_\varkappa$ and $\varepsilon^{\gamma+\varkappa}\gamma\eta v_\gamma w_\varkappa$ can be transformed in the same manner as in 5.7.3. The expression $\gamma\eta^2 w_\gamma w_\varkappa$ can be represented in the form $\eta w_\gamma w_\varkappa + \eta(\eta-1)w_\gamma w_\varkappa$. The second term is supported in $\{x\colon |y|\in [d/2;d]\}$ so $\eta(\eta-1)w_\gamma w_\varkappa$ has the asymptotic expansion in positive powers of ε

$$\sum_{\tau\in Y_\gamma}\sum_{\sigma\in Y_\varkappa} \varepsilon^{\tau+\sigma}\,\eta(\eta-1)r^{-\tau-\sigma}\, w_\gamma^{(\tau)}\, w_\varkappa^{(\sigma)}.$$

We equate the coefficients by the same degrees of ε^γ written in the coordinates $(\xi,z) = (\varepsilon^{-1}y,z)$ and obtain that w_γ satisfies an equation of the form $\Delta(w_\gamma\eta) - \Phi_\gamma\eta = 0$ or, equivalently, $\eta(\Delta w_\gamma - \Phi_\gamma) + [\Delta,\eta]w_\gamma = 0$. The support of the second term on the left is contained in $\{x\colon |y|\in[d/2;d]\}$ and, consequently, the term has the asymptotics

$$\sum_{\tau\in Y_\gamma} \varepsilon^\tau[\Delta,\eta]r^{-\sigma}\, w_\gamma^{(\sigma)}.$$

Thus, v_β satisfies the problem

$$\Delta v_\beta + 2v_0 v_\beta = -\sum_{\{\gamma\in Z:\,\beta-\gamma\in Z\}} v_\gamma v_{\beta-\gamma} - \eta\sum_{A_\beta} r^{-\sigma} w_\varkappa^{(\sigma)} V_\gamma^{(\tau(\sigma,\gamma))}$$
$$-\eta\sum_{B_\beta} r^{-\sigma-\tau} w_\gamma^{(\sigma)} w_\varkappa^{(\tau)} - (\eta-1)\eta\sum_{L_\beta} r^{-\sigma-\tau} w_\gamma^{(\sigma)} w_\varkappa^{(\tau)}$$
$$-\eta\sum_{M_\beta}\Big(P - \sum_{j=0}^{j(\sigma,\gamma)} |y|^{j-1} P_j(y,|y|,z,\partial/\partial y,\partial/\partial z)\Big) r^{-\sigma} w_\gamma^{(\sigma)}$$
$$-\sum_{M_\beta}[\Delta_x,\eta] r^{-\sigma} w_\gamma^{(\sigma)} \quad \text{in} \quad \Omega;$$
$$v_\beta = 0 \quad \text{on} \quad \partial\Omega.$$

Here the sets $A_\beta, B_\beta, L_\beta, M_\beta$, and the numbers $j(\sigma,\gamma)$ and $\tau(\sigma,\gamma)$ are defined by

$$A_\beta = \{\gamma\in Z\cup\{0\}, \varkappa\in Z, \sigma\in Y_\varkappa\colon \varkappa+\gamma+\sigma=\beta\},$$
$$B_\beta = \{\gamma,\varkappa\in Z, \sigma\in Y_\gamma, \tau\in Y_\varkappa\colon \sigma+\tau+\gamma+\varkappa=\beta, \sigma+\tau\le 2\},$$
$$L_\beta = \{\gamma,\varkappa\in Z, \sigma\in Y_\gamma, \tau\in Y_\varkappa\colon \sigma+\tau+\gamma+\varkappa=\beta\},$$
$$M_\beta = \{\gamma,\in Z, \sigma\in Y_\gamma, \colon \sigma+\tau=\beta\},$$
$$j(\sigma,\gamma) = \min\{j\in\mathbb{N}_0\colon j-\sigma\ge 3\},$$
$$\tau(\sigma,\gamma) = \min\{\tau\colon \tau\in X_\gamma, \tau-\sigma+2\ge 0\}.$$

The function w_β is a solution to the following boundary value problem with a parameter $z\in M$:

$$\Delta w_\beta = -\sum_{D_\beta}\Big(w_\gamma w_\varkappa - \sum_{E_{\beta,\gamma,\alpha}} r^{-\sigma-\tau} w_\gamma^{(\sigma)} w_\gamma^{(\tau)}\Big) - \sum_{G_\beta}\rho^\tau v_\gamma^{(\tau)} W_\varkappa^{(\sigma(\tau,\varkappa))}$$
$$-\sum_{N_\beta}|\xi|^{j-1} P_j(\xi/|\xi|, z, |\xi|\partial/\partial\xi, \partial/\partial z) W_\varkappa^{(\Sigma(j,\varkappa))} \quad \text{in} \quad K_\alpha\setminus\overline{\omega(z)};$$
$$w_\beta = -\sum_{H_\beta}\rho^\tau v_\gamma^{(\tau)} \quad \text{on} \quad \partial\big(K_\alpha\setminus\overline{\omega(z)}\big).$$

Here

$$D_\beta = \{\gamma,\varkappa\in Z\colon 2+\gamma+\varkappa=\beta\},$$
$$E_{\beta,\gamma,\varkappa} = \{\sigma\in Y, \tau\in Y_\varkappa\colon \sigma+\tau<2\},$$
$$G_\beta = \{\gamma\in Z\cup\{0\}, \varkappa\in Z, \tau\in X_\varkappa\colon 2+\tau+\gamma+\varkappa=\beta\},$$
$$H_\beta = \{\gamma\in Z\cup\{0\}, \tau\in X_\gamma\colon \tau+\gamma=\beta\},$$
$$N_\beta = \{j\in\mathbb{N}_0, \varkappa\in Z\colon j+1+\varkappa=\beta\},$$
$$\sigma(j,\varkappa) = \Sigma(j,\varkappa) = \min\{\sigma\in Y_\varkappa\colon j-\sigma<3\},$$
$$\sigma(\tau,\varkappa) = \min\{\sigma\colon \sigma\in Y_\varkappa, \tau-\sigma+2<0\}.$$

The next assertion can be obtained in the same way as the inequality (7).

Theorem 12.6.2. *The remainder R_N in (8) (or in (9)) is subject to*

$$\|R_N; \mathbf{C}(\bar\Omega_\varepsilon)\| \le c_\sigma \varepsilon^{N-\sigma},$$

where σ is an arbitrary positive number and the constant c_σ does not depend on ε.

Chapter 13

Asymptotics of Solutions to the Dirichlet Problem for High Order Equations in a Domain with a Thin Tube Excluded

In this chapter we study the asymptotics of solutions to the Dirichlet problem for a strongly elliptic equation of order $2m$ in a subdomain of $\mathbb{R}^n$ from which a small neighborhood of a smooth closed manifold M of codimension $d > 0$ is cut out. We generalize the algorithms described in 12.2 for the Poisson equation in the case $n = 3$, $d = 2$.

The statement of the problem is given in 13.1. The case of "noncritical" d (i.e., d is odd or $d > 2m$) is considered in 13.2, where the asymptotic series

$$\sum_{k=0}^{\infty} \varepsilon^k \{v^{(k)}(x, \log \varepsilon) + \varepsilon^{m-[d/2]} \zeta(x) w^{(k)}(\varepsilon^{-1} y, z, \log \varepsilon)\} \tag{A1}$$

is constructed for a solution. Here ζ is a smooth function with support near M, (y, z) are local coordinates of a point $x \in \operatorname{supp} \zeta$, z being the "tangential" component. The functions $v^{(k)}$ and $w^{(k)}$ are polynomials in $\log \varepsilon$. The coefficients of these polynomials are solutions to uniquely solvable limit problems. The first limit problem is a Sobolev problem in $\Omega \setminus M$, the second one is a Dirichlet problem in the exterior of a domain in $\mathbb{R}^d$ depending on the parameter d.

For the "critical" d (d is even and $d \leq 2m$) the asymptotic expansions are obtained in 13.3 and 13.4. In contrast to the case of noncritical dimension, the second limit problem has nontrivial kernel. To circumvent this difficulty, we expand the class of admissible solutions to the first limit problem. In such a class there is no uniqueness, and the superfluous solutions of the first problem are parametrized by a smooth vector-valued function $\mathbf{c}$ on M. Trying to satisfy the compatibility condition of the second problem, we are led to a system of integral equations of the form

$$(\mathbf{1} \log \varepsilon + B)\mathbf{c} = \mathbf{H}. \tag{A2}$$

The operator B turns out to be bounded only for $d = n$, i.e., in the case that M is a point and (A2) is an algebraic system. Thus, for $d = n$ the system (A2) is uniquely solvable for small ε and the same reason as in 13.2 leads to the following asymptotic series for the solution

$$\sum_{k=0}^{\infty} \varepsilon^k \sum_{j=0}^{k-1} \mu(\varepsilon)^{-j} \{v^{(k,j)}(x, \log \varepsilon) + \varepsilon^{m-n/2} \zeta(x) w^{(k,j)}(\varepsilon^{-1} x, \log \varepsilon)\}, \tag{A3}$$

where $\mu(\varepsilon) = \det \|\mathbf{1} \log \varepsilon + B\|$ (cf.5.3).

If $d < n$, the operator $B = B(z, D_z)$ is unbounded pseudodifferential on M whose symbol increases logarithmically in the second variable. The situation is more complicated because the equation (A2) is generally unsolvable. In 13.5.4 this fact is exemplified by the Dirichlet problem for the Laplace operator in the exterior of a thin torus in $\mathbb{R}^3$. For such a problem the question of solvability admits the complete analysis which shows, in particular, that the inverse operator does not exist for a sequence $\{\varepsilon_k\}_{k\geq 1}$ tending to zero (see also 12.2.2 and 12.4.4).

However, in order to construct the asymptotics of a solution to the original problem, it suffices to find an approximate solution of equation (A2) with error $O(|\log\varepsilon|^{-N})$ or $O(\varepsilon^N)$, if the data of the problem are expanded in powers of $|\log\varepsilon|^{-1}$ or of ε. In the first case we obtain the asymptotics

$$\sum_{k=0}^{\infty} |\log\varepsilon|^{-k}\{v^{(k)}(x) + \varepsilon^{m-d/2}\zeta(x)w^{(k)}(\varepsilon^{-1}y, z)\}, \tag{A4}$$

(see 13.3). In the second case there is an expansion in powers of ε with integral exponents whose coefficients contain terms of the form

$$\prod_{j=1}^{k} P_j R_\varepsilon S_j(g_s, h_s).$$

Here P_j and S_j are operators depending polynomially on $\log\varepsilon$ and R_ε is an asymptotic inverse operator for $\mathbf{1}\log\varepsilon + B$ (see 13.4).

This expansion can be viewed as a generalization of (A3) for the case $d < n$. Note that in order to construct the operator R_ε one needs to invert only a matrix-valued function on M in the case of series (A4) and to invert a pseudodifferential operator on R_ε in the case of the power expansion.

13.1 Statement of the Problem

Let Ω be a subdomain in $\mathbb{R}^n$ with compact closure $\bar\Omega$ and smooth (of class $\mathbf{C}^\infty$) boundary $\partial\Omega$ and let $\varkappa\colon \bar D_\delta^d \times S^{n-d} \to \Omega$ be a diffemorphism of class $\mathbf{C}^\infty$. Here S^{n-d} is the unit $(n-d)$-dimensional sphere for $d < n$, S^0 is a point, and $D_\delta^d = \{y \in \mathbb{R}^d\colon |y| < \delta\}$, $1 < d \leq n$. We denote by $\{\omega(z)\}_{z\in S^{n-d}}$ a family of domains in $\mathbb{R}^d$ that depend smoothly on z and contain the origin having compact closures and smooth boundaries $\partial\omega(z)$.

Introduce the sets $\omega_\varepsilon(z) = \{y \in \mathbb{R}^d\colon \varepsilon^{-1}y \in \omega(z)\}$, $\omega_\varepsilon = \varkappa\{(y,z) \in \mathbb{R}^d \times S^{n-d}\colon y \in \omega_\varepsilon(z)\}$, $M = \varkappa(\{0\} \times S^{n-d})$, $\omega^c = \{(y,z) \in \mathbb{R}^d \times S^{n-d}\colon y \notin \omega(z)\}$, $\Omega_\varepsilon = \Omega \setminus \bar\omega_\varepsilon$, where ε is a sufficiently small positive number.

We consider the boundary value problem

$$\tilde L(x, D_x)\, u(\varepsilon, x) = f(x), \quad x \in \Omega_\varepsilon; \tag{1}$$

$$\nabla_j u(\varepsilon, x) = 0, \quad x \in \partial\Omega_\varepsilon, \quad j = 0, \dots, m-1, \tag{2}$$

where $f \in \mathbf{C}^\infty(\bar\Omega)$ and $\tilde L(x, D_x)$ is an elliptic differential operator with smooth coefficients in $\bar\Omega$. Suppose that for all $v \in \mathbf{C}_0^\infty(\Omega)$ there is the Gårding inequality

$$\operatorname{Re}\int_\Omega \overline{v(x)}\, \tilde L(x, D_x)\, v(x)dx \geq c\|v; \mathbf{W}_2^m(\Omega)\|^2, \tag{3}$$

which provides the unique solvability of problem (1),(2).

In a neighborhood $V = \varkappa(S^{n-d} \times D^d_\delta)$ of the manifold M we write the operator $\tilde{L}$ in coordinates (z, y), where z is a local coordinate system on S^{n-d} and $y \in \mathbb{R}^d$:

$$\tilde{L}(x, D_x) = \sum_{|\alpha|+|\beta|\leq 2m} a_{\alpha\beta}(y, z)\, D^\alpha_y\, D^\beta_z = \tilde{M}(y, z, D_y, D_z). \tag{4}$$

After the change of variables $\eta = \varepsilon^{-1} y$ we expand the operator $\tilde{M}$ in powers of ε. The coefficient by ε^{k-2m} is the differential operator

$$L_k(\eta, z, D_\eta, D_z) = \sum_{|\alpha|+|\beta|\leq 2m} \; \sum_{\{\gamma:\, |\gamma|=k+|\beta|-2m\}} a^{(\gamma)}_{\alpha\beta}(z)\, \eta^\gamma D^\alpha_\eta\, D^\beta_z, \tag{5}$$

where $a^{(\gamma)}_{\alpha\beta}$ are the coefficients in the Taylor series: $a_{\alpha\beta}(z, y) \sim \sum_\gamma a^{(\gamma)}_{\alpha\beta}(z)\, y^\gamma$. In what follows we use the operators

$$\tilde{L}_j = \tilde{L} - \sum_{k=0}^{j} \zeta_0 L_k,$$

where ζ_0 is a function in $\mathbf{C}^\infty_0(\Omega)$ such that $\zeta = 1$ in a neighborhood of M and $\zeta = 0$ outside V.

13.2 The Case of Noncritical Dimension

Let d be either an odd number or an even number that is greater than m. It is natural to seek the principal term $V^{(0)}$ of the asymptotics of the solution u to problem (1), (2) as a solution to the Sobolev boundary value problem

$$\tilde{L}(x, D_x)\, V^{(0)}(x) = f(x), \quad x \in \Omega \setminus M; \tag{1}$$

$$\nabla_j V^{(0)}(x) = 0, \qquad x \in \partial\Omega, \quad j = 0, \dots, m-1; \tag{2}$$

$$\nabla_k V^{(0)}(x) = 0, \qquad x \in M, \quad k < m - d/2. \tag{3}$$

From (3), 13.1, it follows that the problem (1)–(3) is uniquely solvable. In the next assertion we formulate facts on the asymptotics of solutions to the Sobolev problem which will be used in this section.

Theorem 13.2.1. *Suppose that a function $g \in \mathbf{C}^\infty(\bar{\Omega}\setminus M)$ admits in the neighborhood V the asymptotic representation*

$$D^\beta_x\Big(g(x) - \sum_{k=-m-[d/2]}^{N} |y|^k\, g_k(z, \vartheta, \log|y|)\Big) = o(|y|^{N-|\beta|}), \tag{4}$$

where $N \in \mathbb{N}$, β is a multiindex, and $\vartheta = y/|y|$ are local coordinate on S^{d-1} while $g_k(z, \vartheta, t)$ is a polynomial in t whose coefficients are smooth functions in $\mathbf{C}^\infty(S^{n-d} \times S^{d-1})$. There exists a unique solution $v \in \mathbf{C}^\infty(\bar{\Omega} \setminus M)$ to problem (1)–(3) where f is replaced by g. The asymptotic formula

$$D^\beta_x\Big(v(x) - \sum_{k=-m-[d/2]}^{N} |y|^k\, v_k(z, \vartheta, \log|y|)\Big) = o(|y|^{N-|\beta|}) \tag{5}$$

holds, where v_k have the same properties as g_k.

The function $V^{(0)}$ satisfies the equation (1), 13.1, everywhere in Ω_ε, however, the boundary condition (2), 13.1, is fulfilled only on $\partial\Omega$. In the neighborhood V of M we construct a boundary layer to compensate the main part (with respect to ε) of the discrepancy of v_0 in (2),13.1, on $\partial\omega_\varepsilon$. To this end we introduce a function $\eta \to W^{(0)}(\eta, z)$ that smoothly depends on $z \in S^{n-d}$ and satisfies

$$L_0(z, D_\eta)W^{(0)}(\eta, z) = 0, \qquad \eta \in \mathbb{R}^d \setminus \overline{\omega(z)}; \tag{6}$$

$$(\partial/\partial\tau)^j W^{(0)}(\eta, z) = -\delta_{0,j}\Phi^{(0)}(\eta, z), \qquad \eta \in \partial\omega(z), \quad j = 0, \dots, m-1; \tag{7}$$

$$W^{(0)}(\eta, z) = O(|\eta|^{m-1-[d/2]}), \quad |\eta| \to \infty, \tag{8}$$

where τ is the outward normal on $\partial\omega(z)$, $\Phi^{(0)}(\eta, z) = V^{(0)}(0, z)$ if $d > 2m$, and $\Phi^{(0)}(\eta, z) = |\eta|^{m-d/2}\, V^{(0)}_{m-[d/2]}(z, \vartheta)$ if $d < 2m$. Here $V^{(0)}_{m-[d/2]}$ is the function in $\mathbf{C}^\infty(S^{n-d} \times S^{d-1})$ defined by the expansion (5) of the solution $V^{(0)}$ to problem (1)–(3).

We shall use the next theorem on the asymptotic behavior of solutions to the Dirichlet problem (6)–(8) at infinity (cf. Chapters 1 and 3).

Theorem 13.2.2. *Let h be a function in $\mathbf{C}^\infty(\omega^c)$ that smoothly depends on $z \in S^{n-d}$ and admits the representation*

$$D_z^\alpha D_\eta^\beta\Big(h(\eta, z) - \sum_{k=m+[d/2]+1}^{N} |\eta|^{-k}\, h_k(z, \vartheta, \log|\eta|)\Big) = o(|\eta|^{-N-|\beta|}), \tag{9}$$

$|\eta| \to \infty$, where α and β is $(n-d)$- and d-dimensional multiindices, $\vartheta = \eta/|\eta|$. Then for any $\varphi_j \in \mathbf{C}^\infty(\partial\omega^c)$, $j = 0, \dots, m-1$, there exists a solution w to problem (6)–(8) with h and φ_j on the right in (6) and (7), respectively. Moreover,

$$D_z^\alpha D_\eta^\beta\Big(w(\eta, z) - \sum_{k=-m+1+[d/2]}^{N} |\eta|^{-k}\, w_k(z, \vartheta, \log|\eta|)\Big) = o(|\eta|^{-N-|\beta|}) \tag{10}$$

as $|\eta| \to \infty$.

In Ω_ε we define the function

$$U^{(0)}(\varepsilon, x) = \begin{cases} V^{(0)}(x) + W^{(0)}(\varepsilon^{-1}y, z)\,\zeta(x), & \text{if} \quad d > 2m; \\ V^{(0)}(x) + \varepsilon^{m-[d/2]}W^{(0)}(\varepsilon^{-1}y, z)\,\zeta(x), & \text{if} \quad d < 2m, \end{cases}$$

where $\zeta(x)$ is a function in $\mathbf{C}_0^\infty(\Omega)$ that equals 1 in a neighborhood of M and 0 outside the set $\{x \in V \colon \zeta_0(x) = 1\}$, ζ_0 being introduced in 13.1. We substitute $U^{(0)}$ into (1), (2), 13.1, and obtain the discrepancy of order ε on the right in (1), (2), 13.1.

Suppose that the function f on the right-hand side of (1) admits the expansion

$$f(\varepsilon, x) = \sum_{k=0}^{K-1} \varepsilon^{(k)}\big(g^{(k)}(x, \log\varepsilon) + \varepsilon^{-m-[d/2]}h^{(k)}(\varepsilon^{-1}y, z, \log\varepsilon)\zeta(x)\big) + f^{(K)}(\varepsilon, x), \tag{11}$$

where $g^{(k)}(x, t)$ and $h^{(k)}(\eta, z, t)$ are polynomials in t whose coefficients are functions in $\mathbf{C}^\infty(\bar\Omega \setminus M)$ and $\mathbf{C}^\infty(\omega^c)$ and admit representations of the form (4) and (9). Obviously, any function in $\mathbf{C}_0^\infty(\bar\Omega)$ is of the form (11).

We seek an asymptotic solution to problem (1), (2), 13.1, as a formal series

$$\sum_{k=0}^{\infty} \varepsilon^k \big(v^{(k)}(x, \log\varepsilon) + \varepsilon^{m-[d/2]} w^{(k)}(\varepsilon^{-1}y, z, \log\varepsilon)\zeta(x)\big). \tag{12}$$

In (12) the functions $v^{(k)} \in \mathbf{C}^\infty(\bar{\Omega}\backslash M)$ and $w^{(k)} \in \mathbf{C}^\infty(\mathbb{R}^d \times S^{n-d})$ are polynomials in $\log\varepsilon$ and admit expansions of the form (5) and (10) with coefficients $v_j^{(k)}$ and $w_j^{(k)}$, respectively. We substitute (12) into the boundary condition (2), 13.1, on $\partial\omega_\varepsilon$, expand $v^{(k)}$ in asymptotic series and equate the coefficients by the same powers of ε. We obtain

$$\begin{aligned}&(\partial/\partial\tau)^j w^{(q)}(\eta, z, \log\varepsilon)\\ &= -\sum_{s=m-[d/2]}^{m+q-[d/2]} (\partial/\partial\tau)^j \big(|\eta|^s\, v_s^{(q+m-[d/2]-s)}(z, \vartheta, \log|\eta| + \log\varepsilon, \log\varepsilon)\big),\ \eta \in \partial\omega(z).\end{aligned} \tag{13}$$

Proceeding in the same manner with (1), 13.1, we arrive at the formal equality

$$\begin{aligned}&\sum_{k=0}^{\infty} \varepsilon^k \{\tilde{L}(x, D_x) v^{(k)}(x, \log\varepsilon) + \varepsilon^{m-[d/2]}[\tilde{L}(x, D_x), \zeta(x)] w^{(k)}(\varepsilon^{-1}y, z, \log\varepsilon)\\ &+ \varepsilon^{m-[d/2]}\zeta(x)\tilde{M}(\varepsilon\eta, z, \varepsilon^{-1}D_\eta, D_z) w^{(k)}(\eta, z, \log\varepsilon)\}\\ &= \sum_{k=0}^{\infty} \varepsilon^{(k)}\big(g^{(k)}(x, \log\varepsilon) + \varepsilon^{-m-[d/2]}\zeta(x) h^{(k)}(\eta, z, \log\varepsilon)\big).\end{aligned}$$

The operator $\tilde{M}(\varepsilon\eta, z, \varepsilon^{-1}D_\eta, D_z)$ can be expanded in powers of ε with coefficients L_k (see (5), 13.1). Equating the coefficients by the same powers of ε written in the coordinates (η, z), we obtain

$$\begin{aligned}L_0(z, D_\eta) w^{(k)}(\eta, z, \log\varepsilon) = h^{(k)}(\eta, z, \log\varepsilon) - \sum_{j=0}^{k-1} L_{k-j}(\eta, z, D_\eta, D_z)&\\ \times w^{(j)}(\eta, z, \log\varepsilon), \quad \eta \in \mathbb{R}^d \setminus \overline{\omega(z)}.&\end{aligned}$$

From (10) for $w^{(j)}$ it follows that

$$L_{k-j}(\eta, z, D_\eta, D_z)\, w^{(j)}(\eta, z, \log\varepsilon) = O(|\eta|^{k-j-1-[d/2]}), \quad |\eta| \to \infty.$$

Therefore, generally, the condition (8) for $w^{(k)}$ cannot be fulfilled (see Theorem 13.2.2). To avoid terms of order greater than $-(m + [d/2] + 1)$ arising on the right in the second limit problem we apply the procedure of redistribution of discrepancies and represent $L_s w^{(j)}$ in the form

$$\begin{aligned}L_s(\eta, z, D_\eta, D_z) w^{(j,s)}(\eta, z, \log\varepsilon) + \sum_{p=1-m+[d/2]}^{k-m+[d/2]} L_s(\eta, z, D_\eta, D_z)&\\ \times \big(|\eta|^{-p}\, w_p^{(j)}(z, \vartheta, \log|\eta|, \log\varepsilon)\big) = I_1 + I_2.&\end{aligned} \tag{14}$$

Here $w^{(j,s)}$ are defined by

$$w^{(j,s)}(\eta, z, \log\varepsilon) = w^{(j)}(\eta, z, \log\varepsilon) - \sum_{q=1-m+[d/2]}^{s-m+[d/2]} |\eta|^{-q}\, w_q^{(j)}(z, \vartheta, \log\varepsilon). \tag{15}$$

The term I_1 admits the expansion (9) so it can be included in the right-hand side of the boundary value problem for $w^{(k)}$. Thus, the equation for $w^{(q)}$ has the form

$$L_0(z,D_\eta)w^{(q)}(\eta,z,\log\varepsilon) = h^{(q)}(\eta,z,\log\varepsilon) - \sum_{j=1}^{q} L_j(\eta,z,D_\eta,D_z) \times w^{(q-j,j)}(\eta,z,\log\varepsilon), \quad \eta\in\mathbb{R}^d\setminus\overline{\omega(z)}. \tag{16}$$

The boundary value problem (16), (13) with condition at infinity

$$w^{(q)}(\eta,z,\log\varepsilon) = O(|\eta|^{m-[d/2]-1}), \quad |\eta|\to\infty \tag{17}$$

is uniquely solvable by Theorem 13.2.2.

Now we are going to construct $v^{(p)}$. Multiply by ζ the term I_2 in (14), which is large for $|\eta|\to\infty$, and rewrite it in the coordinates x. Taking the commutators $[\tilde L,\zeta]w^{(q)}$ into account we obtain for $v^{(p)}$ the boundary value problem

$$\begin{aligned}\tilde L(x,D_x)\,v^{(p)}(x,\log\varepsilon) &= g^{(p)}(x,\log\varepsilon) - \sum_{k=0}^{p-1}\{[\tilde L(x,D_x),\zeta(x)] + \zeta(x)\tilde L_{p-k}(x,D_x)\}\\ &\quad\times \big(|\eta|^{m+k-p-[d/2]}\,w^{(k)}_{p+[d/2]-k-m}(z,\vartheta,\log|y|-\log\varepsilon,\log\varepsilon)\big),\\ &\quad x\in\Omega\setminus M;\\ \nabla_j v^{(p)}(x,\log\varepsilon) &= 0, \quad x\in\partial\Omega, \quad j=0,\dots,m-1;\\ \nabla_l v^{(p)}(x,\log\varepsilon) &= 0, \quad x\in M, \quad l<m-d/2,\end{aligned} \tag{18}$$

$\tilde L_j$ being defined by (6). 13.1. Note that the right-hand sides of problem (18) as well as of problem (16), (13) are polynomials in $\log\varepsilon$. Therefore $v^{(p)}$ and $w^{(q)}$, too, are polynomials in $\log\varepsilon$.

Thus, the function $v^{(p)}$ is determined by (18) if the functions $w^{(k)}$, $k=0,\dots,q-1$, are known. In its turn, $w^{(q)}$ can be obtained from (16), (13), and (17) if $v^{(j)}$, $j=0,\dots,q$, and $w^{(k)}$, $k=0,\dots,q-1$, have already been found. By Theorems 13.2.1 and 13.2.2, such a sequence of problems is solvable. The validity of asymptotic expansions of the type (5) and (9) for $v^{(p)}$ and $w^{(q)}$ immediately follows from Theorem 13.2.1 and 13.2.2 by induction.

We estimate now the difference between the solution of problem (1), (2), 13.1, and the sum $\tilde S^{(N)}$ of the first N terms of series (12).The function $\tilde R^{(N)} = u - \tilde S^{(N)}$ satisfies the problem

$$\tilde L(x,D_x)\,\tilde R^{(N)}(\varepsilon,x) = f^{(N)}(\varepsilon,x) + \Phi^{(N)}(\varepsilon,x), \quad x\in\Omega_\varepsilon; \tag{19}$$

$$\nabla_j\tilde R^{(N)}(\varepsilon,x) = 0, \quad x\in\partial\Omega, \quad j=0,\dots,m-1; \tag{20}$$

$$(\partial/\partial\tau)^j\,\tilde R^{(N)}(\varepsilon,x) = \sum_{k=0}^{N-1}\varepsilon^k(\partial/\partial\tau)^j v^{(k,N-k)}(x),\ x\in\partial\omega_\varepsilon,\ j=0,\dots m-1, \tag{21}$$

where $f^{(N)}$ is the remainder in (11),

$$\Phi^{(N)} = \varepsilon^{m-[d/2]}\sum_{k=0}^{N-1}\varepsilon^k\{[\tilde L,\zeta] + \zeta\tilde L_{N-1-k}\}w^{(k,N-1-k)}.$$

Let us estimate the norm in $\mathbf{L}_2(\Omega_\varepsilon)$ of $r^m \Phi^{(N)}$,

$$\|r^m \Phi^{(N)}; \mathbf{L}_2(\Omega_\varepsilon)\|^2 \le c\varepsilon^N |\log \varepsilon|^{Q_1} \int_{\Omega_\varepsilon} r^{-2[d/2]} |\log r|^{Q_2} \, dx \le C(\delta)\varepsilon^{N-\delta}, \tag{22}$$

where δ is an arbitrary positive number, Q_1, Q_2 are some positive numbers, r is the distance from M, the constant $C(\delta)$ does not depend on $\varepsilon \in (0,1]$.

Denote by $\Phi_j^{(N)}(\varepsilon, \varepsilon^{-1}y, z)$ the right-hand side of (21) and make the change of variables $y \to \eta = \varepsilon^{-1}y$. The function $(\eta, z) \to \Phi(\varepsilon, \eta, z)$ belongs to $\mathbf{C}^\infty(\partial\omega^c)$, hence, there exists $X^{(N)} \in \mathbf{C}^\infty(\omega^c)$ such that

$$(\partial/\partial\tau)^j X^{(N)} = \Phi_j^{(N)} \quad \text{on} \quad \partial\omega^c.$$

Estimating the norm of $X^{(N)}(\varepsilon, \varepsilon^{-1}y, z)$ in $\mathbf{W}_2^{2m}(\Omega_\varepsilon)$, we obtain

$$\sum_{k=0}^{2m} \|r^{k-m} X^{(N)}; \mathbf{W}_2^k(\Omega_\varepsilon)\| \le C(\sigma)\varepsilon^{N-\sigma}, \tag{23}$$

where the constant $C(\sigma)$ does not depend on $\varepsilon \in (0,1]$.

The main result of the section is the following assertion.

Theorem 13.2.3. *Suppose that the remainder $f^{(N)}$ in* (11) *is subject to*

$$\|r^m f^{(N)}; \mathbf{L}_2(\Omega)\| \le c(\sigma)\varepsilon^{N-\sigma}. \tag{24}$$

Then the partial sum $\tilde{S}^{(N)}$ of series (12) *satisfies*

$$\|\tilde{S}^{(N)} - u; \mathbf{W}_2^m(\Omega_\varepsilon)\| \le C(\sigma)\varepsilon^{N-\sigma}, \tag{25}$$

where u is the solution of problem (1), (2), 13.1, *σ is an arbitrary positive number, and the constant $C(\sigma)$ does not depend on $\varepsilon \in (0,1]$.*

Proof. From (3), 13.1, it follows that the solution of problem (1), (2), 13.1, satisfies the inequality

$$\|u; \mathbf{W}_2^m(\Omega_\varepsilon)\| \le C\|r^m f; \mathbf{L}_2(\Omega_\varepsilon)\|, \tag{26}$$

the constant C being independent of u, f and $\varepsilon \in (0,1]$. Changing the unknown function $\Psi^{(N)} = \tilde{R}^{(N)} + X^{(N)}$ we obtain from (19)–(21) the boundary value problem with homogeneous conditions on $\partial\Omega_\varepsilon$. We apply (26) to its solution $\Psi^{(N)}$, make use of (22)–(24), and arrive at (25). □

13.3 The Case of Critical Dimension (Expansion in $(\log \varepsilon)^{-1}$)

Let d be an even number and $d \le 2m$. We recall some known properties of solutions to the problem

$$L_0(z, D_\eta)\Phi = 0 \quad \text{in} \quad \mathbb{R}^d \setminus \{0\}. \tag{1}$$

According to John [1], the fundamental solution $\Gamma(\eta, z)$ of this equation has the form

$$\Gamma(\eta, z) = S(\eta, z) \log|\eta| + T(\eta, z), \tag{2}$$

where

$$S(\eta, z) = (2\pi i)^{-d}\left((2m-d)!\right)^{-1} \int_{S^{d-1}} (\eta \cdot \vartheta)^{2m-d} L(z, \vartheta)^{-1} ds_\vartheta \tag{3}$$

is a homogeneous polynomial of degree $m - d/2$ and

$$T(\eta, z) = (2\pi i)^{-d}\big((2m-d)!\big)^{-1} \int\limits_{S^{d-1}} (\eta\cdot\vartheta)^{2m-d} \log\big||\eta|^{-1}(\eta\cdot\vartheta)\big| L(z,\vartheta)^{-1} ds_\vartheta$$

is a homogeneous function in η of degree $2m-d$. Clearly, any solution of equation (1) in the Schwartz space $\mathcal{J}'$ is a linear combination of polynomials and the derivatives $D_\eta^\alpha \Gamma$.

We are going to investigate the solution u to problem (1), (2), 13.1. As in 13.2, the solution to problem (1)–(3), 13.2, plays the role of the principal term $V^{(0)}$. The last problem has a unique solution in $\mathbf{C}^{m-d/2}(\bar\Omega)$. However, in the case under study the problem (6), (7), 13.2, is generally unsolvable in the class of functions satisfying (8), 13.2. Indeed, taking account of the above facts on the solutions to equation (1) and of the inequality $\mathrm{Re} L_0(z,\xi) > c|\xi|^{2m}$ with $c = \mathrm{const} > 0$, which results from (3), 13.1, we arrive at the following assertion.

Theorem 13.3.1. *The boundary value problem*

$$L_0(z, D_\eta) w(\eta, z) = h(\eta, z), \quad \eta \in \mathbb{R}^d \setminus \overline{\omega(z)}; \tag{4}$$

$$(\partial/\partial\tau)^j w(\eta, z) = \varphi(\eta, z), \quad \eta \in \partial\omega(z), \quad j = 0, \dots, m-1, \tag{5}$$

where $\varphi_j \in \mathbf{C}^\infty(\partial\omega^c)$ *and*

$$D_z^\alpha D_\eta^\beta h(\eta, z) = o(|\eta|^{-m-d/2-1-|\beta|}), \quad |\eta| \to \infty, \tag{6}$$

has a unique solution $w \in \mathbf{C}^\infty(\partial\omega^c)$ *such that*

$$D_z^\alpha D_\eta^\beta w(\eta, z) = o(|\eta|^{m-d/2-|\beta|}), \quad |\eta| \to \infty. \tag{7}$$

Let $Z^{(\alpha)}$ with arbitrary multiindex α of order $m - d/2$ stand for the solution to the problem

$$L_0^*(z, D_\eta) Z^{(\alpha)}(\eta, z) = 0, \quad \eta \in \mathbb{R}^d \setminus \overline{\omega(z)};$$

$$(\partial/\partial\tau)^j Z^{(\alpha)}(\eta, z) = (\alpha!)^{-1}(\partial/\partial\tau)^j D_\eta^\alpha \Gamma(\eta, z), \quad \eta \in \partial\omega(z); \quad j = 0, \dots, m-1,$$

that satisfies a relation of the form (7). We set

$$\chi^{(\alpha)} = (\alpha!)^{-1} D_\eta^\alpha \bar\Gamma - Z^{(\alpha)} \tag{8}$$

and introduce differential operators $B_k(z, D_\eta)$, $k = 0, \dots, m-1$, of order $2m-k-1$ such that the following Green formula holds:

$$\int\limits_{\mathbb{R}^d\setminus\omega(z)} \big(\bar\psi L_0(z,D_\eta)\varphi - \varphi\overline{L_0^*(z,D_\eta)\psi}\big) d\eta + \int\limits_{\partial\omega(z)} \sum_{k=0}^{m-1} \big((\partial/\partial\tau)^k\varphi\big)\overline{B_k(z,D_\eta)\psi} ds_\eta = 0, \tag{9}$$

where $\varphi, \psi \in \mathbf{C}^\infty(\mathbb{R}^d \setminus \partial\omega(z))$, $\nabla_j \psi = 0$ on $\partial\omega(z)$, $j = 0, \dots, m-1$.

The next assertion follows from Theorems 13.3.1, 3.3.2, and 3.3.9.

Theorem 13.3.2. *i) Let h in (4) admit the estimate (6). Then equation (4) with boundary conditions (5) and the condition at infinity*

$$D_z^\alpha D_\eta^\beta w(\eta, z) = O(|\eta|^{m-d/2-|\beta|-1}), \quad |\eta| \to \infty, \tag{10}$$

is solvable if and only if for any multiindex $\alpha, |\alpha| = m - d/2$,

$$\int_{\mathbb{R}^d \setminus \omega(z)} h\,\overline{\chi^{(\alpha)}}\, d\eta + \int_{\partial\omega(z)} \sum_{k=0}^{m-1} \varphi_k\, \overline{B_k(z, D_\eta)\varphi_k}\, ds_\eta = 0. \tag{11}$$

ii) Let h *satisfy* (9), 13.2, *and let* (11) *be valid. Then there exists a unique solution* w *to problem* (4), (5) *that admits the representation* (10), 13.2.

Remark 13.3.3. From Theorem 13.3.2 it follows that the dimension of the kernel of problem (4), (5), (10) coincides with the dimension T of the linear space of homogeneous polynomials in $y_1, \dots, y_d$ of degree $m - d/2$. A simple calculation leads to

$$T = (m + d/2 - 1)![(d-1)!(m - d/2)!]^{-1}.$$

The algorithm for construction of asymptotics used in 13.2 for the case of noncritical dimension cannot immediately be applied now because the cokernel of the second limit problem is nontrivial. As in Sections 1.3 and 5.3, one should expand the class of solutions to the first limit problem, namely, admit the solutions having order $O(|y|^{m-d/2}|\log|y||)$ as $y \to 0$. Then the kernel of this problem becomes infinite-dimensional containing functions that satisfy the homogeneous equations (1)–(3), 13.2. For any T-tuple of functions in $\mathbf{C}^\infty(M)$ one can construct an element in the kernel. The obtained freedom of choice of a solution to the first limit problem allows one to provide the solvability of the second limit problem in the class of functions satisfying (10). More precisely, the mentioned T-tuple of functions on M can be chosen so that the orthogonality conditions (11) are fulfilled. Thus, at every step of the algorithm we obtain a solvable boundary value problem.

We are going to implement the procedure described.

The next assertion is derived from Theorem 13.3.1 in the same way as Corollary 11.3.5 from Proposition 11.3.4.

Theorem 13.3.4. *For any collection* $\mathbf{c} = \{c_\alpha\}$ *of functions in* $\mathbf{C}^\infty(M)$, *where* $\alpha = (\alpha_1, \dots, \alpha_d)$ *and* $|\alpha| = m - d/2$, *there exists a unique solution* $\tilde{V}_\mathbf{c}$ *to problem* (1)–(3), 13.2, *with* $f = 0$ *such that*

$$\tilde{V}_\mathbf{c}(x) - \sum_{|\beta| = m - d/2} c_\beta(z) D_y^\beta \Gamma(y, z) = O(|y|^{m-d/2}), \quad x \in V. \tag{12}$$

Here Γ *is the fundamental solution of the operator* L_0 *introduced at the beginning of the section.*

We formulate two theorems on the asymptotic behavior of solutions near M that will be of use in what follows (see Section 11.4).

Theorem 13.3.5. *The solution* $\tilde{V}_\mathbf{c}$ *to problem* (1) (3), 13.2, *with* $f = 0$ *satisfying* (12) *admits the asymptotics representation*

$$\begin{aligned}\tilde{V}_\mathbf{c}(x) = \zeta(x) &\sum_{|\beta| = m - d/2} \left(c_\beta(z) D_y^\beta \Gamma(y, z) + \Lambda_\beta(\mathbf{c}) y^\beta\right) \\ &+ \zeta(x) \sum_{k = m - d/2} |y|^k \tilde{V}_\mathbf{c}^{(k)}(z, y/|y|, \log|y|) + \tilde{W}^{(N)}(x),\end{aligned} \tag{13}$$

where Λ_β *is a linear map of* $\mathbf{C}^\infty(M) \times \dots \times \mathbf{C}^\infty(M)$ *into* $\mathbf{C}^\infty(M)$ *(the product consists of* T *factors;* T *is defined in Remark* 13.3.3*);* $\tilde{V}_\mathbf{c}^{(k)}(z, \vartheta, t)$ *are polynomials in* t

depending on $\{c_\alpha\}$ *whose coefficients belong to* $\mathbf{C}^\infty(M)\times\mathbf{C}^\infty(S^{d-1})$; *the remainder in* (13) *satisfies*

$$D_z^\mu D_y^\nu \tilde{W}^{(N)}(x) = o(|y|^{N-|\nu|-\sigma}), \quad |y|\to 0,$$

with arbitrary multiindices μ,ν, *while* N *and* σ *is a positive integer and any positive number.*

Theorem 13.3.6. *Suppose that* $g\in\mathbf{C}^\infty(\bar\Omega\setminus M)$ *admits asymptotic representation* (4), 13.2, *in a neighborhood* V *of the manifold* M. *Then there exists a solution* $v\in\mathbf{C}^\infty(\bar\Omega\setminus M)$ *to problem* (1)–(3), 13.2, *where* f *is replaced by* g, *satisfying* (5), 13.2.

We describe the principal term of the asymptotic expansion of the solution to problem (1)–(2), 13.1. As in 13.2, it is convenient to assume that f on the right in (1), 13.1, has the form (11), 13.2. Denote by $V^{(0)}$ a solution to problem (1)–(3), 13.2, with $f=g^{(0)}$ and set

$$v^{(0)}(x,\log\varepsilon) = V^{(0)}(x,\log\varepsilon) + V_{\mathbf{c}}(x), \tag{14}$$

where $\mathbf{c}$ is a collection of functions in $\mathbf{C}^\infty(M)$ to be determined. As in 13.2, we seek the corrective boundary layer $w^{(0)}$ as the solution to problem

$$L_0(z,D_\eta)w^{(0)}(\eta,z,\log\varepsilon) = h^{(0)}(\eta,z,\log\varepsilon), \quad \eta\in\mathbb{R}^d\setminus\overline{\omega(z)}; \tag{15}$$

$$\begin{aligned}(\partial/\partial\tau)^j w^{(0)}(\eta,z,\log\varepsilon) = -(\partial/\partial\tau)^j\Big(&|\eta|^{m-d/2}V^{(0)}_{m-d/2}(z,\vartheta,\log|\eta|+\log\varepsilon,\log\varepsilon)\\ &+\sum_{|\beta|=m-d/2}\big(c_\beta(z)\varepsilon^{-m+d/2}D_y^\beta\Gamma(\varepsilon\eta,z)+\Lambda_\beta(\mathbf{c})\eta^\beta\big)\Big),\\ &\eta\in\partial\omega(z),\quad j=0,\dots,m-1,\end{aligned} \tag{16}$$

$$w^{(0)}(\eta,z,\log\varepsilon) = o(|\eta|^{m-d/2}), \quad |\eta|\to\infty. \tag{17}$$

Using (2), we obtain

$$\varepsilon^{-m+d/2}D_y^\beta\Gamma(\varepsilon\eta,z) = \log\varepsilon\,\Gamma_1^{(\beta)}(\eta,z)+\Gamma_2^{(\beta)}(\eta,z), \tag{18}$$

where

$$\Gamma_1^{(\beta)}(\eta,z) = D_y^\beta S(\eta,z), \qquad \Gamma_2^{(\beta)}(\eta,z) = D_\eta^\beta\Gamma(\eta,z). \tag{19}$$

Introduce the operator

$$\tilde C\colon (\mathbf{C}^\infty(\partial\omega^c))^m\times\mathbf{C}^\infty(\omega^c)\to(\mathbf{C}^\infty(M))^T$$

by

$$\tilde C(\varphi_1,\dots,\varphi_m,h) = \Big(\sum_{j=0}^{m-1}\int_{\partial\omega(z)}\varphi_j\overline{B_j(z,D_\eta)\chi^{(\alpha)}}\,ds_\eta + \int_{\mathbb{R}^d\setminus\omega(z)} h\overline{\chi^{(\alpha)}}\,d\eta\Big)_{|\alpha|=m-d/2}. \tag{20}$$

We have

$$\begin{aligned}\tilde C\Big(\Big(\big((\partial/\partial\tau)^j c_\beta(z)(\Gamma_1^{(\beta)}(\eta,z)\log\varepsilon+\Gamma_2^{(\beta)}(\eta,z))\big)_{j=0}^m\Big)_{|\beta|=m-d/2}\Big)\\ = (A_1\log\varepsilon+A_2)\mathbf{c}(z),\end{aligned} \tag{21}$$

where A_1 and A_2 are $T\times T$-matrix-valued functions on M. We set

$$\tilde A\mathbf{c} = (\Lambda_\beta\mathbf{c})_{|\beta|=m-d/2}.$$

We rewrite condition (11) for solvability of problem (15)–(17) in the form

$$(A_1 \log\varepsilon + A_2 + \tilde{A})\mathbf{c} = \tilde{C}\Big(\big((\partial/\partial\tau)^j(|\eta|^{m-d/2}V^{(0)}_{m-d/2})\big)_{j=0}^{m-1}, -h^{(0)}\Big). \tag{22}$$

Lemma 13.3.7. *The matrix A_1 is nonsingular.*

Proof. We denote by $a_{\mu\nu}$, $|\mu| = |\nu| = m - d/2$, the entries of A_1. By virtue of (21),

$$a_{\mu\nu} = \int_{\partial\omega(z)} \sum_{k=0}^{m-1} \overline{B_k(z, D_\eta)\chi^{(\mu)}}\,(\partial/\partial\tau)^j\,\Gamma_1^{(\nu)}\,ds_\eta.$$

From the definition (8) of $\chi^{(\mu)}$ and Green's formula (9) it follows that

$$\tilde{C}\Big(\big((\partial/\partial\tau)^j\eta^\beta\big)_{j=0}^{m-1}, 0\Big) = (\delta_{\beta\alpha})_{|\alpha|=m-d/2}. \tag{23}$$

Since $\Gamma_1^{(\nu)}$ is a homogeneous polynomial of degree $m - d/2$, from (23) we obtain

$$(\mu!)^{-1}\,D_\nu^\mu\Gamma_1^{(\nu)} = -\int_{\partial\omega(z)} \sum_{k=0}^{m-1} \overline{B_k(z, D_\eta)\chi^{(\mu)}}\,(\partial/\partial\tau)^j\Gamma_1^{(\nu)}\,ds_\eta.$$

Therefore,

$$\Gamma_1^{(\nu)}(\eta, z) = \sum_{|\mu|=m-d/2} a_{\mu\nu}(z)\,\eta^\mu.$$

Suppose that at a point $z \in M$ the matrix $A_1(z)$ is singular. Then there exists a nonzero vector $\{b_\nu\}$ such that

$$\sum_{|\nu|=m-d/2} b_\nu a_{\mu\nu}(z) = 0.$$

The two last equalities imply

$$\sum_{|\nu|=m-d/2} b_\nu\Gamma_1^{(\nu)}(\eta, z) = 0. \tag{24}$$

We introduce the differential operator

$$P(D_\eta) = \sum_{|\nu|=m-d/2} b_\nu\,D_\eta^\nu.$$

By the definition (19) of $\Gamma_1^{(\nu)}$ the equality (24) is equivalent to

$$P(D_\eta)S(\eta, z) = 0, \tag{25}$$

S being defined by (3). The Fourier transform reduces (25) to the form

$$P(\xi)S(D_\xi, z)\,\delta(\xi) = 0,$$

where δ is Dirac's measure. Hence,

$$S(D_\xi, z)\,P(\xi)\overline{P(\xi)} = 0 \quad \text{for} \quad \xi = 0.$$

Together with (3) this leads to

$$\int_{S^{d-1}} L_0(z, \vartheta)^{-1}(D_\xi\cdot\vartheta)^{2m-d}\big(P(\xi)\overline{P(\xi)}\big)ds_\vartheta = 0 \quad \text{for} \quad \xi = 0$$

or, equivalently,

$$\int_{S^{d-1}} L_0(z,\vartheta)^{-1}|(D_\xi\cdot\vartheta)^{m-d/2}P(\xi)|^2 ds_\vartheta = 0 \quad \text{for} \quad \xi=0.$$

The last equality is impossible because $\operatorname{Re} L_0(z,\vartheta)>0$. The contradiction proves the lemma. □

Now we are in a position to write the formal Neumann series for the system (22). Namely, we have the following evident assertion.

Lemma 13.3.8. *The formal series*

$$K = -\sum_{q=0}^{\infty} |\log\varepsilon|^{-q-1}(A_1^{-1}(A_2+\tilde{A}))^q\, A_1^{-1}\, H \tag{26}$$

with $H\in \left(\mathbf{C}^\infty(M)\right)^T$ *satisfies*

$$(A_1\,\log\varepsilon + A_2+\tilde{A}))K = H. \tag{27}$$

Note that the series (26) is generally not convergent (see 12.2).

Using Lemma 13.3.8, we construct the asymptotic expansion of the solution to problem (1), (2), 13.1, in inverse degrees of $\log\varepsilon$.

Denote by H the right-hand side of (22) and by $\mathbf{k}^{(q)} = (k_\beta^{(q)})_{|\beta|=m-d/2}$ the coefficient by $|\log\varepsilon|^{-q-1}$ in the series (26). We consider the boundary value problem

$$\begin{aligned}
&L_0(z,D_\eta)W_0^{(N)} = h^{(0)} \quad \eta\in\mathbb{R}^d\setminus\overline{\omega(z)};\\
&(\partial/\partial\tau)^j W_0^{(N)} = -(\partial/\partial\tau)^j\Big\{|\eta|^{m-d/2}\,V_{m-d/2}^{(0)} - \sum_{|\beta|=m-d/2}\Big(k_\beta^{(0)}\Gamma_1^{(\beta)} + \sum_{q=0}^{N-1}|\log\varepsilon|^{-q-1}\\
&\qquad\times\big(k_\beta^{(q)}\Gamma_2^{(\beta)} + \Lambda_\beta(\mathbf{k}^{(q)}\eta^\beta - k_\beta^{(q+1)}\Gamma_1^{(\beta)}\big)\Big)\Big\}\\
&\qquad \text{in } \partial\omega(z), \quad j=0,\dots,m-1,\\
&W_0^{(N)}(\eta,z,\log\varepsilon) = o(|\eta|^{m-d/2}), \quad |\eta|\to\infty,
\end{aligned}\tag{28}$$

which is obtained by substituting the partial sum of series (26) into problem (15)–(17).

For any small ε there exists a solution to problem (28). Indeed, the compatibility condition (11) takes the form

$$H - A_1\mathbf{k}^{(0)} + \sum_{q=0}^{N-1}|\log\varepsilon|^{-q-1}(A_2\mathbf{k}^{(q)} + \tilde{A}\mathbf{k}^{(q)} - A_1\mathbf{k}^{(q+1)}) = 0,$$

which is obviously fulfilled by virtue of the definition of the vectors $\mathbf{k}^{(q)}$. The right-hand sides of equations (28) are sums of polynomials in $\log\varepsilon$ and polynomials in $(\log\varepsilon)^{-1}$. So the solution $W_0^{(N)}$, too, is a sum of polynomials in $\log\varepsilon$ and $(\log\varepsilon)^{-1}$. We introduce the function

$$\tilde{S}^{(N)}(\varepsilon,x) = V^{(0)}(x,\log\varepsilon) - \sum_{q=0}^{N}|\log\varepsilon|^{-q-1}\tilde{V}_{k^{(q)}}(x) + \varepsilon^{m-d/2}\zeta(x)W_0^{(N)}(\varepsilon^{-1}y,z,\log\varepsilon).$$

Here $V^{(0)}$ is the solution to problem (1)–(3), 13.2, with $f=g^{(0)}$. The difference $\tilde{R}^{(N)} = u - \tilde{S}^{(N)}$, where u is the solution to problem (1), (2), 13.1, satisfies the boundary value problem

$$\tilde{L}(x, D_x)\,\tilde{R}^{(N)}(x, \log\varepsilon) = f^{(1)}(\varepsilon, x) - \Phi^{(N)}(\varepsilon, x), \quad x \in \Omega_\varepsilon; \tag{29}$$

$$\nabla_j \tilde{R}^{(N)}(x, \log\varepsilon) = 0, \quad x \in \partial\Omega, \quad j = 0, \dots, m-1; \tag{30}$$

$$\begin{aligned}(\partial/\partial\tau)^j \tilde{R}^{(N)}(x, \log\varepsilon) = (\partial/\partial\tau)^j \Big\{ v^{(0,1)}(x, \log\varepsilon) \\ - |\log\varepsilon)|^N \sum_{|\beta| = m-d/2} \big(k_\beta^{(N)} \Gamma_2^{(N)}(y) + \Lambda_\beta(\mathbf{k}^{(N)}) y^\beta\big)\Big\}, \end{aligned} \tag{31}$$

$$x \in \partial\omega_\varepsilon, \quad j = 0, \dots, m-1.$$

The functions $\Phi^{(N)}$, $f^{(1)}$ and $v^{(0,1)}$ are defined by

$$\begin{aligned}\Phi^{(N)}(\varepsilon, x) = \varepsilon^{m-d/2}\{\tilde{L}_0(x, D_x)\big(\zeta(x) W_0^{(N)}(\varepsilon^{-1}y, z, \log\varepsilon)\big) \\ + [\tilde{L}(x, D_x), \zeta(x)] W_0^{(N)}(\varepsilon^{-1}y, z, \log\varepsilon)\},\end{aligned}$$

$$f^{(1)}(\varepsilon, x) = f(\varepsilon, x) - g^{(0)}(x, \log\varepsilon) - \varepsilon^{-m-d/2} h^{(0)}(\varepsilon^{-1}y, z, \log\varepsilon)\zeta(x),$$

$$v^{(0,1)}(x, \log\varepsilon) = v^{(0)}(x, \log\varepsilon) - |y|^{m-d/2} v_0^{(0)}(z, \vartheta, \log|y|, \log\varepsilon),$$

where

$$v^{(0)}(x, \log\varepsilon) = V^{(0)}(x) - \sum_{q=0}^{N} |\log\varepsilon)|^{-q-1} \tilde{V}_{\mathbf{k}^{(q)}}(x).$$

The following estimate on the solution to problem (1), (2), 13.1, results from (3):

$$\|u; \mathbf{W}_2^m(\Omega_\varepsilon)\| \le c\|f r^m(|\log r| + 1); \mathbf{L}_2(\Omega_\varepsilon)\|. \tag{32}$$

We have

$$\|r^m(1 + |\log r|)\Phi^{(N)}; \mathbf{L}_2(\Omega_\varepsilon)\| \le c\varepsilon^{1-\delta}, \quad c = \text{const}, \quad \delta > 0. \tag{33}$$

As in Section 3.2, there exists a function $X^{(N)}$ such that

$$(\partial/\partial\tau)^j X^{(N)} = \Phi_j^{(N)} \quad \text{on} \quad \partial\omega^c,$$

where $\Phi_j^{(N)}$, $j = 0, \dots, m-1$, are the right-hand sides of the equalities (31) written in the coordinates (z, η), $X^{(N)} \in \mathbf{C}_0^\infty(\omega^c)$. Furthermore,

$$\sum_{k=0}^{2m} \|r^{k-m} X^{(N)}\zeta; \mathbf{W}_2^k(\Omega_\varepsilon)\| \le c|\log\varepsilon|^{M_0 - N}, \tag{34}$$

where M_0 is the degree of the polynomial $V_{m-d/2}^{(0)}(z, \vartheta, \log|\eta| + t, t)$ in t.

The next assertion is the main result of the section.

Theorem 13.3.9. *Let*

$$\|r^m(1 + |\log r|)f^{(1)}; \mathbf{L}_2(\Omega_\varepsilon)\| \le c|\log\varepsilon|^{M_0 - N}. \tag{35}$$

Then $\tilde{S}^{(N)}$ satisfies the inequality

$$\|\tilde{S}^{(N)} - u; \mathbf{W}_2^m(\Omega_\varepsilon)\| \le c|\log\varepsilon|^{M_0 - N}, \tag{36}$$

where u is the solution to problem (1), (2), 13.1, *and the constant c does not depend on ε.*

Proof. We set $\Phi^{(N)} = \tilde{R}^{(N)} - X^{(N)}$ in (29)–(31). Applying (32) and using the inequalities (33)–(35), we arrive at (36). □

13.4 The Case of Critical Dimension (Expansion in Powers of ε)

13.4.1 Structure of the asymptotic expansion

We are going to construct a formal asymptotic expansion of the solution to problem (1), (2), 13.1, in degrees of ε. We seek expansion of the form

$$\sum_{k=0}^{\infty} \varepsilon^k \big(v^{(k)}(x, \log\varepsilon) + \varepsilon^{m-d/2}\zeta_x w^{(k)}(\varepsilon^{-1}y, z, \log\varepsilon)\big), \tag{1}$$

where $v^{(k)}(x, \log\varepsilon)$ and $w^{(k)}(\eta, z, \log\varepsilon)$ are formal series in inverse degrees of $\log\varepsilon$. The coefficients of the series $v^{(k)}$ and $w^{(k)}$ belong to $\mathbf{C}^\infty(\bar\Omega \setminus M)$ and $\mathbf{C}^\infty(\omega^c)$ and admit expansions (5) and (10), 13.2, respectively.

We represent the function $v^{(k)}$ in the form

$$V^{(k)}(x, \log\varepsilon) + \tilde V_{\mathbf{c}^{(k)}}(x, \log\varepsilon),$$
$$\mathbf{c}^{(k)} \in (\mathbf{C}^\infty(M))^T.$$

Substitute the formal series (1) into problem (1), (2), 13.1, and as in 13.2 distribute the right-hand sides of the equalities obtained among the right-hand sides of the boundary value problems for $V^{(p)}$ and $w^{(q)}$ in the domains $\Omega\setminus M$ and $\mathbb{R}^d \setminus \overline{\omega(z)}$. Then $V^{(p)}$ satisfies (18), 13.2, and $\tilde V_{\mathbf{c}^{(k)}}$ is the solution to problem (1)–(3), 13.2, with $f = 0$ subject to (7), 13.3, where $c = \mathbf{c}^{(k)}$. Denote by $G^{(q)}$ the right-hand side of (16), 13.2, and introduce

$$\Phi_j^{(q)} = -(\partial/\partial\tau)^j \Big(|y|^{m-d/2} V^{(q)}_{m-d/2} + \sum_{s=m-d/2+1}^{q+m-d/2} |y|^s v_s^{(q+m-d/2-s)} \Big),$$

which are similar to the functions on the right of (13), 13.2. As before in defining the functions $w^{(0)}$, we rewrite the compatibility conditions of problem (16), (17), (13), 13.2, in the form (cf. (22), 13.3)

$$(\log\varepsilon A_1 + A_2 + \tilde A)\big(c_\alpha^{(q)}(z)\big)_{|\alpha|=m-d/2} = \tilde C(\Phi_0^{(q)}, \dots, \Phi_{m-1}^{(q)}, G^{(q)}).$$

From the last equation we obtain the functions $c_\alpha^{(q)}$, $|\alpha| = m - d/2$, (see Lemma 13.3.8) and conclude that the problem with parameter $z \in M$

$$L_0(z, D_\eta) w^{(q)}(\eta, z, \log\varepsilon) = G^{(q)}(\eta, z, \log\varepsilon), \quad \eta \in \mathbb{R}^d \setminus \overline{\omega(z)};$$

$$\begin{aligned}(\partial/\partial\tau)^j w^{(q)}(\eta, z, \log\varepsilon) &= \Phi_j^{(q)}(\eta, z, \log\varepsilon)\\ &\quad - (\partial/\partial\tau)^j \Big\{ \sum_{|\beta|=m-d/2} \big(c_\beta(z)\big(\log\varepsilon\Gamma_1^{(\beta)}(\eta, z) + \Gamma_2^{(\beta)}(\eta, z)\big)\\ &\quad + \Lambda_\beta(\mathbf{c}^{(q)})\eta^\beta\big)\Big\}, \quad \eta \in \partial\omega(z), \quad j = 0, \dots, m-1;\end{aligned}$$

$$w^{(q)}(\eta, z, \log\varepsilon) = o(|\eta|^{m-d/2}), \quad |\eta| \to \infty,$$

is solvable.

Thus, all the terms in (1) are determined. If the series $v^{(k)}$ and $w^{(k)}$ in (1) are convergent then the function

$$\tilde S^{(N)} = \sum_{k=0}^{N-1} \varepsilon^k (v^{(k)} + \varepsilon^{m-d/2}\zeta w^{(k)})$$

is subject to the estimate

$$\|u - \tilde{S}^{(N)}; \mathbf{W}_2^m(\Omega_\varepsilon)\| \leq c\varepsilon^{N-\delta},$$

where u is the solution to problem (1), (2), 13.1, δ being any positive number.

13.4.2 Structure of the formal series (1)

Let us analyze once again the procedure of construction of the coefficients of the series (1). The following operators were used:

A – the inverse operator of problem (1)–(3), 13.2;

$B = B(z)$ – the inverse operator of problem (4)–(7), 13.3, given on functions satisfying the orthogonality condition (11), 13.3;

$\Pi_k = (\Pi_k^{(0)}, \dots, \Pi_k^{(m-1)})$ – the operator that sends a function v having the expansion (5), 13.2, to the vector

$$\Big((\partial/\partial\tau)^j \big(|\eta|^{m-d/2+k}\, v_{m-d/2+k}(z, \vartheta, \log|\eta| + \log\varepsilon)\big)\Big)_{j=0}^{m-1}, \qquad k = 0, 1, 2, \dots;$$

$\tilde{\Pi}_k$ – the operator that sends a function w with expansion (10), 13.2, to

$$|y|^{m-d/2+k}\, w_{d/2-m+k}(z, \vartheta, \log|y| - \log\varepsilon), \quad k = -1, -2, \dots;$$

$R = (A_1 \log\varepsilon + A_2 + \tilde{A})^{-1}$ (assuming that such an operator exists);

$\hat{A}$ – the inverse operator of problem (1), (2), 13.2, (12), 13.3, with $f = 0$.

The functions $v^{(0)}$ and $w^{(0)}$ can be written in the form

$$\begin{aligned} v^{(0)} &= Ag^{(0)} + \hat{A}R\tilde{C}(\Pi_0 Ag^{(0)}, -h^{(0)}), \\ w^{(0)} &= B\big(-\Pi_0 Ag^{(0)} + \Pi_0 \hat{A}R\tilde{C}(\Pi_0 Ag^{(0)}, -h^{(0)}), h^{(0)}\big). \end{aligned} \tag{2}$$

It is convenient to rewrite these equalities in the following manner:

$$(v^{(0)}, w^{(0)}) = (Q + PRS)(g^{(0)}, h^{(0)}). \tag{3}$$

Here Q, P, and S are operators depending polynomially in $\log\varepsilon$.

When defining $v^{(1)}$ and $w^{(1)}$, in the right-hand sides of the problems in the domains Ω and $\mathbb{R}^d \setminus \overline{\omega(z)}$, one obtains the terms

$$\begin{aligned} (T_1 w^{(0)})(x, \log\varepsilon) &= \big([\tilde{L}(x, D_x), \zeta(x)] + \zeta(x)\tilde{L}_1(x, D_x)\big) \\ &\quad \times \big(|y|^{m-[d/2]-1}\, w^{(0)}_{d/2+1-m}(z, \vartheta, \log|y| - \log\varepsilon, \log\varepsilon)\big), \end{aligned}$$

$$(T_2 w^{(0)})(\eta, z, \log\varepsilon) = L_1(\eta, z, D_\eta, D_z)\, w^{(0,1)}(\eta, z, \log\varepsilon)$$

respectively (see (18), 13.2, and (16), 13.2). Thus,

$$\begin{aligned} v^{(1)} &= Ag^{(1)} - ATw^{(0)} + \hat{A}R\tilde{C}\Big(\Pi_0 v^{(0)} + \Pi_0(Ag^{(1)} - AT_1 w^{(0)}), -h^{(1)} + T_2 w^{(0)}\Big), \\ w^{(1)} &= B\big(-\Pi_0 v^{(1)} - \Pi_1 v^{(0)}, h^{(1)} + T_2 w^{(0)}\big). \end{aligned} \tag{4}$$

The representation (4) for $v^{(1)}$ and $w^{(1)}$ can be written in the form

$$\begin{aligned} (v^{(1)}, w^{(1)}) &= \big(Q^{(1,0)} + P_{21}^{(1,0)} RS_{21}^{(1,0)} P_{22}^{(1,0)} RS_{22}^{(1,0)} + P_{11}^{(1,0)} RS_{11}^{(1,0)}\big)\,(g^{(0)}, h^{(0)}) \\ &\quad + \big(Q^{(1,1)} + P_{11}^{(1,1)} RS_{11}^{(1,1)}\big)\,(g^{(1)}, h^{(1)}) \end{aligned}$$

which is similar to (3). Here, as in (3), Q, P, and S stand for operators depending polynomially on $\log\varepsilon$.

Suppose that $v^{(j)}$ and $w^{(j)}, j = 0, \dots, k = 1$ have been determined. By the same reasoning as in 13.4.1 we obtain

$$v^{(k)} = A\Big(g^{(k)} - \sum_{j=0}^{k-1} T_{1j}^{(k)} w^{(j)}\Big)$$
$$+ \hat{A}R\tilde{C}\Big\{\sum_{j=0}^{k-1} \Pi_{k-j} v^{(j)} + \Pi_0 A\Big(g^{(k)} - \sum_{j=0}^{k-1} T_{1j}^{(k)} w^{(j)}\Big), -h^{(k)} + \sum_{j=0}^{k-1} T_{2j}^{(k)} w^{(j)}\Big\}, \tag{5}$$
$$w^{(k)} = B\Big(\sum_{j=0}^{k-1} \Pi_{k-j} v^{(j)}, h^{(k)} - \sum_{j=0}^{k-1} T_{2j}^{(k)} w^{(j)}\Big).$$

Here the operators $T_{1j}^{(k)}$ and $T_{2j}^{(k)}$ are defined by

$$T_{1j}^{(k)} = \{[\tilde{L}(x, D_x), \zeta] + \zeta \tilde{L}_{k-j}(x, D_x)\}\, \tilde{\Pi}_{j-k},$$
$$T_{2j}^{(k)} = L_{k-j}(\eta, z, D_\eta, D_z)\Big(\mathbf{1} - \sum_{p=0}^{k-j+1} \tilde{\Pi}_{-p}\Big). \tag{6}$$

Iterating (5), we arrive at

$$(v^{(k)}, w^{(k)}) = \sum_{s=0}^{k} \Big(Q^{(k,s)} + \sum_{j=1}^{k-s+1} \prod_{q=1}^{j} P_{jq}^{(k,s)} R S_{jg}^{(k,s)}\Big)(g^{(s)}, h^{(s)}),$$

where all the operators except R are polynomials in $\log\varepsilon$.

Thus, the formal series (1) can be written in the form

$$\sum_{k=0}^{\infty} \varepsilon^k \sum_{s=0}^{k} \Bigg(\Big(Q_1^{(k,s)} + \sum_{j=1}^{k-s+1} \prod_{q=1}^{j} P_{jq1}^{(k,s)} R S_{jg1}^{(k,s)}\Big)$$
$$+ \varepsilon^{m-d/2} \zeta \Big(Q_2^{(k,s)} + \sum_{j=1}^{k-s+1} \prod_{q=1}^{j} P_{jq2}^{(k,s)} R S_{jg2}^{(k,s)}\Big)\Bigg)(g^{(s)}, h^{(s)}). \tag{7}$$

In the degenerate case $d = n$ the manifold M is a singleton and the equation (2) is an algebraic system with $T \times T$-matrix,

$$T = (m + n/2 - 1)!/\big((n-1)!\,(m - n/2)!\big),$$

(see Remark 13.3.3). According to Lemma 13.3.7, the matrix is nondegenerate for small ε. Therefore, for $d = n$ the asymptotic solution $\tilde{S}^{(N)}$ takes the form

$$\tilde{S}^{(N)}(\varepsilon, x) = \sum_{k=0}^{N-1} \varepsilon^k \sum_{j=0}^{k+1} \mu(\varepsilon)^{-j} \big(v^{(k,j)}(x, \log\varepsilon) + \varepsilon^{m-n/2} \zeta(x)\, w^{(k,j)}(\varepsilon^{-1}x, \log\varepsilon)\big),$$

where $v^{(k,j)}(x,t)$ and $w^{(k,j)}(\xi,t)$ are polynomials in t and $\mu(\varepsilon) = \det(A_1 \log\varepsilon + A_2 + \tilde{A})$.

13.4.3 Asymptotic inversion of the operator $A_1 \log \varepsilon + A_2 + \tilde{A}$

In order to make the formal series (7) into an asymptotic one it is natural to replace the formal inverse operator R by its asymptotic approximation. Before doing that, we obtain some further information on the operator $\tilde{A}$ in (22), 13.3.

Recall that $\tilde{A}$ is defined by

$$\tilde{A}\mathbf{c} = \left(\Lambda_\beta(\mathbf{c})\right)_{|\beta|=m-d/2}, \tag{8}$$

where Λ_β are the operators in the expansion

$$\tilde{V}_{\mathbf{c}}(x) = \sum_{|\beta|=m-d/2} \left(c_\beta(z) D_y^\beta \Gamma(z,y) + \Lambda_\beta(\mathbf{c}) y^\beta\right) + o(|y|^{m-d/2}), \quad |y| \to 0, \tag{9}$$

of the solution to the problem

$$\begin{aligned}
&\tilde{L}(x, D_x)\, \tilde{V}_{\mathbf{c}}(x) = 0, \quad x \in \Omega \setminus M;\\
&\nabla_j \tilde{V}_{\mathbf{c}} = 0, \quad x \in \partial\Omega, \quad j = 0, \dots, m-1;\\
&\nabla_k \tilde{V}_{\mathbf{c}} = 0, \quad x \in M, \quad k = 0, \dots, m-1-d/2;\\
&\tilde{V}_{\mathbf{c}}(x) = \sum_{|\beta|=m-d/2} \left(c_\beta(z) D_y^\beta \Gamma(z,y) \Lambda_\beta(\mathbf{c}) y^\beta\right) + O(|y|^{m-d/2}), \quad |y| \to 0.
\end{aligned}$$

We consider the Sobolev problem

$$\begin{aligned}
&\tilde{L}(x, D_x)\, W(x) = \sum_{|\beta|<m-d/2} b_\beta(z) D_y^\beta \delta(y), \quad x \in \bar{\Omega};\\
&\nabla_j W(x) = 0, \quad x \in \partial\Omega, \quad j = 0, \dots, m-1;\\
&D_y^\alpha W(x) = \varphi_\alpha(z), \quad x \in M, \quad |\alpha| < m - d/2.
\end{aligned}$$

The problem has a unique solution

$$(W, (b_\alpha)) \in \mathbf{W}_2^m(\Omega) \times \prod_{|\beta|<m-d/2} \mathbf{W}_2^{d/2-m+|\beta|}(M)$$

for any right-hand side $\varphi_\alpha \in \mathbf{W}_2^{m-d/2-|\alpha|}(M)$ (see Theorem 11.4.1). The functions b_β admit the representations

$$b_\beta = \sum_{|\alpha|<m-d/2} K_\beta^\alpha \varphi_\alpha, \tag{10}$$

K_β^α being a classical pseudodifferential operator of order $2m - d - |\alpha| - |\beta|$ on M.

Now we consider the problem

$$\tilde{L}(x, D_x)\, V(x) = \sum_{|\alpha|\le m-d/2} c_\beta D_y^\alpha \delta(y), \quad x \in \Omega; \tag{11}$$

$$\nabla_j V(x) = 0, \quad x \in \partial\Omega, \quad j = 0, \dots, m-1; \tag{12}$$

$$\nabla_k V(x) = 0, \quad x \in M, \quad j = 0, \dots, m - d/2 - 1. \tag{13}$$

From (11) and (12) it follows that

$$V(x) = \int_M \sum_{|\beta|\le m-d/2} c_\beta(\xi) Q_\beta(x,\xi)\, ds_\varsigma, \tag{14}$$

where $\xi = (\eta, \zeta), \zeta$ are local coordinates on M, and Q_β is a linear combination of the derivatives of order $\le |\beta|$ of Green's function of the Dirichlet problem for the operator $\tilde{L}$ in Ω. By virtue of (13),

$$\int_M \sum_{|\gamma|<m-d/2} c_\gamma(\zeta)(D_y^\alpha Q_y)(0,z,0,\zeta)\,ds_\zeta = -\int_M \sum_{|\gamma|=m-d/2} c_\gamma(\zeta)(D_y^\alpha Q_\beta)(0,z,0,\zeta)\,ds_\zeta,$$

where $|\alpha| < m - d/2$. From this and (10) we obtain

$$c_\beta(z) = \sum_{|\alpha|<m-d/2} K_\beta^\alpha \int_M \sum_{|\gamma|<m-d/2} c_\gamma(\zeta)(D_\eta^\alpha Q_y)(0,z,0,\zeta)\,ds_\zeta. \tag{15}$$

Since K_β^α is applied to a pseudodifferential operator of order $-2m+|\alpha|+|\gamma|+d = d/2 - m + |\alpha|$, we have

$$c_\beta = \sum_{|\gamma|=m-d/2} O_\beta^\gamma c_\gamma, \tag{16}$$

where O_β^γ is a psedodifferential operator of order $m - d/2 - |\beta|$.

Represent the function (14) in the form

$$V(x) = \int_M \sum_{|\beta|=m-d/2} c_\beta(z) Q_\beta(x,\xi)\,ds_\zeta + \int_M \sum_{|\beta|=m-d/2} (c_\beta(\zeta) - c_\beta(z)) Q_\beta(x,\xi)\,ds_\zeta,$$
$$+ \int_M \sum_{|\beta|<m-d/2} c_\beta(\zeta) Q_\beta(x,\xi)\,ds_\zeta. \tag{17}$$

The integral

$$I_\beta(x) = \int_M Q_\beta(x,\xi)\,ds_\zeta, \quad |\beta| = m - d/2,$$

satisfies problem (11), (12) with $c_\beta = \delta_{\alpha\beta}$, therefore

$$I_\beta(x) = D_y^\beta \Gamma(y,z) + \sum_{|\alpha|\le m-d/2} (\alpha!)^{-1} y^\alpha \lambda_\alpha^{(\beta)}(z) + o(|y|^{m-d/2}), \quad |y| \to 0.$$

We denote the second and third integrals in (17) by $V_2(x)$ and $V_3(x)$. Since

$$D_y^\alpha Q_\beta(x,\xi) = O(|x-\xi|^{d-n}), \quad \text{for} \quad |\beta| = |\alpha| = m - d/2,$$

and

$$D_y^\alpha Q_\beta(x,\xi) = O(|x-\xi|^{d-n+1}), \quad \text{for} \quad |\alpha| = m - d/2, |\beta| < m - d/2,$$

we obtain that for smooth functions c_β the derivatives $D_y^\alpha V_2$ and $D_y^\alpha V_3$ have the limiting values on M

$$\int_M \sum_{|\beta|=m-d/2} (c_\beta(\zeta) - c_\beta(z))(D_y^\alpha Q_\beta)(0,z,0,\zeta)\,ds_\zeta$$

and

$$\int_M \sum_{|\beta|<m-d/2} c_\beta(\zeta)(D_y^\alpha Q_\beta)(0,z,0,\zeta)\,ds_\zeta.$$

Thus,

$$V(x) = \sum_{|\beta|=m-d/2} \Big\{ c_\beta(z) D_y^\beta \Gamma(y,z) + \sum_{|\alpha|\le m-d/2} (\alpha!)^{-1} y^\alpha \Big(\lambda_\alpha^\beta(z) c_\beta(z) + \int_M \big(c_\beta(\zeta) - c_\beta(z)\big)(D_y^\alpha Q_\gamma)(0,z,0,\zeta)\, ds_\zeta \Big) \Big\}$$
$$+ \sum_{|\beta|<m-d/2} \sum_{|\alpha|<m-d/2} (\alpha!)^{-1} y^\alpha \int_M c_\beta(\zeta)(D_y^\alpha Q_\beta)(0,z,0,\zeta)\, ds_\zeta$$
$$+ o(|y|^{m-d/2}), \quad |y| \to 0. \tag{18}$$

By virtue of the fact that V is a solution to problem (11)–(13), the coefficients by $y^\alpha, |\alpha| < m - d/2$, in (18) equal zero. Hence, $V = \tilde{V}_c$ and

$$\Lambda_\alpha(\mathbf{c}) = \sum_{|\beta|=m-d/2} (\alpha!)^{-1} \Big\{ \lambda_\alpha^\beta(z) c_\alpha(z) + \int_M \big(c_\beta(\zeta) - c_\beta(z)\big)(D_y^\alpha Q_\beta)(0,z,0,\zeta)\, ds_\zeta \Big\}$$
$$+ \sum_{|\beta|<m-d/2} (\alpha!)^{-1} \int_M c_\beta(\zeta)(D_y^\alpha Q_\beta)(0,z,0,\zeta)\, ds_\zeta, \tag{19}$$

where $c_\gamma, |\gamma| < m - d/2$, are connected with $c_\beta, |\beta| = m - d/2$, by the equality (16).

We consider the operator

$$\varphi \to \tilde{E}_\alpha^\beta \varphi = \int_M \varphi(\zeta)(D_y^\alpha Q_\beta)(0,z,0,\zeta) ds_\zeta$$

with $|\beta| < m - d/2$. It is clear that $\tilde{E}_\alpha^\beta$ is a pseudodifferential operator of order $d/2 - m + |\beta|$. From this and (16) we obtain that for $|\beta| < m - d/2$ the mapping

$$(c_\gamma)_{|\gamma|=m-d/2} \to \int_M c_\beta(\zeta)(D_y^\alpha Q_\beta)(0,z,0,\zeta)\, ds_\zeta$$

is a pseudodifferential operator of order zero.

Fix a coordinate neighborhood U on M and denote by k a diffeomorphism $U \to \mathbb{R}^{n-d}$. Let χ_1 and χ_2 be functions in $\mathbf{C}_0^\infty(k(U))$. We introduce the operator

$$\varphi \to \tilde{F}_\beta^\alpha(\varphi) = \chi_1(z) \int_{\mathbb{R}^{n-d}} \big(\varphi(\zeta) - \varphi(z)\big)(D_y^\alpha Q_\beta)(0,z,0,\zeta)\chi_2(\zeta)\, ds_\zeta,$$

where $|\alpha| = |\beta| = m - d/2$. The function $D_y^\alpha Q_\beta$ admits the asymptotic expansion

$$D_y^\alpha Q_\beta(0,z,0,\zeta) \sim q_\beta(z, (\zeta - z)/|\zeta - z|)\, |\zeta - z|^{d-n} + O(|\zeta - z|^{d-n+\delta}),\ |\zeta - z| \to 0,\ \delta > 0,$$

q_β being a smooth function on $\mathbb{R}^{n-d} \times S^{n-d-1}$. Therefore, $\tilde{F}_\beta^\alpha$ is the sum of a pseudodifferential operator of order zero and the operator

$$\varphi \to \tilde{G}_\beta^\alpha \varphi = \chi_1(z) \int_{\mathbb{R}^{n-d}} \big(\varphi(\zeta) - \varphi(z)\big) q_\beta^\alpha(z, (\zeta - z)/|\zeta - z|)|\zeta - z|^{d-n} \chi_2(\zeta)\, d\zeta,$$

which can be written in the form

$$(\tilde{G}_\beta^\alpha \varphi)(z) = (2\pi)^{(d-n)/2} \int_{\mathbb{R}^{n-d}} \exp(i\mu z) \sigma_\beta^\alpha(z,\mu)\hat{\varphi}(\mu) d\mu$$

with

$$\sigma^\alpha_\beta(z,\mu) = \chi_1(z) \int\limits_{\mathbb{R}^{n-d}} \Big(\exp\big(i\mu(\zeta - z)\big) - 1\Big) q^\alpha_\beta\big(z, (\zeta - z)/|\zeta - z|\big)|\zeta - z|^{d-n}\chi_2(\zeta)\, d\zeta$$

$$= \chi_1(z) \int\limits_{\mathbb{R}^{n-d}} (\exp(i\mu\zeta) - 1) q^\alpha_\beta(z, \zeta/|\zeta|)|\zeta|^{d-n}\chi_2(z+\zeta)\, d\zeta.$$

Clearly, for any multiindex γ

$$\nabla_\mu D^\gamma_z \sigma^\alpha_\beta(z,\mu) = O((1+|\mu|)^{-1}). \tag{20}$$

It follows that

$$|D^\gamma_z \sigma^\alpha_\beta(z,\mu)| \le c\,\log(|\mu|+2). \tag{21}$$

From (21) with the help of Remark 3.1 in Kohn and Nirenberg [1] we immediately obtain that the operator $(\log(|D|+2))^{-1}\sigma^\alpha_\beta(\cdot,D)$ is continuous on any space $\mathbf{W}^l_2(\mathbb{R}^{n-d})$ or, equivalently,

$$\|\sigma^\alpha_\beta(\cdot,D)u; \mathbf{W}^l_2(\mathbb{R}^{n-d})\| \le c\|(|D|+1)^l \log(|D|+2)u; \mathbf{L}_2(\mathbb{R}^{n-d})\|. \tag{22}$$

The right-hand side of (22) is the norm in the Hörmander space $\mathbf{B}_{2,k}(\mathbb{R}^{n-d})$, where $k(\mu) = (1+|\mu|)^l \log(|\mu|+2)$ (see Hörmander [1]), § 2, Chapter 3). As in Hörmander [1], § 6, Chapter 2 we introduce the space $\mathbf{B}_{2,k}(M)$ using the norm in $\mathbf{B}_{2,k}(\mathbb{R}^{n-d})$ and a partition of unity. When proving the correctness of the definition of this space, one can make use of the equivalent norm in $\mathbf{B}_{2,k}(\mathbb{R}^{n-d})$

$$\Big(\int\limits_{\mathbb{R}^{n-d}}\int\limits_{\mathbb{R}^{n-d}} |\nabla_l u(z) - \nabla_l u(\zeta)|^2\, |z-\zeta|^{d-n}\log_+|z-\zeta|^{-1} dz d\zeta\Big)^{1/2} + \|u; \mathbf{L}_2(\mathbb{R}^{n-d})\|$$

given in Nagel [1]. From (22) it follows that

$$\|\tilde{A}v; \mathbf{W}^l_2(M)\| \le c\|v; \mathbf{B}_{2,k}(M)\|, \tag{23}$$

where $\tilde{A}$ is the operator (8).

Let Λ be a positive definite selfadjoint operator of second order with smooth coefficients on M. It is known (see Seeley [1] and Shubin [1], Chapter 2) that the following norms are equivalent:

$$\|\Lambda^{l/2}u; \mathbf{L}_2(M)\| \sim \|u; \mathbf{W}^l_2(M)\|. \tag{24}$$

From Strichartz [1] it follows that

$$\|\Lambda^{l/2}\log(\Lambda+1)u; \mathbf{L}_2(M)\| \sim \|u; \mathbf{B}_{2,k}(M)\|. \tag{25}$$

Denote by Ξ a function in $\mathbf{C}^\infty_0(\mathbb{R}^{n-d})$ such that $\Xi(\mu) = 1$ for $|\mu| \le 1$ and $\Xi(\mu) = 0$ for $|\mu| \ge 2$. We introduce the operator $\Xi_\varepsilon(\Lambda) = \Xi(\varepsilon^\delta\Lambda)$ with a small positive δ and set $v = \Xi_\varepsilon(\Lambda)u$ in (23). Then (23) and (25) lead to the estimate

$$\|\tilde{A}\Xi_\varepsilon(\Lambda)u; \mathbf{W}^l_2(M)\| \le c\|\log(\Lambda+1)\Xi_\varepsilon(\Lambda)\Lambda^{l/2}u; \mathbf{L}_2(M)\|.$$

Obviously, the right-hand side does not exceed $C\delta|\log\varepsilon|\|\Lambda^{l/2}u; \mathbf{L}_2(M)\|$, where the constant C does not depend on δ and ε. From this and (24) it follows that

$$\|\tilde{A}\Xi_\varepsilon(\Lambda)u; \mathbf{W}^l_2(M)\| \le c\delta|\log\varepsilon|\|u; \mathbf{W}^l_2(M)\|. \tag{26}$$

In what follows we need the inequality

$$\|\tilde{A}(1-\Xi_\varepsilon(\Lambda))u; \mathbf{W}^l_2(M)\| \le c_N\varepsilon^{\delta(N-1)/2}\|u; \mathbf{W}^{l+N}_2(M)\|, \tag{27}$$

where N is an arbitrary positive integer and the constant c_N is independent of ε. Indeed, the left-hand side does not exceed

$$c\|(1-\Xi_\varepsilon(\Lambda))u; \mathbf{W}_2^{l+1}(M)\| \le c\|(1-\Xi_\varepsilon(\Lambda))\Lambda^{l+1)/2}u; \mathbf{L}_2(M)\|$$
$$\le c(\min\{\lambda_n : \lambda_n \ge \varepsilon^{-\delta}\})^{-(N-1)/2}\|\Lambda^{(l+N)/2}u; \mathbf{L}_2(M)\|,$$

$\{\lambda_n\}$ being the sequence of the eigenvalues of Λ.

We return to equation (27), 13.3. As was noted in 13.4.2, the equation is not necessarily solvable. However, we shall construct a function K_ε that gives the discrepancy $o(\varepsilon^s)$ in (27), 13.3, where s is any positive number. From Lemma 13.3.7 it follows that the matrix (1), 13.3, is nonsingular. By virtue of (26), the operator

$$(\log\varepsilon)^{-1}A_1^{-1}(A_2+\tilde{A}\Xi_\varepsilon(\Lambda))$$

is small with respect to the operator norm in $\mathbf{W}_2^l(M)$ for any $l>0$. Hence, there exists the inverse operator

$$R_\varepsilon = (A_1\log\varepsilon + A_2 + \tilde{A}\Xi_\varepsilon(\Lambda))^{-1}.$$

Set $K_\varepsilon = R_\varepsilon H$. Then

$$(A_1\log\varepsilon + A_2 + \tilde{A})K_\varepsilon = H_\varepsilon,$$

where $H_\varepsilon = H + \tilde{A}(1-\Xi_\varepsilon(\Lambda))R_\varepsilon H$. In view of (27),

$$\|H_\varepsilon - H; \mathbf{W}_2^l(M)\| \le c_N\varepsilon^{\delta(N-1)/2}\|H; \mathbf{W}_2^{l+N}(M)\|. \tag{28}$$

13.4.4 The power asymptotic series for the solution

As before, we seek the function $v^{(0)}$ in the form (14), 13.3, where $V^{(0)}$ is a solution to problem (1)–(3), 13.2, with $f=g^{(0)}$ and $\tilde{V}_{\mathbf{c}^{(0)}}$ is the solution (13), 13.3, to problem (1)–(3), 13.2, with $f=0$, while $\mathbf{c}=\mathbf{c}^{(0)}$ is a collection of smooth functions on M to be determined. The boundary value problem (16)–(18), 13.3, for $w^{(0)}$ must be changed. Namely, the role of the boundary condition (17), 13.3, is played by

$$\begin{aligned}
&(\partial/\partial\tau)^j w^{(0)}(\eta, z, \log\varepsilon)\\
&\quad = -(\partial/\partial\tau)^j\Big\{|\eta|^{m-d/2}\,V^{(0)}_{m-d/2}(z,\vartheta,\log|\eta|+\log\varepsilon,\log\varepsilon)\\
&\quad + \sum_{|\beta|=m-d/2}\big(c_\beta^{(0)}(z)\varepsilon^{-m+d/2}D_y^\beta\Gamma(\varepsilon\eta,z)+\Lambda_\beta(\mathbf{c}^{(0)})+(H_\varepsilon^{(0)}-H^{(0)})_\beta\,\eta^\beta\big)\Big\},\\
&\quad \eta\in\partial\omega(z),\quad j=0,\dots,m-1,
\end{aligned} \tag{29}$$

where

$$H^{(0)} = \tilde{C}\Big(\big((\partial/\partial\tau)^j(|\eta|^{m-d/2}\,V^{(0)}_{m-d/2}(z,\vartheta,\log|\eta|+\log\varepsilon,\log\varepsilon)\big)_{j=0}^{m-1}, -h^{(0)}\Big),$$
$$H_\varepsilon^{(0)} = H^{(0)} + \tilde{A}(1-\Xi_\varepsilon(\Lambda))R_\varepsilon H^{(0)}.$$

The right-hand sides of (17), 13.3, and (29) differ by

$$-(\partial/\partial\tau)^j\sum_{|\beta|=m-d/2}(H_\varepsilon^{(0)}-H^{(0)})_\beta\,\eta^\beta.$$

This term is introduced to provide solvability for the new boundary value problem (16), (18), 13.3, (29).

We denote by $\Phi_0^{(0)}, \dots, \Phi_{m-1}^{(0)}$ the right-hand sides of (29). The compatibility condition for problem (16), (18), 13.3, (29) is the equality

$$\tilde{C}(\Phi_0^{(0)}, \dots, \Phi_{m-1}^{(0)}, -h^{(0)}) = 0. \tag{30}$$

Using the notation introduced in 13.3, we rewrite (30) in the form

$$\begin{aligned} 0 = &-H^{(0)} + (A_1 \log\varepsilon + A_2 + \tilde{A})\mathbf{c}^{(0)} \\ &- \sum_{|\beta|=m-d/2} (H_\varepsilon^{(0)} - H^{(0)})_\beta \tilde{C}\Big(\big((\partial/\partial\tau)^j \eta^\beta\big)_{j=0}^{m-1}, 0\Big). \end{aligned}$$

From this and (23), 13.3, we obtain

$$(A_1 \log\varepsilon + A_2 + \tilde{A})\mathbf{c}^{(0)} = H_\varepsilon^{(0)}.$$

It remains to set $\mathbf{c}^{(0)} = R_\varepsilon H^{(0)}$.

Thus, we have

$$\begin{aligned} v^{(0)} &= Ag^{(0)} + \hat{A}R_\varepsilon \tilde{C}(\Pi_0 Ag^{(0)}, -h^{(0)}) \\ w^{(0)} &= B\Big(- \Pi_0 Ag^{(0)} - \Pi_0 \hat{A} R_\varepsilon \tilde{C}(\Pi_0 Ag^{(0)}, -h^{(0)}) \\ &\quad - \big((\partial/\partial\tau)^j \tilde{A}(1 - \Xi_\varepsilon(\Lambda)) R_\varepsilon(\Pi_0 Ag^{(0)}, -h^{(0)})\big), -h^{(0)}\Big), \end{aligned}$$

where the notation is the same as in 13.4.2.

Represent the solution to problem (1), (2), 13.1, in the form $u = v^{(0)} + \varepsilon^{m-d/2}\zeta w^{(0)} + \hat{R}^{(0)}$. The remainder $\hat{R}^{(0)}$ satisfies the boundary value problem

$$\begin{aligned} &\tilde{L}(x, D_x)\, \hat{R}^{(0)}(x, \log\varepsilon) \\ &\quad = f^{(1)}(\varepsilon, x) - \big(\tilde{L}(x, D_x) - L_0(z, D_\eta)\big)(\zeta(x) w^{(0)}(\varepsilon^{-1}y, z, \log\varepsilon)) \\ &\quad\quad + [L_0(z, D_\eta), \zeta(x)] w^{(0)}(\varepsilon^{-1}y, z, \log\varepsilon), \quad x \in \Omega_\varepsilon; \\ &\nabla_j \hat{R}^{(0)}(x, \log\varepsilon) = 0, \quad x \in \partial\Omega, \quad j = 0, \dots, m-1; \end{aligned} \tag{31}$$

$$\begin{aligned} (\partial/\partial\tau)^k \hat{R}^{(0)}(x, \log\varepsilon) &= -(\partial/\partial\tau)^k \Big\{ v^{(0)}(x) - \varepsilon^{m-d/2}\Big(V_{m-d/2}^{(0)}(y, z, \log\varepsilon) \\ &\quad + \sum_{|\beta|=m-d/2} (H_\varepsilon^{(0)}(z) - H^{(0)}(z))_\beta y^\beta\Big)\Big\}, \\ &\quad x \in \partial\omega_\varepsilon, \quad k = 0, \dots, m-1. \end{aligned} \tag{32}$$

As in 13.3, to estimate the norm of $\hat{R}^{(0)}$ in $\mathbf{W}_2^m(\Omega_\varepsilon)$ we use (32), 13.3. Denote by $F_0^{(0)}, \dots, F_{m-1}^{(0)}$ and $F^{(0)}$ the right-hand sides of (32) and (31), respectively. We have

$$\|r^m(1 + |\log r|)F^{(0)}; \mathbf{L}_2(\Omega_\varepsilon)\| \le c\varepsilon^{1-\delta}, \quad \delta > 0. \tag{33}$$

Furthermore, there exists a function $X^{(0)} \in \mathbf{W}_2^m(\Omega_\varepsilon)$ such that

$$(\partial/\partial\tau)^j X^{(0)} = F_j^{(0)} \quad \text{on} \quad \partial\omega_\varepsilon, \quad j = 0, \dots, m-1,$$

and, by virtue of (28),

$$\|r^{k-m} X^{(0)}; \mathbf{W}_2^k(\Omega_\varepsilon)\| \le c\varepsilon^{1-\delta}, \quad k = 0, \dots, m \tag{34}$$

(cf. (33), (34), 13.3, used in a similar situation in 13.3). From (32), 13.3, (33) and (34) we obtain

$$\|\hat{R}^{(0)}; \mathbf{W}_2^m(\Omega_\varepsilon)\| \le c\varepsilon^{1-\delta}. \tag{35}$$

We now describe the other terms $v^{(k)}$ and $w^{(k)}$ of the asymptotic expansion. We represent the function $v^{(k)}$ in the form $V^{(k)} + \tilde{V}_{\mathbf{c}^{(k)}}$, where $V^{(k)}$ is a solution to the problem

$$\begin{aligned} \tilde{L}V^{(k)} &= g^{(k)} - \sum_{j=0}^{k-1} T_{1j}^{(k)} w^{(j)} \quad \text{in} \quad \Omega \setminus M; \\ \nabla_j V^{(k)} &= 0 \quad \text{on} \quad \partial\Omega, \quad j = 0, \dots, m-1; \\ \nabla_p V^{(k)} &= 0 \quad \text{on} \quad M, \quad p = 0, \dots, m-1-d/2, \end{aligned} \tag{36}$$

and $\tilde{V}_{\mathbf{c}^{(k)}}$ is the solution (13), 13.3, to (1)–(3), 13.2, with $f = 0$, where $\mathbf{c} = \mathbf{c}^{(k)}$ is a collection of functions on M that will be defined in what follows. Guided by the same reasoning as for $w^{(0)}$, we seek $w^{(k)}$ as a solution to the problem

$$\tilde{L}_0 w^{(k)} = h^{(k)} - \sum_{j=0}^{k-1} T_{2j}^{(k)} w^{(j)} \quad \text{on} \quad \mathbb{R}^d \setminus \overline{\omega(z)}; \tag{37}$$

$$\begin{aligned} (\partial/\partial\tau)^p w^{(k)} = -\sum_{j=0}^{k} \Pi_{k-j}^{(p)} v^{(j)} - (\partial/\partial\tau)^p \sum_{|\beta|=m-d/2} (H_\varepsilon^{(k)} - H^{(k)})_\beta \eta^\beta \quad \text{on} \quad \partial\omega(z), \\ p = 0, \dots, m-1. \end{aligned} \tag{38}$$

In (36), (37) and (38) the operators $\Pi_s^{(p)}$ and $T_{gj}^{(k)}$ are the same as in 13.4.2. The vectors $H^{(k)}$ and $H_\varepsilon^{(k)}$ are defined by

$$\begin{aligned} H^{(k)} &= \tilde{C}\Big(\sum_{j=0}^{k-1} \Pi_{k-j} v^{(j)} + \Pi_0 V^{(k)}, -h^{(k)} + \sum_{j=0}^{k-1} T_{2j}^{(k)} w^{(j)}\Big), \\ H_\varepsilon^{(k)} &= H^{(k)} + \tilde{A}(1 - \Xi_\varepsilon(\Lambda)) R_\varepsilon H^{(k)}. \end{aligned}$$

The compatibility condition of problem (37) has the form

$$(A_1 \log\varepsilon + A_2 + \tilde{A})\mathbf{c}^{(k)} = H_\varepsilon^{(k)}.$$

Therefore, $\mathbf{c}^{(k)} = R_\varepsilon H^{(k)}$. The next result is as follows:

$$\begin{aligned} v^{(k)} &= A\Big(g^{(k)} - \sum_{j=0}^{k-1} T_{1j}^{(k)} w^{(j)}\Big) + \hat{A} R_\varepsilon H^{(k)}, \\ w^{(k)} &= B\Big(-\sum_{j=0}^{k} \Pi_{k-j} v^{(j)} \\ &\quad - \Big((\partial/\partial\tau)^p \sum_{|\beta|=m-d/2} (H_\varepsilon^{(k)} - H^{(k)})_\beta \eta^\beta\Big)_{p=0}^{m-1}, h^{(k)} - \sum_{j=0}^{k-1} T_{2j}^{(k)} w^{(j)}\Big). \end{aligned} \tag{39}$$

Now one can schematically write the asymptotic series for u in the form similar to (7):

$$\begin{aligned} \sum_{k=0}^{\infty} \varepsilon^k \sum_{s=0}^{k} \Big\{\Big(Q_1^{(k,s)} + \sum_{j=1}^{k-s+1} \prod_{q=1}^{k-s+1} P_{kq1}^{(k,s)} R_\varepsilon A_{jg1}^{(k,s)}\Big) \\ + \varepsilon^{m-d/2} \zeta_1 \Big(Q_2^{(k,s)} + \sum_{j=1}^{k-s+1} \prod_{q=1}^{j} P_{jq2}^{(k,s)} R_\varepsilon A_{jg2}^{(k,s)}\Big)\Big\}(g^{(s)}, h^{(s)}). \end{aligned} \tag{40}$$

The main result of the section is the next assertion.

Theorem 13.4.1. *Let $S^{(N)}$ be the partial sum of the series (40). Then*

$$\|u - S^{(N)}; \mathbf{W}_2^m(\Omega_\varepsilon)\| \le c(\sigma)\varepsilon^{N-\sigma}, \tag{41}$$

where u is the solution to problem (1), (2), 13.1, σ is an arbitrary positive number, and the constant $c(\sigma)$ does not depend on $\varepsilon \in (0,1]$.

Proof. The difference $\hat{R}^{(N)} = u - S^{(N)}$ satisfies (19), (20), 13.2. The boundary conditions on $\partial\omega_\varepsilon$ has the form

$$(\partial/\partial\tau)^j \hat{R}^{(N)} = \sum_{k=0}^{N-1} \varepsilon^k (\partial/\partial\tau)^j v^{(k,N-k)}$$
$$+ \sum_{k=0}^{N-1} \varepsilon^k (\partial/\partial\tau)^j \sum_{|\beta|=m-d/2} \varepsilon^{m-d/2}(H_\varepsilon^{(k)} - H^{(k)})_\beta y^\beta.$$

Using the inequalities (32), 13.3, and (28) in the same manner as in the proof of Theorem 13.2.3, we arrive at (41). □

13.4.5 More on the Dirichlet problem for the Laplace operator in the exterior of a tube

We return to problem (1), (2), 12.2. The principal term of the asymptotics for its solution was constructed in Theorem 12.2.5. The form of the asymptotic expansion can be obtained from Theorem 13.4.1. We describe the specific representations of the operators $A, \hat{A}, B, \Pi_p, T_{pj}^{(k)}, R_\varepsilon$ in the formulas (39) for the coefficients of the series (40).

Let $n = 3$, $d = 2$, $\tilde{L} = -\Delta$, $M = \{x \in \mathbb{R}^3 : x_3 = 0, x_1^2 + x_2^2 = 1\}$, $\Omega = \mathbb{R}^3$, and $\omega(z)\{y \in \mathbb{R}^2 : r^2 = y_1^2 + y_2^2 < 1\}$. (Although in the previous case the the domain Ω was supposed to be bounded, this requirement was not at the heart of the matter.) In the case under consideration the operator $A_1 \log\varepsilon + A_2 + \tilde{A}$ takes a quite simple form and the question of its invertibility was completely studied in 12.2.2.

The vector-valued function $\tilde{V}_c$ becomes scalar $\mathbf{c} = c$ and $\tilde{V}_c$ satisfies the problem

$$\begin{aligned} -\Delta\tilde{V}_c(x) &= 0, \quad x \in \mathbb{R}^3 \setminus M;\\ \tilde{V}_c(x) &= -c(z)\,(2\pi)^{-1}\log r + O(1), \quad r = |y| \to 0;\\ \tilde{V}_c(x) &= o(1), \quad |x| \to \infty. \end{aligned} \tag{42}$$

According to (11), 12.2,

$$\tilde{V}_c(x) = (4\pi)^{-1} \int_M |x - t|^{-1} c(t)dt. \tag{43}$$

The second limit problem takes the form

$$-\Delta w(\eta) = h(\eta), \quad \eta \in \mathbb{R}^3 \setminus \bar{\omega}; \quad w(\eta) = \varphi(\eta), \quad \eta \in \partial\omega.$$

The equality

$$\int_{\partial\omega} \varphi(\eta)ds - \int_{\mathbb{R}^3\setminus\omega} h(\eta)\log|\eta|d\eta = 0$$

is the compatibility condition for the problem in the class of functions vanishing at infinity. Taking account of the asymptotic formula

$$\tilde{V}_c(x) = \big(-(2\pi)^{-1}\log r + 3(2\pi)^{-1}\log 2\big)\, c(z) + (4\pi)^{-1}\int_M |t-z|^{-1}\big(c(t)-c(z)\big)ds_t + o(1), \quad r\to 0,$$

we obtain the following equation to find the function c:

$$(-\log\varepsilon + 3\log 2)\, c(z) + 2^{-1}\int_M |t-z|^{-1}\big(c(t)-c(z)\big)ds_t = H(z), \tag{44}$$

where

$$H(z) = -\int_{\partial\omega}\varphi(\eta,z)ds + \int_{\mathbb{R}^3\setminus\omega} h(\eta,z)\log|\eta|d\eta.$$

The asymptotic solution to equation (44) constructed in Lemma 12.2.4 is of the form

$$c(\varepsilon,z) = -\sum_{|p|<\varepsilon^{-1/2}} (\log\varepsilon - 3\log 2 + \lambda_{|p|})^{-1} H_p \exp(ipz), \tag{45}$$

where H_p are the Fourier coefficients of the function H and

$$\lambda_p = \sum_{j=0}^{p-1}(2j+1)^{-1}$$

(obviously, in the case of a circle the constant Ξ is equal to zero). According to Lemma 12.2.4, this solution gives the discrepancy $O(\varepsilon^q)$ for any $q>0$.

In the example under discussion all the operators participating in the construction of the series (7) can be written explicitly.

In a small neighborhood of the circle M we introduce the coordinates $y_1=\zeta$, $y_2=\rho-1$, $z=\vartheta$, where (ζ,ρ,ϑ) are cylindrical coordinates with centre at O and axis orthogonal to M. Then

$$-\tilde{L} = \Delta_y + (1+y_2)^{-1}\partial/\partial y_2 + (1+y_2)^{-2}\partial^2/\partial z^2.$$

Hence,

$$L_0 = -\Delta_y, \quad L_k = -(y_2)^{k-1}\partial/\partial y_2 - (k-1)(-y_2)^{k-2}\partial^2/\partial z^2.$$

The inverse operator A of problem (1)–(3), 13.2, is the convolution with $(4\pi|x|)^{-1}$. The problem (4), (5), (7), 13.3, takes the form

$$-\Delta_\eta w(\eta,z) = h(\eta,z), \quad \eta\in\mathbb{R}^2\setminus\{\eta\colon |\eta|\le 1\};$$
$$w(\eta,z) = \varphi(\eta,z), \quad \eta\in\{\eta\colon |\eta|=1\},$$

and its solution $w=B(h,\varphi)$ is given by

$$w(\eta,z) = (2\pi)^{-1}\int_{|\xi|>1}\log(|\eta-\xi|/|\eta\bar{\xi}-1|)h(\xi,z)d\xi_1 d\xi_2 + (2\pi)^{-1}\int_{|\xi|=1}|\eta-\xi|^{-2}(|\eta|^2-1)\varphi(\xi,z)d\tau$$

with $\eta = \eta_1 + i\eta_2, \xi = \xi_1 + i\xi_2 = \rho \exp(i\tau)$. The compatibility condition of problem (4), (5), (10), 13.3, has the form

$$\tilde{C}(\varphi, h; z) = (2\pi)^{-1} \int_{|\xi|=1} \varphi(\xi, z) d\tau - (2\pi)^{-1} \int_{|\xi|>1} \log |\xi| \, h(\xi, z) \, d\xi_1 \, d\xi_2 = 0.$$

The operator $\hat{A}$ is given by

$$\hat{A}c(x) = (4\pi)^{-1} \int_M |x - t|^{-1} c(t) ds_t$$

(see (43)).

The operator R_ε is defined by the equality $R_\varepsilon H(z) = c(\varepsilon, z)$, where c is the function (45).

The operator $\Pi_k = \Pi_k^{(0)}, k = 0, 1, \ldots$ acts on functions that admit the asymptotic representation

$$v(x) \sim \sum_{j=0}^{\infty} |y|^j \, v_j \, (z, \vartheta, \log |y|), \quad |y| \to 0.$$

It is introduced by

$$(\Pi_k v)(\eta) = |\eta|^k \, v_k(z, \vartheta, \log |\eta| + \log \varepsilon).$$

Similarly, the operator $\tilde{\Pi}_k, k = -1, -2, \ldots$ sends a function w with asymptotic expansion

$$w(\eta, z) \sim \sum_{j=1}^{\infty} |\eta|^{-j} \, w_j \, (z, \vartheta, \log |\eta|), \quad |\eta| \to \infty$$

to the function

$$x \to |y|^k w_{-k}(z, \vartheta, \log |y| - \log \varepsilon).$$

The coefficients $v^{(k)}$ and $w^{(k)}$ of the series (1) for the solution to the problem

$$-\Delta u = f \quad \text{in} \quad \mathbb{R}^3 \setminus \bar{\omega}_\varepsilon; \qquad u = 0 \quad \text{on} \quad \partial \omega_\varepsilon;$$
$$u(\varepsilon, x) = O(|x|^{-1}) \quad \text{as} \quad x \to \infty$$

are given by (5) with $f \in \mathbf{C}_0^\infty(\mathbb{R}^3), R = R_\varepsilon$ and

$$T_{1j}^{(k)} = -\Big\{ [\Delta, \zeta] + \zeta \Big(\Delta + \sum_{s=0}^{k-j} L_s \Big) \Big\} \tilde{\Pi}_{j-k}, \quad T_{2j}^{(k)} = L_{k-j} \Big(\mathbf{1} - \sum_{s=0}^{k-j-1} \tilde{\Pi}_{-s} \Big).$$

The two first terms of the series (7) have a particulary simple form. Namely,

$$v^{(0)}(x) = V^{(0)} + 2^{-1} \int_M R_\varepsilon(V^{(0)}|_M) \, |x - t|^{-1} ds_t, \quad w^{(0)}(\eta) = 0,$$

where $V^{(0)} = (4\pi|x|)^{-1} * f$ with $*$ indicating the convolution operator. In order to find $v^{(1)}$ and $w^{(1)}$ we use (5). In the case under consideration we have $g^{(1)} = 0, h^{(1)} = 0, \Pi_1 \, v^{(0)} = \alpha_1(z) y_1 + \alpha_2(z) y_2$, where

$$\alpha_j(z) = \big((\partial / \partial y_j)(V^{(0)}(x) + R_\varepsilon(V^{(0)}|_M) \log r) \big)|_M.$$

Since $\tilde{C}(\Pi_1 v^{(0)}) = 0$, we obtain $v^{(1)} = 0$, $w^{(1)}(\eta, z) = -(\alpha_1(z)\eta_1 + \alpha_2(z)\eta_2)|\eta|^{-2}$. Thus,

$$u(\varepsilon, x) = v^{(0)}(x) + \varepsilon w^{(1)}(\varepsilon^{-1} y, z) \, \zeta(x) + O(\varepsilon^{2-\delta}), \quad \delta > 0.$$

Part VI

Behaviour of Solutions of Boundary Value Problems in Thin Domains

Chapter 14

The Dirichlet Problem in Domains with Thin Ligaments

We consider the Dirichlet problem for the Laplace equation in a domain G_ε depending on a small parameter ε. It is supposed that a part of the domain is located between two close surfaces which are separated by a distance $O(\varepsilon)$. The purpose is to study the asymptotic behavior of solutions as $\varepsilon \to 0$.

In Section 14.1 we obtain the principal term of the asymptotics in the case that one component of the boundary adheres to another at a point (see Fig. 14.1).

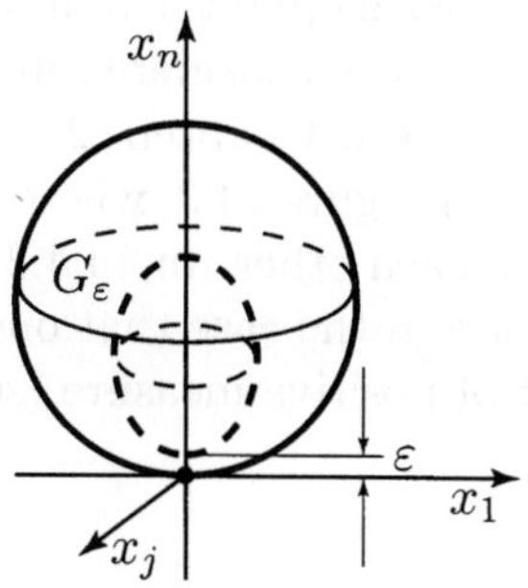

Fig. 14.1

More precisely, in a neighborhood of the origin the domain $G(\varepsilon)$ is given by

$$\gamma_0(y) < x_n < \varepsilon + \gamma_1(y), \tag{A1}$$

where $y = (x_1, \dots, x_{n-1})$ and γ_0, γ_1 are smooth functions vanishing at $y = 0$ while $c_1|y|^m \geq \gamma_1(y) - \gamma_0(y) \geq c_2|y|^m$, $c_j = \text{const} > 0$, with $m = 2, 3, \dots$. Then far from the origin the asymptotic behavior of the solution $u(\varepsilon, \cdot)$ is the same as that of the solution v to the limit problem at $\varepsilon = 0$. Near the origin $u(\varepsilon, x) \sim V(y, t(\varepsilon, x))$, where $t(\varepsilon, x) = (x_n - \gamma_0(y))(\varepsilon + \gamma_1(y) - \gamma_0(y))^{-1}$ and $V(y, t(0, x)) = v(x)$.

In Section 14.2 we construct the complete asymptotic expansion in the domain given by (A1),

$$u(\varepsilon, x) \sim \sum_{k=0}^{\infty} \varepsilon^k v_k(y, t(\varepsilon, x)). \tag{A2}$$

Here v_k are smooth functions that can be found by means of the mentioned limit problem. The expansion of type (A2) is obtained also in the domain G_ε pictured in Figure 14.2.

A more complicated situation is considered in Section 14.3. Here we assume that the derivatives of Dirichlet's data may have power singularities at $y = 0$. Then for the solution in the domain (A1) we obtain an expansion of the form

$$u(\varepsilon, x) \sim \sum_{k=0}^{\infty} \varepsilon^k \big(v_k(y, t(\varepsilon, x)) + \varepsilon^2 w_k(\varepsilon^{-1} y, t(\varepsilon, x))\big).$$

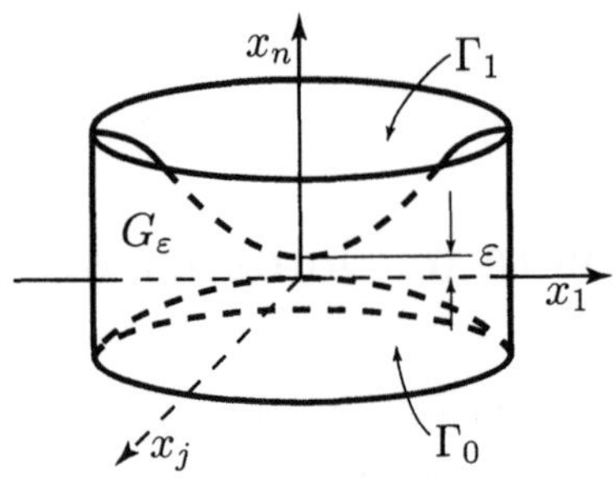

Fig. 14.2

Here v_k are solutions to the first limit problem in the domain G_0 whose derivatives may have singularities at the point 0, and w_k satisfy the second limit problem, i.e., the Dirichlet problem for the Laplace operator in a layer of thickness 1.

Two other types of ligaments are considered in Section 14.4. In 14.4.1, using the algorithms in Chapter 4 (see also Chapters 2 and 6), we study the Dirichlet problem in the domain pictured in Figure 14.3, where the ligament is formed by two conical or corner points close to each other. In 14.4.2 we obtain the principal term of the asymptotics of the solution in the case that one component of the boundary adheres to another at a subset of positive measure (see Fig. 14.4).

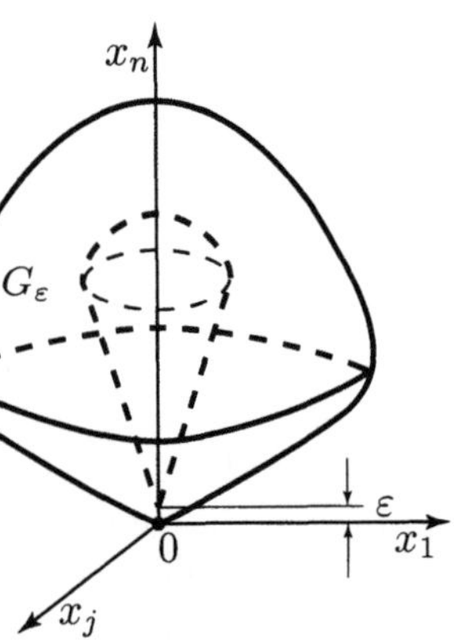

Fig. 14.3

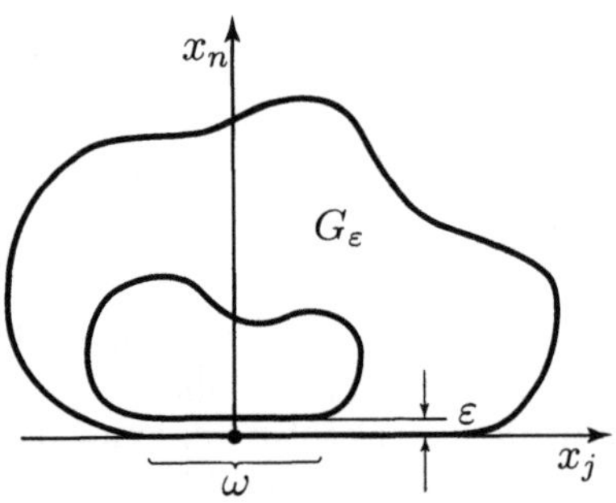

Fig. 14.4

In Section 14.5, as a consequence of the results in 14.1 and 14.2, we derive the asymptotic formulas for the electrostatic capacity of a condenser whose plates are brought into proximity. If in an n-dimensional neighborhood of a domain $\omega \subset \mathbb{R}^{n-1}$ a condenser G_ε is given by

$$0 < x_n < \varepsilon + \alpha \operatorname{dist}(y, \omega)^m, \quad m = 2, 3, \dots,$$

(see Fig. 14.4) then

$$\operatorname{cap} G_\varepsilon = (\varepsilon \mu_n)^{-1}(\operatorname{mes}_{n-1}(\omega) + (\varepsilon/\alpha)^{1/m}(m \sin(\pi/m))^{-1}\pi \operatorname{mes}_{n-2}(\partial\omega) + o(\varepsilon^{1/m})).$$

Here $\mu_n = (n-2)\sigma_{n-1}$ and σ_{n-1} is the measure of the $(n-1)$-dimensional unit sphere while $\sigma_0 = 2$ and $\mu_2 = 2\pi$. Furthermore, an expansion of the form (A2) leads to the complete asymptotic expansion of the capacity of the condenser pictured in Figure 14.1. If near the point 0 the domain G_ε is given by the inequality

$0 < x_n < \alpha|y|^m + \varepsilon$, $m = 2, 4, \dots$, then

$$\operatorname{cap} G_\varepsilon \sim \varepsilon^{(n-1)/m-1} \sum_{q=0}^{\infty} a_q \varepsilon^{q/m} + \sum_{j=0}^{\infty} c_j \varepsilon^j + \varepsilon^{-[(1-n)/m]-1} |\log \varepsilon| \sum_{q=0}^{\infty} b_q \varepsilon^{q/m}. \quad \text{(A3)}$$

The first term in (A3) has the form

$$\begin{cases} (\alpha/\varepsilon)^{1-(n-1)/m} \pi \sigma_{n-2} \{\mu_n \alpha \sin((n-1)\pi/m)\}^{-1} & \text{for } m > n-1, \\ \sigma_{n-2}(\alpha\mu_n)^{-1} |\log \varepsilon| & \text{for } m = n-1, \\ \operatorname{cap} G_0 & \text{for } m < n-1. \end{cases}$$

14.1 The Principal Term in the Asymptotics of Solution

14.1.1 Statement of the problem (the case of two points on smooth surfaces approaching each other)

Let G_ε be a subdomain in $\mathbb{R}^n$ whose boundary is the union of two smooth closed surfaces $\Gamma_\varepsilon^{(0)}$ and $\Gamma_\varepsilon^{(1)}$ so that $\Gamma_\varepsilon^{(0)} \cap \Gamma_\varepsilon^{(1)} = \emptyset$ (see Fig. 14.1). We suppose that $\Gamma_\varepsilon^{(j)}$ smoothly depend on a small positive parameter ε and there exist[1] smooth limiting surfaces $\Gamma_0^{(j)}$ having only one common point 0 (the origin).

Let G_0 stand for the domain bounded by $\Gamma_0^{(0)}$ and $\Gamma_0^{(1)}$. We introduce the ball $D = \{y = (x_1, \dots, x_{n-1})\colon |y| < \alpha_1\}$ and cylinder $V = D \times (-\alpha_1, \alpha_2)$ with positive constants α_j. Suppose that

$$G_\varepsilon \cap V = \{x\colon \gamma_0(y) < x_n < \gamma_1(y) + \varepsilon, y \in D\}, \quad (1)$$

where $\gamma_j \in \mathbf{C}^\infty(\bar{D})$. We set $\delta = \gamma_1 - \gamma_0$ and denote by the same letter a smooth positive continuation of the function $x \to \delta(y)$ from $\bar{V}$ to $\mathbb{R}^n$. Assume that

$$c_1|y|^m \le \delta(y) \le c_2|y|^m, \quad y \in D, \quad (2)$$

where c_j are positive constants and m is an even positive integer. We consider the boundary value problem

$$\Delta u(\varepsilon, x) = 0, \qquad x \in G_\varepsilon; \quad (3)$$

$$u(\varepsilon, x) = \Phi^{(j)}(x), \quad x \in \Gamma_\varepsilon^{(j)}, \quad j = 0, 1, \quad (4)$$

with $\Phi^{(j)} \in \mathbf{C}^\infty(\mathbb{R}^n)$.

14.1.2 The limit boundary value problem

Denote by v the bounded solution to the limit problem

$$\Delta v(x) = 0, \qquad x \in G_\varepsilon; \quad (5)$$

$$v(x) = \Phi^{(j)}(x), \quad x \in \Gamma_0^{(j)} \setminus \{0\}, \quad j = 0, 1. \quad (6)$$

---

[1]This means that there exists a family of diffeomorphisms $X_\varepsilon^{(j)}$ of a neighborhood of $\Gamma_0^{(j)}$ onto itself sending $\Gamma_0^{(j)}$ into $\Gamma_\varepsilon^{(j)}$ while $X_0^{(j)}$ is the identity mapping.

Lemma 14.1.1. *(i) Let $f \in \mathbf{C}^\infty(\bar{G}_0) \setminus \{0\})$ and let the asymptotic formula*

$$f(x) = \delta(y)^{-2}\Big(\sum_{k=0}^{N} f_k(y, t(x)) + R_N(y, t(x))\Big) \tag{7}$$

hold in the neighborhood V, where $N = 0, 1, \dots$, $t(x) = \delta(y)^{-1}(x_n - \gamma_0(y))$, and f_k are positively homogeneous functions in y of degree k that belong to $\mathbf{C}^\infty((\mathbb{R}^{n-1} \setminus \{0\}) \times [0,1])$. The remainder R_N satisfies

$$(\partial/\partial t)^j D_y^\beta R_N(y,t) = O(|y|^{N+1-|\beta|}), \quad |y| \to 0, \tag{8}$$

for all $j = 0, 1, 2, \dots$ and any multiindex β.

Then for the bounded solution to the equation

$$\Delta v(x) = f(x), \quad x \in G_0, \tag{9}$$

with the boundary conditions (6) there holds the representation

$$v(x) = \sum_{k=0}^{N} v_k(y, t(x)) + r_N(y, t(x)). \tag{10}$$

Here v_k and r_N are the same as f_k and R_N, respectively, in (7).

(ii) Under the supplementary condition that in (7) f_k are homogeneous polynomials in y of degree k the functions v_k turn out to be homogeneous polynomials in y of degree k.

Proof. We make the change of variables $x \to (y,t)$ in the neighborhood V, where

$$y = (x_1, \dots, x_{n-1}), \quad t = \delta(y)^{-1}(x_n - \gamma_0(y)). \tag{11}$$

Then $G_0 \cap V$ is mapped into the subdomain $W = D \times (0,1)$ of the layer $\Pi = \{(y,t) \colon y \in \mathbb{R}^{n-1}, t \in (0,1)\}$. The boundary value problem (9), (6) transforms into

$$\begin{aligned} &L(y, t, \nabla_y, \partial/\partial t)\, u(y,t) \\ &\quad = \{\delta(y)^{-2}(\partial/\partial t)^2 + (\nabla_y - \delta(y)^{-1}(t\nabla_y\delta(y) + \nabla_y\gamma_0(y)\partial/\partial t)^2\} u(y,t) \\ &\quad = \delta(y)^{-2}\Big(\sum_{k=0}^{N} f_k(y,t) + R_N(y,t)\Big), \quad (y,t) \in W; \\ &\quad v(y,0) = \Phi_0(y), \quad v(y,1) = \Phi_1(y), \quad y \in D. \end{aligned} \tag{12}$$

We represent the operator L in the form

$$L = \delta^{-2} M_2 + \delta^{-1} M_1 + M_0,$$

where

$$\begin{aligned} M_2 &= (\partial/\partial t)^2 + (t\nabla_y\delta + \nabla_y\gamma_0)^2(\partial/\partial t)^2 + 2\nabla_y\delta \cdot (t\nabla_y\delta + \nabla_y\gamma_0)\partial/\partial t, \\ M_1 &= -2(t\nabla_y\delta + \nabla_y\gamma_0) \cdot \nabla_y \partial/\partial t + (t\Delta_y\delta + \Delta_y\gamma_0)\partial/\partial t, \\ M_0 &= \Delta_y. \end{aligned} \tag{13}$$

Since δ and γ_0 are smooth functions, the operator L admits the formal expansion

$$L(y, t, \nabla_y, \partial/\partial t) \sim \delta^{-2} \sum_{k=0}^{\infty} L^{(k)}(y, t, \nabla_y, \partial/\partial t). \tag{14}$$

Here $L^{(0)} = (1+|\nabla_y\gamma_0(0)|^2)+(\partial/\partial t)^2$ and $L^{(k)}$ are operators of order no greater than 2 whose coefficients are polynomials in t (of degree no greater than 2) and in y (of degree no greater than k) possessing the following property:

$$L^{(k)}(sy,t,s^{-1}\nabla_y,\partial/\partial t) = s^k L^{(k)}(y,t,\nabla_y,\partial/\partial t), \quad s>0.$$

Substituting (7), (10), and (14) into problem (12) and equating the terms of the same degree in y, we obtain the following equations for v_k:

$$\begin{aligned}
&(1+|\nabla_y\gamma_0(0)|^2)(\partial/\partial t)^2 v_k(y,t)\\
&= f_k(y,t) - \sum_{j=1}^{k} L^{(j)}(y,t,\nabla_y,\partial/\partial t)v_{k-j}(y,t), \quad t\in[0,1];\\
&v_k(y,t) = \sum_{|\beta|=k}(\beta!)^{-1}y^\beta(\nabla_y^\beta\Phi_t)(0), \quad t=0,1.
\end{aligned} \tag{15}$$

It is clear that there exists a solution to problem (15) that is a homogeneous function in y of degree k smoothly depending on t. Moreover, if f_k are polynomials in y then v_k, too, are polynomials.

We set $\tilde V = v - \eta(v_1+\cdots+v_N)$, where η is a function in $\mathbf{C}_0^\infty(V)$ equal to zero near the point O. We have

$$\Delta\tilde V = f - [\Delta,\eta]\sum_{k=0}^{\infty} v_k - \eta\sum_{k=0}^{\infty}\Delta v_k. \tag{16}$$

In the coordinates (y,t) the last sum in (16) takes the form

$$\begin{aligned}
\sum_{k=0}^{N} Lv_k &= \sum_{k=0}^{N}\Big(L-\delta^{-2}\sum_{j=0}^{N-k}L^{(j)}\Big)v_k + \delta^{-2}\sum_{p=0}^{N}\sum_{q=0}^{p}L^{(p-q)}v_q\\
&= \sum_{k=0}^{N}\Big(L-\delta^{-2}\sum_{j=0}^{N}L^{(j)}\Big)v_k + \delta^{-2}\sum_{p=0}^{N} f_p.
\end{aligned}$$

Thus, the function $\tilde V$ satisfies

$$\Delta\tilde V = \Psi \quad \text{on} \quad G_0; \quad \tilde V = \varphi_j \quad \text{on} \quad \Gamma_0^{(j)}, \quad j=0,1, \tag{17}$$

where Ψ and φ_j are in $\mathbf{C}^\infty(\bar G_0\setminus\{O\})$ and $\mathbf{C}^\infty(\Gamma_0^{(j)})$, respectively, while in the neighborhood V

$$\begin{aligned}
(\partial/\partial t)^j D_y^\beta(\delta^2\Psi)(y,t) &= O(|y|^{N+1-|\beta|}), \quad |y|\to 0,\\
D_y^\beta\varphi_j(y) &= O(|y|^{N+1-|\beta|}), \quad |y|\to 0, \quad j=0,1.
\end{aligned} \tag{18}$$

Now we estimate the function $\tilde V$. Without loss of generality, one can assume that $\varphi_j = 0$. Let $q_\nu(x) = |x|$ for $|x|\in(\nu,\sigma)$, $q_\nu(x)=\nu$ for $|x|<\nu$, and $q_\nu(x)=\sigma$ for $|x|>\sigma$, where $\nu<\sigma$ and σ is a small number. We multiply the equation in (17) by $q^\mu\bar{\tilde V}$ with $\mu\in\mathbb{R}$. We have

$$\int_{G_0} q_\nu^\mu|\nabla\tilde V|^2\,dx = -\int_{G_0} q_\nu^\mu\Psi\bar{\tilde V}\,dx - \mu\int_{G_0} q_\nu^{\mu-1}\bar{\tilde V}\nabla q_\nu\cdot\nabla V\,dx.$$

It follows that

$$\int_{G_0} q_\nu^\mu |\nabla\tilde{V}|^2 dx \le c\Big(\int_{G_0} q_\nu^{\mu-2}|\nabla q_\nu|^2|\tilde{V}|^2 dx + \Big(\int_{G_0} q_\nu^\mu \delta^2|\Psi|^2 dx\Big)^{1/2}\Big(\int_{G_0} q_\nu^\mu\delta^{-2}|\tilde{V}|^2 dx\Big)^{1/2}\Big).$$

Since for small $|y|$

$$\pi^2\,\delta^{-2}\int_{\gamma_0}^{\gamma_0+\delta}|\tilde{V}(y,z)|^2 dz \le \int_{\gamma_0}^{\gamma_0+\delta}|(\partial/\partial z)\tilde{V}(y,z)|^2 dz,$$

we obtain

$$\int_{G_0} q_\nu^\mu\,\delta^{-2}\,|\tilde{V}|^2\,dx \le c\int_{G_0} q_\nu^\mu\,|\nabla\tilde{V}|^2\,dx,$$

$$\int_{G_0} q_\nu^{\mu-2}\,|\nabla q_\nu|^2\,|\tilde{V}|^2\,dx \le c\max_{\bar{G}_0}(|\nabla q_\nu|^2 q_\nu^{-2}\delta^2)\int_{G_0} q_\nu^\mu\,|\nabla\tilde{V}|^2\,dx.$$

The coefficient of the last integral is not greater than

$$\max_{|x|<\sigma}\{\delta(y)^2\,|x|^{-2}\} = O(\sigma^2).$$

Therefore,

$$\int_{G_0} q_\nu^\mu\,|\nabla\tilde{V}|^2\,dx + \int_{G_0} q_\nu^\mu\,\delta^{-2}\,|\tilde{V}|^2\,dx \le c\int_{G_0} q_\nu^\mu\,\delta^2\,|\Psi|^2\,dx. \tag{19}$$

Using (18) and the condition $\delta(y) \le c|y|^m$, we conclude that the right-hand side in (19) is bounded for $\mu > m - 2N - n - 1$ uniformly with respect to ν. Hence,

$$\int_{G_0} |x|^{\mu-2m}\,|\tilde{V}|^2\,dx < \infty.$$

Let $x \in G_0 \cap V$, $|x| \ne 0$. Denote by $B(x)$ the ball with radius $|y|^m$ and center x. From the known local estimate on solutions to the Dirichlet problem it follows that

$$|D_x^\beta\,\tilde{V}(x)| \le c\,|y|^{m(1-|\beta|+n/2)-\mu/2}\Big(\int_{B(x)\cap G_0}|x|^{\mu-2m}\,|\tilde{V}|^2\,dx\Big)^{1/2} + c\sum_{|\alpha|\le|\beta|}|y|^{m|\alpha|-|\beta|+2}\sup_{B(x)\cap G_0}|D_x^\alpha\,\Psi|.$$

This implies the inequality

$$|D_x^\beta\,\tilde{V}(x)| \le c\,|y|^{K(\beta,N)},$$

where $K(\beta,N) \to +\infty$ for β fixed and $N \to \infty$. □

Corollary 14.1.2. *Assume that the inequalities* (2) *hold and*

$$|\nabla_y \delta(y)| + |\nabla_y \gamma_0(y)| \le c\,|y|^{l-1} \tag{20}$$

with $l = 1, \dots, m$. *Then for the solution* v *to problem* (5), (6) *there is the asymptotic representation*

$$v(x) = \Phi_1(y) + \delta(y)^{-1}\big(x_n - \gamma_0\big)\big(\Phi_1(y) - \Phi_0(y)\big) + r_{m+l-2}\Big(y, \delta(y)^{-1}\big(x_n - \gamma_0(y)\big)\Big). \tag{21}$$

Proof. From (13) and (14), taking (2) and (20) into account, we obtain

$$L^{(k)}(y, t, \nabla_y, \partial/\partial t) = \alpha_k(y,t)(\partial/\partial t)^2, \quad \text{for} \quad k = 1, \dots, m+l-2.$$

Therefore, solving problem (15) for $k = 0, \dots, m+l-2$, we have

$$v_k(y,t) = \sum_{|\beta|=k} (\beta!)^{-1} y^\beta \{\nabla_y^\beta \Phi_0(0) + t((\nabla_y^\beta \Phi_1)(0) - (\nabla_y^\beta \Phi_0)(0))\},$$

so (21) is proven. □

14.1.3 Asymptotics of solutions to the original problem

We introduce the mapping $\sigma_\varepsilon\colon \bar V \cap \bar G_\varepsilon \to \bar W$, where

$$y = (x_1, \dots, x_{n-1}), \quad t = \big(x_n - \gamma_0(y)\big)\big(\delta(y) + \varepsilon\big)^{-1}. \tag{22}$$

In particular, σ_0 is the mapping defined by (11). At a positive distance from the point O the diffeomorphism

$$\varkappa_\varepsilon = \sigma_0^{-1} \circ \sigma_\varepsilon\colon \bar V \cap \bar G_\varepsilon \to \bar V \cap \bar G_0 \tag{23}$$

smoothly depends on $\varepsilon \in [0,1]$ and $\varkappa_0$ is the identity mapping. Therefore, $\varkappa_\varepsilon$ extends to a diffeomorphism $G_\varepsilon \to G_0$ possessing the same properties. We fix such an extension and denote it by the same symbol $\varkappa_\varepsilon$. Notice that on $\bar V \cap \bar G_\varepsilon$

$$\varkappa_\varepsilon(x) = \Big(y, \big(\varepsilon + \delta(y)\big)^{-1}\big(\delta(y) x_n + \varepsilon\gamma_0(y)\big)\Big).$$

Theorem 14.1.3. *Let the functions* δ *and* γ_0 *in* (1) *satisfy the inequalities* (2) *and* (20). *Then for the solution to problem* (3), (4) *there is the asymptotic formula*

$$u(\varepsilon, x) \sim \begin{cases} v(x), & x \in \bar G_0 \setminus V, \\ \tilde V\Big(y, (\delta(y)+\varepsilon)^{-1}(x_n - \gamma_0(y))\Big), & x \in \bar G_0 \cap \bar V. \end{cases} \tag{24}$$

Here v *is the bounded solution to problem* (5), (6) *extended from* $\bar G_0 \setminus V$ *to* $\mathbb{R}^n$ *with the same smoothness;* $\tilde V = v \circ \sigma_0^{-1}$ *is the function* v *written on the set* $G_0 \cap V$ *in the coordinates* (11); $a \sim b$ *means that* $a - b = O(\varepsilon)$ *as* $\varepsilon \to 0$.

Proof. We introduce the functions

$$v_\varepsilon(x) = v(\varkappa_\varepsilon(x)), \quad R(\varepsilon, x) = u(\varepsilon, x) - v_\varepsilon(x).$$

Clearly, to verify (24) it suffices to prove the relation

$$R(\varepsilon, x) = O(\varepsilon). \tag{25}$$

The function R satisfies the boundary value problem

$$\Delta R(\varepsilon, x) = F_\varepsilon(x), \quad x \in G_\varepsilon; \tag{26}$$

$$R(\varepsilon, x) = \varphi_\varepsilon^{(j)}(x), \quad x \in \Gamma_\varepsilon^{(j)}, \quad j = 0, 1. \tag{27}$$

Since the diffeomorphism $\varkappa_0$ is the identity mapping at a positive distance from 0, there hold the relations

$$F_\varepsilon(x) = O(\varepsilon), \quad x \in \bar{G}_\varepsilon \setminus V, \tag{28}$$

$$\varphi_\varepsilon^{(j)}(x) = O(\varepsilon), \quad x \in \Gamma_\varepsilon^{(j)}, \quad j = 0, 1, \tag{29}$$

while one may differentiate them. Moreover,

$$F_\varepsilon = \varepsilon(\delta+\varepsilon)^{-2}\left(M_1(v \circ \sigma_0^{-1})\right) \circ \sigma_\varepsilon + \varepsilon(\delta+2\varepsilon)(\delta+\varepsilon)^{-2}\left(M_0(v \circ \sigma_0^{-1})\right) \circ \sigma_\varepsilon. \tag{30}$$

From (13) it follows that

$$|F_\varepsilon(x)| \le c\,\varepsilon\big(\varepsilon + \delta(y)\big)^{-2}, \quad x \in G_0 \cap V.$$

Let a function $R^{(1)}$ satisfy equation (26) and the boundary conditions

$$R^{(1)}(\varepsilon, x) = 0 \quad x \in \Gamma_\varepsilon^{(0)} \cup \Gamma_\varepsilon^{(1)},$$

and let a function $R^{(2)}$ be harmonic and satisfy (27). From the maximum principle and (29) it follows that

$$|R^{(2)}(\varepsilon, x)| \le c\,\varepsilon \quad x \in \bar{G}_\varepsilon.$$

It remains to verify the similar estimate on $R^{(1)}$. We multiply the equation $\Delta R^{(1)} = F_\varepsilon$ by $(\delta_\sigma + \varepsilon)^2 R^{(1)}$ and integrate over $G_\varepsilon \cap V$. We have

$$\int_{G_\varepsilon} (\delta_\sigma + \varepsilon)^2 |\nabla R^{(1)}|^2 dx + \int_{G_\varepsilon} |R^{(1)}|^2 dx \le c \int_{G_\varepsilon} (\delta_\sigma + \varepsilon)^4 |F_\varepsilon|^2 dx,$$

where $\delta_\sigma(\tau) = \min\{\delta(x), \sigma\}$. Therefore, (28) and (30) imply that

$$\int_{G_\varepsilon} |R^{(1)}|^2 dx \le c\,\varepsilon.$$

From this and the known local estimate of solutions to the Poisson equation it follows that for $|x| \ge \sigma_1$

$$|R^{(1)}(\varepsilon, x)| \le c\,\varepsilon. \tag{31}$$

Now we estimate $R^{(1)}$ in the domain $G_\varepsilon \cap V_\sigma$, where $V_\sigma = \{x \in V\colon |x| < \sigma\}$. Introduce in $G_\varepsilon \cap V_\sigma$ the function

$$B(\varepsilon, x) = \big(\delta(y) + \varepsilon\big)^{-2}(x_n - \gamma_0(y))(x_n - \gamma_0(y) - \delta(y) - \varepsilon).$$

Since

$$\Delta B(\varepsilon, x) = \big(\delta(y) + \varepsilon\big)^{-2}\{2(1 + |\nabla\gamma_0(0)|^2) + O(|y|)\}, \quad |y| \to 0,$$

for sufficiently small σ there exist constants $c_+ > 0$ and $c_- < 0$ such that

$$\Delta\big(\varepsilon c_+ B(\varepsilon, x) - R^{(1)}(\varepsilon, x)\big) \ge 0$$

and

$$\Delta\big(\varepsilon c_- B(\varepsilon, x) - R^{(1)}(\varepsilon, x)\big) \le 0, \quad x \in \bar{G}_\varepsilon \cap \bar{V}_\sigma.$$

Moreover, on $\partial V_\sigma \cap G_\varepsilon$ we have

$$\varepsilon c_\pm B(\varepsilon, x) - R^{(1)}(\varepsilon, x) = O(\varepsilon).$$

Thus, the inequality (31) holds everywhere in $\bar{G}_\varepsilon$. □

The next assertion follows from (21) and (24).

Corollary 14.1.4. *Under the hypotheses of Theorem* 14.1.3 *the solution* $u(\varepsilon, x)$ *to problem* (3), (4) *admits the asymptotic representation*

$$v(\varepsilon,x) = \Phi_0(y) + \big(x_n - \gamma_0(y)\big)(\delta(y)+\varepsilon)^{-1}(\Phi_1(y) - \Phi_0(y)) + O(\varepsilon) \quad \text{for} \quad |y| < c\,\varepsilon^{1/(l+m-2)}.$$

14.2 Complete Asymptotic Expansions of Solutions

14.2.1 The case of a quasicylindrical domain

Let Ω be a subdomain in $\mathbb{R}^{n-1}$ with smooth boundary $\partial\Omega$ and compact closure $\bar{\Omega}$ containing the origin O. We introduce smooth functions γ_0 and γ_1 in $\bar{\Omega}$ such that

$$\gamma_0(y) < \gamma_1(y) \quad \text{for} \quad y \in \bar{\Omega} \setminus \{O\}, \quad \gamma_0(0) = \gamma_1(0).$$

Suppose that

$$\nabla_y \gamma_0(0) = 0; \quad \gamma_j(y) = \gamma_j^0 \quad \text{for} \quad y \in V_1 \cap \bar{\Omega},\ j = 0, 1, \tag{1}$$

where V_1 is a neighborhood of $\partial\Omega$ and γ_j^0 are constants. (This requirement is introduced only for the sake of simplicity of the description.) We also assume that the inequalities (2), 14.1, hold with $\delta = \gamma_1 - \gamma_0$ and $m = 2, 4, \dots$. Denote by G_ε the domain $\{x \in \mathbb{R}^n : y = (x_1, \dots, x_{n-1}) \in \Omega,\ \gamma_0(y) < x_n < \gamma_1(y) + \varepsilon\}$ and by S_ε the set $\{x \in \mathbb{R}^n : y \in \partial\Omega,\ \gamma_0(y) < x_n < \gamma_1^0 + \varepsilon\}$ (see Fig. 14.2). We consider the boundary value problem

$$\Delta u(\varepsilon,x) = f(\varepsilon,x), \qquad x \in G_\varepsilon; \tag{2}$$

$$u(\varepsilon,x) = \Phi^{(j)}(\varepsilon,x), \quad y \in \Omega, \quad x_n = \gamma_j(y) + \delta_{j1}\varepsilon, \quad j = 0, 1; \tag{3}$$

$$u(\varepsilon,x) = 0, \qquad x \in S_\varepsilon. \tag{4}$$

The functions $\Phi^{(j)}$ belong to $\mathbf{C}_0^\infty(\bar{\Omega})$ and smoothly depend on $\varepsilon \in [0,1]$. The function f admits the representation

$$f(\varepsilon,x) = \big(\delta(y)+\varepsilon\big)^{-2}\Big(\sum_{k=0}^{M} \varepsilon^k f^{(k)}(y,t) + Q^{(M)}(\varepsilon,y,t)\Big), \tag{5}$$

where M is an arbitrary positive integer, $t = \big(x_n - \gamma_0(y)\big)(\delta(y)+\varepsilon)^{-1}$, and $f^{(k)}$ satisfies

$$f^{(k)}(y,t) = \sum_{p=0}^{N} f_p^{(k)}(y,t) + R^{(N)}(y,t), \tag{6}$$

N being an arbitrary integer, $f_p^{(k)}(y,t)$ polynomials in y of degree p whose coefficients are smooth functions in $t \in [0,1]$. For the remainders $Q^{(M)}$ and $R^{(N)}$ in (5) and (6) for any multiindex $\beta = (\beta_1, \dots, \beta_{n-1})$ and nonnegative integer $k \in \mathbb{N}_0$ there hold the inequalities

$$\begin{aligned} |D_y^\beta (\partial/\partial t)^k Q^{(M)}(\varepsilon,y,t)| &\le c\varepsilon^{M+1}, \\ |D_y^\beta (\partial/\partial t)^k R^{(N)}(y,t)| &\le c|y|^{N+1-|\beta|}. \end{aligned} \tag{7}$$

Obviously, any function in $\mathbf{C}^\infty(\mathbb{R}^n)$ independent of ε can be represented in the domain G_ε in the form (5), (6).

We make the change of variables $x \to (y,t)$ in G_ε, where

$$y = (x_1, \dots, x_{n-1}), \quad t = (\delta(y)+\varepsilon)^{-1}(x_n - \gamma_0(y)). \tag{8}$$

The mapping (8) sends G_ε into the cylinder $\tilde{C} = \{(y,t)\colon y \in \Omega, t \in (0,1)\}$ whose lateral face will be denoted by S. The boundary value problem (2)–(4) takes the form

$$L(\varepsilon, y, t, \nabla_y, \partial/\partial t)\, u(\varepsilon, y, t) = f(\varepsilon, y, t), \quad (y,t) \in \tilde{C}; \tag{9}$$

$$u(\varepsilon, y, 0) = \Phi^{(0)}(\varepsilon, y), \quad u(\varepsilon, y, 1) = \Phi^{(1)}(\varepsilon, y), \quad y \in \Omega; \tag{10}$$

$$u(\varepsilon, y, t) = 0, \quad (y,t) \in S, \tag{11}$$

where

$$\begin{aligned} &L(\varepsilon, y, t, \nabla_y, \partial/\partial t) \\ &= \big(\delta(y)+\varepsilon\big)^{-2} M_2(y,t,\nabla_y,\partial/\partial t) + \big(\delta(y)+\varepsilon\big)^{-1} M_1(y,t,\nabla_y,\partial/\partial t) + M_0(\nabla_y), \end{aligned} \tag{12}$$

M_j being the operators (13), 14.1. Since $\Phi^{(j)}$ are smooth functions, there is the inequality

$$\Big|\nabla_y^q\Big(\Phi^{(j)}(\varepsilon, y) - \sum_{p=0}^{N} \varepsilon^p \varphi_p^{(j)}(y)\Big)\Big| \le c\varepsilon^{N+1}, \quad y \in \bar{\Omega}, \quad j = 0,1, \tag{13}$$

where N, q, k are nonnegative integers and $\varphi_p^{(j)} \in \mathbf{C}_0^\infty(\Omega)$. We seek the asymptotic solution to problem (2)–(4) (or to problem (9)–(11)) in the form

$$\sum_{k=0}^{\infty} \varepsilon^k u^{(k)}(y,t), \tag{14}$$

where the functions $u^{(k)}$ belong to $\mathbf{C}^{1,\alpha}(\overline{\tilde{C}})$, $0 < \alpha < 1$, and satisfy the boundary value problem

$$\begin{aligned} &L(0, y, t, \nabla_y, \partial/\partial t)\, u^{(k)}(y,t) = \delta(y)^{-2} F_k(y,t), \quad (y,t) \in \tilde{C}; \\ &u^{(k)}(y,0) = \Psi_k^{(0)}(y), \quad u(y,1) = \Psi_k^{(1)}(y), \quad y \in \Omega; \\ &u^{(k)}(y,t) = 0, \quad (y,t) \in S, \end{aligned} \tag{15}$$

with some right-hand sides $F^{(k)}$ and $\Psi_p^{(k)}$. (The existence of such solutions follows from results in MAZ'YA and PLAMENEVSKI [6].)

Note that after the change of variables inverse to (8) for $\varepsilon = 0$ the problem (15) takes the form

$$\begin{aligned} \Delta u^{(k)} &= \delta^{-2} F^{(k)} \quad \text{in} \quad G_0; \\ u^{(k)} &= \Psi_k^{(j)} \quad \text{for} \quad y \in \Omega, \quad x_n = \gamma_j(y), \quad j = 0,1; \\ u^{(k)} &= 0 \quad \text{on} \quad S. \end{aligned} \tag{16}$$

Let us find the right-hand sides $F^{(k)}$ and $\Psi_k^{(j)}$ in (15). To this end we substitute the expansions (13), (14), (5) and (12) into problem (9)–(11). Taking account of the formula

$$\begin{aligned} L(\varepsilon, y, t, \nabla_y, \partial/\partial t) &= \big(\delta(y)+\varepsilon\big)^{-2}\{\delta(y)^2 L(0, y, t, \nabla_y, \partial/\partial t) \\ &\quad + \varepsilon\big(M_1(y,t,\nabla_y,\partial/\partial t) + 2\delta(y) M_0(\nabla_y)\big) + \varepsilon^2 M_0(\nabla_y)\} \end{aligned}$$

and equating the coefficients at the same powers of ε, we obtain for $k = 0, 1, \dots$

$$\begin{aligned} F^{(k)} &= f^{(k)} - (M_1 + 2\delta M_0)u^{(k-1)} - M_0 u^{(k-2)}, \\ \Psi_k^{(j)} &= \varphi_k^{(j)}, \quad j = 0, 1, \quad k \in \mathbb{N}_0. \end{aligned} \tag{17}$$

Here $u^{(-1)}$ and $u^{(-2)}$ equal zero by definition.

Since the functions $f^{(k)}$ admit the expansions (6), Lemma 14.1.1, ii) leads to following assertion.

Lemma 14.2.1. *For the solution $u^{(k)}$ to problem* (15) *(or to problem* (16)*) with right-hand sides* (17) *there holds the representation*

$$u^{(k)}(y,t) = \sum_{p=0}^{M} u_p^{(k)}(y,t) + r_M^{(k)}(y,t),$$

where $u_p^{(k)}$ are polynomials in y of degree p whose coefficients smoothly depend on $t \in [0,1]$, M being an arbitrary positive integer. At a positive distance from S the estimate

$$|D_y^\beta (\partial/\partial t)^j r_M^{(k)}(y,t)| \le c_M |y|^{M+1-|\beta|}$$

is valid with any multiindex β, $k = 0, 1, \dots$.

Theorem 14.2.2. *Let the right-hand sides of problem* (2)–(4) *satisfy* (5)–(7) *and* (13)*. Then the function*

$$\tilde{S}^{(N)}(\varepsilon, x) = \sum_{k=0}^{N} \varepsilon^k u^{(k)}\Big(y, \big(\delta(y) + \varepsilon\big)^{-1}\big(x_n - \gamma_0(y)\big)\Big)$$

is subject to the inequality

$$|u(\varepsilon,x) - \tilde{S}^{(N)}(\varepsilon,x)| \le c_N \varepsilon^{N+1}. \tag{18}$$

Here $u^{(k)}$ are the bounded solutions to problem (15) *with the right-hand sides* (17) *and u is the solution to problem* (2)–(4)*. The constant c_N does not depend on ε.*

Proof. The function $\tilde{R}^{(N)} = u - \tilde{S}^{(N)}$ satisfies the problem

$$\begin{aligned} \Delta \tilde{R}^{(N)}(\varepsilon,x) &= \tilde{F}^{(N)}(\varepsilon,x), \quad x \in G_\varepsilon; \\ \tilde{R}^{(N)}(\varepsilon,x) &= \Xi_j^{(N)}(\varepsilon,y), \quad y \in \Omega, \quad x_n = \gamma_1(y) + \delta_{j1}\varepsilon, \quad j = 0,1; \\ \tilde{R}^{(N)}(\varepsilon,x) &= 0, \qquad\qquad x \in S_\varepsilon, \end{aligned} \tag{19}$$

where

$$\tilde{F}^{(N)} = (\delta+\varepsilon)^{-2}\Big(\varepsilon^{N+1}\big(M_1 + 2\delta M_0\big)u^{(N)} + \varepsilon^{N+1} M_0(\varepsilon u^{(N)} + u^{(N-1)}) - Q^{(N)}\Big)$$

and

$$\Xi_j^{(N)}(\varepsilon,y) = O(\varepsilon^{N+1}), \quad j = 0,1$$

(see (5), (12), (13), and (17)). From this and (28)–(30), 14.1, we obtain (18) in the same manner as in the derivation of (24), 14.1. □

14.2.2 Asymptotics of solutions to the Dirichlet problem

The construction of the asymptotics of solutions to the boundary value problem given in Section 14.2.1 can be carried over almost without changes to the case of the Dirichlet problem

$$\begin{aligned} \Delta u(\varepsilon,x) &= f(\varepsilon,x), && x\in G_\varepsilon;\\ u(\varepsilon,x) &= \Phi^{(j)}(\varepsilon,x), && x\in\Gamma^{(j)}_\varepsilon,\quad j=0,1, \end{aligned} \tag{20}$$

in the domain G_ε defined in 14.1.1. We suppose that the functions $\Phi^{(j)}$ smoothly depend on x and ε. The function f possesses the same property outside any neighborhood of the point O while in V it admits the expansion (5).

We introduce a diffeomorphism $\chi_\varepsilon\colon G_\varepsilon\to H$ of the domain G_ε onto a domain H independent of ε. Assume that on $V\cap G_\varepsilon$ the mapping χ_ε is defined by

$$\chi_\varepsilon(x)=\Big(y,(x_n-\gamma_0(y))\big(\delta(y)+\varepsilon\big)^{-1}\Big) \tag{21}$$

and at a positive distance from O the diffeomorphism χ_ε smoothly depends on ε. To obtain χ_ε it suffices to define $\chi_\varepsilon^{(1)}$ by the equality (21) everywhere in G_ε while γ_0 and δ are extended to $\mathbb{R}^{n-1}$ with $\delta(y)>0$ for $y\neq 0$. The image $\chi_\varepsilon^{(1)}(G_\varepsilon\cap V)$ coincides with $D\times(0,1)$ and at a positive distance from the origin $\chi_\varepsilon^{(1)}$ smoothly depends on ε. One can eliminate the dependence of $\chi_\varepsilon^{(1)}(G_\varepsilon)$ on ε by means of an almost identical mapping.

After the change of variables $x\to\zeta=\chi_\varepsilon(x)$ we obtain the asymptotic expansion

$$\big(\delta(y)+\varepsilon\big)^2\Delta_x\sim\sum_{k=0}^{\infty}\varepsilon^k A_k(\zeta,\nabla_\zeta),$$

where A_k are second order differential operators with coefficients smooth in $\bar H$ and A_0 is elliptic. Moreover, in V the operator Δ_x admits the representation (12).

We seek the asymptotics of the solution u to problem (20) in the form

$$u(\varepsilon,x)\sim\sum_{k=0}^{\infty}\varepsilon^k u^{(k)}(\chi_\varepsilon(x)) \tag{22}$$

(cf. (14)). The functions $u^{(k)}$ are solutions of the boundary value problem

$$A_0u^{(k)}=f^{(k)}-\sum_{p=1}^{k}A_p u^{(k-p)}\quad\text{in}\quad H;\quad u^{(k)}=\Phi^{(k)}\quad\text{on}\quad\partial H$$

(cf. (16) and (17)). One can estimate the remainder in the same manner as in 14.2.1. Obviously, the asymptotics (22) can be rewritten in the form

$$u(\varepsilon,x)\sim\begin{cases}\sum_{k=0}^{\infty}\varepsilon^k v^{(k)}(x), & x\in G_\varepsilon\setminus V,\\ \sum_{k=0}^{\infty}\varepsilon^k V^{(k)}\Big(y,\big(\delta(y)+\varepsilon\big)^{-1}(x_n-\gamma_0(y))\Big), & x\in G_\varepsilon\cap V.\end{cases} \tag{23}$$

14.3 Asymptotics of Solutions for Nonsmooth Right-hand Side Terms

14.3.1 The second limit problem

We consider the boundary value problem (2)–(4), 14.2, in the quasicylindrical domain described in Section 14.2.1. Assume that the right-hand side f in (2), 14.2, satisfies (5) and (6), 14.2, where $f_p^{(k)}$ are functions in $\mathbf{C}^\infty((\mathbb{R}^{n-1} \setminus \{O\}) \times [0,1])$ positively homogeneous in y of degree k (we supposed before that $f_p^{(k)}$ are polynomials in y). The algorithm of the construction of the asymptotics of solutions to problem (2) - (4), 14.2, given in the preceding section, does not apply in this more complicated case. Indeed, looking for the asymptotic solution in the form (14), 14.2, we obtain for the coefficients $u^{(k)}$ the asymptotic estimates

$$u^{(k)}(y,t) = O(|y|^{-k}) \quad \text{for} \quad |y| \to 0, \quad k \in \mathbb{N}_0$$

(see (17), 14.2, and Lemma 14.1.1). Thus, the coefficients of the series (14), 14.2, have intolerable singularity at the point O. As usual, such a situation requires an additional limit problem to describe the asymptotics of u near the singular point. Changing variables

$$(y,t) \to (\xi,t) = (\varepsilon^{-1}y, t)$$

and passing to $\varepsilon = 0$, we transform the domain G_ε into the layer $\Pi = \mathbb{R}^{n-1} \times (0,1)$. The problem (9)–(11) takes the form

$$\begin{aligned} \Delta_{\xi,t} w(\xi,t) &= g(\xi,t), \qquad (\xi,t) \in \Pi; \\ w(\xi,0) &= \Psi^{(0)}(\xi), \quad w(\xi,1) = \Psi^{(1)}(\xi), \quad \xi \in \mathbb{R}^{n-1}, \end{aligned} \tag{1}$$

Here we used the representation (12), 14.2, for the operator L and the relation (1), 14.2. The problem (1) is called the second limit problem. There exists a unique solution to the problem in the class of functions of no more than power growth provided the right-hand sides $\Psi^{(0)}$, $\Psi^{(1)}$, and g belong to this class. One can verify this with the help of the energy inequality

$$\int_\Pi (1+|\xi|)^\mu \left(|\nabla w|^2 + |w|^2\right) d\xi\, dt \le c_\mu \int_\Pi (1+|\xi|)^\mu\, |g|^2\, d\xi\, dt, \tag{2}$$

where $\mu \in \mathbb{R}$, $w \in \mathbf{C}^1(\Pi)$ and $w = 0$ on $\partial\Pi$. The inequality can be proved in the same way as (19), 14.1.

Lemma 14.3.1. *Suppose that the right-hand sides $g \in \mathbf{C}^\alpha(\bar{\Pi})$ and $\Psi^{(0)}$, $\Psi^{(1)} \in \mathbf{C}^{2,\alpha}(\mathbb{R}^{n-1})$ $(0 < \alpha < 1)$ of problem (1) satisfy*

$$\begin{aligned} g(\xi,t) &= \sum_{k=-1}^{N-1} |\xi|^{-k} g_k(\xi/|\xi|, t) + G_N(\xi,t), \\ \Psi^{(j)}(\xi,t) &= \sum_{k=-1}^{N-1} |\xi|^{-k} \Psi_k^{(j)}(\xi/|\xi|) + T_N^{(j)}(\xi), \end{aligned} \tag{3}$$

where $g_k \in \mathbf{C}^\infty(S^{n-2} \times [0,1])$, $\Psi_k^{(j)} \in \mathbf{C}^\infty(S^{n-2})$. The remainders G_N and $T_N^{(j)}$ are subject to the inequalities

$$\begin{aligned} |\nabla_\xi^k (\partial/\partial t)^p G_N(\xi,t)| &\le c_{pk} |\xi|^{-N-k}, \quad t \in [0,1], \quad |\xi| > 0; \\ |\nabla_\xi^k T_N^{(j)}(\xi)| &\le c_k |\xi|^{-N-k}, \quad |\xi| > 0, \end{aligned} \tag{4}$$

for all $k, p \in \mathbb{N}_0$. *Then there exists a unique solution* $w \in \mathbf{C}^{2,\alpha}(\bar{\Pi})$ *to problem* (3) *that admits the representation*

$$\begin{aligned} &w(\xi,t) = \sum_{k=-1}^{N-1} |\xi|^{-k} w_k(\xi/|\xi|,t) + W^{(N)}(\xi,t), \\ &|\nabla_\xi^k (\partial/\partial t)\, W^{(N)}(\xi,t)| \le c_{pk} |\xi|^{-N-k}, \quad t \in [0,1], \quad |\xi| > 0. \end{aligned} \tag{5}$$

Proof. We use the same scheme as in the proof of Lemma 14.1.1. It suffices to consider the homogeneous boundary conditions. We find the coefficients w_k in (5) from the sequence of the boundary value problems

$$(\partial/\partial t)^2 w_k(\xi/|\xi|,t) = g_k(\xi/|\xi|,t) - |\xi|^k \Delta_\xi \big(|\xi|^{2-k} w_{k-2}(\xi/|\xi|,t)\big), \quad t \in [0,1];$$
$$w_k(\xi/|\xi|,t) = 0, \quad t = 0,1.$$

Here $k = -1, 0, 1, \ldots$ while w_{-3} and w_{-2} are equal to zero.

The function

$$R(\xi,t) = w(\xi,t) - \chi(\xi) \sum_{k=-1}^{N-1} |\xi|^{-k} w_k(\xi/|\xi|,t),$$

where $\chi \in \mathbf{C}^\infty(\mathbb{R}^{n-1})$ and $\chi(\xi) = 1$ for $|\xi| \ge 1$, has to be substituted into problem (1). After that we use the energy inequality (2) and the local estimate of the derivatives. □

14.3.2 Asymptotics of solutions

To proceed further it is convenient to rewrite the expansions (5) and (6), 14.2, of f using both the coordinate systems (y,t) and $(\varepsilon^{-1}y,t)$. Namely,

$$f(\varepsilon,y,t) = \big(\delta(y)+\varepsilon\big)^{-2} \Big\{ \sum_{k=0}^{M-1} \varepsilon^k \big(f^{(k)}(y,t)) + \varepsilon^2 g^{(k)}(\varepsilon^{-1}y,t)\big) + Q^{(M)}(\varepsilon,y,t) \Big\}, \tag{6}$$

where

$$f^{(k)}(y,t) = \sum_{p=3}^{N-1} |y|^p f_p^{(k)}(y/|y|,t) + \tilde{F}_N^{(k)}(y,t), \tag{7}$$

$$g^{(k)}(\xi,t) = \sum_{p=-2}^{N-1} |\xi|^{-p} g_p^{(k)}(\xi/|\xi|,t) + \tilde{G}_N^{(k)}(\xi,t), \tag{8}$$

and $f_p^{(k)}$, $g_p^{(k)} \in \mathbf{C}^\infty(S^{n-2} \times [0,1])$. (Estimates for the remainders are given below.) Similarly,

$$\begin{aligned} \Phi^{(j)}(\varepsilon,y) &= \sum_{k=0}^{M-1} \varepsilon^k \big(\Phi^{(j,k)}(y) + \varepsilon^2 \Psi^{(j,k)}(\varepsilon^{-1}y)\big) + R^{(j,M)}(\varepsilon,y), \quad j = 0,1; \\ \Phi^{(j,k)}(y,t) &= \sum_{p=3}^{N-1} |y|^p \Phi_p^{(j,k)}(y/|y|) + \tilde{\Phi}_N^{(j,k)}(y), \quad j = 0,1; \\ \Psi^{(j,k)}(\xi) &= \sum_{p=-2}^{N-1} |\xi|^{-p} \Psi_p^{(j,k)}(\xi/|\xi|) + \tilde{\Psi}_N^{(j,k)}(\xi), \quad j = 0,1, \end{aligned} \tag{9}$$

and $\Phi_p^{(j,k)}$, $\Psi_p^{(j,k)} \in \mathbf{C}^\infty(S^{n-2})$, $j = 0, 1$. For $q, s = 0, 1, \dots$ there hold the inequalities

$$\begin{aligned} &|(\partial/\partial t)^s \nabla_y^q Q^{(M)}(\varepsilon, y, t)| + |\nabla_y^q R^{(j,M)}(\varepsilon, y)| \le c^M |y|^{-q}(\varepsilon + |y|)^2; \\ &|(\partial/\partial t)^s \nabla_y^q \tilde{F}_N^{(k)}(y,t)| + |\nabla_y^q \tilde{\Phi}_N^{(j,k)}(y)| \le c|y|^{N-q}, \quad j = 0, 1; \\ &|(\partial/\partial t)^s \nabla_\xi^q \tilde{G}_N^{(k)}(\xi,t)| + |\nabla_\xi^q \tilde{\Psi}_N^{(j,k)}(\xi)| \le c|\xi|^{-N-q}, \quad j = 0, 1. \end{aligned} \tag{10}$$

We seek an asymptotic solution to problem (9)–(11), 14.2, in the form

$$u(\varepsilon, x) \sim \sum_{k=0}^{\infty} \varepsilon^k \big(v^{(k)}(y,t) + \varepsilon^2 w^{(k)}(\varepsilon^{-1}y, t)\big), \tag{11}$$

where $v^{(k)}$ and $w^{(k)}$ admit the expansions

$$v^{(k)}(y,t) = \sum_{p=3}^{N-1} |y|^p v_p^{(k)}(y/|y|, t) + V_N^{(k)}(y,t), \tag{12}$$

$$w^{(k)}(\xi,t) = \sum_{p=-2}^{N-1} |\xi|^{-p} w_p^{(k)}(\xi/|\xi|, t) + W_N^{(k)}(\xi,t) \tag{13}$$

and $v_p^{(k)}$, $w_p^{(k)} \in \mathbf{C}^\infty(S^{n-2} \times [0,1])$. The remainders $V_N^{(k)}$ and $W_N^{(k)}$ are subject to the inequalities

$$\begin{aligned} &|(\partial/\partial t)^s \nabla_y^q V_N^{(k)}(y,t)| \le c|y|^{N-q}, \\ &|(\partial/\partial t)^s \nabla_\xi^q W_N^{(k)}(\xi,t)| \le c|\xi|^{N-q}, \quad s, q \in \mathbb{N}_0. \end{aligned}$$

In what follows it will be shown that the coefficient $v^{(k)}$ in the series (11) satisfies the problem

$$\begin{aligned} &L_0(y, t, \nabla_y, \partial/\partial t)\, v^{(k)}(y,t) = \delta(y)^{-2} \hat{F}^{(k)}(y,t), \quad (y,t) \in \tilde{C}; \\ &v^{(k)}(y, 0) = \hat{H}_0^{(k)}(y), \quad v^{(k)}(y,1) = \hat{H}_1^{(k)}(y), \quad y \in \Omega; \\ &v^{(k)}(y,t) = \hat{H}_2^{(k)}(y,t), \quad (y,t) \in S, \end{aligned} \tag{14}$$

where $\hat{F}^{(k)}$ and $\hat{H}_p^{(k)}$ are some functions determined by means of $v^{(s)}$, $w^{(s)}$, $s = 0, \dots, k-1$, and $f^{(k)}$, $\Phi^{(q,k)}$, $q = 0, 1$.

The coefficient $w^{(k)}$ is a solution to the problem

$$\begin{aligned} &\Delta_{\xi,t} w^{(k)}(\xi,t) = \hat{G}^{(k)}(\xi,t), \quad (\xi,t) \in \Pi; \\ &w^{(k)}(\xi,0) = \hat{B}_0^{(k)}(\xi), \quad w^{(k)}(\xi,1) = \hat{B}_1^{(k)}(\xi), \quad \xi \in \mathbb{R}^{n-1}. \end{aligned} \tag{15}$$

In order to describe the right-hand sides $\hat{F}^{(k)}$, $\hat{H}_p^{(k)}$, $\hat{G}^{(k)}$, $\hat{B}_q^{(k)}$ of problems (14) and (15) we need various representations of the operators

$$(\delta + \varepsilon)^2 L = \tilde{M}_2 + \varepsilon \tilde{M}_1 + \tilde{M}_0$$

with $\tilde{M}_2 = M_2 + \delta M_1 + \delta^2 M_0$, $\tilde{M}_1 = M_1 + 2\delta M_0$ and $\tilde{M}_0 = M_1$ (see (13), 14.1). Since γ_0 and δ are smooth functions and the relations (2), 14.1, are fulfilled, the operators $\tilde{M}_2$ and $\tilde{M}_1$ admit the expansions

$$\tilde{M}_j(y, t, \nabla_y, \partial/\partial t) \sim \sum_{k=0}^{\infty} \mathbf{M}_j^{(k)}(y, t, \nabla_y, \partial/\partial t), \tag{16}$$

where $\mathbf{M}_2^{(0)} = (\partial/\partial t)^2$ while $\mathbf{M}_j^{(k)}$ are operators of order no greater than 2 whose coefficients are polynomials in t (of degree no greater than 2) and in y (of degree no greater than k) satisfying

$$\mathbf{M}_j^{(k)}(sy, t, s^{-1}\nabla_y, \partial/\partial t) = s^k \mathbf{M}_j^{(k)}(y, t, \nabla_y, \partial/\partial t), \quad s > 0. \tag{17}$$

We introduce the operators

$$\hat{M}_j^{(k)} = \tilde{M}_j - \sum_{p=0}^{k-1} \mathbf{M}_j^{(k)}, \quad j = 1, 2. \tag{18}$$

The derivation of the asymptotics (11) of the solution u to problem (2)–(4), 14.2, is an iterative process. At the pth step we define the functions $w^{(p)}$ and $v^{(p)}$ compensating the discrepancies of order $O(\varepsilon^p)$ on the right in (2)–(4), 14.2, and distribute the new discrepancies (which are generated by $v^{(p)}$ and $w^{(p)}$) between the right-hand sides $\hat{F}^{(q)}$ and $\hat{G}^{(q)}$ of the limit problems for the next terms of the series. To illustrate this we consider the expressions $\tilde{X}^{(1)} = \varepsilon^{p+2}(\delta + \varepsilon)^{-2}\tilde{M}_0 v^{(p)}$ and $\tilde{X}^{(2)} = \varepsilon^{p+3}(\delta + \varepsilon)^{-2}\tilde{M}_1 w^{(p)}$.

The function $X^{(1)}$ is written in the coordinates (y, t) and admits the expansion

$$\tilde{X}^{(1)}(\varepsilon, y, t) = \big(\delta(y) + \varepsilon\big)^{-2}\tilde{K}(y, t) \sim \big(\delta(y) + \varepsilon\big)^{-2}\sum_{q=1}^{\infty} |y|^q \tilde{K}^{(q)}(t).$$

By Lemma 14.1.1, the solution to problem

$$L_0 v = \delta^{-2}|y|\tilde{K}^{(1)} \quad \text{on} \quad \tilde{C}; \qquad v = 0 \quad \text{on} \quad \partial\tilde{C} \setminus S; \qquad v = 0 \quad \text{on} \quad S$$

satisfies $v(y, t) = O(|y|)$ as $|y| \to 0$. Therefore, the function v cannot be represented in the form (12) and is not contained in a coefficient of the series (11). We single out the terms of order $O(|y|)$ and $O(|y|^2)$ in $\tilde{K}(y, t)$ and rewrite them in the coordinate (ξ, t), i.e.,

$$\begin{aligned}\tilde{X}^{(1)}(\varepsilon, y, t) = \big(\delta(y) + \varepsilon\big)^{-2}\Big\{&\varepsilon^{p+2}\Delta_y V_5^{(p)}(y, t) \\ &+ \varepsilon^{p+3}\Delta_\xi\big(|\xi|^3 v_3^{(p)}(\xi/|\xi|, t)\big) + \varepsilon^{p+4}\Delta_\xi\big(|\xi|^4 v_4^{(p)}(\xi/|\xi|, t)\big)\Big\}.\end{aligned}$$

Here $V_5^{(p)}$, $v_3^{(p)}$, and $v_4^{(p)}$ are defined by (12) with $N = 5$. The function $\Delta_y V_5^{(p)}$ admits the representation

$$\sum_{k=3}^{\infty} |y|^k \tilde{V}_k(t)$$

(cf. (7)). We move it to the right-hand side of problem (14) to find $v^{(p+2)}$. The terms

$$\Delta_\xi\big(|\xi|^3 v_3^{(p)}(\xi/|\xi|, t)\big) = O(|\xi|), \qquad \Delta_\xi\big(|\xi|^4 v_4^{(p)}(\xi/|\xi|, t)\big) = O(|\xi|^2)$$

(cf.(8)) must be moved to the right-hand sides of problems (15) to find $w^{(p+1)}$ and $w^{(p+2)}$, respectively.

The expression $\tilde{X}^{(2)}$ contains the products of functions written in the coordinates (y, t) and (ξ, t). We are going to represent it in the form

$$\tilde{X}^{(2)}(\varepsilon, y, t) \sim \varepsilon^{p+1}\big(\delta(y) + \varepsilon\big)^{-2}\Big\{\sum_{q=0}^{\infty} \varepsilon^q\big(\tilde{T}^{(q)}(y, t) + \varepsilon^2\tilde{N}^{(q)}(\xi, t)\big)\Big\}, \tag{19}$$

where

$$\tilde{T}^{(q)}(y,t) \sim \sum_{k=3}^{\infty} |y|^k \tilde{T}_k^{(q)}(t), \qquad \tilde{N}^{(q)}(\xi,t) \sim \sum_{k=-2}^{\infty} |\xi|^{-2} \tilde{N}^{(q)}(t) \tag{20}$$

(cf. (6)–(8)). To this end we substitute the series (16) for $\tilde{M}_1$ and (13) for $w^{(p)}$ into $\tilde{K}^{(2)} = \varepsilon^{p+3}\tilde{M}_1 w^{(p)}$. The formal double series obtained contains the terms

$$\varepsilon^{p+3}\mathbf{M}_1^{(k)}(y,t,\nabla_y,\partial/\partial t)\big(|\xi|^{-q} w_q^{(p)}(\xi/|\xi|,t)\big) = O\big(\varepsilon^{p+q+3}|y|^{k-q}\big).$$

If $k-q>2$, we make the change of variables $\xi \to x = \varepsilon\xi$. In the case $k-q\le 2$, we write the operator $\mathbf{M}_1^{(k)}$ in the coordinates (ξ,t) using (17). Adding the coefficients by degrees of ε written in the same coordinates, we arrive at the relation

$$\begin{aligned}\tilde{K}^{(2)}(\varepsilon,y,t) \sim \varepsilon^{p+1}\sum_{q=0}^{\infty}\varepsilon^q\big\{&\hat{M}_1^{(q+1)}(y,t,\nabla_y,\partial/\partial t)\big(|y|^{2-q}w_{q-2}^{(p)}(y/|y|,t)\big)\\ &+\varepsilon^2\mathbf{M}_1^{(q)}(\xi,t,\nabla_\varepsilon,\partial/\partial t)W_{q-2}^{(p)}(\xi,t)\big\},\end{aligned} \tag{21}$$

where $\hat{M}_1^{(q+1)}$ and $W_{q-2}^{(p)}$ are defined by (13) and (18). The formulas (19) and (20) follow from (21). Therefore, one can send the terms $\tilde{T}^{(q)}$ to the right-hand sides of problems (14) (to define $v^{(p+q+1)}$) and the terms $\tilde{N}^{(q)}$ to those of problems (15) (to define $w^{(p+q+1)}$), $q=0,1,2,\ldots$.

We substitute the representations (6), (9), (11) into problem (9)–(11), 14.2, and write every term in the series obtained using either the coordinates (y,t) or (ξ,t) according to the above rule. Equating the coefficients at the same powers of ε in the same coordinates, we find the right-hand sides of problems (14) and (15):

$$\begin{aligned}
\hat{F}^{(k)}(y,t) &= f^{(k)}(y,t) - \tilde{M}_1(y,t,\nabla_y,\partial/\partial t)v^{(k-1)}(y,t) - \tilde{M}_0(\nabla_y)V_5^{(k-2)}(y,t)\\
&\quad-\sum_{j=1}^{2}\sum_{p=-2}^{k+j+4}\hat{M}_j^{(p+3)}(y,t,\nabla_y,\partial/\partial t)\big(|y|^{-p}w_p^{(k+j-2-p)}(y/|y|,t)\big),\\
\hat{G}^{(k)}(\xi,t) &= g^{(k)}(\xi,t) - \Delta_\xi\big(|\xi|^3 v_3^{(k-1)}(\xi/|\xi|,t) + |\xi|^4 v_4^{(k-2)}(\xi/|\xi|,t)\big)\\
&\quad-\sum_{j=1}^{2}\sum_{q=j-1}^{k+j-2}\mathbf{M}_j^{(q)}(\xi,t,\nabla_\varepsilon,\partial/\partial t)W_{q+2}^{(k+j-2+q)}(\xi,t),\\
\hat{H}_j^{(k)}(y) &= \Phi^{(j,k)}(y),\quad \hat{B}_j^{(k)}(\xi) = \Psi^{(j,k)}(\xi),\quad j=0,1,\\
\hat{H}_2^{(k)}(t) &= -\sum_{q=0}^{k}|y|^{q-2}w_{q-2}^{(k-q)}(y/|y|,t).
\end{aligned} \tag{22}$$

Here $k=0,1,2,\ldots$ and the functions with negative superscripts are equal to zero. Note that $\hat{G}^{(k)}$ and $\hat{B}_j^{(k)}$ must first be found. Thereupon one can define $v^{(0)},\ldots,v^{(k-1)}$ and $w^{(0)},\ldots,w^{(k-1)}$. To obtain $\hat{F}^{(k)}$ and $\hat{H}_q^{(k)}$ we first determine $v^{(0)},\ldots,v^{(k-1)}$ and $w^{(0)},\ldots,w^{(k)}$.

The next assertion follows by induction from (22), (7), (8), Lemma 14.1.1 (ii), and Lemma 14.3.1.

Lemma 14.3.2. *For the solutions $v^{(k)} \in \mathbf{C}^{1,\alpha}(\bar{\tilde{C}})$ and $w^{(k)} \in \mathbf{C}^{2,\alpha}(\bar{\Pi})$ $(0<\alpha<1)$ to problems* (14) *and* (15) *with right-hand sides* (22) *there hold the representations* (12) *and* (13).

The next theorem justifies the asymptotic expansion (11) of the solution to problem (2)–(4), 14.2.

Theorem 14.3.3. *Let the relations* (6)–(10) *be valid for the right-hand sides of problem* (9)–(11), 14.2. *Suppose that u is a solution to problem* (2)–(4), 14.2, *and*

$$\tilde{S}^{(N)}(\varepsilon,x)=\sum_{k=0}^{N}\varepsilon^k\Big\{v^{(k)}\Big(y,\big(\delta(y)+\varepsilon\big)^{-1}(x_n-\gamma_0(y))\Big)$$
$$+\varepsilon^2 w^{(k)}\Big(\varepsilon^{-1}y,\big(\delta(y)+\varepsilon\big)^{-1}(x_n-\gamma_0(y))\Big)\Big\},$$

where $v^{(k)}$ and $w^{(k)}$ are solutions to problems (14) *and* (15) *with right-hand sides* (22) *admitting the expansions* (12) *and* (13). *Then*

$$|u(\varepsilon,x)-\tilde{S}^{(N)}(\varepsilon,x)|\le c\varepsilon^{N-1}. \tag{23}$$

Proof. The function $\tilde{R}^{(N)}=u-\tilde{S}^{(N)}$ satisfies problem (19), 14.2, where

$$\tilde{F}^{(N)}=(\delta+\varepsilon)^{-2}\{Q^{(N+1)}+\varepsilon^{N+1}(\tilde{M}v^{(N)}+\Delta_y(v^{(N-1)}+\varepsilon v^{(N)}))$$
$$+\sum_{k=0}^{N}\varepsilon^k(\varepsilon\hat{M}_1^{(N-k)}W_{N-k-2}^{(k)}+\hat{M}_2^{(N-k+1)}W_{N-k+1}^{(k)})\Big\};$$

$$\Xi_j^{(N)}=R^{(j,N+1)},\quad j=0,1;\qquad \Xi_2^{(N)}=\sum_{k=0}^{N}\varepsilon^{k+2}W_{N-k-1}^{(k)},$$

the functions $Q^{(N+1)}$ and $R^{(j,N+1)}$ being defined by (6) and (9). From (10) we obtain

$$\tilde{F}^{(N)}(\varepsilon,x)=O\big(\varepsilon^{N-1}(|y|+\varepsilon)^2\,(\delta(y)+\varepsilon)^{-2}\big),$$
$$\Xi_j^{(N)}(\varepsilon,y)=O\big(\varepsilon^{N+1}(|y|+\varepsilon)^2\big),\qquad j=0,1,$$
$$\Xi_2^{(N)}(\varepsilon,x)=O(\varepsilon^{N+1}).$$

Now the estimate (23) can be obtained in the same way as (18), 14.2, in Theorem 14.2.2 and (24), 14.1, in Theorem 14.1.3. □

14.4 Ligaments of a Different Form

14.4.1 Two close conical points

The restriction $m\ge 2$ was essential in (2), 14.1. In this section we consider problem (3), (4), 14.1, for $m=1$ (see Fig. 14.3). We deal with a special case of the situation studied in Chapter 4 where the complete asymptotic expansions were constructed for solutions to general elliptic boundary value problems in domains perturbed near conical points. In contrast to Section 14.1, where the inclusions $\gamma_j\in\mathbf{C}^\infty(\bar{D})$ were supposed, we require the asymptotic representations

$$\gamma_j(y)=|y|\gamma_j^0(y/|y|)+O(|y|^2),\quad j=0,1,$$

which can be differentiated, γ_j^0 being smooth functions on the sphere S^{n-2}

As was shown in Chapter 4, the asymptotics of the solution u to the original problem (3), (4), 14.1, can be obtained with the help of the bounded solutions to the following limit problems.

The limit problem in the domain G_0:

$$\Delta v = 0 \quad \text{in} \quad G_0; \quad v = \Phi^{(j)} \quad \text{on} \quad \Gamma_0^{(j)}, \quad j = 0, 1.$$

The limit problem in the domain

$$\tilde{K} = \{x \in \mathbb{R}^n \colon |y|\gamma_0^0(y/|y|) < x_n < |y|\gamma_1^0(y/|y|)|y| + 1\} \ :$$

$$\Delta w = 0 \quad \text{in} \quad \tilde{K}; \quad w = \Phi^{(j)}(0) \quad \text{on} \quad \partial_j \tilde{K}, \quad j = 0, 1,$$

where $\partial_1 \tilde{K}$ and $\partial_0 \tilde{K}$ are the upper and lower components of the boundary $\partial \tilde{K}$. Assume that w is continued as an odd function with respect to x to some conical neighborhoods of $\partial_0 \tilde{K}$ and $\partial_1 \tilde{K}$. Note that in the case $n = 2$ such a continuation is harmonic. For $n \geq 3$ there is the estimate

$$|\Delta w(x)| \leq c d_1(x)^{-1}, \tag{1}$$

where $d_1(x)$ is the distance from the point x to the nearest of two points 0 and $(0, \dots, 0, 1)$.

By K we denote the cone $\{x \colon \gamma_0^0(y/|y|) < x_n/|y| < \gamma_1^0(y/|y|)\}$ and by h a positively homogeneous function of order zero that is smooth in $\mathbb{R}^n \setminus \{0\}$ and harmonic in K and equal to 1 and 0 on the upper and lower components of ∂K, respectively. In particular, for $n = 2$ the function h linearly depends on the angular variable ϑ and

$$(\partial/\partial\vartheta)h = \alpha_\pm^{-1} \quad \text{for} \quad \pm y > 0, \tag{2}$$

where $\alpha_\pm$ is the opening of the angle $K \cap \{x \colon \pm y > 0\}$.

Set

$$H(x) = \Phi^{(0)}(0) + h(x)\big(\Phi^{(1)}(0) - \Phi^{(0)}(0)\big).$$

Obviously, for the solutions to the limit problems there are the asymptotic formulas

$$\begin{aligned} v(x) &= H(x) + O(|x|^\Lambda), \quad |x| \to 0, \\ w(x) &= H(x) + O(|x|^{-1}), \quad |x| \to \infty, \end{aligned} \tag{3}$$

where $\Lambda = \min\{1, \lambda\}$; λ is the first eigenvalue of the boundary value problem

$$\delta U + \lambda(\lambda + n - 2)U = 0 \quad \text{for} \quad S^{n-1} \cap K; \quad U = 0 \quad \text{for} \quad S^{n-1} \cap \partial K,$$

δ being the Beltrami operator on the $(n-1)$-dimensional unit sphere S^{n-1} (see Section 1.6).

Theorem 14.4.1. *There holds the representation*

$$u(\varepsilon, x) = \big(1 - \zeta(\varepsilon^{-1}y)\big)v(x) + \zeta(y)w(\varepsilon^{-1}x) - \big(1 - \zeta(\varepsilon^{-1}y)\big)\zeta(y)H(x) + R(\varepsilon, x),$$

where ζ is a function in $\mathbf{C}_0^\infty(D)$ equal to 1 near the point 0 while

$$|R(\varepsilon, x)| \leq c\varepsilon^\Lambda.$$

Proof. The function R satisfies

$$\Delta R(\varepsilon, x) = F(\varepsilon, x), \quad x \in G_\varepsilon; \quad R(\varepsilon, x) = \varphi^{(j)}(\varepsilon, x), \quad x \in \Gamma_\varepsilon^{(j)}, \quad j = 0, 1, \tag{4}$$

with

$$\begin{aligned} F(\varepsilon, x) = {} & \big(\zeta(\varepsilon^{-1}y) - 1\big)\Delta v(x) + [\Delta, \zeta(\varepsilon^{-1}y)](v - H)(x) \\ & - [\Delta, \zeta(y)](w - H)(\varepsilon^{-1}x) - \zeta(y)\Delta w(\varepsilon^{-1}x) + \zeta(y)\big(1 - \zeta(\varepsilon^{-1}y)\big)\Delta H(x). \end{aligned} \tag{5}$$

Further,

$$\begin{aligned}\varphi^{(j)}(\varepsilon,x) &= -\big(1-\zeta(\varepsilon^{-1}y)\big)\big(\Phi^{(j)}(x)+O(\varepsilon)\big)-\zeta(y)\big(\Phi^{(j)}(0)+O(\varepsilon)\big)\\ &\quad+\big(1-\zeta(\varepsilon^{-1}y)\big)\zeta(y)\big(\Phi^{(j)}(0)+O(\varepsilon)\big)+\Phi^{(j)}(x)\\ &= \big(\Phi^{(j)}(x)+\Phi^{(j)}(0)\big)\zeta(\varepsilon^{-1}y)+O(\varepsilon)=O(\varepsilon).\end{aligned}\tag{6}$$

It follows from (3) that

$$\Big|\Big(1-\zeta\big((\alpha\varepsilon)^{-1}y\big)\Big)\,F(\varepsilon,x)\Big|\le c\varepsilon^{\Lambda}\,(|y|+\varepsilon)^{-2},\tag{7}$$

α being a small constant. For $n=2$ the extension of w is harmonic and $\zeta\Delta w=0$. Hence, $F(\varepsilon,x)=0$ for $|y|<\alpha\varepsilon$. In the case $n\ge 3$ we obtain from (1) that

$$\Big|\zeta\big((\alpha\varepsilon)^{-1}y\big)\,F(\varepsilon,x)\Big|=\Big|\zeta\big((\alpha\varepsilon)^{-1}y\big)\,\Delta w(\varepsilon^{-1}x)\Big|\le C\varepsilon d_\varepsilon(x)^{-1},$$

where d_ε is the distance from x to the conic vertices O and $(0,\dots,0,\varepsilon)$.

Let $R^{(1)}$ stand for the solution to the problem

$$\begin{aligned}\Delta R^{(1)}(\varepsilon,x)&=\zeta\big((\alpha\varepsilon)^{-1}y\big)\,F(\varepsilon,x), && x\in G_\varepsilon;\\ R^{(1)}(\varepsilon,x)&=0,\quad x\in\Gamma_\varepsilon^{(j)}, && j=0,1\end{aligned}$$

($R^{(1)}=0$ for $n=2$). Assume that $n\ge 3$. Since Green' function $\tilde G(x,\xi)$ is $O(|x-\xi|^{-1})$, we have

$$|R^{(1)}(\varepsilon,x)|\le c\varepsilon\int_{G_\varepsilon} d_\varepsilon(\xi)^{-1}\,|x-\xi|^{-1}d\xi\le C\varepsilon.$$

After the change of variables $x\to X=(Y,X_n)=\varepsilon^{-1}x$ we obtain

$$\begin{aligned}\Delta_X(R-R^{(1)})&=\varepsilon^2\zeta(\alpha^{-1}Y)\,F && \text{for}\quad \varepsilon X\in G_\varepsilon,\\ R-R^{(1)}&=\varphi^{(j)} && \text{for}\quad \varepsilon X\in\Gamma_\varepsilon^{(j)},\quad j=0,1.\end{aligned}$$

From (6), (7), and the maximum principle it follows that

$$|R(\varepsilon,x)-R^{(1)}(\varepsilon,x)|=O(\varepsilon^{\Lambda}).$$

□

14.4.2 Components of the boundary adhering at a large set

In Sections 14.1–14.3, it was supposed that the limiting surfaces $\Gamma_0^{(0)}$ and $\Gamma_0^{(1)}$ had the only common point. Here we briefly consider the case when the set $\omega=\{y\colon \delta(y)=0\}$ is compact in D with nonempty interior and smooth boundary $\partial\omega$ (see Fig. 14.4). In a neighborhood of $\partial\omega$ we introduce the coordinate $(\nu,\tau)=(\nu,\tau_1,\dots,\tau_{n-2})$, where ν is the distance to $\partial\omega$ and τ any local coordinate system on $\partial\omega$. Suppose that

$$\begin{aligned}&\gamma_j\in\mathbf{C}^\infty(\bar D\setminus\omega)\cap\mathbf{C}^\infty(\bar\omega),\quad j=0,1;\\ &c_1\nu^m\le\delta(y)\le c_2\nu^m,\quad y\in\bar D\setminus\omega;\quad \delta(y)=0,\quad y\in\bar\omega,\end{aligned}\tag{8}$$

where, as in (2), 14.1, c_j is a positive constant and m is an even positive integer.

The limit problem has the form

$$\Delta v=0\quad\text{in}\quad G_0;\quad v=\Phi^{(j)}\quad\text{on}\quad \Gamma_0^{(j)}\setminus\bar\Omega,\quad j=0,1,\tag{9}$$

with $\Omega=\{x\in\Gamma_0^{(0)}: y\in\omega\}$. A slight change in the proof of Lemma 14.1.1 justifies the following assertion on the asymptotics of the solution to problem (9) near $\partial\omega$.

Lemma 14.4.2. *Let $\Phi^{(j)} \in \mathbf{C}^\infty(\Gamma_0^{(j)} \setminus \Omega)$, $j = 0, 1$, and let (8) be valid. Then for the bounded solution v to problem (9) there is the asymptotic formula*

$$v(x) = \sum_{k=0}^{N} v_k\Big(\nu, \tau, \delta(y)^{-1}\big(z - \gamma_0(y)\big)\Big) + r_N(x), \quad N \in \mathbb{N}_0,$$

on the set $\bar{G}_0 \cap V$. Here $v_k(\nu, \tau, t)$ are homogeneous polynomials in ν of degree k whose coefficients are smooth functions in $\tau \in \partial\omega$ and $t \in [0, 1]$. The remainder r_N admits the estimate

$$(\partial/\partial\nu)^k (\partial/\partial t)^j\, D_\tau^\beta\, r_N(\nu, \tau, t) = O(\nu^{N+1-k}), \quad \nu \to 0, \tag{10}$$

where $k, j = 0, 1, \dots$ and β is an arbitrary multiindex.

Corollary 14.4.3. *Let the hypotheses of Lemma 14.4.2 be fulfilled and let*

$$|\nabla_\nu \delta(\nu)| + |\nabla_\nu \gamma_0(\nu)| \le c\nu^{l-1}, \tag{11}$$

where $l = 1, \dots, m$. Then the solution v to problem (9) admits the representation

$$v(x) = \Phi_1(y) + \delta(y)^{-1}\big(x_n - \gamma_1(y)\big)\big(\Phi_1(y) - \Phi_0(y)\big) + r_{m+l-2}(x), \quad x \in \bar{G}_0 \cap V,$$

with $\Phi_j = \Phi^{(j)}(y, \gamma_j(y))$. The remainder r_{m+l-2} is subject to (10) with $N = m+l-2$.

Using Lemma 14.4.2, one can prove the main result of the section in the same way as Theorem 14.1.3.

Theorem 14.4.4. *Let the relations (8) and (9) hold and let v be the bounded solution to problem (9) extended from $\bar{G}_0$ to $\mathbb{R}^n \setminus V$ with the same smoothness. Then the solution u to problem (3), (4) has the asymptotics*

$$u(\varepsilon, x) \sim \begin{cases} v(x), & x \in \bar{G}_\varepsilon \setminus V; \\ \tilde{V}\Big(y, \big(\delta(y) + \varepsilon\big)^{-1}\big(x_n - \gamma_1(y)\big)\Big), & y \in D \setminus \omega; \\ \Phi_0(y) + \big(\delta(y) + \varepsilon\big)^{-1}\big(x_n - \gamma_1(y)\big)\big(\Phi_1(y) - \Phi_0(y)\big), & y \in \omega, \end{cases} \tag{12}$$

where $\tilde{V}$ (as in Theorem 14.1.3) coincides with the solution v to problem (40), 14.1, written in $\bar{G}_0 \cap V$ in the coordinates

$$(y, t) = \Big(x_1, \dots, x_{n-1}, \delta(y)^{-1}\big(x_n - \gamma_0(y)\big)\Big)$$

and $a \sim b$ means $a - b = O(\varepsilon)$.

Note that by Theorem 14.4.4 and Corollary 14.4.3 there is the asymptotic representation

$$\begin{aligned} u(\varepsilon, x) &= \Phi_0(y) + \big(\delta(y) + \varepsilon\big)^{-1}\big(x_n - \gamma_0(y)\big)\big(\Phi_1(y) - \Phi_0(y)\big) + O(\varepsilon) \\ &\text{for}\quad \operatorname{dist}(y, \omega) \le c\varepsilon^{1/(l+m-2)}. \end{aligned}$$

14.5 Asymptotics of the Condenser Capacity

We apply the asymptotic formulas obtained in Sections 14.1, 14.2, and 14.4 to calculate harmonic capacity. We consider the domain G_ε as the condenser with plates $\Gamma_\varepsilon^{(0)}$ and $\Gamma_\varepsilon^{(1)}$. Then the solution to problem (3), (4), 14.1, for $\Phi^{(0)} = 0$, $\Phi^{(1)} = 1$ is the electrostatic potential and the capacity of the condenser is defined by

$$\operatorname{cap} G_\varepsilon = \mu_n^{-1} \int\limits_{\Gamma_\varepsilon^{(0)}} (\partial/\partial N) u(\varepsilon, x) ds, \tag{1}$$

where $\mu_n = (n-2)|S^{n-1}|$ for $n > 2$ and $\mu_2 = 2\pi$ while N is the inward normal.

14.5.1 Components of the boundary adhering at one point

To calculate the asymptotics we need the following assertion.

Lemma 14.5.1. *Let the domain G_ε satisfy the hypotheses of Theorem* 14.1.3. *Then*

$$\operatorname{cap} G_\varepsilon = \mu_n^{-1} \int\limits_{\Gamma_\varepsilon^{(0)}} (\partial/\partial N)(v \circ \varkappa_\varepsilon) ds + \rho_n^{(\varepsilon)},$$

where $\varkappa_\varepsilon$ is the diffeomorphism (23), 14.1, *and*

$$\rho_n^{(\varepsilon)} = \begin{cases} O(\varepsilon^{n/m}) & \text{for} \quad m > n, \\ O(\varepsilon |\log \varepsilon|) & \text{for} \quad m = n, \\ O(\varepsilon) & \text{for} \quad m < n. \end{cases}$$

Proof. By the known local estimate of solutions to the Poisson equation and by (25)–(30), 14.1, we obtain

$$\begin{aligned} |\nabla R(\varepsilon, x)| \le c\Big\{ \big(\delta(x) + \varepsilon\big)^{-1} &\sup_{\xi \in D_{\delta+\varepsilon}(x) \cap G_\varepsilon} |R(\varepsilon, \xi)| \\ + \big(\delta(x) + \varepsilon\big)^{-1} &\sup_{\xi \in D_{\delta+\varepsilon}(x) \cap G_\varepsilon} |F(\varepsilon, \xi)| \\ + \sum_{j=0}^{1} &\sup_{\xi \in D_{\delta+\varepsilon}(x) \cap \Gamma_\varepsilon^{(j)}} |\nabla \varphi^{(j)}(\varepsilon, \xi)| \Big\} \le C\varepsilon \big(\delta(x) + \varepsilon\big)^{-1}, \end{aligned}$$

where $D_{\delta+\varepsilon}(x)$ is the ball with radius $\delta(x) + \varepsilon$ and center x. Hence,

$$|\rho_n^{(\varepsilon)}| \le C\varepsilon \int\limits_{\Gamma_\varepsilon^{(0)}} \big(\delta(x) + \varepsilon\big)^{-1} ds \le C\varepsilon \Big(1 + \int\limits_{|y| \le 1} (|y|^m + \varepsilon)^{-1} dy\Big). \qquad \square$$

The next theorem is the main result of this section.

Theorem 14.5.2. *Let the domain G_ε satisfy the hypotheses of Theorem* 14.1.3 *and let*

$$\delta_0(y) = \sum_{|\gamma| = m} (\gamma!)^{-1} y^\gamma \, (\nabla_y^\gamma \delta)(0).$$

Then

$$\operatorname{cap} G_\varepsilon = \begin{cases} \operatorname{cap} G_0 + O(\varepsilon) \quad \text{for} \quad m<n-1, \\ \mu_n^{-1}(1+|\nabla_y \gamma_0(0)|^2)|\log\varepsilon| \int\limits_{S^{n-2}} \delta_0(\vartheta)^{-1} d\vartheta + O(1) \quad \text{for} \quad m=n-1, \\ (1+|\nabla_y\gamma_0(0)|^2)\mu_n^{-1}\varepsilon^{-1+(n-1)/m} \int\limits_{\mathbb{R}^{n-1}} \big(\delta_0(y)+1\big)^{-1} dy + \beta_\varepsilon, \\ \qquad\qquad \text{for} \quad m>n-1, \end{cases}$$

where ϑ is a point of the sphere S^{n-2}, $\beta_\varepsilon = O(\varepsilon^{n/m-1})$ for $n < m$ and $\beta_\varepsilon = O(|\log\varepsilon|)$ for $n = m$.

Proof. In the case $m < n-1$ it suffices to verify the relation

$$\int\limits_{\Gamma_\varepsilon^{(0)}\cap V} (\partial/\partial N)(v\circ \varkappa_\varepsilon) ds = \int\limits_{\Gamma_\varepsilon^{(0)}\cap V} (\partial/\partial N) v\, ds + o(1).$$

Since in $\bar{G}_\varepsilon \cap V$ the mapping $\varkappa_\varepsilon$ is given by

$$\varkappa_\varepsilon(x) = \Big(y, \big(\delta(y)\, x_n + \varepsilon\gamma_0(y)\big)\, \big(\delta(y)+\varepsilon\big)^{-1}\Big),$$

we have

$$|(\partial/\partial N)(v\circ\varkappa_\varepsilon) - (\partial/\partial N) v \circ \varkappa_\varepsilon| \le c\varepsilon\big(\delta(x)+\varepsilon\big)^{-1}.$$

It remains to note that for $m < n-1$

$$\int\limits_{\Gamma_\varepsilon^{(0)}\cap V} \big(\delta(y)+\varepsilon\big)^{-1} ds = O(1).$$

Now let $m \ge n-1$. By Theorem 14.1.3 and Lemma 14.1.1,

$$\begin{aligned}(\partial/\partial N)(v\circ\varkappa_\varepsilon) &= (\partial/\partial N)\Big(\big(x_n - \gamma_0(y)\big)\big(\delta(y)+\varepsilon\big)^{-1}\Big) + O(|y|^{m+l-3}) \\ &= \big(\delta(y)+\varepsilon\big)^{-1}\Big(\big(1+|\nabla_y\gamma_0(0)|^2\big)^{1/2} + O(|y|)\Big) + O(|y|^{m+l-3}).\end{aligned}$$

Further, by Lemma 14.5.1,

$$\begin{aligned}\operatorname{cap} G_\varepsilon = \mu_n^{-1}\big(1+|\nabla_y\gamma_0(0)|^2\big)^{1/2} &\int\limits_{\Gamma_0^{(0)}\cap V} \big(\delta(y)+\varepsilon\big)^{-1} ds \\ + O\Big(1 + &\int\limits_{\Gamma_0^{(0)}\cap V} |y|\, \big(\delta(y)+\varepsilon\big)^{-1} ds\Big).\end{aligned}$$

It is clear that

$$\begin{aligned}\int\limits_{\Gamma_0^{(0)}\cap V} \big(\delta(y)+\varepsilon\big)^{-1} ds &= \big(1+|\nabla_y\gamma_0(0)|^2\big)^{1/2} \int\limits_D \big(\delta_0(y)+\varepsilon\big)^{-1} dy \\ + O\Big(\int\limits_{\Gamma_0^{(0)}\cap V} |y|\, \big(\delta(y)+\varepsilon\big)^{-1} ds\Big), &\quad \int\limits_{\Gamma_0^{(0)}\cap V} |y|\, \big(\delta(y)+\varepsilon\big)^{-1} ds = O(\beta_\varepsilon).\end{aligned}$$

Since

$$\int_D \left(\delta_0(y)+\varepsilon\right)^{-1} dy = \begin{cases} |\log\varepsilon| \int_{S^{n-2}} \delta_0(\vartheta)^{-1}\, d\vartheta + O(1) & \text{for} \quad m = n-1, \\ \varepsilon^{-1+(n-1)/m} \int_{\mathbb{R}^{n-1}} \left(\delta_0(y)+1\right)^{-1} dy + \beta_\varepsilon & \text{for} \quad m > n-1, \end{cases}$$

we obtain the desired result. □

14.5.2 Other condensers

We seek the asymptotics of capacity in the case of close conic vertices considered in 14.4.1 (see Fig. 14.3).

Theorem 14.5.3. *Let the domain G_ε satisfy the hypotheses of Theorem* 14.4.1. *Then*

$$\begin{aligned} \operatorname{cap} G_\varepsilon &= \operatorname{cap} G_0 + O(\varepsilon^\Lambda), & n \geq 3, \\ \operatorname{cap} G_\varepsilon &= (\alpha_+^{-1} + \alpha_-^{-1})|\log\varepsilon| + O(1), & n = 2. \end{aligned}$$

Proof. In view of the homogeneity of the function H, from Theorem 14.4.1 it follows that

$$\begin{aligned} (\partial/\partial N)u(\varepsilon,x) &= (\partial/\partial N)\{(1-\zeta(\varepsilon^{-1}y))v(x)\} \\ &\quad + (\partial/\partial N)\{\zeta(y)(w(\varepsilon^{-1}x) - (1-\zeta(\varepsilon^{-1}y))H(\varepsilon^{-1}x))\} \\ &\quad + (\partial/\partial N)R(\varepsilon,x). \end{aligned}$$

By virtue of (1), 14.4, and (3), 14.4,

$$\begin{aligned} |w(\varepsilon^{-1}x) - (1-\zeta(\varepsilon^{-1}y))H(\varepsilon^{-1}x)| &\leq c\varepsilon(|y|+\varepsilon)^{-1}, \\ |\nabla\left(w(\varepsilon^{-1}x) - (1-\zeta(\varepsilon^{-1}y))H(\varepsilon^{-1}x)\right)| &\leq c\varepsilon d_\varepsilon(x)^{-1}(|y|+\varepsilon)^{-1}, \end{aligned} \tag{2}$$

where $d_\varepsilon(x)$ is the distance from x to the conic vertices O and $(0,\dots,0,\varepsilon)$. Since the right-hand sides admit the estimates (5), 14.4, and (6), 14.4, and $R(\varepsilon,x) = O(\varepsilon^\Lambda)$, we have

$$\begin{aligned} |\nabla R(\varepsilon,x)| &\leq c\Big(\max_{|x|/2\leq|z|\leq 2|x|}\{|F(\varepsilon,z)|\,|x| + |R(\varepsilon,z)|\,|x|^{-1}\} \\ &\qquad + \sum_{j=0}^{1} \max_{|x|/2\leq|z|\leq 2|x|} |\nabla\varphi^{(j)}(\varepsilon,z)|\Big) \\ &\leq c\varepsilon^\Lambda\left((|y|+\varepsilon)^{-2}|x| + |x|^{-1}(|y|+\varepsilon)^{-1}\right) \leq c\varepsilon^\Lambda(|y|+\varepsilon)^{-1}. \end{aligned}$$

Hence,

$$\int_{\Gamma_\varepsilon^{(0)}} |(\partial/\partial N)R(\varepsilon,x)|ds \leq c\varepsilon^\Lambda\Big(\int_D (|y|+\varepsilon)^{-1}dy + 1\Big) \leq c\varepsilon^\Lambda. \tag{3}$$

Moreover,

$$|v(x)| + |y|\,|(\partial/\partial N)v(x)| \leq c, \qquad |(\partial/\partial N)\zeta(\varepsilon^{-1}y)| \leq c(|y|+\varepsilon)^{-1}. \tag{4}$$

Thus, in the case $n \geq 3$, from (2)–(4) we obtain

$$\begin{aligned} &\mu_n \operatorname{cap} G_\varepsilon \\ &= \int_{\Gamma_0^{(0)}} (\partial/\partial N)v(x)ds - \int_{\Gamma_0^{(0)}} (\zeta(\varepsilon^{-1}y)(\partial/\partial N)v(x) + v(x)(\partial/\partial N)\zeta(\varepsilon^{-1}y))ds + O(\varepsilon^\Lambda) \\ &= \mu_n \operatorname{cap} G_0 + O(\varepsilon^\Lambda). \end{aligned}$$

Now let $n = 2$. By (2)–(4),

$$2\pi \operatorname{cap} G_\varepsilon = \int\limits_{\Gamma_0^{(0)} \cap \{y:\, |y| \geq \alpha_1 \varepsilon\}} (\partial/\partial N) v(x) ds - \int\limits_{\Gamma_0^{(0)} \cap \{y:\, |y| \geq \alpha_1 \varepsilon\}} \Big\{ (\partial/\partial N)\Big(\big(1 - \zeta(\varepsilon^{-1} y)\big) v(x) \Big) \Big\} ds + O(\varepsilon^\Lambda + \varepsilon |\log \varepsilon|). \tag{5}$$

Here α_1 is a constant such that $\operatorname{supp} \zeta \subset \{y:\, |y| \leq \alpha_1\}$. Since $|v(x)| \leq \text{const}$ and

$$\operatorname{supp} \Big\{ (\partial/\partial N)\Big(\big(1 - \zeta(\varepsilon^{-1} y)\big) v(x) \Big) \Big\} \subset \{y:\, |y| \geq \alpha_1 \varepsilon\},$$

the second integral on the right in (5) is $O(1)$. Taking account of the equality

$$(\partial/\partial N) v(x) = (\alpha_\pm |x|^{-1}) + O(1) \quad \text{for} \quad \pm y > 0$$

on $\Gamma_0^{(0)} \cap V$ (see (2), 14.4), we arrive at

$$\begin{aligned} \operatorname{cap} G_\varepsilon &= (2\pi)^{-1} \int\limits_{\Gamma_0^{(0)} \cap \{y:\, |y| \geq \alpha_1 \varepsilon\}} (\partial/\partial N) v(x) ds + O(1) \\ &= (2\pi)^{-1} (\alpha_+^{-1} + \alpha_-^{-1}) |\log \varepsilon| + O(1). \end{aligned}$$

□

We turn to the case when one component of the boundary adheres to another one at a "large" set (see 14.4.2 and Fig. 14.4).

Theorem 14.5.4. *Let the hypotheses of Theorem 14.4.4 be fulfilled. Then*

$$\begin{aligned} \operatorname{cap} G_\varepsilon = (\mu_n \varepsilon)^{-1} \Big(\int\limits_\omega (1 + |\nabla_y \gamma_0(y)|^2) dy + (\varepsilon/m!)^{1/m} \big(m \sin(\pi/m)\big)^{-1} \pi \\ \times \int\limits_{\partial\omega} (1 + |\nabla_y \gamma_0(\tau, 0)|^2) \big((\partial/\partial\nu)^m \, \delta(\tau, 0) \big)^{1/m} ds_\tau \Big) + \beta_\varepsilon, \end{aligned}$$

where $\beta_\varepsilon = O(|\log \varepsilon|)$ for $m = 2$ and $\beta_\varepsilon = O(\varepsilon^{-1+2/m})$ for $m > 2$.

Proof. The scheme of proof is the same as that in Theorem 14.5.2. One should substitute the representation (12), 14.4, of u into (1). By (12), 14.4, with $\Phi_0 = 0$ and $\Phi_1 = 1$, on the set Ω

$$(\partial/\partial N) u(\varepsilon, x) = \varepsilon^{-1} (1 + |\nabla_y \gamma_0(y)|^2)^{1/2} + O(1).$$

Therefore,

$$\begin{aligned} \int\limits_\Omega (\partial/\partial N) u(\varepsilon, x) ds &= \int\limits_\omega (1 + |\nabla_y \gamma_0(y)|^2)^{1/2} (\partial/\partial N) u(\varepsilon, x) dy \\ &= \varepsilon^{-1} \int\limits_\omega (1 + |\nabla_y \gamma_0(y)|^2) ds + O(1). \end{aligned}$$

Besides, from (12), 14.4, it follows that

$$\begin{aligned} (\partial/\partial N) u(\varepsilon, x) = \big(\delta(\tau, \nu) + \varepsilon\big)^{-1} (1 + |\nabla_y \gamma_0(\tau, 0)|^2)^{1/2} \\ + O\Big(\varepsilon \big(\delta(\tau, \nu) + \varepsilon\big)^{-1} \Big), \quad x \in (\Gamma_0^{(0)} \cap V) \setminus \Omega. \end{aligned}$$

Hence,

$$\int\limits_{(\Gamma_0^{(0)}\cap V)\setminus\Omega} (\partial/\partial N)u(\varepsilon,x)ds$$

$$=\int\limits_{\partial\omega}\left(1+|\nabla_y\gamma_0(\tau,0)|^2\right)\int\limits_0^1\left(\delta(\tau,\nu)+\varepsilon\right)^{-1}d\nu ds_\tau+O(\beta_\varepsilon)$$

$$=\varepsilon^{-1+1/m}\int\limits_{\partial\omega}\left(1+|\nabla_y\gamma_0(\tau,0)|^2\right)\int\limits_0^\infty\left((m!)^{-1}\nu^m(\partial/\partial\nu)^m\delta(\tau,0)+1\right)^{-1}d\nu ds_\tau+O(\beta_\varepsilon).$$

□

To conclude the section we formulate the corresponding assertion for the case $m=1$ (see Fig. 14.5). Its proof does not require essentially new arguments.

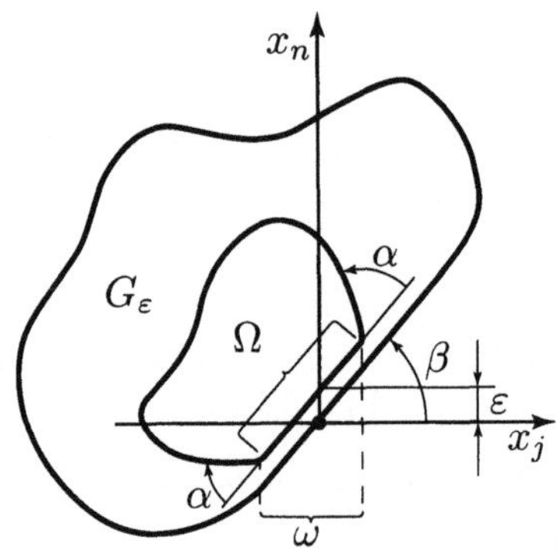

Fig. 14.5

Theorem 14.5.5. *Let the domain G_ε be the same as in* 14.4.2 *and let $m=1$ (see* (8), 14.4*). Then*

$$\operatorname{cap} G_\varepsilon=(\varepsilon\mu_n)^{-1}\int\limits_\Omega|\cos\beta|^{-1}ds+\mu_n^{-1}|\log\varepsilon|\int\limits_{\partial\Omega}\alpha^{-1}dl+O(1),$$

where α is the opening of the angle between the surfaces $\Gamma_0^{(0)}$ and $\Gamma_0^{(1)}$, and β is the angle between the normal to Ω and the axis x_n.

14.5.3 The complete asymptotics of capacity

One can derive the asymptotic representation (A3) for the capacity of the condenser G_ε from (22), 14.2, and (23), 14.2. Substituting (22), 14.2, into (1), we obtain

$$\operatorname{cap} G_\varepsilon\sim\mu_n^{-1}\int\limits_{\Gamma_\varepsilon^{(0)}}\sum_{j=0}^\infty\varepsilon^j(\partial/\partial N)\Big(u^{(j)}\big(\chi_\varepsilon(x)\big)\Big)ds. \tag{6}$$

Let us write the integral on the right in (6) as the sum of two integrals where the first one is over $\Gamma_\varepsilon^{(0)}\setminus V$. By virtue of (23), 14.2,

$$\int\limits_{\Gamma_\varepsilon^{(0)}\cap V}\sum_{j=0}^\infty\varepsilon^j(\partial/\partial N)\Big(u^{(j)}\big(\chi_\varepsilon(x)\big)\Big)ds\sim\sum_{j=0}^\infty\varepsilon^q\sigma_q. \tag{7}$$

Since on $\Gamma_\varepsilon^{(0)} \cap V$

$$\partial/\partial N = (1 + |\nabla_y \gamma_0(y)|^2)\Big\{(\delta(y)+\varepsilon)^{-1}\partial/\partial t$$
$$+ \nabla_y\gamma_0(y)\cdot\Big(\nabla_y - (\delta(y)+\varepsilon)^{-2}\Big((\delta(y)+\varepsilon)\nabla_y\gamma_0(y) + \gamma_0(y)\nabla_y\delta(y)\Big)\Big)\partial/\partial t\Big\}$$

and $\gamma_0 = 0$ (see (1), 14.2), the integral over $\Gamma_\varepsilon^{(0)} \cap V$ takes the form (in the coordinates (y,t))

$$\sum_{j=0}^{\infty} \varepsilon^j \int_D k_j(y)\big(\delta(y)+\varepsilon\big)^{-1} dy,$$

where k_j are smooth functions on $\bar{D}$. It remains to find the asymptotics of the integral

$$\int_0^c \big(\delta(r\vartheta)+\varepsilon\big)^{-1} k_j(r\vartheta) r^{n-2} dr$$

with small constant c and $\vartheta \in S^{n-2}$. Setting $z^m = \delta(r\vartheta)\delta(\vartheta)^{-1}$, we rewrite the integral in the form

$$\int_0^1 (z^m+\varepsilon)^{-1} l_j(z,\vartheta) z^{n-2} dz$$

l_j being a smooth function. Now (A3) results from (7) and the following assertion.

Lemma 14.5.6. *There exists the asymptotic expansion*

$$\int_0^1 (z^m+\varepsilon)^{-1} k(z) z^{n-2} dz \sim \varepsilon^{-1+(n-1)/m} \sum_{q=0}^{\infty} A_q \varepsilon^{q/m} + \varepsilon^{-1-[(n-1)/m]} |\log\varepsilon|$$
$$\times \sum_{q=0}^{\infty} B_q \varepsilon^{q/m} + \sum_{j=0}^{\infty} C_j \varepsilon^j. \quad ^2 \tag{8}$$

where k is a smooth function on $[0,1]$ while A_q, B_q, and C_j are constant.

Proof. After the change of variables $z \to \eta = \varepsilon^{-1/m} z$, we obtain

$$\int_0^\varepsilon (z^m+\varepsilon)^{-1} k(z) z^{n-2} dz = \varepsilon^{-1+(n-1)/m} \int_0^1 (\eta^m+\varepsilon)^{-1} k(\varepsilon^{1/m}\eta)\eta^{n-2} d\eta$$
$$\sim \varepsilon^{-1+(n-1)/m} \sum_{q=0}^{\infty} K_q \varepsilon^{q/m}.$$

[2] $[z]$ stands for the integral part of z and $\{z\} = z - [z]$.

Moreover, using the Taylor expansions of the functions $k(z)$ and $z \to (z^m+\varepsilon)^{-1}$, we have

$$\begin{aligned}\int_\varepsilon^1 (z^m+\varepsilon)^{-1}\, k(z) z^{n-2} dz &= \int_\varepsilon^1 \sum_{j=0}^{J} (-\varepsilon z^{-m})^j\, z^{n-2-m} \\ &\quad \times \Big(k(z) - \sum_{p=0}^{m(j+1)-n+1} (j!)^{-1} z^j (d/dz)^j k(0)\Big) dz \\ &\quad + \int_\varepsilon^1 \sum_{p=0}^{m(J+1)-n+1} (p!)^{-1} z^{p+n-2-m} (d/dz)^p k(0) \\ &\quad \times \Big((1+\varepsilon z^{-m})^{-1} - \sum_{0\le j\le (p+n-1)/m-2} (-\varepsilon z^{-m})^j\Big) dz \\ &\quad + \int_\varepsilon^1 z^{n-2-m}\Big((1+\varepsilon z^{-m}) - \sum_{j=0}^{J} (-\varepsilon z^{-m})^j\Big) \\ &\quad \times \Big(k(z) - \sum_{p=0}^{m(J+1)-n+1} (p!)^{-1} z^p (d/dz)^p k(0)\Big) dz. \end{aligned} \tag{9}$$

Denote the integrals on the right in (9) by I_1, I_2, and I_3. It is clear that $I_3 = O(\varepsilon^{J+1})$. The integral I_2 (after the change $z \to \eta = \varepsilon^{-1} z)$) and the integral I_1 can be rewritten in the form

$$\varepsilon^{-1+(n-1)/m} \int_1^{1/\varepsilon} \sum_{p=0}^{m(J+1)-n+1} \varepsilon^{p/n} \eta^{\{(p+n-1)/m\}m-1} \lambda_p(\eta^{-1}) d\eta$$

and

$$\int_\varepsilon^1 \sum_{j=0}^{J} \varepsilon^j \mu_j(z) dz,$$

respectively. Here λ_p and μ_j are smooth functions on $[0,1]$. Thus, I_1 is represented by the last series on the right in (8) and I_2 by the sum of the two first series. □

Chapter 15

Boundary Value Problems of Mathematical Physics in Thin Domains

Chapters 15 and 16 are devoted to the asymptotics of solutions to elliptic boundary value problems in thin domains. Some dimensions of such a domain are negligibly small in comparison with the others. In this chapter we consider a number of problems in mathematical physics. The general boundary value problems are studied in the next chapter.

Applications of the asymptotics are many and varied. In one way or another, any approximation of three-dimensional equations by two- or one-dimensional ones is connected with the asymptotics.

The first section in the chapter is auxiliary. Considering the Dirichlet problem and the mixed boundary value problem for the Poisson equation in a thin rectangle, we show some devices that allow us to find the asymptotics. In Section 15.2 we deal with a self-adjoint system of second order equations that is elliptic in the sense of Petrovski in an n-dimensional cylinder of small height. We obtain the principal term of the asymptotics of solutions. These are precisely the problems that arise in the majority of applications; the corresponding examples are collected in Section 15.3.

In the two last sections we study boundary value problems in noncylindrical domains. Section 15.4 contains the asymptotics of the solution to the Neumann problem in a thin rectangle notched lengthwise and in a three-dimensional cylinder of small height having a two-dimensional crack in the middle plane. In Section 15.5 we construct the principal term of the asymptotics of the solution to the Neumann problem in a domain with two close cuts.

15.1 Boundary Value Problems for the Laplace Operator in a Thin Rectangle

15.1.1 Statements of the problems

We consider the two simplest boundary value problems for the Poisson equation in the thin rectangle $Q_\varepsilon = (-1,1)\times(-\varepsilon/2,\varepsilon/2)$; namely, the Dirichlet problem

$$\Delta u(\varepsilon,x) = f(\varepsilon,x), \qquad x\in Q_\varepsilon; \quad u(\varepsilon,x_1,\pm\varepsilon/2) = \varphi^\pm(\varepsilon,x_1), \quad |x_1|<1; \quad (1)$$

$$u(\varepsilon,\pm 1,x_2) = \psi^\pm(\varepsilon,x_2), \qquad |x_2|<\varepsilon/2, \quad (2)$$

and the mixed boundary value problem

$$\Delta u(\varepsilon,x) = f(\varepsilon,x), \quad x\in Q_\varepsilon; \quad (\partial/\partial x_2)u(\varepsilon,x_1,\pm\varepsilon/2) = \varphi^\pm(\varepsilon,x_1), \quad |x_1|<1; \quad (3)$$

$$u(\varepsilon,\pm 1,x_2) = \psi^\pm(\varepsilon,x_2), \quad |x_2|<\varepsilon/2. \quad (4)$$

After the change of variable $x \to (x_1, \varepsilon^{-1}x_2)$ the domain Q_ε converts into $Q_1 = (-1,1) \times (-1/2, 1/2)$ and the Laplace operator takes the form

$$\Delta = (\partial/\partial x_1)^2 + \varepsilon^{-2}(\partial/\partial y)^2.$$

Therefore, the limit problem for (1) is the boundary value problem for the ordinary differential equation in the segment $[-1/2, 1/2]$.

15.1.2 Asymptotics of solution to the Dirichlet problem

Suppose that the right-hand sides f and $\varphi^\pm$, $\psi^\pm$ in (1), (2) admit the expansions

$$f(\varepsilon, x) = \varepsilon^{-2}\Big(\sum_{j=0}^{J} \varepsilon^j f_j(x_1, \varepsilon^{-1}x_2) + \tilde{f}_J(\varepsilon, x)\Big), \tag{5}$$

$$\psi^\pm(\varepsilon, x_2) = \sum_{j=0}^{J} \varepsilon^j \psi_j^\pm(\varepsilon^{-1}x_2) + \tilde{\psi}_J^\pm(\varepsilon, x_2), \tag{6}$$

$$\varphi^\pm(\varepsilon, x_1) = \sum_{j=0}^{J} \varepsilon^j \varphi_j^\pm(x_1) + \tilde{\varphi}_J^\pm(\varepsilon, x_1), \tag{7}$$

where $f_j \in \mathbf{C}^\infty(\bar{Q}_1)$, $\varphi_j^\pm \in \mathbf{C}^\infty[-1,1]$, and $\psi_j^\pm \in \mathbf{C}^\infty[-1/2,1/2]$ while the remainders $\tilde{f}_J$, $\tilde{\psi}_J^\pm$, and $\tilde{\varphi}_J^\pm$ are subject to the inequality

$$\max_{x \in \bar{Q}_\varepsilon} |\tilde{f}_J(\varepsilon, x)| + \max_{x_2 \in [-1/2,1/2]} |\tilde{\psi}_J^\pm(\varepsilon, x_2)| + \max_{x_1 \in [-1,1]} |\tilde{\varphi}_J^\pm(\varepsilon, x_1)| \le c\varepsilon^{J+1}. \tag{8}$$

To provide the existence of a continuous solution to problem (1), (2) in $\bar{Q}_\varepsilon$ the Dirichlet data must satisfy the compatibility condition at the corner points

$$\varphi^\pm(\varepsilon, 1) = \psi^\pm(\varepsilon, \pm\varepsilon/2), \quad \varphi^\pm(\varepsilon, -1) = \psi^\pm(\varepsilon, \pm\varepsilon/2). \tag{9}$$

We seek an asymptotic solution to equations (1) in the form

$$v(\varepsilon, x) \sim \sum_{j=0}^{\infty} \varepsilon^j v_j(x_1, \varepsilon^{-1}x_2). \tag{10}$$

Substituting (5)–(7) and (10) into (1) and equating the coefficients by the same degrees of ε, we obtain the sequence of the boundary value problems for v_j:

$$\begin{aligned} &(\partial/\partial y)^2 v_j(x_1, y) = f_j(x_1, y) - (\partial/\partial x_1)^2 v_{j-2}(x_1, y), \quad y \in [-1/2, 1/2]; \\ &v_j(x_1, \pm 1/2) = \varphi_j^\pm(x_1), \quad j \in \mathbb{N}_0, \quad v_{-2} = v_{-1} = 0. \end{aligned} \tag{11}$$

It is clear that the problems (11) are uniquely solvable.

The series (10) formally satisfies (1) and produces a discrepancy in (2). To compensate the discrepancy we construct boundary layers near the sides $\Gamma^\pm = \{x \colon x_1 = \pm 1, |x_2| < \varepsilon/2\}$ of the rectangle Q_ε. To this end we introduce the new coordinates

$$\eta^+ = \varepsilon^{-1}(1 - x_1, x_2), \quad \eta^- = \varepsilon^{-1}(1 + x_1, x_2). \tag{12}$$

The coordinate η^+ corresponds to the boundary layer near Γ^+. Since both boundary layers can be constructed in the same way, in what follows we consider only Γ^- and set $\eta^- = \eta$.

Substitute the series (10) and the series

$$w^{-}(\varepsilon, x) \sim \sum_{j=0}^{\infty} \varepsilon^{j} w_{j}^{-}\left(\varepsilon^{-1}(1+x_{1}), \varepsilon^{-1} x_{2}\right) \tag{13}$$

into problem (1), (2) and make use of the coordinates η. The domain Q_ε is mapped to $\Pi = \{\eta \in \mathbb{R}^2 \colon \eta_1 > 0, |\eta_2| < 2\}$. Therefore, we obtain the boundary value problem

$$\begin{aligned} &\Delta_\eta w_j^-(\eta) = 0, \quad \eta \in \Pi; \quad w_j^-(\eta_1, \pm 1/2) = 0, \quad \eta_1 \in (0, \infty); \\ &w_j^-(0, \eta_2) = \psi_j^-(\eta_2) - v_j(-1, \eta_2), \quad |\eta_2| < 1/2. \end{aligned} \tag{14}$$

This problem is called the second limit problem.

The next assertions follow from 15.1.1.

Proposition 15.1.1. *Let*

$$\psi_j^-(\pm 1/2) = v_j(-1, \pm 1/2). \tag{15}$$

There exists a unique solution w_j^- to problem (14) with finite Dirichlet's integral. The solution and its first derivatives are continuous in $\bar{\Pi}$ and

$$|w_j^-(\eta)| \le \text{const} \exp(-\pi \eta_2). \tag{16}$$

Note that (15) follows from the expansions (6) and (7) of the right-hand sides $\varphi^\pm$ and $\psi^\pm$ of problem (1), (2), from the definition (10) of v_j, and assumption (9). Thus, the exponentially decaying boundary layer has been constructed. The justification of the obtained asymptotic representation of the solution u to problem (1), (2) as the sum of series $v + w^- + w^+$ can easily be derived from the maximum principle. We formulate the corresponding assertion.

Proposition 15.1.2. *Let the right-hand sides of problem (1), (2) be subject to (6)–(9). Then for the solution u to problem (1), (2) the inequality*

$$\max_{x \in \bar{Q}_\varepsilon} \left| u(\varepsilon, x) - \sum_{j=0}^{J} \varepsilon^j \Big(v_j(x_1, \varepsilon^{-1} x_2) + \sum_{\pm} w_j^\pm\left(\varepsilon^{-1}(1 \pm x_1), \varepsilon^{-1} x_2\right) \Big) \right| \le \text{const} \cdot \varepsilon^{J+1}$$

holds with any positive integer J. Here v_j satisfies (11), $w_j^\pm$ is the exponentially decaying solution to problem (14). Replacing "$-$" by "$+$" and $v_j(-1, \eta_2)$ by $v_j(1, \eta_2)$ in (14), we obtain the problems for w_j^+.

15.1.3 Asymptotics of solution to the mixed problem

Let the right-hand sides f and $\psi^\pm$ of problem (3), (4) admit the expansions (6) and let

$$\varphi^\pm(\varepsilon, x_1) = \varepsilon^{-1} \Big(\sum_{j=0}^{J} \varepsilon^j \, \varphi_j^\pm(x_1) + \tilde{\varphi}_J^\pm(\varepsilon, x_1) \Big), \tag{17}$$

where the notation is similar to that in (7); the remainders in (6) and (17) are subject to (8) and

$$\max_{x_2 \in [-\varepsilon/2, \varepsilon/2]} |(\partial/\partial x_2) \tilde{\psi}_J^\pm(\varepsilon, x_2)| \le c \cdot \varepsilon^J. \tag{18}$$

One cannot find a formal asymptotic solution to problem (3) in the form (10). Indeed, the first limit problem

$$(d/dy)^2 \, V(y) = F(y), \quad y \in [-1/2, 1/2]; \quad (d/dy) \, V(\pm 1/2) = \Phi^\pm \tag{19}$$

is not uniquely solvable. Problem (17) has a solution only if

$$\int_{-1/2}^{1/2} F(y)\,dy = \Phi^+ - \Phi^- \tag{20}$$

and its solution is determined up to a constant term. From now on by a solution to problem (19) with right-hand sides F and $\Phi^\pm$ subject to (20) we mean the function whose mean value on the segment $[-1/2, 1/2]$ is equal to zero. To obtain the asymptotics of the solution to problem (3), (4) we make use of the obvious fact that for a function z depending only on x_1 there is the equality

$$\Delta z(x_1) = (d/dx_1)^2 z(x_1).$$

A formal asymptotic solution to equations (3) is sought as the sum of series (10) and the series

$$z(\varepsilon, x_1) \sim \varepsilon^{-2} \sum_{j=0}^{\infty} \varepsilon^j z_j(x_1). \tag{21}$$

Choosing the coefficients z_j, we provide the solvability of the first limit problem at every step of the algorithm.

We replace the function u in (3) by the sum of series (10) and (21) and the right-hand sides f and $\varphi^\pm$ by their expansions (6) and (17). Equating the coefficients by the same degrees of ε, we obtain the equations

$$\begin{aligned}(\partial/\partial y)^2 v_j(x_1, y) =& f_j(x_1, y) - (\partial/\partial x_1)^2 v_{j-2}(x_1, y) - (d/dx_1)^2 z_j(x_1),\\ & y \in [-1/2, 1/2];\\ & (\partial/\partial y)^2 v_j(x_1, \pm 1/2) = \varphi_j^\pm(x_1).\end{aligned} \tag{22}$$

Taking account of the condition

$$\int_{-1/2}^{1/2} v_q(x_1, y)\,dy = 0, \quad q = \mathbb{N}_0, \tag{23}$$

we find that problem (22) has a solution only in the case that

$$(d/dx_1)^2 z_j(x_1) = \int_{-1/2}^{1/2} f_j(x_1, y)\,dy + \varphi_j^-(x_1) - \varphi_j^+(x_1), \quad x_1 \in [-1, 1]. \tag{24}$$

Suppose that the functions z_j satisfy (24). Then problem (22) takes the form

$$\begin{aligned}(\partial/\partial y)^2 v_j(x_1, y) =& f_j(x_1, y) - \int_{-1/2}^{1/2} f_j(x_1, \xi)\,d\xi + \varphi_j^+(x_1) - \varphi_j^-(x_1)\\ & - (\partial/\partial x_1)^2 v_{j-2}(x_1, y), \quad y \in [-1/2, 1/2];\\ & (\partial/\partial y) v_j(x_1, \pm 1/2) = \varphi_j^\pm(x_1)\end{aligned} \tag{25}$$

and has the solution v_j.

Now we are going to construct a boundary layer. As in 15.1.2, we first seek a boundary layer near Γ^- in the form (13). Substitute the sum of series (10), (21), (13) for u in (3), (4) and make the change of variables $x \to \eta^-$ (see (12)). Then, using (6) and (17), we equate the coefficients at ε^q, $q = -2, -1, 0, 1, \ldots$. For $q = -2$ and $q = -1$ from the boundary condition (4) we obtain

$$z_0(-1) = 0, \quad z_1(-1) = 0. \tag{26}$$

For the rest q we find

$$w_q^-(0, \eta_2) = \psi_q^-(\eta_2) - v_q(-1, \eta_2) - z_{q+2}(-1), \quad \eta_2 \in (-1/2, 1/2). \tag{27}$$

From (3) we obtain the relations

$$\Delta_\eta w_q^-(\eta) = 0, \quad \eta \in \Pi; \quad (\partial/\partial\eta_2) w_q^-(\eta_1, \pm 1/2) = 0, \quad \eta_1 \in (0, \infty), \tag{28}$$

which together with (27) for $q \in \mathbb{N}_0$ give the boundary value problems for the coefficients w_j^- of the series (13).

The next assertion can be derived from Proposition 1.1.4 in the same way as Lemma 1.4.1 from Proposition 1.2.4.

Proposition 15.1.3. *Let $\Psi \in \mathbf{C}^\infty[-1/2, 1/2]$. Then there exists a unique solution to the problem*

$$\Delta_\eta W(\eta) = 0, \quad \eta \in \Pi;$$
$$(\partial/\partial\eta_2) W(\eta_1, \pm 1/2) = 0, \quad \eta_1 > 0; \quad W(0, \eta_2) = \Psi(\eta_2), \quad |\eta_2| < 1/2,$$

which is continuous in $\bar{\Pi}$ and has the finite Dirichlet integral. For this solution there holds the asymptotic formula

$$W(y) = c_1 + O\big(\exp(-\pi\eta_2)\big) \quad \text{with} \quad c_1 = \int_{-1/2}^{1/2} \Psi(\xi)\, d\xi.$$

Thus, the functions w_q^- satisfy (28) and (27); they are exponentially decaying only in the case

$$\int_{-1/2}^{1/2} \big(\Psi_q^-(\xi) - v_q(-1, \xi) - z_{q+2}(-1)\big) d\xi = 0, \quad q \in \mathbb{N}_0. \tag{29}$$

By virtue of (23), relations (29) can be rewritten in the form

$$z_{q+2}(-1) = \int_{-1/2}^{1/2} \psi_q^-(\xi) d\xi, \qquad q \in \mathbb{N}_0. \tag{30}$$

Similar analysis of the boundary layer near Γ^+ enables us to obtain the equations

$$z_0(1) = 0, \quad z_1(1) = 0, \quad z_{q+2}(1) = \int_{-1/2}^{1/2} \psi_q^+(\xi) d\xi, \quad q \in \mathbb{N}_0. \tag{31}$$

Therefore, the boundary value problems for $w_j^{\pm}$ take the form

$$\begin{aligned}
&\Delta_\eta w_j^{\pm}(\eta) = 0, \quad \eta \in \Pi;\\
&(\partial/\partial\eta_2) w_j^{\pm}(\eta_1, 1/2) = (\partial/\partial\eta_2) w_j^{\pm}(\eta_1, -1/2) = 0, \quad \eta_1 > 0,\\
&w_j^{\pm}(0,\eta_2) = \psi_j^{\pm}(\eta_2) - \int_{-1/2}^{1/2} \psi_j^{\pm}(\xi)\,d\xi - v_j(\pm 1, \eta_2), \quad |\eta_2| < 1/2.
\end{aligned} \tag{32}$$

From (23) and Proposition 15.1.3 it follows that problems (23), $j \in \mathbb{N}_0$, have solutions subject to (16).

We have constructed the formal asymptotics. To justify the asymptotics one can use the next assertion, which follows from the energy inequality.

Proposition 15.1.4. *Let the right-hand sides f, $\varphi^{\pm}$, and $\psi^{\pm}$ of problem* (3), (4) *be subject to* (6), (8), (17), *and* (18). *Then for the solution u to problem* (3), (4) *and the partial sum*

$$U^{(J)}(\varepsilon, x) = \varepsilon^{-2} z_0(x_1) + \sum_{j=0}^{J} \varepsilon^j \Big(v_j(x_1, \varepsilon^{-1} x_2) + \varepsilon^{-1} z_{j+1}(x_1) + \sum_{\pm} w_j^{\pm}\big(\varepsilon^{-1}(1 \pm x_1), \varepsilon^{-1} x_2\big)\Big).$$

there holds the inequality

$$\|u - U^{(J)};\, \mathbf{W}_2^1(Q_\varepsilon)\| \le \text{const} \cdot \varepsilon^{J-1/2}, \quad J \in \mathbb{N}.$$

Here J is an arbitrary positive integer, v_j being the solutions to the ordinary differential equations (25) *subject to* (23). *The functions z_q are the solutions to problems* (24), (26), (30), *and* (31). *Finally, $w_j^{\pm}$ are the exponentially decaying solutions to problems* (32).

15.2 The Principal Term in Asymptotics of the Solution to a Boundary Value Problem for a System of Second Order Equations in a Cylinder of Small Height

In this section we study asymptotics of the solution to the mixed boundary value problem for a self-adjoint system of second order differential equations in an n-dimensional cylinder $\tilde{C}_\varepsilon = \Omega \times (-\varepsilon/2, \varepsilon/2)$. The system is supposed to be elliptic in the sense of Petrovski. We describe only the third limit problem in Ω (in 15.1.2 the role of such a problem was played by the ordinary differential equation (24), 15.1, with boundary conditions (30), 15.1). In so doing we obtain the principal term of the asymptotics.

15.2.1 Statement of the problem

Let $\Omega \in \mathbb{R}^{n-1}$ be a domain in $\mathbf{C}^\infty$ with compact closure Ω and $\mathbf{C}^\infty$ smooth boundary $\partial\Omega$. We denote by $\tilde{C}_\varepsilon$ the cylinder $\Omega \times (-\varepsilon/2, \varepsilon/2)$ of a small height ε (see

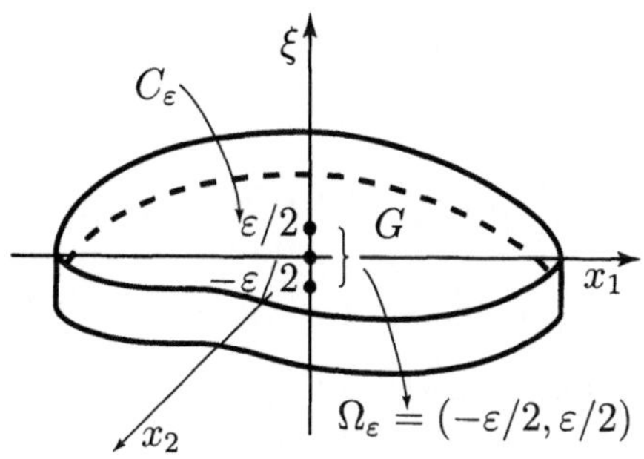

Fig. 15.1

Fig. 15.1) and consider the strongly elliptic (in the Petrovski sense) self-adjoint $T \times T$-system of second order equations

$$\sum_{k=1}^{T} L_{lk}(\varepsilon, x, \partial/\partial x)\, u_k(\varepsilon, x) = f_l(\varepsilon, x), \quad x \in \tilde{C}_\varepsilon, \quad l = 1, \dots, T, \tag{1}$$

where

$$\begin{aligned} L_{lk}(\varepsilon, x, \partial/\partial x) = -&\sum_{j=1}^{n} (\partial/\partial x_j) \Big\{ \sum_{i=1}^{n} a_{lk}^{(j,i)}(\varepsilon, x)\, (\partial x_j) + a_{lk}^{(j)}(\varepsilon, x) \Big\} \\ +&\sum_{i=1}^{n} a_{lk}^{(j)}(\varepsilon, x)\, (\partial/\partial x_j) + a_{lk}(\varepsilon, x), \end{aligned} \tag{2}$$

$a_{lk}^{(j,i)}$, $a_{lk}^{(j)}$, and a_{lk} are smooth real functions in $\varepsilon \in [0,1]$ and $x \in \overline{\tilde{C}_1}$,

$$a_{lk}^{(j,i)}(\varepsilon, x) = a_{kl}^{(i,j)}(\varepsilon, x), \quad a_{lk}(\varepsilon, x) = a_{kl}(\varepsilon, x), \quad i, j = 1, \dots, n;\ k, l = 1, \dots, T. \tag{3}$$

Denote by $B_{lk}^{(j)}(\varepsilon, x, \partial/\partial x))$ the differential equations in braces in (2) and introduce the boundary conditions

$$\begin{aligned} &\sum_{k=1}^{T} B_{lk}^{(n)}(\varepsilon, x', \pm\varepsilon/2, \partial/\partial x)\, u(\varepsilon, x', \pm\varepsilon/2) = g_l^{\pm}(\varepsilon, x'), \\ &\quad x' \in \Omega, \quad l = 1, \dots T; \end{aligned} \tag{4}$$

$$u_l(\varepsilon, x) = 0, \quad x \in S_\varepsilon = \partial\Omega \times (-\varepsilon/2, \varepsilon/2), \quad l = 1, \dots, T, \tag{5}$$

with $(x' = x_1, \dots, x_{n-1})$.

We introduce vector-valued functions $u, f, g^{\pm}$ with components $u_k, f_k, g_k^{\pm}$, $k = 1, \dots, T$ and the $T \times T$-matrix differential operators $L(\varepsilon, x, \partial/\partial x)$ and $B^{(n)}(\varepsilon, x', \pm\varepsilon/2, \partial/\partial x)$ with entries $L_{lk}(\varepsilon, x, \partial x)$ and $B_{lk}^{(n)}(\varepsilon, x', \pm\varepsilon/2, \partial/\partial x)$. For ease of notation, we write the system (1), (4), (5) in the vector form

$$\begin{aligned} &L(\varepsilon, x, \partial/\partial x)\, u(\varepsilon, x) = f(\varepsilon, x), \quad x \in \tilde{C}_\varepsilon, \quad u(\varepsilon, x) = 0, \quad x \in S_\varepsilon; \\ &B^{(\pm)}(\varepsilon, x', \partial/\partial x)\, u(\varepsilon, x', \pm\varepsilon/2) = g^{\pm}(\varepsilon, x'), \quad x' \in \Omega. \end{aligned} \tag{6}$$

Assume that for any vector-valued function $u\in\mathbf{W}^1_{2,0}(\tilde{C}_\varepsilon;S_\varepsilon)=\{u\in\mathbf{W}^1_2(\tilde{C}_\varepsilon): u=0\,\mathrm{on}\,S_\varepsilon\}$ there holds the inequality of Korn's type

$$E(u;\tilde{C}_\varepsilon)\geq c_\delta^2\sum_{k=1}^{T}\varepsilon^{2\alpha_k-2+\delta}\Big(||\rho_\varepsilon(x')^{1-\alpha_k}(\partial/\partial x')u_k;\mathbf{L}_2(\tilde{C}_\varepsilon)||^2 + ||\rho_\varepsilon(x')^{-\alpha_k}u_k;\mathbf{L}_2(\tilde{C}_\varepsilon)||^2+\varepsilon^2||\rho_\varepsilon(x')^{-\alpha_k}(\partial/\partial x_n)u_k;\mathbf{L}_2(\tilde{C}_\varepsilon)||^2\Big), \tag{7}$$

where $\delta>0$, $\rho_\varepsilon(x')=\max\{\varepsilon,\ \mathrm{dist}(x',\partial\Omega)\}$, and $\alpha_1,\ \dots,\ \alpha_T$ are some integers. The constant c_δ depends on δ being independent of ε. Finally,

$$\begin{aligned}E(u;\tilde{C}_\varepsilon)=\sum_{l,k=1}^{T}\int_{\tilde{C}_\varepsilon}\Big\{\sum_{j=1}^{n}\Big(\sum_{i=1}^{n}a_{lk}^{(j,i)}(\varepsilon,x)\,(\partial/\partial x_i)\,u_k(\varepsilon,x)\\ +a_{lk}^{(j)}(\varepsilon,x)\,u_k(\varepsilon,x)+a_{kl}^{(j)}(\varepsilon,x)\Big)(\partial/\partial x_j)u_l(\varepsilon,x)\\ +a_{lk}(\varepsilon,x)\,u_k(\varepsilon,x)\,u_l(\varepsilon,x)\Big\}dx.\end{aligned} \tag{8}$$

Lemma 15.2.1. *(i) For any right-hand sides $f\in\mathbf{L}_2(\tilde{C}_\varepsilon)$ and $g^\pm\in\mathbf{L}_2(\Omega)$ there exists a unique generalized solution $u\in\mathbf{W}^1_{2,0}(\tilde{C}_\varepsilon;S_\varepsilon)$ and*

$$E(u;\tilde{C}_\varepsilon)^{1/2}\leq c_\delta\sum_{l=1}^{T}\varepsilon^{1-\alpha_l-\delta}\Big(||\rho_\varepsilon(x')^{\alpha_l}f_l;\mathbf{L}_2(\tilde{C}_\varepsilon)||+\varepsilon^{1/2}\sum_{\pm}||\rho_\varepsilon(x')^{\alpha_l}g_l^\pm;\mathbf{L}_2(\Omega)||\Big). \tag{9}$$

(ii) The numbers α_l are positive for $l=1,\dots,T$.

Proof. (i) One can prove the assertion using the standard scheme (see, for example, LIONS and MAGENES [1], LADYSZHENSKAYA [1]).

(ii) We substitute a vector-valued function $w\in\mathbf{W}^1_{2,0}(\Omega;\partial\Omega)$ with nonzero components into (7). Since $E(w;\tilde{C}_\varepsilon)=O(\varepsilon)||w;\mathbf{W}^1_2(\Omega)||$, we have

$$\sum_{k=1}^{T}\varepsilon^{\alpha_k}||w_k;\mathbf{L}_2(\Omega)||=O(\varepsilon).$$

If $\alpha_l\leq 0$, we let $\varepsilon\to 0$ and obtain the equality $||w_l;\mathbf{L}_2(\Omega)||=0$, which is impossible. The contradiction completes the proof. □

Corollary 15.2.2. *From (ii) and Lemma* 15.2.1 *it follows that* (7) *and* (9) *are still valid in the case $\rho_\varepsilon(x')=1$.*

15.2.2 Auxiliary constructions

As in Section 15.1, with the help of the change of variables $x\to(y,z)=(x',\varepsilon^{-1}x_n)$ we reduce the thin domain $\tilde{C}_\varepsilon$ to the standard $\tilde{C}_1$. The operators L and $B^\pm$ in (6) can be decomposed into the following formal series :

$$\begin{aligned}L(\varepsilon,y,\partial/\partial y,\varepsilon^{-1}\partial/\partial z)&\sim\varepsilon^{-2}\sum_{m=0}^{\infty}\varepsilon^m L_m(y,z,\partial/\partial y,\partial/\partial z),\\ B^\pm(\varepsilon,y,\partial/\partial y,\varepsilon^{-1}\partial/\partial z)&\sim\varepsilon^{-1}\sum_{m=0}^{\infty}\varepsilon^m B_m^\pm(y,\partial/\partial y,\partial/\partial z).\end{aligned} \tag{10}$$

Here L_m and $B_m^\pm$ are differential operators of order no greater than 2 and 1, respectively; the coefficients of L_m are polynomials in z of degrees not greater than m; finally,

$$L_0(y,z,\partial/\partial y,\partial/\partial z) = L_0(y,\partial/\partial z) = -\big(a_{lk}^{(n,n)}(0,y,0)\,(\partial/\partial z)^2\big)_{l,k=1}^{T},$$
$$B_0^\pm(y,\partial/\partial y,\partial/\partial z) = B_0^\pm(y,\partial/\partial z) = -\big(a_{lk}^{(n,n)}(0,y,0)\,(\partial/\partial z)^2\big)_{l,k=1}^{T}.$$

By virtue of (3) and the strong ellipticity of L the matrix $A(y)=\big(a_{lk}^{(n,n)}(0,y,0)\big)_{l,k=1}^{T}$ is self-adjoint and positive definite. Therefore, with the help of a unitary matrix-valued function $U(y)$ it reduces to the diagonal form $U(y)^{-1}A(y)U(y)=\operatorname{diag}\big(A_1(y),\dots,A_T(y)\big)$ with $A_k>0$ in $\bar\Omega$. We assume that all the necessary changes from $u,f,g^\pm$ to $Uu,Uf,Ug^\pm$ have been fulfilled in (1), (4)–(6) from the very beginning. Also suppose that the right-hand sides in (6) admit the asymptotic expansions

$$f_k(\varepsilon,x)\sim\varepsilon^{-2}\sum_{m=0}^{\infty}\varepsilon^m f_k^{(m)}(x',\varepsilon^{-1}x_n),\quad g_k^\pm(\varepsilon,x')\sim\varepsilon^{-1}\sum_{m=0}^{\infty}\varepsilon^m g_k^{(m)\pm}(x'),\tag{11}$$

where $f_k^{(m)}\in\mathbf{C}^\infty(\overline{\tilde C_1})$ and $g_k^{(m)}\in\mathbf{C}^\infty(\bar\Omega)$, $k=1,\dots,T$.

According to the terminology of Section 15.1, the first limit problem for (6) is the Neumann problem

$$\begin{aligned}&L_0(y,\partial/\partial z)w^{(0)}(y,z)=f^{(0)}(y,z),\quad z\in(-1/2,1/2);\\&B_0^\pm(y,\partial/\partial z)w^{(0)}(y,\pm1/2)=g^{(0)\pm}(y)\end{aligned}\tag{12}$$

with parameter y. The compatibility condition for problem (12) is of the form

$$g_l^{(0)-}(y)-g_l^{(0)+}(y)+\int_{-1/2}^{1/2}f_l^{(0)}(y,\eta)d\eta=0,\quad l=1,\dots,T.\tag{13}$$

Therefore, as was shown in 15.1.2, one has to construct the third limit problem (the second limit problem is the problem on the boundary layer near S_ε, see 15.1). The method of the construction consists in an attempt to solve the homogeneous problem (6). Denote by $v_1,\dots,v_T$ some functions in $\mathbf{C}^\infty(\bar\Omega)$ and consider the series

$$V^{(k)}(\varepsilon,x,\varepsilon^{-1}x_n)=e^{(k)}v_k(x')+\sum_{r=1}^{\infty}\varepsilon^r v^{(k,r)}(x',\varepsilon^{-1}x_n),\tag{14}$$

where $e^{(1)},\dots,e^{(T)}$ are the basis vectors in $\mathbb{R}^T$ and the functions $v^{(k,r)}$ have mean values (in z) on $(-1/2,1/2)$ equal to zero. Since $A_k(\partial/\partial z)v_k=0$, from (10) and the formal relations $LV^{(k)}=0$ and $B^\pm V^{(k)}=0$ it follows that

$$\begin{aligned}&L_0(y,\partial/\partial z)v^{(k,r)}(y,z)=-\sum_{m=1}^{r}L_m(y,z,\partial/\partial y,\partial/\partial z)v^{(k,r-m)}(y,z),\ z\in(-1/2,1/2);\\&B_0^\pm(y,\partial/\partial z)v^{(k,r)}(y,\pm1/2)=-\sum_{m=1}^{r}B_m^\pm(y,\partial/\partial y,\partial/\partial z)v^{(k,r-m)}(y,\pm1/2)\end{aligned}\tag{15}$$

with $v^{(k,0)} = e^{(k)} v_k$, $r \in \mathbb{N}$. Denote the right-hand sides of problems (15) by $F^{(k,r)}$ and $G^{(k,r)\pm}$ and set

$$X^{(k,r)}(y; v_k) = G^{(k,r)-}(y) - G^{(k,r)+}(y) + \int_{-1/2}^{1/2} F^{(k,r)}(y, \eta) d\eta. \tag{16}$$

As was mentioned, problems (15) can be unsolvable. We change them. Namely, we assume that the mean values (in z) of the functions $v^{(k,1}, v^{(k,2)}, \dots$ are equal to zero and the functions satisfy the equations

$$\begin{aligned} &L_0(y, \partial/\partial z) v^{(k,r)}(y, z) = F^{(k,r)}(y, z) - X^{(k,r)}(y; v_k), \quad z \in (1 - 1/2, 1/2); \\ &B_0^{\pm}(y, \partial/\partial z) v^{(k,r)}(y, \pm 1/2) = G^{(k,r)\pm}(y). \end{aligned} \tag{17}$$

By virtue of (16), the compatibility condition of type (13) is fulfilled for every problem (17). Therefore, $v^{(k,r)}$ exist and are given by

$$\begin{aligned} v_l^{(k,r)}(y, z) = -A_l(y)^{-1} \Big(&\int_{-1/2}^{z} (z - \eta) F_l^{(k,r)}(y, \eta) d\eta + z G_l^{(k,r)}(y) \\ - 2^{-1} &\int_{-1/2}^{1/2} (4^{-1} - \eta + \eta^2) F_l^{(k,r)}(y, \eta) - 2^{-1}(z^2 + z - 1/12) X^{(k,r)}(y; v_k) \Big). \end{aligned} \tag{18}$$

Lemma 15.2.3. *(i) The vectors $X^{(k,s)}(y; v_k)$ defined by (16) admit the representations*

$$X^{(k,s)}(y; v_k) = M^{(s,k)}(y, \partial/\partial y) v_k(y), \quad s \in \mathbb{N}, \tag{19}$$

where v_k are differential operators of order no greater than s.

(ii) The vector-valued functions (18) are polynomials in z of degree no greater than $2s$ and linearly depend on the derivatives of v_k of order no greater than s.

Proof. The argument is by induction on s. For $s = 0$ the above assertions are obviously fulfilled. Let i) and ii) hold for $s = 0, 1, \dots, r-1$. Since $L_m(y, z, \partial/\partial y, \partial/\partial z)$ and $B_m^{\pm}(y, \partial/\partial y, \partial/\partial z)$ are polynomials in z and $\partial/\partial y$ of degree no greater than m (see (10)), by the induction hypothesis $F^{(k,r)}$ is a polynomial in z of degree $2r - 2$ while $F^{(k,r)}$ and $G^{(k,r)\pm}$ are linear combinations of the derivatives of v_k of order no greater than r. This implies i). Solving ordinary differential equations with constant coefficients and right-hand sides $F^{(k,r)}$, we arrive at ii). □

15.2.3 Asymptotics of the solution

Far from S_ε we seek the asymptotics of the solution to problem (1), (4), (5) in the form

$$u_k(\varepsilon, x) \sim \sum_{m=0}^{\infty} \varepsilon^r \Big(w_k^{(r)}(x', \varepsilon^{-1} x_n) + \varepsilon^{-\alpha_{\max}} \sum_{\tau=1}^{T} \varepsilon^{-\alpha_r} \sum_{s=0}^{r} v_k^{(\tau,s,r-s)}(\varepsilon, x', \varepsilon^{-1} x_n) \Big). \tag{20}$$

Here $v^{(\tau,0,\rho)}(\varepsilon, x', \varepsilon^{-1} x_n) = e^{(\tau)} v_\tau^{(\rho)}(\varepsilon, x')$ and the vectors $v^{(\tau,s,\rho)}$ can be determined by means of $v_\tau^{(\rho)}$ in the same way as $v^{(\tau,s)}$ by means of v_τ in 15.2.2.

Substituting (20) into (1) and (4), we obtain the following equations for $w^{(r)}$:

$$L_0(y,\partial/\partial z)w^{(r)}(y,z) = f^{(r)}(y,z) - \sum_{m=1}^{r} L_m(y,z,\partial/\partial y,\partial/\partial z)w^{(r-m)}(y,z) - H^{(r)}, \qquad (21)$$

$$z \in (-1/2,1/2);$$

$$B_0^{\pm}(y,\partial/\partial z)w^{(r)}(y,\pm 1/2) = g^{(r,\pm)}(y) - \sum_{m=1}^{r} B_m^{\pm}(y,\partial/\partial y,\partial/\partial z)w^{(r-m)}(y,\pm 1/2).$$

Here the vector-valued function $H^{(r)} \in \mathbf{C}^\infty(\bar\Omega)$ is chosen so that the compatibility condition (21) is fulfilled, i.e.,

$$H^{(r)}(y) = g^{(r)-}(y) - g^{(r)+}(y) + \int_{-1/2}^{1/2} f^{(r)}(y,\eta)d\eta$$

$$- \sum_{m=1}^{r} \Big(B_m^{-}(y,\partial/\partial y,\partial/\partial z)\, w^{(r-m)}(y,1/2) - B_m^{+}(y,\partial/\partial y,\partial/\partial z)\, w^{(r-m)}(y,-1/2)$$

$$+ \int_{-1/2}^{1/2} L_m(y,\eta,\partial/\partial y,\partial/\partial \eta)\, w^{(r-m)}(y,\eta)d\eta \Big). \qquad (22)$$

The solution $w^{(r)}(y,z)$ to problem (21) whose mean value (in z) on $(-1/2,1/2)$ equals zero is given by a formula similar to (18).

The series (20) satisfies the boundary conditions (4) but in the equations (1) it produces the discrepancy

$$\sum_{r=0}^{\infty} \varepsilon^{r-2}\Big(H_l^{(r)}(y) + \varepsilon^{-\alpha_{\max}} \sum_{\tau=0}^{T} \varepsilon^{-\alpha_\tau} \sum_{s=0}^{r} M_l^{(s,\tau)}(y,\partial/\partial y)v_\tau^{r-s}(\varepsilon,y)\Big), \qquad (23)$$

which is a function in $y \in \Omega$ independent of x_n. We introduce the matrix operator

$$M^0(\varepsilon,y,\partial/\partial y) = -\left(\sum_{s=1}^{\alpha_k+\alpha_l} \varepsilon^{s-\alpha_k-\alpha_l} M_l^{(s,k)}(y,\partial/\partial y) \right)_{k,l=1}^{M}, \qquad (24)$$

which is called the operator of third limit problem. The series (23) is formally equal to zero if

$$M^0(\varepsilon,y,\partial/\partial y)v^{(m)}(y) = \big(H_l^{(m+\alpha_l-\alpha_{\max})}(y)\big)_{l=1}^{T} - \sum_{r=1}^{m} M^{(r)}(y,\partial/\partial y)v^{(m-r)}(y), \quad y \in \Omega, \quad m \in \mathbb{N}_0, \qquad (25)$$

with $v^{(m)} = (v^{(m)}, \dots, v_T^{(m)})$, $M^{(r)} = \Big(M_l^{(r+\alpha_l+\alpha_k,k)}\Big)_{k,l=1}^{T}$. In (25) the functions $H_l^{(h)}$ with negative h are equal to zero.

Remark 15.2.4. One cannot take the series (20) as asymptotics of the solution to problem (6) everywhere in the cylinder $\tilde C_\varepsilon$, because its terms do not satisfy the boundary conditions (5) at the lateral area of the cylinder. As in Section 15.1, the corresponding discrepancies have to be compensated for with the help of a boundary layer. Here, studying only the principal term of the asymptotics, we do not consider the boundary layer. In a more general situation its construction will be given in Chapter 16.

15.2.4 Properties of the limit operator

The first assertion guarantees that the operator $M^0(\varepsilon, y, \partial/\partial y)$ differs from zero.

Theorem 15.2.5. *For any function $\zeta_\tau \in \mathbf{C}_0^\infty(\Omega)$ there holds the inequality*

$$\varepsilon^{-\alpha_\tau-\delta}\sum_{\nu=1}^{T}||M^0_{\tau\nu}(\varepsilon, y, \partial/\partial y)\zeta_\tau; \mathbf{L}_2(\Omega)|| \geq C||\zeta_\tau; \mathbf{L}_2(\Omega)|| + O(\varepsilon) \tag{26}$$

with constant C independent of ε and ζ_τ.

Proof. We construct the vector-valued functions $\zeta^{(\tau,r)}$ for ζ_τ in the same way as $v^{(k,r)}$ for v_k in 15.2.2. Let

$$Z_\nu^{(\tau)}(\varepsilon, x) = \varepsilon^{-\alpha_\tau}\Big(\delta_{\tau\nu}\zeta_\tau(x') + \sum_{r=1}^{\alpha_\tau+\alpha_{\max}} \varepsilon^r \zeta_\nu^{(\tau,r)}(x', \varepsilon^{-1}x_n)\Big), \quad \nu = 1, \dots, T. \tag{27}$$

We have

$$||Z_\tau^{(\tau)}; \mathbf{L}_2(\tilde{C}_\varepsilon)|| = \varepsilon^{-\alpha_\tau+1/2}(||\zeta_\tau; \mathbf{L}_2(\Omega)|| + O(\varepsilon)),$$

$$\sum_{l,k=1}^{T}\varepsilon^{1-\alpha_l}\Big(||L_{lk}\, Z_k^{(\tau)}; \mathbf{L}_2(\tilde{C}_\varepsilon)|| + \varepsilon^{1/2}\sum_{\pm}||B_{lk}^{\pm}\, Z_k^{(\tau)}; \mathbf{L}_2(\Omega)||\Big)$$

$$= \varepsilon^{-1/2}\Big(\sum_{\nu=1}^{T}||M^0_{\nu\tau}(\varepsilon, y, \partial/\partial y)\zeta_\tau; \mathbf{L}_2(\Omega)|| + O(\varepsilon)\Big).$$

By virtue of (9) and Corollary 15.2.2,

$$\varepsilon^{\alpha_\tau-1}||Z^{(\tau)}; \mathbf{L}_2(\tilde{C}_\varepsilon)||$$

$$\leq c\sum_{l,k=1}^{T}\varepsilon^{1-\alpha_l-\delta}\Big(||L_{lk}\, Z_k^{(\tau)}; \mathbf{L}_2(\tilde{C}_\varepsilon)|| + \sum_{\pm}\varepsilon^{1/2}||B_{lk}^{\pm}\, Z_k^{(\tau)}; \mathbf{L}_2(\Omega)||\Big).$$

The above relations lead to (27). □

Theorem 15.2.6. *The matrix operator $M^0(\varepsilon, y, \partial/\partial y)$ is formally self-adjoint.*

Proof. Let $\zeta_\tau, \xi_\tau \in \mathbf{C}_0^\infty(\Omega)$, $\tau = 1, \dots, T$. Using (27) and the similar formula for ξ_τ, we introduce the vector-valued functions $Z^{(\tau)}$ and $\Xi^{(\tau)}$. It is clear that $Z^{(\tau)}$ and $\Xi^{(\tau)}$ together with all their derivatives vanish on S_ε. We consider the bilinear form

$$\Lambda(Z, \Xi) = \int_{\tilde{C}_\varepsilon}\Xi(\varepsilon, x)\cdot L(\varepsilon, x\partial/\partial x)\, Z(\varepsilon, x)\, dx$$

$$+\sum_{\pm}\pm\int_{\Omega}\Xi(\varepsilon, x', \pm\varepsilon/2)\cdot B^{\pm}(\varepsilon, x', \partial/\partial x)\, Z(\varepsilon, x', \pm\varepsilon/2)\, dx'.$$

Since problem (6) is self-adjoint, we have $\Lambda(Z^{(\tau)},\Xi^{(\nu)})=\Lambda(\Xi^{(\tau)},Z^{(\nu)})$. Moreover,

$$\Lambda(Z^{(\tau)},\Xi^{(\nu)})=\sum_{q=1}^{J}\varepsilon^{q-2-\alpha_1}\sum_{l=1}^{T}\int_{\tilde{C}_\varepsilon}\Xi_\tau^{(\nu)}(\varepsilon,x)\,M_l^{(\tau,q)}(x',\partial/\partial x')\zeta_\tau(x')dx+O(\varepsilon^{J-\alpha_\tau-\alpha_\nu}). \tag{28}$$

The second factor in every integral in (28) does not depend on x_n. Therefore, in the first factor the terms whose mean values on $(-\varepsilon/2,\varepsilon/2)$ are equal to zero may be dropped. Thus, if $J=\alpha_\tau+\alpha_\nu$, then

$$\Lambda(Z^{(\tau)},\Xi^{(\nu)})=\varepsilon^{-1}\int_\Omega \zeta_\nu(y)\,M^0_{\nu\tau}(\varepsilon,y,\partial/\partial y)\,\xi_\tau(y)dy+O(1).$$

Similarly,

$$\Lambda(\Xi^{(\nu)},Z^{(\tau)})=\varepsilon^{-1}\int_\Omega \zeta_\tau(y)\,M^0_{\tau\nu}(\varepsilon,y,\partial/\partial y)\,\xi_\nu(y)dy+O(1).$$

Comparing the two last relations, we find that $M^0(\varepsilon,y,\partial/\partial y)=(M^0)^*(\varepsilon,y,\partial/\partial y)$ in any strictly interior subdomain Ω' of Ω for sufficiently small ε. It remains to note that the coefficients of M_0 are polynomials in powers of ε^{-1} whose coefficients are smooth in Ω. □

15.2.5 Unique solvability of the limit problem

We restrict ourselves to the study of the situation where the limit problem is independent of ε and elliptic.

Let the operator $M^0(\varepsilon,y,\partial/\partial y)$ be independent of ε and the orders of its entries M^0_{lk} be equal to $t_k+t_l-t_{\max}$, where $t_r=\alpha_r+\alpha_{\max}$, $\alpha_{\max}=\max\{\alpha_1,\dots,\alpha_T\}$, $t_{\max}=\max\{t_1,\dots,t_T\}$. We suppose that the operator M^0 is elliptic in the Douglis-Nirenberg sense in $\bar\Omega$ while $s_l=t_l-t_{\max}$ and the $m\times T$-system of boundary operators

$$\tilde D(y,\partial/\partial y)=\{e^{(k)}(\partial/\partial N)^p:\quad k=1,\dots,T,\quad p=0,\dots,\alpha_k-1\} \tag{29}$$

together with M^0 form the elliptic boundary value problem in Ω (see Section 3.1). Here N is the outward normal to $\partial\Omega$ and $m=2(\alpha_1+\dots+\alpha_T)$.

Theorem 15.2.7. *The boundary value problem*

$$M^0(y,\partial/\partial y)\,v(y)=F(y),\quad y\in\Omega;\quad \tilde D(y,\partial/\partial y)\,v(y)=0,\quad y\in\partial\Omega, \tag{30}$$

is uniquely solvable, i.e., the mapping

$$M^0\colon \prod_{k=1}^{T}(\mathbf{W}_2^{q+t_k}(\Omega)\cap \mathbf{W}^{\alpha_k}_{2,0}(\Omega;\partial\Omega))\to\prod_{l=1}^{T}\mathbf{W}_2^{q+t_{\max}-t_l}(\Omega)$$

is an isomorphism for any $q\in\mathbb{N}_0$.

Proof. By Theorem 15.2.6, problem (30) is self-adjoint and therefore it suffices to verify the absence of nontrivial solutions to the homogeneous equations (30). Let v^0 satisfy (30) with $F=0$. Since the boundary $\partial\Omega$ is smooth, from Theorem 3.1.8 it follows that the components v^0_k belong to $\mathbf{C}^\infty(\bar\Omega)$ and for any $q\in\mathbb{N}_0$

$$|\nabla^q_{x'}v^0_k(x')|<\text{const}\cdot\rho(x')^{(\alpha_k-q)_+}, \tag{31}$$

where $\nabla^q_{x'} v^0_k$ is the vector of all the derivatives of v^0_k of order q. Let χ_0 be a function in $\mathbf{C}^\infty(\mathbb{R})$ equal to 0 for $|t| \le 1/2$ and 1 for $|t| \ge 1$ while $\chi_\sigma(\varepsilon, x') = \chi_0(\varepsilon^{-\rho}\sigma(x'))$. It is clear that for sufficiently small ε the function χ_σ is smooth in $\bar{\Omega}$, $\operatorname{supp}(\nabla^q_{x'}\chi_\sigma) \subset \{x' \in \bar{\Omega} \colon \rho(x') \le \varepsilon^\sigma\}$ for $q \ge 1$ and

$$|\nabla^q_{x'} \chi_\sigma(\varepsilon, x')| \le c_q \varepsilon^{-\sigma q}. \tag{32}$$

We introduce the functions $\zeta_k = \chi_\sigma v^0_k$ and the vectors $Z^{(k)}$ by (27). All the derivatives of ζ_k and $Z^{(k)}$ vanish on $\partial\Omega$. Therefore, as in the proof of Theorem 2.6, we obtain

$$\Lambda(Z^{(k)}, Z^{(l)}) = \sum_{r=1}^{\alpha_k+\alpha_l} \varepsilon^{r-\alpha_k-\alpha_l-1} \int_\Omega \zeta_l(\varepsilon, x')\, M_l^{(k,r)}(x', \partial/\partial x')$$
$$\times\, \zeta_k(\varepsilon, x')dx' + O(\varepsilon^{-\sigma(\alpha_k+\alpha_l+2\alpha_{\max}+2)}).$$

Thus, according to (24), (31), and (32), we have

$$\sum_{r=1}^{\alpha_k+\alpha_l} \varepsilon^{r-\alpha_k-\alpha_l-1} \int_\Omega \zeta_l(\varepsilon, y)\, M_l^{(k,r)}(y, \partial/\partial y)\, \zeta_k(\varepsilon, y) dy$$
$$= \varepsilon^{-1} \int_\Omega \zeta_l(\varepsilon, y) \{\chi_\sigma(\varepsilon, y)\, M^0_{lk}(y, \partial/\partial y) + [M^0_{lk}(y, \partial/\partial y), \chi_\sigma(\varepsilon, y)]\}\, v^0_k(y)\, dy$$
$$= \varepsilon^{-1} O\Big(\int_0^{\varepsilon^\sigma} \rho^{\alpha_t} \sum_{q=0}^{\alpha_k+\alpha_l} \varepsilon^{-\sigma q} \rho^{(q-\alpha_l)} + d\rho\Big) = O(\varepsilon^{\sigma-1}).$$

We set $Z = Z^{(1)} + \cdots + Z^{(T)}$. For small σ

$$E(Z; \tilde{C}_\varepsilon) = \Lambda(Z, Z) = \sum_{k,l=1}^{T} \Lambda(Z^{(k)}, Z^{(l)}) = O(\varepsilon^{\sigma-1}).$$

By virtue of the inequalities (31), (32), and (7) and Corollary 15.2.2,

$$E(Z; \tilde{C}_\varepsilon)^{1/2} \ge c_\delta \sum_{k=1}^{T} \varepsilon^{\alpha_k-1+\delta} ||Z_k; \mathbf{L}_2(\tilde{C}_\varepsilon)||$$
$$= c_\delta \varepsilon^{\delta-1/2} \sum_{k=1}^{T} \delta_{\alpha_k \alpha_{\max}} (||\zeta_k; \mathbf{L}_2(\Omega)|| + O(\varepsilon^{1-\sigma}))$$
$$= c_\delta \varepsilon^{\delta-1/2} \sum_{k=1}^{T} \delta_{\alpha_k \alpha_{\max}} (||v^0_k; \mathbf{L}_2(\Omega)|| + O(\varepsilon^{1-\sigma})).$$

Because v^0_k is independent of ε, we obtain $||v^0_k; \mathbf{L}_2(\Omega)|| = 0$ for k subject to $\alpha_k = \alpha_{\max}$. Thus, $v^0_k = 0$ in $\bar{\Omega}$ provided $\alpha_k = \alpha_{\max}$. Then

$$\sum_{k=1}^{T} \varepsilon^{\alpha_k-1} ||Z_k; \mathbf{L}_2(\tilde{C}_\varepsilon)|| = c_\delta \varepsilon^{\delta-1/2} \sum_{k=1}^{T} \delta_{\alpha_k \alpha_{\max}-1} (||\zeta_k; \mathbf{L}_2(\Omega)|| + O(\varepsilon^{1-\sigma}))$$

and, consequently, $v^0_k = 0$ in $\bar{\Omega}$ for $\alpha_k = \alpha_{\max} - 1$. We continue in this fashion and find that all the components of v^0 are equal to zero. □

15.2.6 Justification of the asymptotic expansion of the solution

Let all assumptions in 15.2.1 and 15.2.5 be fulfilled. To derive an estimate on the difference between the solution to problem (1), (4), (5) and its approximation we need the following assertion.

Lemma 15.2.8. *If $v_k \in \mathbf{W}_2^{t_k+1}(\Omega) \cap \mathbf{W}_{2,0}^k(\Omega; \partial\Omega)$, then*

$$\begin{aligned} &||\big(|\log \rho(y)| + 1\big)^{-1} \rho(y)^{-(\alpha_k - q)_+ - 1/2} \nabla_y^q v_k(y); \mathbf{L}_2(\Omega)|| \\ &\leq \text{const}\, ||v_k; \mathbf{W}_2^{t_k+1}(\Omega)||, \quad q = 0, \dots, t_k. \end{aligned} \tag{33}$$

Proof. One can obtain (33) making use of the Hardy inequality (see Lemma 1.4.2). □

Let v be the solution to problem (30) with right-hand side F given by the formulas

$$\begin{aligned} &F_t(y) = g_l^{(0)-}(y) - g_l^{(0)+}(y) + \int_{-1/2}^{1/2} f_l^{(0)}(y, \eta) : = H_l^{(0)}(y), \quad \alpha_l = \alpha_{\max}, \\ &F_t(y) = 0, \quad \alpha_l < \alpha_{\max}, \qquad l = 1, \dots, T, \end{aligned} \tag{34}$$

while

$$g^{(0)\pm} \in \mathbf{W}_2^1(\Omega), f^{(0)}, (\partial/\partial x') f^{(0)} \in \mathbf{L}_2(\tilde{C}_1). \tag{35}$$

Since $F_l \in \mathbf{W}_2^{t_{\max} - t_1 + 1}(\Omega)$, $l = 1, \dots, T$, we have $v_k \in \mathbf{W}_2^{t_k+1}(\Omega) \cap \mathbf{W}_{2,0}^{\alpha_k}(\Omega; \partial\Omega)$, $k = 1, \dots, T$, by Theorem 15.2.7. We set

$$\begin{aligned} &V(\varepsilon, x) = \\ &\chi_1(\varepsilon, x')\Big(w^{(0)}(x', \varepsilon^{-1} x_n) + \sum_{\tau=1}^{T} \varepsilon^{-\alpha_{\max} - \alpha_\tau} \sum_{r=0}^{\alpha_\tau + \alpha_{\max}} \varepsilon^r v^{(\tau, r)}(x', \varepsilon^{-1} x_n)\Big), \end{aligned} \tag{36}$$

where $v^{(k,r)}$ are the vector-valued functions in (15), $w^{(0)}$ the solution to problem (21), and $r = 0$. By Lemma 15.2.3 and what was said on $v_1, \dots, v_T$, the vector V belongs to the space $\mathbf{W}_{2,0}^1(\tilde{C}_\varepsilon; S_\varepsilon)$. We estimate the norm in $\mathbf{W}_2^1(\tilde{C}_\varepsilon)$ of the difference $U = V - u$, where u is the solution to problem (6). To this end we consider

$$\begin{aligned} &\sum_{k=1}^{T} L_{lk}(\varepsilon, x, \partial/\partial x)\, V_k(\varepsilon, x) = \chi_1(\varepsilon, x')\varepsilon^{-2}\{f_l^0(x', \varepsilon^{-1} x_n) + H_l^{(0)}(x')\} \\ &+ \sum_{k=1}^{T} \{L_{lk}(\varepsilon, x, \partial/\partial x) - \delta_{lk} A_{ll}(x')(\partial/\partial x_n)^2\}\{\chi_1(\varepsilon, x') w^{(0)}(x', \varepsilon^{-1} x_n)\} \\ &+ \sum_{\tau=1}^{T} \varepsilon^{-\alpha_{\max} - \alpha_\tau} \sum_{r=0}^{\alpha_\tau + \alpha_{\max}} \varepsilon^r \sum_{k=1}^{T} [L_{lk}(\varepsilon, x, \partial/\partial x), \chi_1(\varepsilon, x')] v_k^{(\tau, r)}(x', \varepsilon^{-1} x_n) \end{aligned}$$

$$+\chi_1(\varepsilon,x')\sum_{\tau=1}^{T}\varepsilon^{-\alpha_{\max}-\alpha_\tau}\sum_{r=0}^{\alpha_\tau+\alpha_{\max}}\varepsilon^{r-2}\Big\{\sum_{m=0}^{r}(L_m)_{lk}(x',\varepsilon^{-1}x_n,\partial/\partial x',\varepsilon\partial/\partial x_n)$$
$$\times v_k^{(\tau,r-m)}(x',\varepsilon^{-1}x_n)\Big\}+\chi_1(\varepsilon,x')\sum_{k=1}^{T}\sum_{\tau=0}^{T}\varepsilon^{-\alpha_{\max}-\alpha_\tau}\sum_{r=0}^{\alpha_\tau+\alpha_{\max}}\varepsilon^{r}\Big\{L_{lk}(\varepsilon,x,\partial/\partial x)$$
$$-\sum_{m=0}^{\alpha_\tau+\alpha_{\max}-r}\varepsilon^{m-2}(L_m)_{lk}(x',\varepsilon^{-1}x_n,\partial/\partial x',\varepsilon\partial/\partial x_n)\Big\}v_k^{(\tau,r)}(x',\varepsilon^{-1}x_n). \tag{37}$$

Rewrite the right-hand side in (37) in the form $I_1+\cdots+I_5$. For I_1 we have

$$\begin{aligned}\int_{\tilde C_\varepsilon} & I_1(\varepsilon,x)\,U_l(\varepsilon,x)dx\\ &=\varepsilon^{-2}\int_{\tilde C_\varepsilon}\big(f_l^0(x',\varepsilon^{-1}x_n)-H_l^{(0)}(x')\big)\,U_l(\varepsilon,x)dx\\ &\quad+O\big(\varepsilon^{(\alpha_l-2)/2}\|\rho_\varepsilon^{-\alpha_l}U_l;\mathbf{L}_2(\tilde C_\varepsilon)\|\Xi\big)\end{aligned} \tag{38}$$

with

$$\Xi=\sum_{\pm}\|g^{(0)\pm};\mathbf{W}_2^1(\Omega)\|+\|f^{(0)};\mathbf{L}_2(\tilde C_1)\|+\|\partial f^{(0)}/\partial x';\mathbf{L}_2(\tilde C_1)\|.$$

Since the operator L_0 is reduced to the diagonal form (see 15.2.2), the coefficient by $(\partial/\partial x_n)^2$ in the operator $L_{lk}-\delta_{lk}A_{ll}(\partial/\partial x_n)^2$ is $O(\varepsilon)$ and the remaining coefficients are $O(1)$. Therefore,

$$\begin{aligned}\int_{\tilde C_\varepsilon} & I_2(\varepsilon,x)\,U_l(\varepsilon,x)dx\\ &\le c\Big(\varepsilon\int_{\tilde C_\varepsilon}\chi_1(\varepsilon,x')|(\partial/\partial x_n)^2w^{(0)}(x',\varepsilon^{-1}x_n)||U_l(\varepsilon,x)|dx\\ &\quad+\int_{\tilde C_\varepsilon}|(\partial/\partial x')\chi_1(\varepsilon,x')w^{(0)}(x',\varepsilon^{-1}x_n)|\,|(\partial/\partial x')U_l(\varepsilon,x)|dx\Big)\\ &\le c\varepsilon^{-1/2}\Xi\big(\|U_l;\mathbf{L}_2(\tilde C_\varepsilon)\|+\|\partial U_l/\partial x';\mathbf{L}_2(\tilde C_\varepsilon)\|\big).\end{aligned} \tag{39}$$

We consider I_3. Taking account of the fact that the support of $[L_{lk},\chi_1]v_k^{(\tau,r)}$ is located in the set $D_\varepsilon=\{x\in\tilde C_\varepsilon\colon \varepsilon>\rho(x')>\varepsilon/2\}$ and using Lemmas 15.2.3 and 15.2.9, we find

$$\begin{aligned}\int_{\tilde C_\varepsilon} & I_3(\varepsilon,x)\,U_l(\varepsilon,x)dx\\ &\le c\sum_{\tau=1}^{T}\varepsilon^{-\alpha_{\max}-\alpha_\tau}\sum_{r=0}^{\alpha_\tau+\alpha_{\max}}\varepsilon^{r}\sum_{\mu=0}^{1}\varepsilon^{\mu-2}\int_{D_\varepsilon}|\nabla_{x'}^{r}v_\tau(x)|\,|\nabla_{x'}^{\mu}U_l(\varepsilon,x)|dx\end{aligned}$$

$$\leq c\sum_{\tau=1}^{T}\sum_{r=0}^{\alpha_\tau+\alpha_{\max}} |\log\varepsilon|\varepsilon^{-\alpha_{\max}+\alpha_1+(\alpha_k-r)_++r-\alpha_k-1}$$
$$\times\int\limits_{D_\varepsilon}\varepsilon^{-1/2}\big(|\log\rho(x')|+1\big)\,\rho(x')^{-(\alpha_\tau-r)_+-1/2} \tag{40}$$
$$\times|\nabla_{x'}^r v_\tau(x')|\,\rho(x')^{-\alpha_l}\big(|U_l(\varepsilon,x)|+\rho(x')^{-1}|(\partial/\partial x')U_l(\varepsilon,x')|\big)dx$$
$$\leq c|\log\varepsilon|\varepsilon^{-\alpha_{\max}+\alpha_l-1}\Xi\big(||\rho_\varepsilon^{-\alpha_l}U_l;\mathbf{L}_2(\tilde{C}_\varepsilon)||+||\rho_\varepsilon^{1-\alpha_l}\partial U_l/\partial x';\mathbf{L}_2(\tilde{C}_\varepsilon)||\big).$$

According to (17) and (24),

$$\int\limits_{\tilde{C}_\varepsilon} I_4(\varepsilon,x)\,U_l(\varepsilon,x)dx=\sum_{\tau=1}^{T}\varepsilon^{\alpha_l-\alpha_{\max}-2}\int\limits_{\tilde{C}_\varepsilon}M_l^0(x',\partial/\partial x')\,v_\tau(x')\,U_l(\varepsilon,x)dx$$
$$+O\big(\varepsilon^{\alpha_l-\alpha_{\max}-1}\Xi||\rho_\varepsilon^{-\alpha_l}U_l;\mathbf{L}_2(\tilde{C}_\varepsilon)||\big). \tag{41}$$

Finally, in the same manner as in estimating the expression that contains I_2, we obtain

$$\int\limits_{\tilde{C}_\varepsilon} I_5(\varepsilon,x)\,U_l(\varepsilon,x)dx\leq c\varepsilon^{-1/2}\Xi\big(||U_l;\mathbf{L}_2(\tilde{C}_\varepsilon)||+||\partial U_l/\partial x';\mathbf{L}_2(\tilde{C}_\varepsilon)||\big). \tag{42}$$

Thus, from (34) and (37)–(42) it follows that

$$\sum_{k=1}^{T}\int\limits_{\tilde{C}_\varepsilon} L_{lk}(\varepsilon,x,\partial/\partial x)\,V_k(\varepsilon,x)dx$$
$$=\varepsilon^{-2}\int\limits_{\tilde{C}_\varepsilon}\big(f_l^{(0)}(x',\varepsilon^{-1}x_n)+(\delta_{\alpha_1\alpha_{\max}}-1)H_l^{(0)}(x')\big)\,U_l(\varepsilon,x)dx \tag{43}$$
$$+O\Big(\varepsilon^{\alpha_l-\alpha_{\max}-1}|\log\varepsilon|\Xi\big(||\rho_\varepsilon^{-\alpha_l}U_l;\mathbf{L}_2(\tilde{C}_\varepsilon)||+||\rho_\varepsilon^{1-\alpha_l}\partial U_l/\partial x';\mathbf{L}_2(\tilde{C}_\varepsilon)||\big)\Big).$$

Similarly,

$$\sum_{k=1}^{T}\int\limits_{\Omega} B_{lk}^{(\pm)}(\varepsilon,x',\partial/\partial x)\,V_k(\varepsilon,x',\pm\varepsilon/2)U_l(\varepsilon,x',\pm\varepsilon/2)dx'$$
$$=\varepsilon^{-1}\int\limits_{\Omega} g_l^{(0)\pm}(x')U_l(\varepsilon,x',\pm\varepsilon/2)dx'$$
$$+O\Big(\varepsilon^{\alpha_l-\alpha_{\max}-3/2}|\log\varepsilon|\Xi\big(||\rho_\varepsilon^{-\alpha_l}U_l;\mathbf{L}_2(\tilde{C}_\varepsilon)||+||\rho_\varepsilon^{1-\alpha_l}\partial U_l/\partial x';\mathbf{L}_2(\tilde{C}_\varepsilon)||\big)\Big). \tag{44}$$

We admit that the right-hand sides in (1), (4), and (6) are subject to the inequalities

$$||\rho_\varepsilon^{\alpha_l}(f_l-\varepsilon^{-2}f_l^{(0)});\mathbf{L}_2(\tilde{C}_\varepsilon)||\leq K\varepsilon^{\alpha_l-\alpha_{\max}-1}|\log\varepsilon|,$$
$$||\rho_\varepsilon^{\alpha_l}(g_l^{(0)\pm}(x')-\varepsilon^{-1}g_l^{(0)\pm});\mathbf{L}_2(\Omega)||\leq K\varepsilon^{\alpha_l-\alpha_{\max}-3/2}|\log\varepsilon|. \tag{45}$$

Then, combining (43)–(45), we obtain

$$E(U;\tilde{C}_\varepsilon)=\sum_{l=1}^{T}\Big\{\int_{\tilde{C}_\varepsilon}\Big(\sum_{k=1}^{T}L_{lk}(\varepsilon,x,\partial/\partial x)\,V_k(\varepsilon,x)-f_l(\varepsilon,x)\Big)\,U_l(\varepsilon,x)\,dx$$

$$+\sum_{\pm}\pm\int_{\Omega}\Big(\sum_{k=1}^{T}B^{\pm}_{lk}(\varepsilon,x',\partial/\partial x)\,V_k(\varepsilon,x',\pm\varepsilon/2)-g^{\pm}_l(\varepsilon,x')\Big)$$

$$\times\, U_l(\varepsilon,x',\pm\varepsilon/2)dx'\Big\}$$

$$=O\Big(\varepsilon^{-\alpha_{\max}}|\log\varepsilon|(\Xi+K)\sum_{l=1}^{T}\varepsilon^{\alpha_l-1}(\|\rho_\varepsilon^{-\alpha_l}U_l;\mathbf{L}_2(\tilde{C}_\varepsilon)\|$$

$$+\|\rho_\varepsilon^{1-\alpha_l}\partial U_l/\partial x';\mathbf{L}_2(\tilde{C}_\varepsilon)\|)\Big).$$

Taking (7) into account, we arrive at the estimate

$$\sum_{k=1}^{T}\varepsilon^{\alpha_k}(\|\rho_\varepsilon^{1-\alpha_k}\partial U_k/\partial x';\mathbf{L}_2(\tilde{C}_\varepsilon)\|+\varepsilon\|\rho_\varepsilon^{-\alpha_k}\partial U_k/\partial x_n;\mathbf{L}_2(\tilde{C}_\varepsilon)\|+\|\rho_\varepsilon^{-\alpha_k}U_k;\mathbf{L}_2(\tilde{C}_\varepsilon)\|)$$
$$\le C_\sigma(\Xi+K)\varepsilon^{1-\alpha_{\max}-\sigma},\tag{46}$$

σ being an arbitrary positive number. We have obtained the main result of the section.

Theorem 15.2.9. *Let all the hypotheses in* 15.2.1 *and* 15.2.5 *be fulfilled and let u be the solution to problem* (1), (2), *and* (5) *with right-hand sides subject to* (45) *and* (35). *Finally, let V stand for the vector-valued function* (36). *Then for the difference $U=V-u$ there holds the inequality* (46), *where the constant C is independent of $f,g,$ and ε; K is the constant in* (45) *and Ξ is the sum of the norms of the functions in* (35).

Corollary 15.2.10. *Under the assumptions of Theorem 2.9, from* (36) *it follows that for $U_k^0(\varepsilon,x)=u_k(\varepsilon,x)-\varepsilon^{-\alpha_{\max}-\alpha_k}\,v_k(x')$ there exists estimate*

$$\sum_{k=1}^{T}\varepsilon^{\alpha_k}(\|\rho_\varepsilon^{1-\alpha_k}\partial U_k^0/\partial x';\mathbf{L}_2(\tilde{C}_\varepsilon)\|+\varepsilon\|\rho_\varepsilon^{-\alpha_k}\partial U_k^0/\partial x_n;\mathbf{L}_2(\tilde{C}_\varepsilon)\|+\|\rho_\varepsilon^{-\alpha_k}U_k^0;\mathbf{L}_2(\tilde{C}_\varepsilon)\|)$$
$$\le C_\sigma(\Xi+K)\varepsilon^{1-\alpha_{\max}-\sigma}.\tag{47}$$

Remark 15.2.11. Since

$$\|\rho_\varepsilon^{1-\alpha_k}\partial v_k/\partial x;\mathbf{L}_2(\tilde{C}_\varepsilon)\|+\|\rho_\varepsilon^{-\alpha_k}v_k;\mathbf{L}_2(\tilde{C}_\varepsilon)\|=O(\varepsilon^{1/2}),$$

the inequality (47) shows that in the sense of mean convergence

$$u_k(\varepsilon,x)=\varepsilon^{-\alpha_{\max}-\alpha_k}\big(v_k(x')+O(\varepsilon^{1/2-\sigma})\big),\quad k=1,\dots,T.\tag{48}$$

15.3 Applications of Theorem 15.2.9 to Particular Boundary Value Problems

In this section, with the help of the algorithm given in Section 15.2, we find the additional limit problems for some special problems in mathematical physics.

Example 15.3.1. Let $l_\pm$ be positive smooth functions on a segment $[-a,a]$. We introduce the thin domain $Q_\varepsilon = \{\xi \in \mathbb{R}^2 : |\xi_1| < a, -l_-(\xi_1) < \varepsilon^{-1}\xi_2 < l_+(\xi_1)\}$ and consider the boundary value problem

$$\begin{aligned} &-\Delta U(\varepsilon,\xi) = \tilde{F}(\varepsilon,\xi), \quad \xi \in Q_\varepsilon;\\ &U(\varepsilon,\pm a,\xi_2) = 0, \quad -l_-(\pm a) < \varepsilon^{-1}\xi_2 < l_+(\pm a);\\ &(\partial/\partial\nu)\,U\big(\varepsilon,\xi_1,\pm\varepsilon l_\pm(\xi_1)\big) = \tilde{G}_\pm(\varepsilon,\xi_1), \quad |\xi_1| < a, \end{aligned} \tag{1}$$

ν being the outward normal. The problem describes the antiplane deformation of a thin trapezium with curvilinear sides. With the help of the change of variables $\xi \to x = \big(\xi_1, I(\xi_1)\xi_2 + J(\xi_1)\big)$, $I(t) = \big(l_+(t) + l_-(t)\big)^{-1}$, $J(t) = I(t)\big(l_-(t) - l_+(t)\big)/2$ problem (1) reduces to problem (6), 15.2, in the thin rectangle $\tilde{C}_\varepsilon = (-a,a) \times (-\varepsilon/2, \varepsilon/2)$ with $n=2, T=1$ and

$$\begin{aligned} &a_{11}^{(1,1)}(\varepsilon,x) = I(x_1)^{-1}, \quad a_{11}^{(1,2)}(\varepsilon,x) = a_{11}^{(2,1)}(\varepsilon,x) = -(\partial/\partial x_1)\,\xi_2(x),\\ &a_{11}^{(2,2)}(\varepsilon,x)^2 = \big(a_{11}^{(2,1)}(\varepsilon,x)^2\,a_{11}^{(1,1)}(\varepsilon,x)^2 + 1\big)\,a_{11}^{(1,1)}(\varepsilon,x)^{-1},\\ &a_{11}^{(1)}(\varepsilon,x) = a_{11}^{(2)}(\varepsilon,x) = a_{11}(\varepsilon,x) = 0 \end{aligned}$$

The energy inequality and Hardy's inequality (see Lemma 1.4.2) imply (7), 15.2, with $\alpha_1 = 1$. The functions $v^{(1,r)}$ are given by

$$v^{(1,1)}(x_1) = 0, \quad v^{(1,2)}(x_1,z) = \big(z - J(x_1)\big)\,I(x_1)^{-1},$$

and the operators $M_1^{(s,l)}$ by

$$\begin{aligned} &M_1^{(1,1)}(x_1,\partial/\partial x_1) = 0\\ &M_1^{(1,2)}(x_1,\partial/\partial x_1) = -(\partial/\partial x_1)\,\big(l_+(x_1) + l_-(x_1)\big)\,(\partial/\partial x_1). \end{aligned}$$

Thus, the role of limit problem (30), 15.2, for (1) is played by the Dirichlet problem for the ordinary differential operator $M^{(0)}(x_1,\partial/\partial x_1) = -(\partial/\partial x_1)(l_+(x_1) + l_-(x_1))(\partial/\partial x_1)$. This example shows, in particular, how to apply the algorithm described in order to find the limit problem in noncylindrical domains, too.

Example 15.3.2. We consider a three-dimensional problem in elasticity theory. Let $\tilde{C}_\varepsilon = \Omega \times (-\varepsilon/2, \varepsilon/2)$ be a thin plate, where Ω is a plane domain with smooth boundary. The Lamé system in $\tilde{C}_\varepsilon$ is of the form (2), 15.2, where

$$\begin{aligned} a_{ii}^{(i,i)} &= (1+\nu)^{-1}E(1-2\nu)^{-1}(1-\nu),\\ a_{ij}^{(i,j)} &= a_{ii}^{(j,j)} = \big(2(1+\nu)\big)^{-1}\,E,\\ a_{ij}^{(i,j)} &= \big(2(1+\nu)\big)^{-1}\,E(1-2\nu)^{-1}\nu \end{aligned}$$

for $i,j = 1,2,3$, $j \neq i$, the remaining coefficients equal zero, E stands for Young's module and ν for Poisson's coefficient. If the lateral side S_ε is rigidly fixed and the external force is applied to the bases, then the corresponding boundary conditions take the form (4), (5), 15.2. The estimate (7), 15.2, coincides with the modified Korn inequality while $\alpha_1 = \alpha_2 = 1, \alpha_3 = 2$ (see SHOIKHET [1]). The iterative procedures in 15.3.3 give the following expressions for the components of the auxiliary functions $v^{(k,r)}$:

$$\begin{aligned} &v_3^{(p,1)}(x',\zeta) = (\nu-1)^{-1}\nu\zeta(\partial/\partial x_p)\,v_p(x'),\\ &v_p^{(3,1)}(x',\zeta) = -\zeta(\partial/\partial x_p)\,v_3(x'),\\ &v_3^{(3,2)}(x',\zeta) = (\nu-1)^{-1}\nu(\zeta^2/2-1/24)\Delta v_3(x'),\\ &v_p^{(3,3)}(x',\zeta) = ((1-\nu)^{-1}(2-\nu)\zeta^3/6-(-1)^{-1}\zeta/4)(\partial/\partial x_p)\,\Delta v_3(x'). \end{aligned} \tag{2}$$

The other components vanish and p takes values 1 and 2. Only the following operators $M_l^{(s,k)}(x',\partial/\partial x')$ differ from zero:

$$\begin{aligned} M_q^{(2,p)}(x',\partial/\partial x') &= \left(2(1+\nu)\right)^{-1} E\left(\delta_{pq}\Delta-(1-\nu)^{-1}(1+\nu)(\partial/\partial x_p)(\partial/\partial x_q)\right)\\ & M_{qp}^{(0)}(\partial/\partial x'),\quad p,q=1,2;\\ & M_3^{(4,3)}(x',\partial/\partial x') = \left(12(1-\nu^2)\right)^{-1} E\Delta_{x'}^2, = M_{33}^{(0)}(\partial/\partial x'). \end{aligned}$$

The corresponding limit problem (30), 15.2, coincides with the problem on the plane stress state for the vector (v_1, v_2) (the two first lines) and the Sophie Germain equation for the bend v_3 (the third line).

Example 15.3.3. We consider a boundary value problem for an elliptic system of three second order equations in a rectangle $\tilde{C}_\varepsilon = (-1,1)\times(-\varepsilon/2,\varepsilon/2)$ whose limit problem contains a differential operator of sixth order. More precisely, the limit system is elliptic in the sense of Douglis-Nirenberg with $t_1 = 2$, $t_2 = 4$, $t_3 = 6$ (see 15.2.5). In the rectangle $\tilde{C}_\varepsilon$ we consider the system

$$\begin{aligned} &-\Delta u_1-(\partial/\partial x_1)(\partial/\partial x_2)u_2 = f_1,\\ &-\Delta u_2-(\partial/\partial x_1)(\partial/\partial x_2)u_1-(\partial/\partial x_1)(\partial/\partial x_2)u_3 = f_2,\\ &-\Delta u_3-(\partial/\partial x_1)(\partial/\partial x_2)u_2 = f_3 \end{aligned} \tag{3}$$

with Dirichlet conditions on the vertical sides of the rectangle and with boundary conditions

$$\begin{aligned} &\partial u_1/\partial x_2+\partial u_2/\partial x_1 = g_1^\pm,\quad \partial u_2/\partial x_2+\partial u_3/\partial x_1 = g_2^\pm,\\ &\partial u_3/\partial x_3 = g_3^\pm \qquad \text{for}\quad x_2=\pm\varepsilon/2. \end{aligned} \tag{4}$$

One can verify inequality (7) with $\alpha_1 = 1, \alpha_2 = 2, \alpha_3 = 3$. It turns out that the auxiliary functions $v_l^{(k,r)}$ are given by the equalities

$$\begin{aligned} v_1^{(2,1)} &= -\zeta\partial v_2/\partial x_1, v_1^{(2,3)} = (\zeta^3/6-\zeta/8)\partial^3 v_2/\partial x_1^3,\\ v_2^{(3,1)} &= -\zeta\partial v_3/\partial x_1, v_1^{(3,2)} = (\zeta^2/2-1/24)\partial^2 v_3/\partial x_1^2,\\ v_1^{(3,4)} &= -(\zeta^4/24-\zeta^2/48+7/5760)\partial^4 v_3/\partial x_1^4,\\ v_2^{(3,5)} &= (\zeta^5/120-\zeta^3/144-\zeta/484)\partial^5 v_3/\partial x_1^5. \end{aligned} \tag{5}$$

The operators $M_l^{(s,k)}(\partial/\partial x_1)$ are of the form

$$\begin{aligned} &M_1^{(2,1)}(\partial/\partial x_1) = -(\partial/\partial x_1)^2,\quad M_2^{(4,2)}(\partial/\partial x_1) = 12^{-1}(\partial/\partial x_1)^4,\\ &M_3^{(6,3)}(\partial/\partial x_1) = -720^{-1}(\partial/\partial x_1)^6. \end{aligned} \tag{6}$$

As usual, in (5) and (6) we represent only the expressions that differ from zero. Thus, the limit problem for (3) and (4) is as follows:

$$-(\partial/\partial x_1)^2 v_1(x_1) = F_1(x_1), \quad x_1 \in (-1,1); \quad v_1(\pm 1) = 0;$$
$$12^{-1}(\partial/\partial x_1)^4 v_2(x_1) = F_2(x_1), \quad x_1 \in (-1,1); \quad v_2(\pm 1) = (\partial/\partial x_1) v_2(\pm 1) = 0;$$
$$-720^{-1}(\partial/\partial x_1)^6 v_3(x_1) = F_3(x_1), \quad x_1 \in (-1,1);$$
$$v_3(\pm 1) = (\partial/\partial x_1) v_3(\pm 1) = (\partial/\partial x_1)^2 v_3(\pm 1) = 0.$$

Example 15.3.4. Let $\tilde{C}_\varepsilon$ be the same cylinder as in Example 15.3.2. We consider the problem in micropolar elasticity theory (the Cosserat continuum)

$$(\mu+\alpha)\Delta u(x) + (\lambda+\mu-\alpha)\,\mathrm{grad}\ \mathrm{div}\, u(x) + 2\alpha \mathrm{rot}\, \omega(x) = X(x),$$
$$(\gamma+\varepsilon)\Delta\omega(x) + (\beta+\gamma-e)\,\mathrm{grad}\ \mathrm{div}\, \omega(x) - 4\alpha\omega(x) + 2\alpha \mathrm{rot}\, u(x) = Y(x), \tag{7}$$
$$x \in \tilde{C}_\varepsilon;$$
$$u(x) = 0, \quad \omega(x) = 0 \quad x \in S_\varepsilon; \tag{8}$$
$$\sigma_{3j}(u,\omega;y,\pm\varepsilon/2) = (\mu+\alpha)(\partial/\partial x_3)\, u_j(y,\pm\varepsilon/2) + (\mu-\alpha)(\partial/\partial x_j)\, u_3(y,\pm\varepsilon/2)$$
$$+\,\delta_{j3}\lambda\, \mathrm{div}\, u(y,\pm\varepsilon/2) + 2\alpha\big(\delta_{j2}\omega_1(y,\pm\varepsilon/2) - \delta_{j1}\omega_2(y,\pm\varepsilon/2)\big)$$
$$= p_j^{\pm}(y), \quad y \in \Omega, \quad j = 1,2,3;$$
$$\mu_{3j}(u,\omega;y,\pm\varepsilon/2) = (\gamma+e)(\partial/\partial x_3)\omega_j(y,\pm\varepsilon/2) + (\gamma-e)(\partial/\partial x_j)\omega_3(y,\pm\varepsilon/2)$$
$$+\,\delta_{j3}\, \mathrm{div}\, \omega(y,\pm\varepsilon/2) = m_j^{\pm}(y), \quad y \in \Omega, \quad j = 1,2,3 \tag{9}$$

(see COSSERAT and COSSERAT [1]), AERO and KUVSHINSKI [1], NOWACKI [1]). Here $x = (x_1, x_2, x_3) = (y, x_3)$; u and ω are displacement and rotation vectors; X and Y vectors of mass forces and momenta applied to the bases $\Gamma^{\pm}$ of $\tilde{C}_\varepsilon$; λ and μ the Lamé coefficients; α, β, γ, e the moment constants ($\alpha \neq 0, e \neq 0$); σ_{jk}, μ_{jk} the stresses and the moment stresses. To be in agreement with the previous notation we introduce the vector $u = (u_1, \dots, u_6) = (u_1, u_2, u_3, \omega_1, \omega_2, \omega_3)$. Since the energy integral for problem (7)–(9) is equal to

$$E(u,\omega) = \int_{\tilde{C}_\varepsilon} \Big\{2^{-1} \sum_{j,k=1}^{3} \big(\mu(\partial u_j/\partial x_k + \partial u_k/\partial x_j)^2 + \gamma(\partial\omega_j/\partial x_k + \partial\omega_k/\partial x_j)^2$$
$$+\, e(\partial\omega_j/\partial x_k - \partial\omega_k/\partial x_j)^2\big) + \lambda|\mathrm{div}\ u|^2 + \beta|\mathrm{div}\ \omega|^2$$
$$+\,\alpha\big(\partial u_1/\partial x_2 - \partial u_2/\partial x_1 - 2\omega_3)^2 + (\partial u_2/\partial x_3 - \partial u_3/\partial x_2 - 2\omega_1)^2$$
$$+\,(\partial u_3/\partial x_1 - \partial u_1/\partial x_3 - 2\omega_2)^2\big)\Big\}\, dx,$$

in the case $e \neq 0$ from the one-dimensional Hardy inequality it follows that (7) holds with $\alpha_\tau = 1, \tau = 1, \dots, 6$. Therefore, all the results in 15.2.1–15.2.6 are valid for problem (7)–(9). We carry out the corresponding calculation.

Define the vectors (14), 15.2, and set

$$V(x') = \big(v_k(x')\big)_{k=1}^{6}, \quad V^{(r)}(\varepsilon, x) = \sum_{k=1}^{6} v^{(k,r)}(x', \varepsilon^{-1}x_3), \quad r = 1,2,$$

where $v^{(k,r)}$ are the solutions to problems (15), 15.2, and v_k are some functions (approximations for the components u_τ of the solution u to problem (7)–(9)). In what follows it will be convenient to write $(v_1, v_2) = (w_1, w_2) = w$, $(v_4, v_5) = (\vartheta_1, \vartheta_2) = \vartheta$, $v_3 = W$, $v_6 = \Theta$. Solving the mentioned problems, we find

$$V^{(1)} = \zeta \begin{pmatrix} (\alpha+\mu)^{-1}\,(\alpha-\mu)\,\partial W/\partial y_1 \;-2(\mu+\alpha)^{-1}\,\alpha\,\vartheta_2 \\ (\alpha+\mu)^{-1}\,(\alpha-\mu)\,\partial W/\partial y_2 \;+2(\mu+\alpha)^{-1}\,\alpha\vartheta_1 \\ -2(\mu+\lambda)^{-1}\,\lambda\,\mathrm{div}_y\, w \\ -(\gamma+e)^{-1}\,(\gamma-e)\,\partial\Theta/\partial y_1 \\ -(\gamma+e)^{-1}\,(\gamma-e)\,\partial\Theta/\partial y_2 \\ -(2\gamma+\beta)^{-1}\,\beta\,\mathrm{div}_y\,\vartheta \end{pmatrix},$$

and

$$V^{(2)} = (\zeta^2/2 - 1/24) \begin{pmatrix} \begin{pmatrix} (2\mu+\lambda)^{-1}\lambda(\mu+\alpha)^{-1}(\mu-\alpha)\partial/\partial y_1 \mathrm{div}_y\, w \\ \qquad - 2(\mu+\alpha)^{-1}\alpha(\gamma+e)^{-1}(\gamma-e)\partial\Theta/\partial y_2 \end{pmatrix} \\ \begin{pmatrix} (2\mu+\lambda)^{-1}\lambda(\mu+\alpha)^{-1}(\mu-\alpha)\partial/\partial y_2 \mathrm{div}_y\, w \\ \qquad + 2(\mu+\alpha)^{-1}\alpha(\gamma+e)^{-1}(\gamma-e)\partial\Theta/\partial y_1 \end{pmatrix} \\ \begin{pmatrix} (2\mu+\lambda)^{-1}\,\lambda\big((\mu+\alpha)^{-1}\,(\mu-\alpha)\,\Delta_y W \\ \qquad + 2(\mu+\alpha)^{-1}\,\alpha(\partial\vartheta_1/\partial y_2 - \partial\vartheta_2/\partial y_1)\big) \end{pmatrix} \\ (2\gamma+\beta)^{-1}\,\beta(\gamma+e)^{-1}\,(\gamma-e)\,\partial/\partial y_1\,\mathrm{div}_y\,\vartheta \\ (2\gamma+\beta)^{-1}\,\beta(\gamma+e)^{-1}\,(\gamma-e)\,\partial/\partial y_2\,\mathrm{div}_y\,\vartheta \\ (2\gamma+\beta)^{-1}\,\beta(\gamma+e)^{-1}\,(\gamma-e)\,\Delta_y\Theta \end{pmatrix}.$$

The operators $M^{1,k}$ in (19), 15.2, are equal to zero and

$$\sum_{k=1}^{6} M^{(2,k)}(y, \partial/\partial y) V(y)$$

$$= \begin{pmatrix} (\mu+\alpha)\Delta_y w_1 + (\Lambda+\mu-\alpha)\partial/\partial y_1 \mathrm{div}_y\, w + 2\alpha\partial\Theta/\partial y_2 \\ (\mu+\alpha)\Delta_y w_2 + (\Lambda+\mu-\alpha)\partial/\partial y_2 \mathrm{div}_y\, w - 2\alpha\partial\Theta/\partial y_1 \\ A\Delta_y W - A(\partial\vartheta_1/\partial y_2 - \partial\vartheta_2/\partial y_1) \\ (\gamma+e)\Delta_y\vartheta_2 + (B+\gamma-e)\partial/\partial y_2 \mathrm{div}_y\vartheta - A(\vartheta_2 + \partial W/\partial y_1) \\ (\gamma+e)\Delta_y\vartheta_1 + (B+\gamma-e)\partial/\partial y_1 \mathrm{div}_y\vartheta - A(\vartheta_1 - \partial W/\partial y_2) \\ \Gamma\Delta_y\Theta - 2\alpha(\partial w_1/\partial y_2 - \partial w_2/\partial y_1 + 2\Theta) \end{pmatrix}.$$

Here the constants A, B, Γ, Λ are defined by

$$A = (\mu+\alpha)^{-1}4\mu\lambda, \quad B = (2\gamma+\beta)^{-1}2\gamma\beta, \quad \Gamma = (\gamma+e)^{-1}4\gamma e,$$
$$\Lambda = (2\mu+\lambda)^{-1}2\mu\lambda.$$

Thus, the system of equations in Ω in the third limit problem (30), 15.2, decomposes into the two following: the planar problem in Cosserat's moment theory

$$\begin{aligned} &(\mu+\alpha)\Delta_y w_1(y) + (\Lambda+\mu-\alpha)\partial/\partial y_1 \mathrm{div}_y\, w(y) + 2\alpha(\partial/\partial y_2)\,\vartheta(y) = F_1(y), \\ &(\mu+\alpha)\Delta_y w_2(y) + (\Lambda+\mu-\alpha)\partial/\partial y_2 \mathrm{div}_y\, w(y) - 2\alpha(\partial/\partial y_1)\,\vartheta(y) = F_2(y), \quad (10) \\ &\Gamma\Delta_y\vartheta(y) - 2\alpha\big((\partial/\partial y_2)v_1(y) - (\partial/\partial y_1)v_2(y) + 2\vartheta(y)\big) = F_6(y), \quad y \in \Omega, \end{aligned}$$

and the problem in the moment theory of thin plates

$$\begin{aligned} &(\gamma+e)\Delta_y\vartheta_1(y) + (B+\gamma-e)\partial/\partial y_1 \mathrm{div}_y\vartheta(y) - A\big(\vartheta_1(y) - (\partial/\partial y_2)\,W(y)\big) = F_5(y), \\ &(\gamma+e)\Delta_y\vartheta_2(y) + (B+\gamma-e)\partial/\partial y_2 \mathrm{div}_y\vartheta(y) - A\big(\vartheta_2(y) + (\partial/\partial y_1)\,W(y)\big) = F_4(y), \\ &A\Delta_y W(y) - A\big((\partial/\partial y_2)\vartheta_1(y) - (\partial/\partial y_1)\vartheta_2(y)\big) = F_3(y), \quad y \in \Omega. \end{aligned} \quad (11)$$

The corresponding boundary conditions are of the form

$$w_1(y) = w_2(y) = \Theta(y) = 0, \quad y \in \partial\Omega; \tag{12}$$

$$\vartheta_1(y) = \vartheta_2(y) = W(y) = 0, \quad y \in \partial\Omega. \tag{13}$$

Note that, according to Theorem 15.2.7, problems (10), (12) and (11), (13) are uniquely solvable.

Remark 15.3.5. The system of equations (9) can be reduced to Reissner's system (Reissner [1], Timoshenko and Woinowsky-Krieger [1]). Indeed, make the change of coordinates $y \to (x, y) = (y_1, -y_2)$ and introduce the new functions $\vartheta_1 = \vartheta_y, \vartheta_2 = \vartheta_x$. Then (9) takes the form

$$\begin{aligned}
&(\gamma + e)\Delta\vartheta_x - (B + \gamma - e)\,(\partial/\partial y)(\partial\vartheta_y/\partial x - \partial\vartheta_x/\partial y) - A(\vartheta_x + \partial W/\partial x) = 0,\\
&(\gamma + e)\Delta\vartheta_y - (B + \gamma - e)\,(\partial/\partial x)(\partial\vartheta_x/\partial y - \partial\vartheta_y/\partial x) - A(\vartheta_y + \partial W/\partial y) = 0,\\
&\Delta W + \partial\vartheta_x/\partial x - \partial\vartheta_y/\partial y = -(\varepsilon A)^{-1}\, p,
\end{aligned} \tag{14}$$

p being the transversal force. If the constants γ, e, B, A satisfy

$$\alpha = 5\mu/19, \quad \gamma + e = (1 - \nu^2)^{-1}\, E\varepsilon^2/12 = \varepsilon^{-1} D,$$
$$(2\gamma + \beta)^{-1}\gamma(\gamma + \beta) = \varepsilon^{-1}(1 - \nu)D/8,$$

where ε is the thickness of the plate, D its cylindrical stiffness, E Young's modulus, and ν Poisson's coefficient, then (14) precisely coincides with Reissner's system.

15.4 Antiplanar Shear and Flow of an Ideal Fluid in a Thin Domain with a Longitudinal Cut

In this section we consider the Neumann problem for the Laplace operator in a cylinder of small height with a longitudinal cut. The problem can be interpreted as describing the potential flow of an ideal fluid; in the two-dimensional case we deal with the antiplanar deformation or torsion of a rod. The purpose is to show how one should change the iterative procedures (in comparison with Section 15.1) to construct the asymptotics of the solution to the problem in a thin noncylindrical domain. Therefore, more attention is focused on description of the algorithm itself and its justification is dropped.

15.4.1 Statement of the problem

Let Γ_1, Γ_2, and Γ_3 be simple smooth closed contours while Γ_1 encompasses the contour Γ_2, and Γ_2 the contour Γ_3 (it is possible that Γ_3 is empty). We denote by Ω and ω the domains bounded by the curves $\Gamma_1 \cup \Gamma_3$ and $\Gamma_2 \cup \Gamma_3$, respectively. Introduce a cylinder Q_ε and the set $\{(x, z) \colon x = (x_1, x_2) \in \Omega, |z| < \varepsilon/2\}$. Let $G_\varepsilon = Q_\varepsilon \setminus M$ (see Fig. 15.2).

In the domain G_ε we consider the Neumann problem

$$-\Delta_{x,z} u(\varepsilon, x, z) = \varepsilon^{-1} f(x, \varepsilon^{-1} z), \quad (x, z) \in G_3; \tag{1}$$

$$(\partial/\partial z)u(\varepsilon, x, \pm\varepsilon/2) = \varphi^{\pm}(x), \quad x \in \Omega; \quad (\partial/\partial z)u(\varepsilon, x, \pm 0) = \psi^{\pm}(x), \quad x \in \omega; \tag{2}$$

$$(\partial/\partial n)u(\varepsilon, x, z) = \varepsilon^{-1} g(x, \varepsilon^{-1} z), \quad (x, z) \in S^1_\varepsilon \cup S^3_\varepsilon, \tag{3}$$

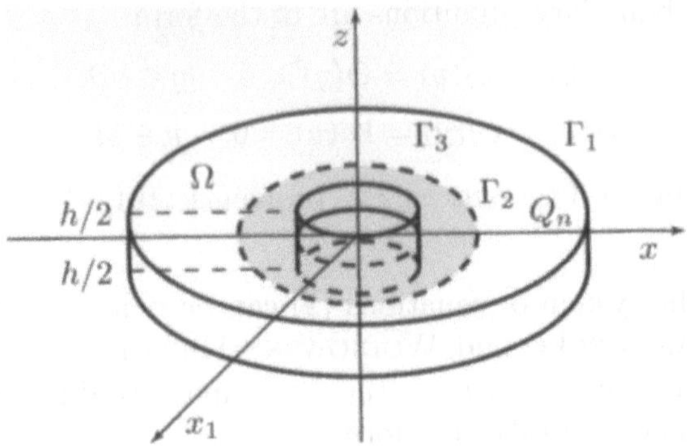

Fig. 15.2

where $\Delta_{x,z} = \Delta + (\partial/\partial z)^2$, $\Delta = (\partial/\partial x_1)^2 + (\partial/\partial x_2)^2$ is the outward normal to $\partial\Omega$, and $f, \varphi^{\pm}, \psi^{\pm}, g$ are smooth functions. We assume that there holds the compatibility condition of problem (1)–(3):

$$\varepsilon^{-1} \int\limits_{Q_\varepsilon} f(x, \varepsilon^{-1} z)dxdz + \varepsilon^{-1} \int\limits_{S_\varepsilon^1 \cup S_\varepsilon^3} g(x, \varepsilon^{-1} z)ds_x dz$$
$$+ \sum_{\pm} \pm \Big(\int\limits_{\Omega} \varphi^{\pm}(x)dx - \int\limits_{\omega} \psi^{\pm}(x)dx \Big) = 0. \tag{4}$$

Note that the right-hand sides of (1) and (3) contain ε^{-1} in such a manner that every integral in (4) is $O(1)$ as $\varepsilon \to 0$.

Near the edge ∂M the solution to problem (1), (2) admits the representation

$$u(\varepsilon, x, z) = C_0(\varepsilon, s) + C(\varepsilon, s)\, r^{1/2} \cos(\vartheta/2) + O(r), \quad \text{as} \quad r \to 0 \tag{5}$$

(see Proposition 11.3.3), where (r, ϑ) are the polar coordinates in the planes orthogonal to ∂M, and C_0 and C are smooth functions in $s \in \partial M$. We seek the asymptotics of the coefficient $C(\varepsilon, s)$.

15.4.2 The two-dimensional case

Let Q_s be a rectangle $(b_-, b_+) \times (-\varepsilon/2, \varepsilon/2)$ and let $M = \{(x, z) \colon a_- < x < a_+, z = 0\}$ with $b_- < a_- < a_+ < b_+$ (see Fig. 15.3). The planar version of problem (1)–(3) is of the form

$$-\big((\partial/\partial x)^2 + (\partial/\partial z)^2\big)\, u(\varepsilon, x, z) = \varepsilon^{-1} f(x, \varepsilon^{-1} z), \quad x \in G_\varepsilon = Q_\varepsilon \setminus M; \tag{6}$$

$$(\partial/\partial z)u(\varepsilon, x, \pm\varepsilon/2) = \varphi^{\pm}(x), \quad x \in (b_-, b_+);$$
$$(\partial/\partial z)u(\varepsilon, x, \pm 0) = \psi^{\pm}(x), \quad x \in (a_-, a_+); \tag{7}$$
$$(\partial/\partial x)u(\varepsilon, b_{\pm}, x) = \varepsilon^{-1} g^{\pm}(\varepsilon^{-1} z), \quad |z| < \varepsilon/2. \tag{8}$$

Let us try to find the principal term of the asymptotics of the solution u to problem (6), (7) in the form $\varepsilon w(x, \varepsilon^{-1} z)$. Then, separating out the leading terms of orders ε^{-1} and ε^0 in (6) and (7), we obtain

$$-(\partial/\partial\eta)^2 w(x, \eta) = f(x, \eta), \quad \eta \in (-1/2, 1/2); \quad (\partial/\partial\eta)w(x, \pm 1/2) = \varphi^{\pm}(x) \tag{9}$$

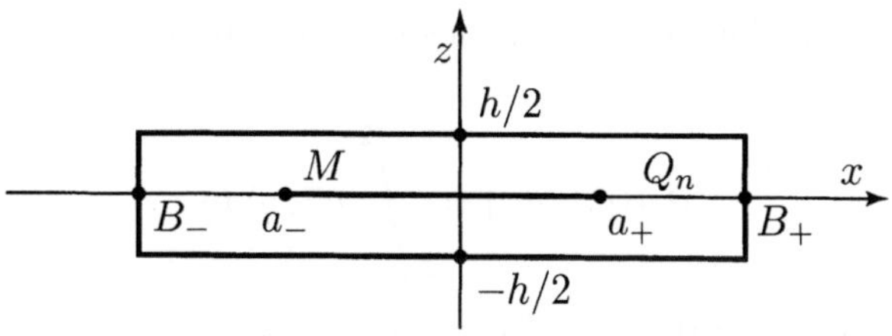

Fig. 15.3

for $x \in (b_-, a_-) \cup (a_+, b_+)$ and $\eta = \varepsilon^{-1} z$. The compatibility condition for problem (9) is as follows:

$$-\int_{-1/2}^{1/2} f(x,\eta)d\eta = \varphi^+(x) - \varphi^-(x). \tag{10}$$

Thus, it does not always happen that the right-hand sides of (6) and (7) can be compensated for by means of $\varepsilon w(x, \varepsilon^{-1}z)$.

As in 15.1.2, one should take an approximation for u in the form $\varepsilon^{-1}v(x) + \varepsilon\, w(x, \varepsilon^{-1}z)$. Calculating the coefficient by ε^{-1} in (6) and by ε^0 in (7), we obtain the equations

$$\begin{aligned} -(\partial/\partial\eta)^2 w(x,\eta) &= f(x,\eta) + (\partial/\partial x)^2 v(x), \quad \eta \in (-1/2, 1/2); \\ (\partial/\partial\eta) w(x, \pm 1/2) &= \varphi^\pm(x) \end{aligned} \tag{11}$$

in the intervals (b_-, a_-) and (b_+, a_+). The equality

$$(\partial/\partial x)^2 v(x) = -\int_{-1/2}^{1/2} f(x,\eta)d\eta - \varphi^+(x) + \varphi^-(x), \quad x \in (b_-, a_-) \cup (a_+, b_+), \tag{12}$$

is the compatibility condition for problem (11) (cf. (10)), which enables us to find $v(x)$. Similarly, for the functions w and $v^\pm$ forming the approximation $\varepsilon^{-1}v^\pm(x) + w(x, \varepsilon^{-1}z)$ for the solution u for $x \in (a_-, a_+)$ and $\pm z > 0$ we obtain

$$\begin{aligned} -(\partial/\partial\eta)^2 w(x,\eta) &= f(x,\eta) + (\partial/\partial x)^2 v^\pm(x), \quad \eta \in A_\pm; \\ (\partial/\partial\eta) w(x, \pm 1/2) &= \varphi^\pm(x), \quad (\partial/\partial\eta) w(x, \pm 0) = \psi^\pm(x); \end{aligned} \tag{13}$$

$$(\partial/\partial x)^2 v^\pm(x) = -2\int_{A_\pm} f(x,\eta)d\eta \pm 2(\varphi^\pm(x) - \psi^\pm(x)), \quad x \in (a_+, b_+), \tag{14}$$

where $A_+ = (0, 1/2)$ and $A_- = (-1/2, 0)$.

15.4.3 Boundary layers

Now we find the boundary conditions for the principal terms $\varepsilon^{-1}v$ and $\varepsilon^{-1}v^\pm$ in the asymptotics of the solution to problem (6)–(8). As in Section 15.1, a boundary layer arises near every vertical side $\{x = \pm 1, |z| < \varepsilon/2\}$ of the rectangle Q_ε. To describe these layers we introduce the variables $\eta = \varepsilon^{-1}z$ and $\xi = \varepsilon^{-1}(b_\pm \mp x)$. Letting $\varepsilon \to 0$, from (6)–(8) we obtain the boundary value problem

$$\begin{aligned} -\Delta_{\xi,\eta} V_0^\pm(\xi,\eta) &= 0, \quad (\xi,\eta) \in (0,\infty) \times (-1/2, 1/2); \\ (\partial/\partial\eta) V_0^\pm(\xi, 1/2) &= (\partial/\partial\eta) V_0^\pm(\xi, -1/2) = 0, \\ (\partial/\partial\xi) V_0^\pm(0,\eta) &= \pm((\partial/\partial x) v(b_\pm) - g^\pm(\eta)), \quad |\eta| < 1/2. \end{aligned} \tag{15}$$

By Proposition 15.1.3, there exist solutions to problem (15) exponentially decaying as $\xi \to \infty$ provided

$$(\partial/\partial x)v(b_\pm) = \int_{-1/2}^{1/2} g^\pm(\eta)d\eta. \tag{16}$$

Boundary layers arise near the tips of the cut M, too. Let us consider, for instance, the left endpoint $(a_-, 0)$ of M. After the change of variables $(x, z) \to \zeta = (\zeta_1, \zeta_2) = \varepsilon^{-1}(x - a_-, z)$, equations (6) and (7) give the Neumann problem in the strip $\Pi = \{\mathbb{R} \times (-1/2, 1/2)\} \setminus \{\zeta\colon \zeta_1 \geq 0, \zeta_2 = 0\}$ with semi-infinite cut.

Lemma 15.4.1. *Every solution to the homogeneous Neumann problem in the domain* Π *whose growth is at most polynomial is a linear combination* $c_1 + c_2\zeta + c_3 Y(\zeta)$, *where* c_j *is constant,* Y *an odd function in* ζ_2 *harmonic in* Π *that has zero Neumann data on* $\partial\Pi$ *and admits the representation*

$$\begin{aligned} &Y(\zeta) = O\big(\exp(\pi\zeta_1/2)\big) \quad \text{as} \quad \zeta_1 \to -\infty; \\ &Y(\zeta) = \pm\zeta_1 \pm \pi^{-1}\log 2 + O\big(\exp(-\pi\zeta_1/4)\big) \quad \text{as} \quad \zeta_1 \to +\infty, \quad \pm\zeta_2 > 0. \end{aligned} \tag{17}$$

Proof. It is clear that a linear function in ζ_1 is such a desired solution to the Neumann problem in Π. Moreover, with the help of the conformal mapping $\zeta = \zeta_1 + i\zeta_2 \to y = \big(1 - \exp(-2\pi\zeta)\big)^{1/2}$, which maps Π into the half-plane, we obtain the explicit formula for Y:

$$Y(\zeta) = (2\pi)^{-1}\log\big|\big(y(\zeta_1 + i\zeta_2) + 1\big)^{-1}\big(y(\zeta_1 + i\zeta_2) - 1\big)\big|.$$

It follows that

$$Y(\zeta) = (2\rho/\pi)^{1/2}\cos(\vartheta/2) + O(\rho) \quad \text{as} \quad \rho = \varepsilon^{-1}r \to 0 \tag{18}$$

in a neighborhood of the singular point $\zeta = 0$.

By Proposition 15.1.5, for any aforementioned solution $\hat{H}$ to the Neumann problem in π there are the asymptotic formulas

$$\begin{aligned} &\hat{H}(\zeta) = \hat{A} + \hat{B}\zeta_1 + O\big(\exp(\pi\zeta_1/2)\big) \quad \text{as} \quad \zeta_1 \to -\infty, \\ &\hat{H}(\zeta) = \hat{A}_\pm + \hat{B}_\pm\zeta_1 + O\big(\exp(-\pi\zeta_1/4)\big) \quad \text{as} \quad \zeta_1 \to \infty, \quad \pm\zeta_2 > 0. \end{aligned}$$

We substitute H and one of the functions $1, \zeta_1$ or Y into Green's formula for the set $\{\zeta \in \Pi\colon |\zeta_1| < R\}$. Using the representations of these functions as $\zeta_1 \to \pm\infty$ and letting $R \to \infty$, we derive the equalities

$$2\hat{B} = \hat{B}_+ + \hat{B}_-, \quad 2\hat{A} = \hat{A}_+ + \hat{A}_-, \quad \hat{B}_+\pi^{-1}\log 2 - \hat{A}_+ = \hat{B}_-\pi^{-1}\log 2 - \hat{A}_-.$$

From this and from the fact that the solution to the Neumann problem in Π vanishing at infinity is equal to zero, we obtain that there exist only three linearly independent solutions to the problem whose growth at infinity is at most polynomial. □

Let us describe the construction of the boundary layer. We denote by $\varepsilon^{-1}V(\varepsilon, x, z)$ the function coinciding with $\varepsilon^{-1}v(x)$ for $x \in (b_-, a_-) \cup (a_+, b_+)$ and with $\varepsilon^{-1}v_\pm(x)$ for $x \in (a_-, a_+)$, $\pm z > 0$. Generally, the function has discontinuities at $x = a_\pm$. We multiply it by cut-off functions vanishing near these points x:

$$\varepsilon^{-1}\Big(1 - \chi\big(\varepsilon^{-1}(x - a_+)\big)\Big)\Big(1 - \chi\big(\varepsilon^{-1}(x - a_-)\big)\Big)V(\varepsilon, x, z), \tag{19}$$

where $\chi \in \mathbf{C}_0^\infty(-1,1)$, $\chi(t) = 1$ for $|t| < 1/2$. The expression (19) gives the discrepancy in (6)–(8), which partially originates from the cut-off functions (in the ε-neighborhoods of the points $x = b_\pm$). We pass to the coordinates ζ, separate out the terms of orders $\varepsilon^{-3}, \varepsilon^{-2}$ and $\varepsilon^{-2}, \varepsilon^{-1}$ in (6) and (7), respectively, and obtain that the principal terms of the aforementioned discrepancy can be compensated for through the use of the solution to the problem

$$\Delta_\zeta W(\varepsilon,\zeta) = F(\varepsilon,\zeta), \quad \zeta \in \Pi; \quad (\partial/\partial\zeta_2)W(\varepsilon,\zeta) = 0, \quad \zeta \in \partial\Pi, \tag{20}$$

where

$$F(\varepsilon,\zeta) = [\Delta_\zeta, \chi(\zeta_1)]\{v(a_-) + \varepsilon\zeta_1(\partial/\partial x)v(a_-)\} \quad \text{for} \quad \zeta_1 < 0, \quad \zeta_2 \in (-1/2, 1/2);$$
$$F(\varepsilon,\zeta) = [\Delta_\zeta, \chi(\zeta_1)]\{v^\pm(a_-) + \varepsilon\zeta_1(\partial/\partial x)v^\pm(a_-)\} \quad \text{for} \quad \zeta_1 < 0, \quad \zeta_2 \in A_\pm.$$

Note that F is a smooth function. The solution to the problem for $\zeta_1 < 0$ is of the form

$$\begin{aligned}&\chi_1(\zeta_1)\big(v(a_-) + \varepsilon\zeta_1(\partial/\partial x)v(a_-)\big)\\ &\quad + 2^{-1}\big((\partial/\partial x)v^+(a_-) - (\partial/\partial x)v^-(a_-)\big)Y(\zeta),\end{aligned} \tag{21}$$

where $\chi_1(t) = \chi(t)$ for $t < 0$ and $\chi_1(t) = 1$ for $t \geq 1$. In the case $\zeta_1 > 0, \pm\zeta_2 > 0$ the functions

$$\big(1 - \chi_1(\zeta_1)\big)\big(v^\pm(a_-) + \varepsilon\zeta_1(\partial/\partial x)v^\pm(a_-)\big) \tag{22}$$

must be added to the expression (21). The constructed solution to problem (21) vanishes for $\zeta_1 < -1$ and, according to (17), admits the representation

$$\begin{aligned}W(\varepsilon,\zeta) &= \varepsilon\{(\partial/\partial x)v(a_-) - 2^{-1}\varepsilon\big((\partial/\partial x)v^+(a_-) + (\partial/\partial x)v^-(a_-)\big)\}\zeta_1\\ &\quad + v(a_-) - v^\pm(a_-) \pm (2\pi)^{-1}\varepsilon\big((\partial/\partial x)v^+(a_-) + (\partial/\partial x)v^-(a_+)\big)\log 2\\ &\quad + O\big(\exp(-\pi\zeta_1/4)\big)\end{aligned} \tag{23}$$

with $\varepsilon^{-1}v(x)$ for $x \in (b_-, a_-) \cup (a_+, b_+)$ and with $\varepsilon^{-1}v(x)_\pm(x)$ for $x \in (a_, a_+), \pm z > 0$ for $\zeta_1 > 0$ and $\pm\zeta_2 > 0$.

15.4.4 A supplementary limit problem and asymptotics of the intensity factor

The assumption $W(\varepsilon,\zeta) = O(\varepsilon)$ for $\zeta_1 = O(\varepsilon^{-1})$ leads to the equations

$$v(a_-) = v^+(a_-) = v^-(a_-), \quad 2(\partial/\partial x)v(a_-) = (\partial/\partial x)v^+(a_-) + (\partial/\partial x)v^-(a_-). \tag{24}$$

Similarly, when constructing the boundary layer near the point $(a_+, 0)$, we obtain

$$v(a_+) = v^+(a_+) = v^-(a_+), \quad 2(\partial/\partial x)v(a_+) = (\partial/\partial x)v^+(a_+) + (\partial/\partial x)v^-(a_+). \tag{25}$$

Thus, we have the boundary value problem (12)–(14), (16), (24), and (25) to find v and $v^\pm$. Note that the problem has a solution by virtue of the compatibility condition (4) of the original problem (6)–(8).

Proposition 15.4.2. *Let $\hat{U}$ be the sum of a smooth solution of the form* (19) *constructed by means of the solutions $v, v^\pm$ to the limit problem* (12)–(14), (16), (24), (25) *and the boundary layers*

$$W_\pm(\varepsilon, \pm\varepsilon^{-1}(x - a_\pm), \varepsilon^{-1}z) + \chi(x - b_\pm)V_0^\pm\big(\varepsilon(b_\pm \pm x), \varepsilon^{-1}z\big)$$

(see (20)–(23) *and* (15)*). Then*

$$\|\nabla(\hat{U}-u); \mathbf{L}_2(Q_\varepsilon \setminus M)\| \le c\varepsilon^{1/2},$$

where u is the solution to problem (6)–(8) *and the constant c is independent of ε.*

Near the left endpoint of M the constructed approximation to the solution u coincides with the boundary layer $\varepsilon^{-1}W(\varepsilon, \zeta)$. This enables us to find an asymptotic formula for the intensity factor $C_-(\varepsilon)$ in the expansion of u near the singular point $(a_-, 0)$:

$$u(\varepsilon, x, z) = \text{const} + C^-(\varepsilon) r^{1/2}\cos(\vartheta/2) + O(r) \quad \text{as} \quad r \to 0, \tag{26}$$

(see Proposition 1.4.6). Taking (21), (22) and (18) into account, we obtain

$$C_-(\varepsilon) = (2\pi\varepsilon)^{-1/2}\big((\partial/\partial x)v^+(a_-) - (\partial/\partial x)v^-(a_-)\big) + O(\varepsilon^{1/2}), \tag{27}$$

where $v^\pm$ is the solution to problem (12)–(14), (16), (24), and (25). Replacing a_- by a_+ in (27), we arrive at the corresponding formula for $C_+(\varepsilon)$.

15.4.5 The three-dimensional case

Asymptotics of the solution to problem (1)–(4) and the asymptotics of the coefficient in the expansion (5) of the solution near the boundary ∂M of a planar cut M can be constructed essentially in the same way as in 15.4.2 and 15.4.3. Note only that in constructing the boundary layers near the contours Γ_1, Γ_2, and Γ_3 one should make the change of variables $(x, z) \to (s, \xi_1, \xi_2) = (s, \varepsilon^{-1}n, \varepsilon^{-1}z)$, where s and n are the normal and tangent coordinates near Γ_j.

Far from Γ_j the main approximation for u is $\varepsilon^{-1}v(x)$ (for $x \in \Omega \setminus \bar{\omega}$) and $\varepsilon^{-1}v^\pm(x)$ (for $x \in \omega, \pm z > 0$) which satisfy the following boundary value problem:

$$-\Delta v(x) = \int_{-1/2}^{1/2} f(x, \eta)\,d\eta + \varphi^+(x) - \varphi^-(x), \quad x \in \Omega \setminus \bar{\omega};$$

$$-\Delta v(x) = 2\int_{A_\pm} f(x, \eta)\,d\eta \pm 2\big(\varphi^\pm(x) - \psi^\pm(x)\big), \quad x \in \omega;$$

$$(\partial/\partial n)v(x) = \int_{-1/2}^{1/2} g(x, \eta)\,d\eta, \quad x \in \Gamma_1;$$

$$(\partial/\partial n)v^\pm(x) = \int_{A_\pm} g(x, \eta)\,d\eta, \quad x \in \Gamma_3;$$

$$v(x) = v^+(x) = v^-(x), \quad 2(\partial/\partial n)v(x) = (\partial/\partial n)v^+(x) + (\partial/\partial n)v^-(x), \quad x \in \Gamma_2. \tag{28}$$

From now on in the section the normal n to Γ_1 and Γ_2 is outward with respect to $\Omega \setminus \bar{\omega}$ and the normal n to Γ_3 is outward for ω. Note that problem (28) is solvable by virtue of (4).

The boundary layer near ∂M can be constructed as in 15.4.2. In an ε-neighborhood of ∂M the solution to problem (1)–(3) admits the representation

$$u(\varepsilon, x, z) = \varepsilon^{-1}\big(J_0(s) + \varepsilon J_1(s) + \varepsilon\xi J_2(s) + \varepsilon J_3(s)Y(\xi)\big) + O(\varepsilon),$$

where $\xi = \varepsilon^{-1}(n, z)$, Y is the function in Lemma 15.4.1, J_k are smooth functions on ∂M, and $J_0 = v|_{\partial M}$, $J_2 = (\partial v/\partial n)|_{\partial M}$, $J_3 = 2^{-1}(\partial v^+/\partial n - \partial v^-/\partial n)|_{\partial M}$. The asymptotics of the coefficient $C(\varepsilon, s)$ in the representation (5) of the solution to problem (1)–(3) is of the form

$$C(\varepsilon, s) = (2\pi\varepsilon)^{-1/2}\big((\partial v^+/\partial n)|_{\partial M} - (\partial v^-/\partial n)|_{\partial M}\big)(s) + O(\varepsilon^{1/2}) \quad \text{as} \quad \varepsilon \to 0. \tag{29}$$

15.4.6 Examples

Relying on the results of preceding sections, we find asymptotics of solutions to some particular problems.

Example 15.4.3. We consider the problem on the antiplanar shear of a rectangle Q_ε with crack M (see 15.4.2). Suppose that there are no mass forces, the surfaces of the crack and the vertical sides are free from stresses, and shear forces with intensity $\tau(x)$ are given on the horizontal sides. All this means that in (6)–(8)

$$f(x, \eta) = 0, \quad \psi^\pm(x) = 0, \quad g^\pm(x) = 0, \quad \varphi^\pm(x) = \mu^{-1}\tau(x),$$

where μ is the shear modulus. In that case $v = 0$ and

$$v^\pm(x) = \pm 2\mu^{-1}\Big((a_+ - a_-)^{-1}(x - a_+)\int_{a_-}^{x}(y - a_-)\tau(y)dy$$

$$+ (a_+ - a_-)^{-1}(x - a_-)\int_{x}^{a_+}(y - a_+)\tau(y)dy\Big).$$

Then

$$(\partial/\partial x)v^\pm(a_-) = \pm 2(a_+ - a_-)^{-1}\mu^{-1}\int_{a_-}^{a_+}(y - a_+)\tau(y)dy,$$

and by (27) the intensity factor at the left tip $(a_-, 0)$ of the crack M is equal to

$$(8/\pi\varepsilon)^{1/2}(a_+ - a_-)^{-1}\mu^{-1}\int_{a_-}^{a_+}(y - a_+)\tau(y)dy + O(\varepsilon^{1/2}), \quad \varepsilon \to 0. \tag{30}$$

We observe that for $\tau > 0$ and small ε the coefficient $C_-(\varepsilon)$ is negative.

Example 15.4.4. We consider the potential flow of an ideal fluid in a three-dimensional circular annulus of small height ε containing an impenetrable absolutely rigid membrane (see Section 15.4.4). Let Γ_j be the circle of radius $R_j (R_1 > R_2 > R_3)$. Suppose that the fluid is uniformly supplied through the upper part of the side wall $S^3_{\varepsilon,+} = \{(x, z)\colon |x| = R_3, z \in (0, \varepsilon/2)\}$ and evacuated through the lower base; the rate of flow is equal to $\pi q R_3$. The other walls and the membrane are inpenetrable. Therefore, in (1)–(3)

$$f(x, e^{-1}z) = 0,\ \varphi^+(x) = 0,\ \varphi^\pm(x) = 0;\ g(x, e^{-1}z) = 0 \text{ for } (x, z) \in S^1_\varepsilon \cup S^3_{\varepsilon,-},$$
$$g(x, e^{-1}z) = -\varepsilon^{-1}\pi q/R_3 \quad \text{for} \quad (x, z) \in S^3_{\varepsilon,+}, \quad \varphi^-(x) = -\pi q/(R_1^2 - R_3^2).$$

In that case

$$\begin{aligned} v(x) &= 2^{-1}\pi(R_1^2 - R_3^2)^{-1}q(R_1^2\log r - r^2/2),\\ v^+(x) &= \pi q\log r + 2^{-1}(R_1^2 - R_3^2)^{-1}q((2R_3^2 - R_1^2)\log R_2 - R^2/2),\\ v^-(x) &= (R_1^2 - R_3^2)^{-1}\pi q(R_3^2\log r - r^2/2) + 2^{-1}\pi(R_1^2 - R_3^2)^{-1}q\\ &\quad\times\big((R_1^2 - 2R_3^2)\log R_2 + R_2^2/2\big). \end{aligned}$$

According to (30), for the coefficient $C(\varepsilon, s)$ in (5) by the singularity of the velocity ∇u at the edge ∂M there holds the asymptotic formula

$$C(\varepsilon, s) = (\pi/2\varepsilon)^{1/2}R_2^{-1}q(R_1^2 - R_3^2)^{-1}(R_1^2 + R_2^2 - 2R_3^2) + O(\varepsilon^{1/2}), \quad \varepsilon \to 0.$$

For small ε the coefficient $C(\varepsilon, s)$ has the same sign as q.

Example 15.4.5. The intensity factors can be approximately calculated by the above procedure in more general situations, too. For instance, we discuss the antiplanar shear of a thin curvilinear trapezoid with a crack. Let l be a smooth positive function on $[b_-, b_+]$ and $Q_\varepsilon = \{(x, z)\colon x \in (b_-, b_+), |z| < \varepsilon l(x)\}$. We consider the problem

$$\begin{aligned} -\Delta_{x,z}u(\varepsilon, x, z) &= 0, \quad (x, z) \in Q_\varepsilon \setminus M;\\ (\partial/\partial x)u(\varepsilon, b_\pm, z) &= 0, \quad |z| < \varepsilon l(b_\pm);\\ (\partial/\partial n)u\big(\varepsilon, x, \pm\varepsilon l(x)\big) &= \pm\mu^{-1}\tau(x), \quad x \in (b_-, b_+);\\ (\partial/\partial z)u(\varepsilon, x, \pm 0) &= 0, \quad x \in (a_-, a_+), \end{aligned} \tag{31}$$

where n is the outward normal to ∂Q_ε (see Example 15.4.3).

Taking account of the equality $n(\varepsilon, x) = \big(1 + \varepsilon^2 l'(x)^2\big)^{-1/2}\big(-\varepsilon l'(x), \pm 1\big)$ for $z = \pm\varepsilon l(x)$ and substituting the approximation $\varepsilon^{-1}v^\pm(x) + \varepsilon w(x, \varepsilon^{-1}z), x \in (a_-, a_+), \pm z \in \big(0, \varepsilon l(x)\big)$, into problem (31), we obtain

$$\begin{aligned} (\partial/\partial\eta)^2 w(x, \eta) &= -(\partial/\partial x)^2 v^\pm(x), \quad \pm\eta \in \big(0, l(x)\big);\\ (\partial/\partial\eta)w\big(x, \pm l(x)\big) &= \pm l'(x)(\partial/\partial x)v^\pm(x) \pm \mu^{-1}\tau(x),\\ (\partial/\partial\eta)w(x, 0) &= 0. \end{aligned} \tag{32}$$

The compatibility condition of the boundary value problem (32) for the ordinary equation with parameter x is of the form

$$l(x)(\partial/\partial x)^2 v^\pm(x) + l'(x)(\partial/\partial x)v^\pm(x) = \pm\mu^{-1}\tau(x), \quad x \in (a_-, a_+). \tag{33}$$

In a similar way, one can derive the equation for v

$$2l(x)(\partial/\partial x)^2 v(x) + 2l'(x)(\partial/\partial x)v(x) = 0, \quad x \in (b_-, a_-) \cup (a_+, b_+). \tag{34}$$

The boundary layers near the points $(b_\pm, 0)$ and $(a_\pm, 0)$ are constructed in the same manner as in Section 15.4.3. We obtain the boundary conditions and the transmission conditions

$$\begin{aligned} (\partial/\partial x)v(b_\pm) = 0, \quad v(a_\pm) &= v^+(a_\pm) = v^-(a_\pm),\\ 2(\partial/\partial x)v(a_\pm) &= (\partial/\partial x)v^+(a_\pm) + (\partial/\partial x)v^-(a_\pm). \end{aligned} \tag{35}$$

Besides, we derive asymptotic representations of the solution to problem (31) in the zones $|x-a_\pm|<\varepsilon, |z|<\varepsilon l(x)$ located near the tips of the crack M

$$u(\varepsilon,x,z)=\varepsilon^{-1}v(a_\pm)+L_\pm\Big(\big(2\varepsilon l(a_\pm)\big)^{-1}(a_\pm-x)\Big)+l(a_\pm)\big((\partial/\partial x)v^+(a_\pm) \\ -(\partial/\partial x)v^-(a_\pm)\big)Y\Big(\big(2\varepsilon l(a_\pm)\big)^{-1}\big(\pm(a_\pm-x),z\big)\Big)+O(\varepsilon), \tag{36}$$

where $L_\pm$ is a linear function and Y the function in Lemma 15.4.1. Solving equations (33)–(35) and combining (36), (18) and (26), we obtain the asymptotics of the intensity factor

$$C_-(\varepsilon)=-2\big(\pi\varepsilon l(a_-)\big)^{-1/2}\Big(\mu\int_{a_-}^{a_+}\big(l(y)\big)^{-1}dy\Big)\int_{a_-}^{a_+}\int_{a_-}^{y}\tau(t)dt\big(l(y)\big)^{-1}dy+O(\varepsilon^{1/2}),$$

$\varepsilon\to 0$.

Note that for $l(x)=1/2$ the last formula coincides with (30).

15.5 Intensity Factors for Close Parallel Cracks

In this section we study the antiplanar deformation of a domain with two parallel cracks. The cracks have different lengths; the smaller is over the large or is displaced with respect to the latter (see Figs. 15.4 and 15.5). The distance between the cracks is a small parameter of the problem.

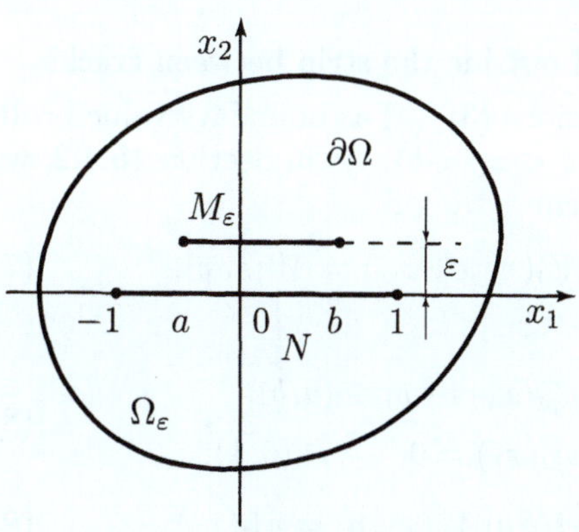

Fig. 15.4

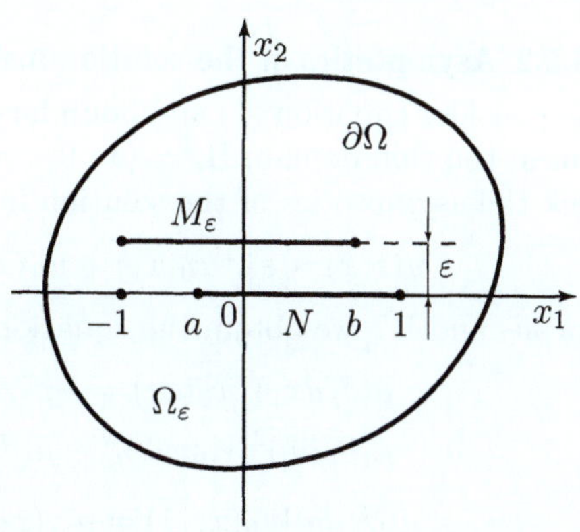

Fig. 15.5

We consider this problem as that in a domain with thin ligaments. Such singular perturbations for the Dirichlet problem were studied in Chapter 14. When constructing asymptotics of the solution to the Neumann problem in a domain with thin ligaments, we again invoke the expedients used for problems in thin domains (see Section 15.1) and in domains with small variation of the boundary near a corner point (see Chapter 2).

15.5.1 Statement of the problem

Let Ω be a domain in $\mathbb{R}^2$ with smooth boundary $\partial\Omega$ containing the segment $N=\{x\in\mathbb{R}^2\colon x^2=0,-1\le x_1\le 1\}$. We introduce $M_\varepsilon=\{x\colon x_2=\varepsilon, a\le x_1\le b\}$ and $\Omega_\varepsilon=\Omega\setminus(N\cup M_\varepsilon)$ with small positive ε (see Fig. 15.4). Here a and b are numbers in the interval $(-1,1)$ and $\varepsilon \ll \min\{1-b, 1+a, a+b\}$. We consider the problem of antiplanar deformation

$$\begin{aligned}
\mu\Delta u(\varepsilon,x) &= 0, && x \in \Omega_\varepsilon; && (1)\\
\mu(\partial/\partial n)u(\varepsilon,x) &= q(x), && x \in \partial\Omega; && (2)\\
\mu(\partial/\partial x_2)u(\varepsilon,x_1,\varepsilon\pm 0) &= \mp p_M^\pm(x_1), && a < x_1 < b; && (3)\\
\mu(\partial/\partial x_2)u(\varepsilon,x_1,\pm 0) &= \mp p_N^\pm(x_1), && -1 < x_1 < 1, && (4)
\end{aligned}$$

μ being shear modulus, n the outward normal to $\partial\Omega$, and $q, p_M^\pm$, and $p_N^\pm$ smooth external forces applied to $\partial\Omega$ and to the sides $M_\varepsilon^\pm$ and $N^\pm$ of the cuts M_ε and N. Suppose that the forces at the boundary $\partial\Omega$ are self-balanced, i.e., there holds the compatibility condition of problem (1)–(4):

$$\int_{\partial\Omega} q(x)ds + \sum_\pm \Big(\int_a^b p_M^\pm(x_1)dx_1 + \int_{-1}^1 p_N^\pm(x_1)dx_1 \Big) = 0. \qquad (5)$$

Let r_b, ϑ_b be the polar coordinates based on the tip (b,ε) of M_ε so that $M_\varepsilon^\pm$ are given by $\vartheta_b = \pm\pi$. In a small neighborhood of the point (b,ε) the solution u of problem (1)–(4) admits the representation

$$u(x,\varepsilon) = \text{const } + K_b(\varepsilon)r_b^{1/2}\sin(\vartheta/2) + O(r_b|\log r_b|), \qquad (6)$$

where $K_b(\varepsilon)$ is the intensity factor. Similar formulas hold near the endpoints (a,ε) and $(\pm 1, 0)$ of M_ε and N, too. Let $K_a(\varepsilon)$ and $K_\pm(\varepsilon)$ stand for the corresponding intensity factors.

15.5.2 Asymptotics of the solution inside and outside the strip between cracks

We consider equation (1) and boundary conditions (3), (4) as boundary value problem in the thin domain $\Pi_\varepsilon = \{x\colon 0 < x_2 < \varepsilon, a < x_1 < b\}$. As in Section 15.4.2, we seek the asymptotics of the solution in the form

$$u(\varepsilon,x) \sim \varepsilon^{-1}w_0(x_1) + w_1(x_1) + \varepsilon W_0(x_1, \varepsilon^{-1}x_2) =: W(\varepsilon,x). \qquad (7)$$

For w_0 and W_0 we obtain the equations

$$\left.\begin{aligned}
\mu(d/dx_1)^2 w_0(x_1) &= -p_M^-(x_1) - p_N^+(x_1), \quad x_1 \in (a,b);\\
\mu(\partial/\partial\eta)^2 W_0(x_1,\eta) + \mu(d/dx_1)^2 w_0(x_1) &= 0, \quad \eta \in (0,1);
\end{aligned}\right. \qquad (8)$$

$$\mu(\partial/\partial\eta)W_0(x_1,1) = p_M^-(x_1), \quad \mu(\partial/\partial\eta)W_0(x_1,0) = p_N^+(x_1) \qquad (9)$$

with $\eta = \varepsilon^{-1}x^2$. The required boundary conditions for (8) will be found when studying the boundary layers near the points (a,ε) and (b,ε). The equation for w_1 in (7) is similar to (8). However, we need not use such a function when constructing the principal term of the asymptotics of u. Note that the right-hand side of the mentioned equation is equal to zero, i.e., w_1 is a linear function.

In case $\varepsilon = 0$ the domain Ω_ε converts into the domain Ω_0 with the only crack N and problem (1)–(4) takes the form

$$\mu\Delta v_0(x) = 0, \quad x \in \Omega_0; \quad \mu(\partial/\partial n)v_0(x) = q(x), \quad x \in \partial\Omega; \qquad (10)$$

$$\begin{aligned}
\mu(\partial/\partial x_2)v_0(x_1,+0) &= -p_N^+(x_1), \quad x_1 \in (-1,a)\cup(b,1);\\
\mu(\partial/\partial x_2)v_0(x_1,+0) &= -p_M^+(x_1), \quad x_1 \in (a,b);\\
\mu(\partial/\partial x_2)v_0(x_1,-0) &= -p_N^-(x_1), \quad x_1 \in (-1,1). \qquad (11)
\end{aligned}$$

Problem (10), (11) may have no bounded solution because, by virtue of (5),

$$\mu \int_{\partial\Omega_0} (\partial/\partial n) v_0(x) ds = -I, \quad I = \int_a^b (p_N^+(x_1) + p_M^-(x_1)) dx_1. \tag{12}$$

Therefore, as in Section 15.2, it is necessary to expand the class of solutions and allow the function u to have logarithmic singularities at the points $(a, +0)$ and $(b, +0)$. Then the boundary value problem turns out to be solvable and its solution is determined up to a linear combination of two functions that satisfy the homogeneous problem. One of those functions is identically equal to 1 and the other is the Neumann function whose poles are located at the above points. We recall that G satisfies the equations

$$\Delta G(x) = 0, \quad x \in \Omega_0; \quad (\partial/\partial n) G(x) = 0, \quad x \in \partial\Omega; \tag{13}$$

$$G(x) = -\pi^{-1} \log r_b + G_b + O(r_b), \quad x_2 > 0, \quad r_b \to 0; \tag{14}$$

$$G(x) = \pi^{-1} \log r_a + G_a + O(r_a), \quad x_2 > 0, \quad r_a \to 0, \tag{15}$$

where r_d is the distance from x to the point $(d, 0)$, and G_a and G_b are some constants.

Thus, as the asymptotics of u we take a linear combination

$$u(\varepsilon, x) \sim c_v + V_0(x) + A_0(\varepsilon) G(x) =: V(\varepsilon, x), \tag{16}$$

where c_v is an arbitrary constant (the rigid displacement), $A_0(\varepsilon)$ is to be determined, and the function V_0 is bounded outside any neighborhood of the point $(b, 0)$ and satisfies (10), (11), and

$$V_0(x) = \pi^{-1} I \mu^{-1} \log r_b + V_b^0 + O(r_b |\log r_b|), \quad x_2 > 0, \quad r_b \to 0. \tag{17}$$

15.5.3 Boundary layers near tips of the cut

In Section 15.5.2, we found formal asymptotics of u inside and outside Π_ε. In order to match those representations and, in doing so, to eliminate arbitrariness of some constants, we study the behavior of the solution to problem (1)–(4) near the points (a, ε) and (b, ε). As in the preceding sections, there are boundary layers in such zones. Due to the symmetry of the problem, it suffices to consider only one tip of the crack M_ε, say, the point (b, ε). We make the change of variables $x \to \xi = \varepsilon^{-1}(x_1, -b, x_2)$. Letting $\varepsilon \to 0$ and considering $\xi_2 > 0$, we obtain the boundary value problem

$$\mu \Delta z(\xi) = 0, \quad \xi \in \Xi; \quad \mu(\partial/\partial \xi_2) z(\xi) = 0, \quad \xi \in \partial\Xi, \tag{18}$$

with $\Xi = \mathbb{R}^2_+ \setminus \{\xi \in \mathbb{R}^2 \colon \xi_2 = 1, \xi_1 < 0\}$. The domain Ξ has two "outlets" to infinity: the one is the corner Ξ_+ and the other is the semistrip Ξ_-.

Lemma 15.5.1. *Every solution to problem* (18) *that is of polynomial growth in* Ξ_- *and is* $O(|\log|\xi||^N)$ *in* Ξ_+ *turns out to be a linear combination* $a + bY(\varepsilon)$, *where* a *and* b *are constants and*

$$Y(\xi) = \log|\gamma(\xi_1 + i\xi_2)| \tag{19}$$

while $\gamma \colon \xi_1 + i\xi_2 \to \eta_1 + i\eta_2$ *is the inverse for*

$$\eta_1 + i\eta_2 \to \xi_1 + i\xi_2 = \pi^{-1}(e^{-1}(\eta_1 + i\eta_2) + \log(\eta_1 + i\eta_2)).$$

There hold the representations

$$Y(\xi) = \pi\xi_1 + O\big(\exp(\pi\xi_1)\big) \quad \text{as} \quad \xi_1 \to -\infty \quad \text{in} \quad \Xi_-; \tag{20}$$
$$Y(\xi) = \log|\xi| + 1 + \log\pi + O\big(|\xi|^{-1}|\log|\xi||\big) \quad \text{as} \quad |\xi| \to \infty \quad \text{in} \quad \Xi_+. \tag{21}$$

Proof. We list the solutions to problem (18) that are of polynomial growth in Ξ_- and are $O(|\log|\xi||^N)$ in Ξ_+. The solution $Y_0(\xi) \equiv 1$ is evident. Propositions 15.1.1 and 15.1.4 imply that any solution Y_* possessing the mentioned properties admits the representations

$$Y_*(\xi) = c_1\xi_1 + c_2 + O\big(\exp(\pi\xi_1)\big) \quad \text{as} \quad \xi_1 \to -\infty \quad \text{in} \quad \Xi_-, \tag{22}$$
$$Y_*(\xi) = c_3\log|\xi| + c_4 + O\big(|\xi|^{-1}|\log|\xi||\big) \quad \text{as} \quad \xi_1 \to \infty \quad \text{in} \quad \Xi_+, \tag{23}$$

where c_j are some constants. We substitute the functions Y_* and Y_0 into the Green formula for the domain $\Xi_R = \{\xi \in \Xi\colon |\xi| < R \text{ for } \xi \in \Xi_+ \text{ and } \xi_1 > -R \text{ for } \xi \in \Xi_-\}$ with large positive R. Then

$$\begin{aligned} 0 &= \int\limits_{\xi_R} \big(Y_*(\xi)\Delta Y_0(\xi) - Y_0(\xi)\Delta Y_*(\xi)\big)d\xi \\ &= \int\limits_{\partial\xi_R} \big(Y_*(\xi)(\partial/\partial n)Y_0(\xi) - Y_0(\xi)(\partial/\partial n)Y_*(\xi)\big)ds. \end{aligned} \tag{24}$$

In the last integral in (24) the integrand differs from zero only for the integral I_+ over the arc $\{\pi - \arcsin(R^{-1}) > \vartheta > 0, |\xi| = R\}$ and for the integral I_- over the segment $\{\xi_1 = -R, 0 < \xi_2 < 1\}$. Using (22) and (23), we find that

$$I_+ = -\int\limits_0^{\pi-\arcsin(R^{-1})} \big(c_3 + O(R^{-1}|\log R|)\big)d\vartheta = -\pi c_3 + O(R^{-1}|\log R|), \tag{25}$$

$$I_- = \int\limits_0^1 \Big(c_1 + O\big(\exp(-\pi R)\big)\Big)d\xi_2 = c_1 + O\big(\exp(-\pi R)\big). \tag{26}$$

Thus, letting $R \to \infty$ and taking (25) and (26) into account, we derive the equality $c_1 = \pi c_3$ from (24).

Suppose that a solution to problem (18) has asymptotics (22), (23) with $c_1 = c_3 = 0$. Then from Green's formula (24) for Y_* and Y it follows that $c_2 = c_4$. Moreover, the solution with zero constants in (22) and (23) vanishes at infinity and has finite Dirichlet's integral. Hence, the solution is trivial. Thus, Y_0 and Y form a basis in the space of solutions to problem (18). □

Remark 15.5.2. (i) In what follows we make use of a representation of the type (6)

$$Y(\xi) = 1 + (2\pi\rho)^{1/2}\sin(\varphi/2) + O(\rho) \tag{27}$$

for the function Y near the tip of the cut, where (ρ, φ) are the polar coordinates based on $(0, 1)$ so that the sides of the cut are given by $\varphi = \pm\pi$ and $\rho = \varepsilon^{-1}(r + O(\varepsilon))$ and $\varphi = \vartheta + O(\varepsilon)$ (see (6)).

(ii) Allowing Y_* to be of linear growth in Ξ_+, we obtain the new solution $Y_1(\xi) = \xi_1$ to the homogeneous problem (18). Now we turn to matching the asymptotics (7) and (16) with the help of a boundary layer. We seek an approximation to the solution u of problem (1)–(4) near the points (a, ε) and (b, ε) in the form

$$u(\varepsilon, x) \sim c_a(\varepsilon) + B_a(\varepsilon)Y\big(\varepsilon^{-1}(a - x_1), \varepsilon^{-1}x_2\big) = Z_a(\varepsilon, x), \tag{28}$$

$$u(\varepsilon, x) \sim c_b(\varepsilon) + B_b(\varepsilon)Y\big(\varepsilon^{-1}(x_1 - b), \varepsilon^{-1}x_2\big) = Z_b(\varepsilon, x). \tag{29}$$

As was said, the role of the boundary layers (28) and (29) is to match the constructed approximations of u. We find $c_a(\varepsilon), c_b(\varepsilon), B_a(\varepsilon), B_b(\varepsilon)$, and $A_0(\varepsilon)$ in (28), (29), and (16) equating the asymptotic expansions (7) and (16) in intermediate zones.

For points x such that $0 < x_2 < \varepsilon$ and $x_1 \sim b - \varepsilon^{1/2}$ we obtain

$$\begin{aligned} W(\varepsilon, x) = {} & \varepsilon^{-1}w_0(0) + w_1(0) + \varepsilon^{-1}(x_1 - b)(d/dx)w_0(b) \\ & + (2\varepsilon)^{-1}(x_1 - b)^2(d/dx_1)^2 w_0(b) + O(\varepsilon^{1/2}) \end{aligned}$$

or, in the coordinates ξ,

$$\begin{aligned} W(\varepsilon, (b,0) + \varepsilon\xi) = {} & \varepsilon^{-1}w_0(0) + w_1(0) + \xi_1(d/dx_1)w_0(b) \\ & + 2^{-1}\varepsilon_1^2(d/dx_1)^2 w_0(b) + O(\varepsilon^{1/2}). \end{aligned} \tag{30}$$

From (20) it follows that

$$Z_b(\varepsilon, x) \sim c_b(\varepsilon) + B_b(\varepsilon)\pi\xi_1 + O\big(\exp(\pi\xi_1)\big). \tag{31}$$

Combining (30) and (31), we derive the equalities

$$c_b(\varepsilon) = \varepsilon^{-1}w_0(0) + w_1(0), \quad B_b(\varepsilon) = B_b = \pi^{-1}(d/dx_1)w_0(b). \tag{32}$$

If $x_2 > \varepsilon$ and $r_b \sim \varepsilon^{1/2}$, then according to (14)–(17) we have

$$V(\varepsilon, x) = c_v + \pi^{-1}I\mu^{-1}\log r_b + V_b^0 + A_0(\varepsilon)(-\pi^{-1}\log r_b + G_b) + O(\varepsilon^{1/2}). \tag{33}$$

Moreover, by virtue of (21) in the same zone

$$Z_b(\varepsilon, x) = c_b(\varepsilon) + B_b\big(\log(r_b\varepsilon^{-1}) + \log\pi + 1\big) + O(\varepsilon^{1/2}). \tag{34}$$

The asymptotics (33) and (34) must coincide. Therefore,

$$\mu^{-1}I - A_0(\varepsilon) = \pi B_b, \quad c_v + V_b^0 + A_0(\varepsilon)G_b = c_b(\varepsilon) + B_b\big(\log(\varepsilon^{-1}\pi) + 1\big). \tag{35}$$

In the same manner as in studying the boundary layer Z_a we derive

$$c_a(\varepsilon) = \varepsilon^{-1}w_0(a) + w_1(a), \quad B_a(\varepsilon) = B_a = -\pi^{-1}(d/dx_1)w_0(a), \tag{36}$$

$$A_0(\varepsilon) = \pi B_a, \quad c_v + V_a^0 + A_0(\varepsilon)G_a = c_a(\varepsilon) + B_a\big(\log(\varepsilon^{-1}\pi) + 1\big). \tag{37}$$

Solving the algebraic system (32), (35)–(37), we find the deficient boundary conditions for w_0 satisfying (8)

$$w_0(b) - w_0(a) = 0, \quad (d/dx_1)w_0(b) - (d/dx_1)w_0(a) = I\mu^{-1}, \tag{38}$$

the expressions for the constants A_0, B_a, and B_b

$$\begin{aligned} A_0(\varepsilon) = A_0 = -\pi^{-1}(d/dx_1)w_0(a), \quad B_a &= -\pi^{-1}(d/dx_1)w_0(a) \\ B_b &= -\pi^{-1}(d/dx_1)w_0(b) \end{aligned} \tag{39}$$

and, finally, one boundary condition for w_1

$$\begin{aligned} w_1(b) - w_1(a) = {} & V_b^0 - V_a^0 - \pi^{-1}(d/dx_1)w_0(a)(G_b - G_a) \\ & + \pi^{-1}\big(1 + \log(\pi/\varepsilon)\big)\big((d/dx_1)w_0(b) + (d/dx_1)w_0(a)\big). \end{aligned}$$

The second boundary condition and the equation for w_1 can be found in constructing next terms in the asymptotics.

Note that problem (8), (38) is solvable due to the definition (3) of I and the condition (5).

15.5.4 Estimate on the remainder in asymptotics

Using the asymptotics (7), (16), (28), and (29), we construct a global approximation to the solution u of problem (1)–(4). Let $\chi \in \mathbf{C}_0^\infty(-1,1)$ and let $\chi(t) = 1$ for $|t| < 1/2$. We set

$$X(\varepsilon,x) = \big(1-\chi(r_a\varepsilon^{-1/2})\big)\big(1-\chi(r_b\varepsilon^{-1/2})\big)$$

and introduce the function

$$U(\varepsilon,x) = \begin{cases} V(\varepsilon,x), & x\in\Omega_\varepsilon, x_2<0;\\ X(\varepsilon,x)V(\varepsilon,x)+\chi(r_a\varepsilon^{-1/2})Z_a(\varepsilon,x)+\chi(r_b\varepsilon^{-1/2})Z_b(\varepsilon,x), & x\in\Omega_\varepsilon\setminus\overline{\Pi_\varepsilon}, x_2\in 0;\\ \Big(1-\chi\big((x_1-a)\varepsilon^{-1/2}\big)\Big)\Big(1-\chi\big((x_1-b)\varepsilon^{-1/2}\big)\Big)W(\varepsilon,x) & \\ \quad +\chi\big((x_1-a)\varepsilon^{-1/2}\big)Z_a(\varepsilon,x)+\chi\big((x_1-b)\varepsilon^{-1/2}\big)Z_b(\varepsilon,x), & x\in\Pi_\varepsilon, \end{cases} \tag{40}$$

(see (7), (16), (23), and (24)). The function U satisfies (2)–(4) on M_ε^- and N, producing small discrepancies on M_ε^+ and in the equation (1). Therefore, we arrive at the next assertion.

Proposition 15.5.3. *There exists the inequality*

$$||\nabla(u-U);\mathbf{L}_2(\Omega_\varepsilon)|| \le c\varepsilon^{1/4}|\log\varepsilon|, \tag{41}$$

where u is the solution to problem (1)–(4), *U is the function* (40), *the constant c being independent of ε.*

Remark 15.5.4. The construction (40) obtained by the method of matched asymptotic expansions is inexact and gives the error $O(\varepsilon^{1/4}|\log\varepsilon|)$ (see (41)). By the same argument as in Proposition 2.1.5, one could obtain a lesser error $O(\varepsilon^{1/2}|\log\varepsilon|)$. However, the corresponding approximation is too unwieldy and we do not describe it.

15.5.5 Asymptotics of the intensity factors

The representations (40) of the solution to problem (1)–(4) lead to the asymptotic formulas for the intensity factors at the tips of the cracks M_ε and N (see (6)). We introduce the coefficients $k_V^\pm$ and $k_G^\pm$ in the expansions

$$V_0(x) = \mathrm{const}_\pm + k_V^\pm r_\pm^{1/2}\sin(\vartheta_\pm/2) + O(r_\pm|\log r_\pm|), \tag{42}$$

$$G(x) = \mathrm{const}_\pm + k_G^\pm r_\pm^{1/2}\sin(\vartheta_\pm/2) + O(r_\pm|\log r_\pm|), \tag{43}$$

of the harmonic functions V_0 and G (see (17) and (13)–(15)). Here $r_\pm$ and $\vartheta_\pm$ are the polar coordinates based on $(\pm1,0)$ so that the sides $N^\pm$ of the cut N are given by $\vartheta_\pm = \pi$ and $\vartheta_\pm = -\pi$.

Combining (6), (16), (42), (43), and (39), we find

$$K_\pm(\varepsilon) = k_V^\pm - \pi^{-1}(d/dx_1)w_0(a)k_G^\pm + O(\varepsilon|\log\varepsilon|). \tag{44}$$

Similarly, from (28), (29), and (27) we derive the relations

$$\begin{aligned} K_a(\varepsilon) &= (2\pi\varepsilon)^{1/2}(d/dx_1)w_0(a) + O(\varepsilon^{1/2}|\log\varepsilon|), \\ K_b(\varepsilon) &= -(2\pi\varepsilon)^{1/2}(d/dx_1)w_0(b) + O(\varepsilon^{1/2}|\log\varepsilon|). \end{aligned} \tag{45}$$

The asymptotic formulas contain the solution w_0 of the ordinary differential equation, see (8) and (38). If $w_0 = \text{const}$ then the representations (45) are of little use. This is the case, for example, if the shores of cracks are free from forces, i.e., $p_M^{\pm} = p_N^{\pm} = 0$ in (3) and (4). We consider such a situation constructing the second term in the asymptotics.

Far from Π_ε we seek the asymptotics of u in the form

$$u(\varepsilon, x) \sim v_0(x) + \varepsilon v_1(x) + \varepsilon A_1 G(x), \tag{46}$$

A_1 being a constant. The function v_0 satisfies (10) and the boundary condition

$$\mu(\partial/\partial x_2)v_0(x) = 0, \quad x \in N^{\pm}. \tag{47}$$

The function v_1 is a solution to the boundary value problem

$$\Delta v_1(x) = 0, \quad x \in \Omega_0; (\partial/\partial n)v_1(x) = 0, \quad x \in \partial\Omega; \tag{48}$$

$$\begin{aligned} &(\partial/\partial x_2)v_1(x_1, +0) = (\partial/\partial x_1)^2 v_0(x_1, +0), \quad x_1 \in (a, b); \\ &(\partial/\partial x_2)v_1(x) = 0, \quad x \in N^- \cup (N^+ \setminus M_0^+). \end{aligned} \tag{49}$$

We explain the choice of the right-hand sides in (49). Using Taylor's expansion for the function on the right in (46), we find that for $x \in M_\varepsilon^+$ (or for $x_1 \in (a, b), x_2 = \varepsilon + 0$)

$$\begin{aligned} (\partial/\partial x_2)u(\varepsilon, x_1, \varepsilon + 0) &\sim (\partial/\partial x_2)v(x_1, +0) + \varepsilon((\partial/\partial x_2)v_1(x_1, +0) \\ &+ A_1(\partial/\partial x_2)G(x_1, +0) + (\partial/\partial x_2)^2 v_0(x_1, +0)) + O(\varepsilon^2). \end{aligned} \tag{50}$$

Since v_0 and G satisfy the homogeneous Neumann condition on N^+, the coefficient by ε in (50) vanishes provided

$$\begin{aligned} (\partial/\partial x_2)v_1(x_1, +0) &= -(\partial/\partial x_2)^2 v_0(x_1, +0) \\ &= (\partial/\partial x_1)^2 v_0(x_1, +0) - \Delta v_0(x_1, +0) = (\partial/\partial x_1)^2 v_0(x_1, +0). \end{aligned}$$

There exists a solution to problem (48), (49) that is bounded outside any neighborhood of the point $(b, +0)$ and admits the representation

$$v_1(x) = \pi^{-1} I_1 \log r_b + V_b^1 + O(r_b|\log r_b|), \quad r_b \to 0, \quad x_2 > 0 \tag{51}$$

(cf. (17)), where

$$\begin{aligned} I_1 &= -\int_{\partial\Omega} (\partial/\partial n)v_1(x)ds = -\int_a^b (\partial/\partial x_1)^2 v_0(x_1, +0))dx_1 \\ &= (\partial/\partial x_1)v_0(a, +0) - (\partial/\partial x_1)v_0(b, +0). \end{aligned} \tag{52}$$

Inside Π_ε the solution $u(\varepsilon, u)$ is approximated by $w_1(x_1)$, which is a linear function $\alpha x_1 + \beta$. The boundary layers (28) and (29) are of the form

$$Z_a(\varepsilon, x) = c_a^0 + \varepsilon\Big(c_a^1 + B_a Y(\varepsilon^{-1}(a - x_1), \varepsilon^{-1}x_2) + D_a Y_1(\varepsilon^{-1}(a - x_1), \varepsilon^{-1}x_2)\Big), \tag{53}$$

$$Z_b(\varepsilon, x) = c_b^0 + \varepsilon\Big(c_b^1 + B_b Y(\varepsilon^{-1}(x_1 - b), \varepsilon^{-1}x_2) + D_b Y_1(\varepsilon^{-1}(x_1 - b), \varepsilon^{-1}x_2)\Big). \tag{54}$$

Matching the above representations, we obtain, as in Section 15.4.3, the algebraic equations

$$\begin{aligned}&\alpha b+\beta=v_0(b,+0),\quad \alpha a+\beta=v_0(a,+0),\quad \pi B_b=I_1-A_1,\quad \pi B_a=A_1,\\ &\qquad\alpha=\pi B_b+D_b=-\pi B_a-D_a,\quad D_b=(\partial/\partial x_1)v_0(b,+0),\\ &\qquad\qquad D_a=-(\partial/\partial x_1)v_0(a,+0).\end{aligned}\tag{55}$$

The system (55) is overdetermined but solvable by virtue of (52). Solving it, we find that

$$\begin{aligned}&A_1=\pi B_a=I_1-\pi B_b,\quad \alpha=\big(v_0(b,+0)-v_0(a,+0)\big)/(b-a),\\ &B_b=\pi^{-1}\big(\alpha-(\partial/\partial x_1)v_0(b,+0)\big),\quad B_a=\pi^{-1}\big((\partial/\partial x_1)v_0(a,+0)-\alpha\big).\end{aligned}\tag{56}$$

Thus, from (46), (58), (54), and (56) we derive the following asymptotic formulas for the intensity factors of the solution to problem (1)–(4) for $p_M^{\pm}=p_N^{\pm}=0$:

$$\begin{aligned}K_{\pm}(\varepsilon)=k_V^{\pm}+\varepsilon\{k_{1,v}^{\pm}+k_G^{\pm}\Big(&(\partial/\partial x_1)v_0(a,+0)-(b-a)^{-1}\big(v_0(b,+0)-v_0(a,+0)\big)\Big)\}\\ &+O(\varepsilon^2|\log\varepsilon|),\end{aligned}\tag{57}$$

$$\begin{aligned}K_b(\varepsilon)=(2\varepsilon/\pi)^{1/2}\{&(b-a)^{-1}\big(v_0(b,+0)-v_0(a,+0)\big)-(\partial/\partial x_1)v_0(b,+0)\}\\ &+O(\varepsilon^{3/2}|\log\varepsilon|),\end{aligned}\tag{58}$$

where $k_{1,v}^{\pm}$ are the coefficients in representations of the form (42), (43) for the solution v_1 to problem (48), (49). The asymptotics of $K_a(\varepsilon)$ takes the form (58) with the sign reversed and a and b interchanged.

In the case $\Omega=\mathbb{R}^2$ the formulas (57) and (58) are more explicit. In particular, for the harmonic function u in $\Omega_\varepsilon=\mathbb{R}^2\setminus(N\cup M_\varepsilon)$ subject to the relations

$$\mu(\partial/\partial x_2)u(\varepsilon,x)=0,\quad x\in N\cup M_\varepsilon;\quad \mu(\partial/\partial x_2)u(\varepsilon,x)=q_0+O(1),\quad |x|\to\infty,\tag{59}$$

the intensity factors are given by

$$\begin{aligned}K_+(\varepsilon)=2^{1/2}\mu^{-1}q_0\{1+(2\pi)^{-1}\varepsilon\Big(&4^{-1}\log\big((1+a)^{-1}(1-b)^{-1}(1+b)(1-a)\big)\\ &+2^{-1}(1-a)^{-1}(1-b)^{-1}(b-a)\\ &+\big((1+b)/(1-b)\big)^{1/2}\big((1-b^2)^{-1/2}b-(1-a^2)^{-1/2}a\big)\\ &+\Big(\big((1+b)/(1-b)\big)^{1/2}\big((1+a)(1-a)\big)^{1/2}\Big)\\ &\times\big(\big((1-a^2)^{1/2}+(1-b^2)^{1/2}\big)^{-1}(a+b)-(1-a^2)^{-1/2}a\big)\Big)\}+O(\varepsilon^2|\log\varepsilon|),\end{aligned}\tag{60}$$

$$\begin{aligned}K_b(\varepsilon)=\mu^{-1}q_0(2\varepsilon/\pi)^{1/2}(1-b^2)^{-1/2}&\big(1-ab+(1-a^2)^{1/2}(1-b^2)^{-1/2}\big)^{-1}(b-a)\\ &+O(\varepsilon^{3/2}|\log\varepsilon|).\end{aligned}$$

15.5.6 Shifted cracks

We consider a location of cracks different from that in Sections 15.5.1–15.5.5. We keep the same notation a,b,ε, and Ω as in 15.5.1. Set $M_\varepsilon=\{x\in\mathbb{R}^2: x_2=\varepsilon,-1\leq x_1\leq b\}$, $N_1=\{x\in\mathbb{R}^2: x_2=0, a\leq x_2\leq 1\}$, and $\Omega_\varepsilon=\Omega\setminus(N_1\cup M_\varepsilon)$ (Fig. 15.5). We consider problem (1)–(4) with $p_M^{\pm}=0$ and $p_N^{\pm}=0$. The asymptotics of the solution can be constructed in the same manner as before. The only distinction consists in defining the function v_1. In the case under consideration the discrepancy $\partial^2 v_0/\partial x_1^2$

arises on the set $M_\varepsilon^+ \cup \{x \in M_\varepsilon^- : \ -1 \le x_1 \le a\}$. Therefore, the right-hand sides of boundary conditions of type (49) have the inadmissible growth $O(r_-^{-3/2})$ as $r_- \to 0$. This is because the image $(-1, 0)$ of the left tip of M_ε in the limit problem is shifted from the original position and the function v_0 does not satisfy the boundary conditions near $(-1, 0)$. It necessitates taking an approximation for u in the form

$$v^*(\varepsilon, x) = X_\delta(x) v_0(x_1, x_2 + \varepsilon) + \big(1 - X_\delta(x)\big) v_0(x), \tag{61}$$

where $X_\delta(x) = \chi\big(\delta^{-1}(x_1 + 1)\big)\chi(\delta^{-1}x_2)$, χ is the cut-off function introduced in 15.5.4, and δ is a small number such that the support of X_δ intersects neither $\partial\Omega$ nor N_1. Note that outside a neighborhood of the point $(-1, \varepsilon)$ the function (61) has an expansion in nonnegative integer powers of ε, which can be obtained by Taylor's formula for the first term on the right in (61).

The principal term of the discrepancy of v^* in the boundary condition (3) is of the form $\varepsilon\mu\varphi_1^\pm$, where

$$\begin{aligned}
&\varphi_1^+(x_1) = \chi\big(\delta^{-1}(x_1+1)\big)(\partial/\partial x_1)^2 v_0(x_1, +0), \quad x_1 \in (-1, b),\\
&\varphi_1^-(x_1) = 0, \quad x_1 \in (a, b),\\
&\varphi_1^-(x_1) = x\big(\delta^{-1}(x_1+1)\big)(\partial/\partial x_1)^2 v_0(x_1, -0), \quad x_1 \in (-1, a).
\end{aligned}$$

Moreover, there arises a discrepancy in the equation (1) whose principal term coincides with $\varepsilon\mu\psi_1(x)$,

$$\psi_1(x) = -(\partial/\partial x_2)v_0(x)\Delta X_\delta(x) - 2\nabla(\partial/\partial x_2)v_0(x)\cdot\nabla X_\delta(x).$$

Since the function v_0 is harmonic,

$$\begin{aligned}
\int_{\Omega_0}\psi_1(x)dx &= \int_{\Omega_0}(\partial/\partial x_2)X_\delta(x)(\partial/\partial x_1)^2 v_0(x) - (\partial/\partial x_1)X_\delta(x)(\partial/\partial x_1)(\partial/\partial x_2)v_0(x)dx\\
&= \sum_\pm \pm\int_{-1}^{a}(\partial/\partial x_1)\chi\big(\delta^{-1}(x_1+1)\big)(\partial/\partial x_1)v_0(x_1, \pm 0)dx.
\end{aligned}$$

Hence, there exists a solution v_1 to the boundary value problem

$$\begin{aligned}
\Delta v_1(x) &= \psi_1(x), && x \in \Omega_0\\
(\partial/\partial x_2)v_1(x) &= \pm\varphi^\pm(x_1), && x \in N^\pm;\\
(\partial/\partial n)v_1(x) &= 0, && x \in \partial\Omega,
\end{aligned} \tag{62}$$

which admits the representations (51) and (52) with $(a, -0)$ replaced by $(a, +0)$.

The rest of construction runs as in Section 15.5.3. As a result, for the intensity factors we obtain the asymptotic formulas (10) and (11) where $(a, +0)$ must be replaced by $(b, +0)$, $k_{1,v}^\pm$ stand for the coefficients in the asymptotics of the solution to problem (62), and $k_G^\pm$ are the coefficients in the expansion (43) of Neumann's function with poles $(a, -0)$ and $(b, +0)$ (cf. (14) and (15). In the case of planar problem (1), (54), the formulas (60) convert into

$$\begin{aligned}K_+(\varepsilon) = 2^{1/2}\mu^{-1}q_0\{1 + (\varepsilon/2\pi)\Big(&4^{-1}\log\left((1+a)^{-1}(1-b)^{-1}(1+b)(1-a)\right) \\ &+ \big(2(1-a)(1-b)\big)^{-1}(a+3b-4ab) \\ &- \Big(\big((1+b)/(1-b)\big)^{1/2} + \big((1+a)/(1-a)\big)^{1/2}\Big)(b-a)^{-1} \\ &\times \Big(\big((1-b^2)^{1/2} + (1-a^2)^{1/2}\big)\Big)\} + O(\varepsilon^2|\log\varepsilon|),\end{aligned}$$

$$\begin{aligned}K_b(\varepsilon) = \mu^{-1}q_0(2\varepsilon/\pi)^{1/2}(b-a)^{-1}(1-b^2)^{-1/2}\big(1-ab+(1-a^2)^{1/2}(1-b^2)^{1/2}\big) \\ + O(\varepsilon^{3/2}|\log\varepsilon|).\end{aligned}$$

We conclude the section with the formulation of the following result. Let us consider the cracks shown in Figure 15.5 in the case that the lengths of the cracks equal 1 and the sizes of the rectangle between the cracks are $l \in (0,1)$ and ε. At the right tip of the upper crack M_ε in problem (1), (59) there is the relation

$$K(\varepsilon, l) = q_0\mu^{-1}(2\varepsilon/\pi)^{1/2}\varkappa(l) + O(\varepsilon^{3/2}|\log\varepsilon|),$$

where

$$\varkappa(l) = l^{-1}2(1-l)^{1/2} + 2^{-1}(1-l)^{-1/2}\,l = \big(2l((1-l)^{1/2}\big)^{-1}(4-4l+l^2).$$

The value $\varkappa(l)$ tends to ∞ as $l \to 0$ or $l \to 1$. The formula for $K(\varepsilon, l)$ loses its exactness for l and $1-l$ close to ε because the assumption that ε is sufficiently small turns out to be violated (see Section 15.5.1). The function $\varkappa$ (and, consequently, the intensity factor, too) attains its minimum at $l = l_* = 2(2^{1/2}-1)$, which is a root of the equation $l^3 + 2l^2 - 12l + 8 = 0$.

Chapter 16

General Elliptic Problems in Thin Domains

In this chapter we study general elliptic boundary value problems in a thin domain of arbitrary dimension and apply the algorithms described for a number of special problems in Chapter 15. As before, the basic iterative process for construction of asymptotics consists in successivly solving two limit problems in the sections of the thin domain. If these problems are not uniquely solvable we introduce the third limit problem. Choosing its solution, we provide compatibility conditions for the two first problems. In the general case, the role of the third limit problem can be played by a problem with small parameter by the derivatives of higher order, algebraic or differential equations on the boundary of a section, etc. (Section 16.3 contains the corresponding examples.)

In Sections 16.1.1–16.1.3 we describe the structure of domains and differential operators and formulate the ellipticity condition. The two first limit problems are defined in Sections 16.1.4 and 16.1.5; there are given the related decompositions of the operators of the original problem into a series in powers of ε. The algorithm for the construction of the limit problem based on the iterative procedures in Chapter 4 is contained in Section 16.4.

Construction and justification of asymptotic expansions for the solution to the original problem are given in Section 16.2. In Section 16.2.1 we prove the Fredholm property for the problem and formulate the assumption on the estimate of solutions uniform with respect to the parameter. The simple case that the two first limit problems are uniquely solvable is considered in 16.2.2. We formulate needed properties of solutions to the third limit problem in Section 16.2.3. Finally, Section 16.2.4 contains the construction and justification of the asymptotics of the solution to the original problem.

In Section 16.3 we discuss a few particular problems for which the third limit problems are of various types. In addition, in Section 16.4 we turn once again to the Lamé system (see Example 15.3.2). In particular, Section 16.4.4 contains a detailed analysis of the boundary layer near the edge of a plate.

16.1 Limit Problems

16.1.1 Statement of the problem

Let G and Ω be domains in $\mathbb{R}^n$ and $\mathbb{R}^\nu$ with compact closures. We suppose that the boundaries ∂G and $\partial\Omega$ are smooth $(n-1)$- and $(n-\nu)$-dimensional submanifolds in $\mathbb{R}^n$ and $\mathbb{R}^\nu$. We denote by s the local coordinates in ∂G and by τ the distance from $x \in \mathbb{R}^n$ to ∂G. In a small d-neighborhood $\tilde{O}_d$ of ∂G the coordinates (s,τ) define the systems of local coordinates and the intersection $\tilde{O}_d \cap G$ is given by $0 < \tau < d$. We introduce a domain $\omega \subset \mathbb{R}^{\nu+1}$ with smooth boundary $\partial\omega$ such that

$$\{(\tau,\xi) \in \mathbb{R}^{\nu+1} : \xi \in \Omega, \tau > 1\} \subset \omega \subset \{(\tau,\xi) \in \mathbb{R}^{\nu+1} : \xi \in \Omega, \tau > 0\}$$

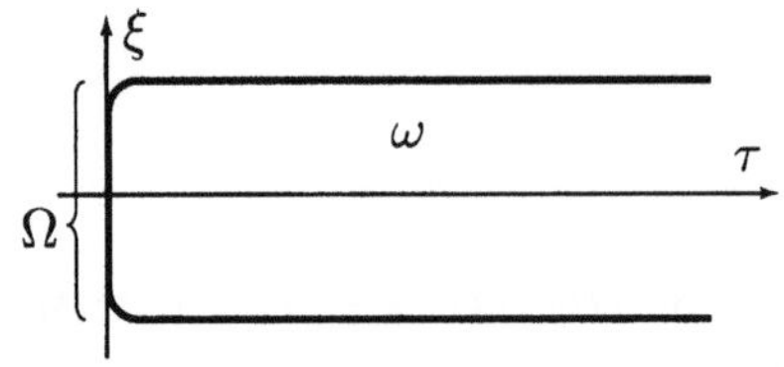

Fig. 16.1

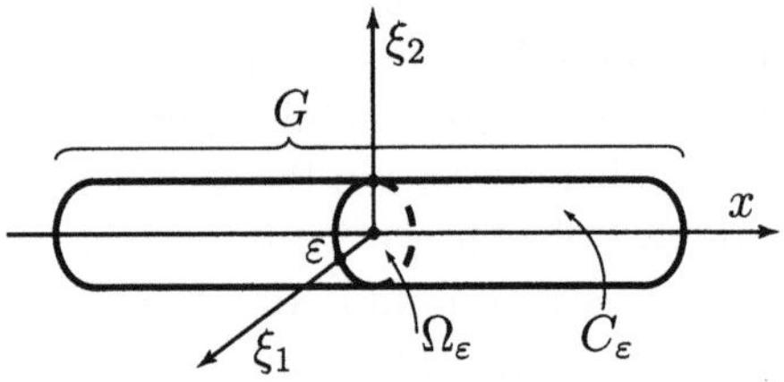

Fig. 16.2

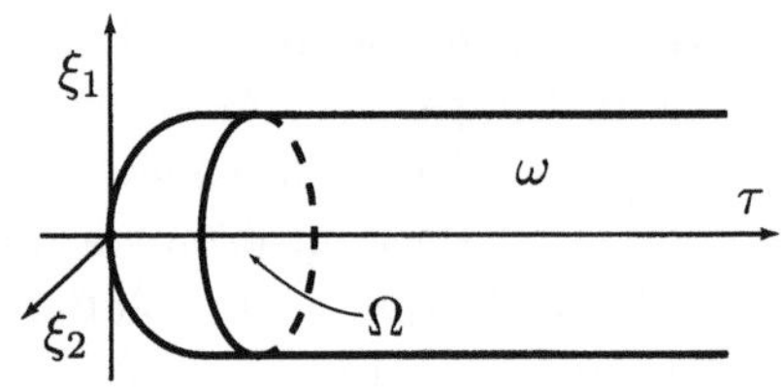

Fig. 16.3

(see Figs. 15.1, 16.1–16.3). We define the set $\tilde{C}_\varepsilon$ depending on a small positive parameter ε. Outside the cylinder $\tilde{O}_d \times \mathbb{R}^\nu$ it coincides with $G \times \Omega_\varepsilon$ while

$$\tilde{C}_\varepsilon \cap (\tilde{O}_d \times \mathbb{R}^\nu) = \{(x, \xi) : x = (s, \tau), (\tau, \xi) \in \omega_\varepsilon\}.$$

Here

$$\Omega_\varepsilon = \{\xi \in \mathbb{R}^\nu : \varepsilon^{-1}\xi \in \Omega\}$$

and

$$\omega_\varepsilon = \{(\tau, \xi) \in \mathbb{R}^{\nu+1} : \varepsilon^{-1}(\tau, \xi) \in \omega\}.$$

Let us fix a neighborhood $\hat{O}$ in $\mathbb{R}^{n+\nu}$ of the set $G \times \{0\}$. Clearly, the boundary $\partial\tilde{C}_\varepsilon$ is smooth for small ε and $\partial\tilde{C}_\varepsilon \subset \hat{O}$.

We consider the boundary value problem

$$\tilde{L}(\varepsilon, x, \xi, D_x, D_\xi)\, u(\varepsilon, x, \xi) = f(\varepsilon, x, \xi), \quad (x, \xi) \in \tilde{C}_\varepsilon, \tag{1}$$

$$\partial\tilde{B}(\varepsilon, x, \xi, D_x, D_\xi)\, u(\varepsilon, x, \xi) = g(\varepsilon, x, \xi), \quad (x, \xi) \in \partial\tilde{C}_\varepsilon, \tag{2}$$

with $u = (u_1, \dots, u_k)$, $f = (f_1, \dots, f_k)$, and $g = (g_1, \dots, g_m)$.

16.1.2 Structure of differential operators

Let $\tilde{L}$ and $\tilde{B}$ be matrix differential operators of sizes $k \times k$ and $m \times k$ whose entries $\tilde{L}_{hj}$ and $\tilde{B}_{qj}$ are of orders $s_h + t_j$ and $r_q + t_j$, respectively. Here $\mathbf{s} = (s_h), \mathbf{t} = (t_j)$, and $\mathbf{r} = (r_q)$ are collections of integers such that $t_j > 0$, $s_h \le 0$, and $t_1 + s_1 + \ldots + t_k + s_k = 2m$. We suppose that coefficients of the differential operators $\tilde{L}$ and $\tilde{B}$ are smooth functions of the variables $(\varepsilon, x, \xi) \in (0, 1] \times \hat{O}$. We also assume that on the set $\overline{\hat{O}} \setminus (\tilde{O}_\varepsilon \times \mathbb{R}^\nu)$ the entries $\tilde{L}_{hj}$ and $\tilde{B}_{qj}$ admit the representations

$$\begin{aligned} \tilde{L}_{hj}(\varepsilon, x, \xi, D_x, D_\xi) &= \sum_{|\alpha| \le s_h + t_j} \varepsilon^{|\alpha| - s_h - t_j} \tilde{L}^{(\alpha)}_{hj}(\varepsilon, x, \varepsilon^{-1}\xi, \varepsilon D_\xi) D^\alpha_x, \\ \tilde{B}_{qj}(\varepsilon, x, \xi, D_x, D_\xi) &= \sum_{|\alpha| \le r_q + t_j} \varepsilon^{|\alpha| - r_q - t_j} \tilde{B}^{(\alpha)}_{qj}(\varepsilon, x, \varepsilon^{-1}\xi, \varepsilon D_\xi) D^\alpha_x. \end{aligned} \tag{3}$$

Here $L^{(\alpha)}_{hj}(\varepsilon, x, \eta, D_\eta)$ and $B^{(\alpha)}_{qj}(\varepsilon, x, \eta, D_\eta)$ are differential operators of orders $s_h + t_j - |\alpha|$ and $r_q + t_j - |\alpha|$ whose coefficients are smooth functions in the variables

$(\varepsilon, x, \eta) \in [0,1] \times \bar{G} \times \bar{\Omega}$. Hence, there hold the inequalities

$$\begin{aligned} &\left\| L_{hj}^{(\alpha)}(\varepsilon, x, \eta, D_\eta) - \sum_{p=0}^{N} \varepsilon^p L_{hj}^{(\alpha,p)}(x, \eta, D_\eta) : \mathbf{W}_2^{l+s_h+t_j-|\alpha|}(\Omega) \to \mathbf{W}_2^l(\Omega) \right\| \\ &+ \left\| B_{qj}^{(\alpha)}(\varepsilon, x, \eta, D_\eta) - \sum_{p=0}^{N} \varepsilon^p B_{qj}^{(\alpha,p)}(x, \eta, D_\eta) : \mathbf{W}_2^{l+r_q+t_j-|\alpha|}(\Omega) \to \mathbf{W}_2^l(\Omega) \right\| \\ &\le c_{Nl}\, \varepsilon^{N+1}, \end{aligned} \tag{4}$$

where $L_{hj}^{(\alpha,p)}$ and $B_{qj}^{(\alpha,p)}$ are some differential operators whose coefficients smoothly depend on $(x, \eta) \in \bar{G} \times \bar{\Omega}$, and the constants $N, l \in \mathbb{N}_0, c_{Nl}$ are independent of $\varepsilon \in [0,1]$ and $x \in \bar{G}$.

In the coordinates $(s, \tau, \xi) \in \bar{\hat{O}} \cap (\bar{\hat{O}}_{2\varepsilon} \times \mathbb{R}^n)$ the operators $\tilde{L}_{hj}$ and $\tilde{B}_{qj}$ take the form

$$\begin{aligned} \tilde{L}_{hj}(\varepsilon, x, \xi, D_x, D_\xi) &= \sum_{\mu=0}^{s_h+t_j} \sum_{|\gamma| \le \mu} \varepsilon^{|\gamma|-\mu} \tilde{P}_{hj}^{(\mu,\gamma)}(\varepsilon, s, \varepsilon^{-1}\tau, \varepsilon^{-1}\xi, D_s) D_\tau^{\gamma_1} D_\xi^{\gamma'}, \\ \tilde{B}_{qj}(\varepsilon, x, \xi, D_x, D_\xi) &= \sum_{\mu=0}^{r_q+t_j} \sum_{|\gamma| \le \mu} \varepsilon^{|\gamma|-\mu} \tilde{Q}_{qj}^{(\mu,\gamma)}(\varepsilon, s, \varepsilon^{-1}\tau, \varepsilon^{-1}\xi, D_s) D_\tau^{\gamma_1} D_\xi^{\gamma'}, \end{aligned} \tag{5}$$

where $\gamma = (\gamma_1, \dots, \gamma_{\nu+1}) = (\gamma_1, \gamma')$; $\tilde{P}_{hj}^{(\mu,\gamma)}(\varepsilon, s, t, \eta, D_s)$ and $\tilde{Q}_{qj}^{(\mu,\gamma)}(\varepsilon, s, t, \eta, D_s)$ are differential operators on ∂G with $\operatorname{ord} \tilde{P}_{hj}^{(\mu,\gamma)} = s_h + t_j - \mu$ and $\operatorname{ord} \tilde{Q}_{qj}^{(\mu,\gamma)} = r_q + t_j - \mu$. Suppose that the coefficients of the operators are smooth functions in the variables $(\varepsilon, s, t, \eta) \in [0,1] \times \partial G \times [0,2] \times \bar{\Omega}$. Then there are valid inequalities

$$\begin{aligned} &\left\| \tilde{P}_{hj}^{(\mu,\gamma)}(\varepsilon, s, t, \eta, D_s) - \sum_{p=0}^{N} \varepsilon^p \tilde{P}_{hj}^{(\mu,\gamma,p)}(s, t, \eta, D_s) : \mathbf{W}_2^{l+s_h+t_j-\mu}(\partial G) \to \mathbf{W}_2^l(\partial G) \right\| \\ &+ \left\| \tilde{Q}_{qj}^{(\mu,\gamma)}(\varepsilon, s, t, \eta, D_s) - \sum_{p=0}^{N} \varepsilon^p \tilde{Q}_{qj}^{(\mu,\gamma,p)}(s, t, \eta, D_s) : \mathbf{W}_2^{l+r_q+t_j-\mu|}(\partial G) \to \mathbf{W}_2^l(\partial G) \right\| \\ &\le c_{Nl} \varepsilon^{N+1}, \end{aligned} \tag{6}$$

the constants c_{Nl} being independent of $\varepsilon \in [0,1]$, $t \in [0,2]$, and $\eta \in \bar{\Omega}$. It is obvious that any matrix operators $\tilde{L}$ and $\tilde{B}$ with smooth coefficients admit the representations (3) and (5) subject to (4) and (6).

16.1.3 The ellipticity condition

We introduce matrix differential operators $L_0, B_0, \tilde{P}_0$, and $\tilde{Q}_0$ with entries

$$(L_0)_{hj} = \sum_{|\alpha| \le s_h+t_j} L_{hj}^{(\alpha,0)}(x, \eta, D_\eta) D_x^\alpha, \qquad (B_0)_{qj} = \sum_{|\alpha| \le r_q+t_j} B_{qj}^{(\alpha,0)}(x, \eta, D_\eta) D_x^\alpha,$$

$$(\tilde{P}_0)_{hj} = \sum_{\mu=0}^{s_h+t_j} \sum_{|\gamma| \le \mu} \tilde{P}_{hj}^{(\mu,\gamma,0)}(s, \zeta, D_s) D_\zeta^\gamma, \quad (\tilde{Q}_0)qj = \sum_{\mu=0}^{r_q+t_j} \sum_{|\gamma| \le \mu} \tilde{Q}_{qj}^{(\mu,\gamma,0)}(s, \zeta, D_s) D_\zeta^\gamma.$$

The operators L_0, B_0 and $\tilde{P}_0, \tilde{Q}_0$ at a point x^0 are obtained from $\varepsilon^{\mathbf{s}} \tilde{L} \varepsilon^{\mathbf{t}}$ and $\varepsilon^{\mathbf{r}} \tilde{B} \varepsilon^{\mathbf{t}}$ by changing the variables $(x, \xi) \to (y, \eta) = \varepsilon^{-1}(x - x^0, \xi)$ and letting $\varepsilon = 0$.

Assume that the operator L_0 is uniformly elliptic in the Douglis-Nirenberg sense on the set $\bar{G} \times \bar{\Omega}$ and together with the boundary operator ∂B_0 satisfies the ellipticity condition on the set $\bar{G} \times \partial\Omega$. We also suppose that the operator $\tilde{P}_0$ is uniformly elliptic in the Douglis-Nirenberg sense on the set $\{(s,\zeta) \in \partial G \times \mathbb{R}^{\nu+1} : \zeta = (t,\eta) \in \bar{\omega}, t \le 2\}$ and together with the boundary operator $\partial\tilde{Q}_0$ satisfies the ellipticity condition on the set $\{(s,\zeta) \in \partial G \times \mathbb{R}^{\nu+1} : \zeta \in \partial\omega, t \le 2\}$ (see Section 3.1.6).

16.1.4 The first limit problem

We make the change of variables $\xi \to \eta = \varepsilon^{-1}\xi$ and set $u = \varepsilon^{\mathbf{t}} v$. Then we multiply the h-th equation of system (1) by ε^{s_h} and the q-th equation of system (2) by ε^{r_q}. Letting $\varepsilon \to 0$, we obtain the boundary value problem in Ω with parameter $x \in \bar{G}$:

$$\begin{aligned} L^{(0,0)}(x,\eta,D_\eta)\,v(\eta,x) &= h(\eta,x), \quad \eta \in \Omega; \\ \partial B^{(0,0)}(x,\eta,D_\eta)\,v(\eta,x) &= k(\eta,x), \quad \eta \in \partial\Omega, \end{aligned} \tag{7}$$

where $L^{(0,0)}$ and $B^{(0,0)}$ are the matrix operators with entries $L^{(0,0)}_{hj}$ and $B^{(0,0)}_{qj}$ (see (4)). Note that

$$L^{(0,0)}(x,\eta,D_\eta) = \{\varepsilon^s \tilde{L}(\varepsilon,x,\varepsilon\eta,D_x,\varepsilon^{-1}D_\eta)\varepsilon^t\}|_{\varepsilon=0},$$

$$B^{(0,0)}(x,\eta,D_\eta) = \{\varepsilon^r B(\varepsilon,x,\varepsilon\eta,D_x,\varepsilon^{-1}D_\eta)\varepsilon^t\}|_{\varepsilon=0}.$$

Lemma 16.1.1. *The mapping*

$$A_\Omega(x) = (L^{(0,0)}, \partial B^{(0,0)}) : \mathbf{DW}_2^l(\Omega) \to \mathbf{RW}_2^l(\Omega) \tag{8}$$

is Fredholm and $\operatorname{Ind} A_\Omega(x) = 0$. *Here the spaces* $\mathbf{DW}_2^l(\Omega)$ *and* $\mathbf{RW}_2^l(\Omega)$ *are defined by*

$$\mathbf{DW}_2^l(\Omega) = \prod_{j=1}^{k} \mathbf{W}_2^{l+t_j}(\Omega),$$

$$\mathbf{RW}_2^l(\Omega) = \prod_{h=1}^{k} \mathbf{W}_2^{l-s_h}(\Omega) \times \prod_{q=0}^{m} \mathbf{W}_2^{l-r_q-1/2}(\partial\Omega).$$

Proof. Since the operator $(L_0, \partial B_0)$ is elliptic in the Douglis-Nirenberg sense, the operator $\big(L_0(x^0,\eta,\lambda,\dots,\lambda,D_\eta), B_0(x^0,\eta,\lambda,\dots,\lambda,D_\eta)\big)$ with complex parameter λ is elliptic in the sense of Section 3.1.5. By Theorem 3.1.9, such an operator is Fredholm with index 0 in the pair $\mathbf{DW}_2^l(\Omega), \mathbf{RW}_2^l(\Omega)$ for any λ, in particular, for $\lambda = 0$. □

Denote by K the dimension of the kernel (or cokernel) of problem (7) and suppose that it does not depend on x. Let $\varphi_1, \dots, \varphi_K$ and $\psi_1, \dots, \psi_K$ be bases in the subspaces $\ker A_\Omega(x)$ and $\operatorname{coker} A_\Omega(x)$. Because the coefficients of $L^{(0,0)}$ and $B^{(0,0)}$ and the boundary of Ω are smooth we have $\varphi_j \in \mathbf{C}^\infty(\bar{G} \times \bar{\Omega})$.

The operator of problem (1), (2) admits the formal expansion

$$\Big(\varepsilon^{\mathbf{s}} \tilde{L}(\varepsilon,x,\varepsilon\eta,D_x,\varepsilon^{-1}D_\eta)\varepsilon^{\mathbf{t}}, \varepsilon^{\mathbf{r}}\partial\tilde{B}(\varepsilon,x,\varepsilon\eta,D_x,\varepsilon^{-1}D_\eta)\varepsilon^{\mathbf{t}}\Big) = A_\Omega(x) - \sum_{p-1}^{\infty} \varepsilon^p A_\Omega^{(p)}(x), \tag{9}$$

where (see representations (3) and (4))

$$A_{\Omega}^{(p)}(x) = -\Big(\Big(\sum_{|\alpha|\le\min\{t_j+s_h,p\}} L_{hj}^{(\alpha,p-|\alpha|)}(x,\eta,D_\eta)D_x^\alpha\Big)_{h,j=1}^{k},$$
$$\partial\Big(\sum_{|\alpha|\le\min\{t_j+r_q,p\}} B_{qj}^{(\alpha,p-|\alpha|)}(x,\eta,D_\eta)D_x^\alpha\Big)_{q=1,j=1}^{m,k}\Big), \quad p\in\mathbb{N}. \tag{10}$$

16.1.5 The second limit problem

In a neighborhood $\tilde{O}_d$ of ∂G the operator D_x^α can be written in the form

$$D_x^\alpha = \sum_{p\le|\alpha|} \tilde{D}_p^\alpha(s,\tau,D_s)D_\tau^p, \tag{11}$$

where $\tilde{D}_p^\alpha(s,\tau,D_s)$ is a differential operator on ∂G of order $|\alpha|-p$ with coefficients smoothly depending on $s\in\partial G$ and $\tau\in[0,d)$. We substitute (11) into (3) and expand the coefficients of $\tilde{L}_{hj}$ and $\tilde{B}_{qj}$ into the series in degrees of τ. After the change of variables

$$(x,\xi)\to(s,\zeta)=(s,t,\eta)=(s,\varepsilon^{-1}\tau,\varepsilon^{-1}\xi), \tag{12}$$

the relations (3) enable us to decompose the operator $(\tilde{L},\partial\tilde{B})$ on the set $\{(x,\xi)\in\tilde{C}_\varepsilon : \tau\in[\varepsilon,d]\}$ into the formal series

$$\Big(\varepsilon^{\mathbf{s}}\tilde{L}(\varepsilon,x(s,\varepsilon t),\varepsilon\eta,D_x,\varepsilon^{-1}D_\eta)\varepsilon^{\mathbf{t}},\varepsilon^{\mathbf{r}}\partial\tilde{B}(\varepsilon,x(s,\varepsilon t),\varepsilon\eta,D_x,\varepsilon^{-1}D_\eta)\varepsilon^{\mathbf{t}}\Big)$$
$$= A_\omega(s) - \sum_{\mu=1}^{\infty}\varepsilon^\mu A_\omega^{(\mu)}(s) = (P^{(0)},\partial Q^{(0)}) - \sum_{\mu=1}^{\infty}\varepsilon^\mu (P^{(\mu)},\partial Q^{(\mu)}), \tag{13}$$

where $P^{(0)}, Q^{(0)}, P^{(\mu)}$, and $Q^{(\mu)}$ are the matrix differential operators with entries

$$P_{hj}^{(0)}(s,\zeta,D_\zeta) = \sum_{|\alpha|\le s_h+t_j} L_{hj}^{(\alpha,0)}(x(s,0),\eta,D_\eta)\tilde{D}_{|\alpha|}^\alpha(s,0,0)D_t^{|\alpha|},$$

$$Q_{qj}^{(0)}(s,\zeta,D_\zeta) = \sum_{|\alpha|\le r_q+t_j} B_{qj}^{(\alpha,0)}(x(s,0),\eta,D_\eta)\tilde{D}_{|\alpha|}^\alpha(s,0,0)D_t^{|\alpha|},$$

$$P_{hj}^{(\mu)}(s,\zeta,D_s,D_\zeta) = -\sum_{|\alpha|\le s_h+t_j}\sum_{m=0}^{\mu}\sum_{p=\max\{0,\mu-m-|\alpha|\}}(m!)^{-1}t^m$$
$$\times\Big((\partial/\partial\tau)^m L_{hj}^{(\alpha,\mu+p-|\alpha|-m)}(x(s,\tau),\eta,D_\eta)\tilde{D}_p^\alpha(s,\tau,D_s)\Big)|_{\tau=0}D_t^p, \tag{14}$$

$$Q_{qj}^{(\mu)}(s,\zeta,D_s,D_\zeta) = -\sum_{|\alpha|\le r_q+t_j}\sum_{m=0}^{\mu}\sum_{p=\max\{0,\mu-m-|\alpha|\}}(m!)^{-1}t^m$$
$$\times\Big((\partial/\partial\tau)^m B_{gj}^{(\alpha,\mu+p-|\alpha|-m)}(x(s,\tau),\eta,D_\eta)\tilde{D}_p^\alpha(s,\tau,D_s)\Big)|_{\tau=0}D_t^p.$$

On the other hand, by virtue of (5) and (6), the operator $(\tilde{L}, \partial\tilde{B})$ on the set $\tilde{C}_\varepsilon \cap (\tilde{O}_2 \times \mathbb{R}^\nu)$ and $P^{(0)}, Q^{(0)}, P^{(\mu)}, Q^{(\mu)}$ are of the form

$$\begin{aligned} P^{(0)}_{hj} &= \sum_{|\gamma|\le s_h+t_j} P^{(s_h+t_j,\gamma,0)}_{hj}(s,\zeta)D^\gamma_\zeta, \\ Q^{(0)}_{qj} &= \sum_{|\gamma|\le r_q+t_j} \tilde{Q}^{(r_q+t_j,\gamma,0)}_{qj}(s,\zeta)D^\gamma_\zeta, \\ P^{(\mu)}_{hj} &= \sum_{\lambda=\max\{0,s_h+t_j-\mu\}}^{s_h+t_j} \sum_{|\gamma|\le\lambda} \tilde{P}^{(\lambda,\gamma,\mu+\lambda-s_h-t_j)}_{hj}(s,\zeta,D_s)D^\gamma_\zeta, \\ Q^{(\mu)}_{qj} &= \sum_{\lambda=\max\{0,r_q+t_j-\mu\}}^{r_q+t_j} \sum_{|\gamma|\le\lambda} \tilde{Q}^{(\lambda,\gamma,\mu+\lambda-r_q-t_j)}_{qj}(s,\zeta,D_s)D^\gamma_\zeta. \end{aligned} \tag{15}$$

Since the coefficients of $\tilde{L}$ and $\tilde{B}$ are smooth, the right-hand sides of the corresponding equalities in (14) and (15) coincide for $t \in [1,2]$. Hence, one can consider that $P^{(\mu)}$ and $Q^{(\mu)}$ are matrix differential operators on ω with smooth coefficients. Moreover, the operators (14) polynomially depend on t.

We make the change of variables (12) in problem (1), (2) and set $u = \varepsilon^{\mathbf{t}}v$. Then we multiply the equation with number h in (1) by ε^{s_h} and that with number q in (2) by ε^{r_q}. Setting $\varepsilon = 0$, we obtain the boundary value problem depending on a parameter $s \in \partial G$:

$$\begin{aligned} P^{(0)}(s,\zeta,D_\zeta)\, w(\zeta,s) &= \tilde{H}(\zeta,s), \quad \zeta \in \omega; \\ \partial Q^{(0)}(s,\zeta,D_\zeta)\, w(\zeta,s) &= \tilde{K}(\zeta,s), \quad \zeta \in \partial\omega. \end{aligned} \tag{16}$$

We shall consider the problem in the spaces $\mathbf{W}^h_{2,\beta}(\omega)$ of functions w with norm

$$||w; \mathbf{W}^h_{2,\beta}(\omega)|| = \left(\sum_{j=0}^h || \exp(\beta t)\, w; \mathbf{W}^j_2(\omega)||^2\right)^{1/2}.$$

As usual (see Section 3.2.1) we introduce the space of traces $\mathbf{W}^{h-1/2}_{2,\beta}(\partial\omega)$ of functions in $\mathbf{W}^h_{2,\beta}(\omega), h \ge 1$, and the spaces of vector-valued functions

$$\mathbf{DW}^l_{2,\beta}(\omega) = \prod_{j=1}^k \mathbf{W}^{l+t_j}_{2,\beta}(\omega),$$

$$\mathbf{RW}^l_{2,\beta}(\omega) = \prod_{h=1}^k \mathbf{W}^{l-s_h}_{2,\beta}(\omega), \times \prod_{q=1}^m \mathbf{W}^{l-r_q-1/2}_{2,\beta}(\partial\omega).$$

According to the two first formulas in (8), for $t > 1$ the coefficients of $P^{(0)}$ and $Q^{(0)}$ are independent of t, i.e., $P^{(0)}(s,\zeta,D_\zeta) = \mathbf{P}(s,\xi,D_t,D_\xi)$ and $Q^{(0)}(s,\zeta,D_\zeta) = \mathbf{Q}(s,\xi,D_t,D_\xi)$. We introduce the differental operator on Ω with complex parameter λ and parameter s:

$$\hat{A}(\lambda,s) = \Big(\mathbf{P}(s,\xi,\lambda,D_\xi), \partial\mathbf{Q}(s,\xi,\lambda,D_\xi)\Big) : \mathbf{DW}^l_2(\Omega) \to \mathbf{RW}^l_2(\Omega).$$

By Theorems 3.2.1 and 3.3.1, the mapping

$$A_\omega(s) : \mathbf{DW}^l_{2,\beta}(\omega) \to \mathbf{RW}^l_{2,\beta}(\omega) \tag{17}$$

is Fredholm if and only if the line $\mathrm{Im}\lambda = \beta$ contains no eigenvalues of $\hat{A}(\lambda, s)$. Suppose that there exists a positive number β_0 such that the strip $\{\lambda \in \mathbb{C} \colon \mathrm{Im}\lambda \in (0, \beta_0]\}$ is free from the eigenvalues of the family of the operators $\hat{A}(\lambda, s)$ with $s \in \partial G$ and the kernel $\ker A_\omega(s)$ and cokernel $\mathrm{coker} A_\omega(s)$ have dimensions K^k and K^c independent of $s \in \partial G$. We denote by $\Phi_1, \dots, \Phi_{K^k}$ and $\Psi_1, \dots, \Psi_{K^c}$ bases in $\ker A_\omega(s)$ and $\mathrm{coker} A_\omega(s)$. Since the boundary $\partial\omega$ and the coefficients of $P^{(0)}$ and $Q^{(0)}$ are smooth, $\Phi_j \in \mathbf{C}^\infty(\partial G \times \bar{\omega}), j = 1, \dots, K^k$. Moreover, from Theorem 3.3.17 we obtain the following assertion.

Theorem 16.1.2. *The kernel (cokernel) of the mapping* (17) *does not depend on* β *for* $\beta \in (0, \beta_0]$.

Remark 16.1.3. If ∂G consists of several connected components $\Gamma_j, j = 1, \dots, J$, the above condition $K^k(s) = \mathrm{const}$, $K^c(s) = \mathrm{const}$ can be replaced by the requirement:

$$K^k(s) = K^k_j, \quad K^c(s) = K^c_j, \quad s \in \Gamma_j, \quad j = 1, \dots, J.$$

16.1.6 The third limit problem

Suppose that at least one of the subspaces $\mathrm{coker} A_\Omega(x)$ and $\mathrm{coker} A_\omega(s)$ is nontrivial. In this case the limit problems (7) and (6) may have no solutions and in the general situation one cannot find an asymptotic series for the solution u to problem (1), (2) with help of two limit problems only (see Section 15.1.3). The purpose of this section is to describe an algorithm for the construction of an additional limit problem that is similar to that of the construction of solutions to homogeneous problems given in Sections 4.3.4 and 4.4.4.

We denote by $\tilde{\Pi}_\Omega$ and $\tilde{\Pi}_\omega$ the projections

$$\tilde{\Pi}_\Omega(x) : \mathbf{RW}^l_2(\Omega) \to \mathrm{im}\, A_\Omega(x),$$
$$\tilde{\Pi}_\omega(s) : \mathbf{RW}^l_{2,\beta}(\omega) \to \mathrm{im}\, A_\omega(s),$$

and by $A^{-1}_\Omega(x)$ and $A^{-1}_\omega(s)$ the right inverse operators for $A_\Omega(x)$ and $A_\omega(s)$ sending $\mathrm{im}\, A_\Omega(x)$ and $\mathrm{im}\, A_\omega(s)$ onto the direct complements of the subspaces $\ker A_\Omega(x)$ and $\ker A_\omega(s)$, respectively. One can consider that the finite dimensional operators $1 - \tilde{\Pi}_\Omega(x)$ and $1 - \tilde{\Pi}_\omega(s)$ project into the subspaces $(\Pi \mathbf{W}^{l-s_h}_2 \Omega)) \times \{0\}$ and $(\Pi \mathbf{W}^{l-s_h}_{2,\beta} \omega)) \times \{0\}$ (i.e., the boundary value problems $A_\Omega y = (1 - \tilde{\Pi}_\Omega)\tilde{F}$ and $A_\omega y = (1 - \tilde{\Pi}_\omega)\tilde{F}$ have zero right-hand sides in the boundary conditions).

Let $z_1, \dots, z_K$ and $r_1, \dots, r_{K^k}$ be some collections of functions in $\mathbf{C}^\infty(\bar{G})$ and $\mathbf{C}^\infty(\partial\bar{G})$. We introduce

$$Z(x, \eta) = \sum_{j=1}^{K} z_j(x)\varphi_j(x, \eta), \quad R(s, \zeta) = \sum_{j=1}^{K^k} r_j(s)\Psi_j(s, \zeta). \tag{18}$$

Let us find a system of differential equations in G and ∂G to find z_j and r_p. First we consider the vector $z = (z_1, \dots, z_K)$ and try to find an "asymptotic solution" to problem (1), (2).

As the principal approximation, we take the function Z satisfying $A_\Omega Z = 0$ and almost in the same way as in Section 4.4.4 we construct "the next approximating terms." The arising boundary value problems cannot be solved. Therefore, at every step we project their right-hand sides into A_Ω. The remaining discrepancy belongs to the subspace $(\mathrm{coker}\, A_\Omega)^\perp$. We expand it into the elements of a basis in the subspace. The coefficients of such an expansion are linear differential operators applied to $z = (z_1, \dots, z_K)$.

The sum of N first terms in the mentioned "homogeneous solution" $\tilde{Z}$ and its discrepancy $\tilde{X}$ have the form

$$\tilde{Z} = Z + \sum_{j=1}^{K} \varepsilon^j A_\Omega^{-1} \tilde{\Pi}_\Omega \tilde{X}_j, \quad \tilde{X} = \sum_{j=1}^{N} \varepsilon^j (1 - \tilde{\Pi}_\Omega) \tilde{X}_j, \tag{19}$$

where

$$\tilde{X}_j = \sum_{\gamma_1 + \dots + \gamma_\varkappa = j} A_\Omega^{(\gamma_1)} A_\Omega^{-1} \tilde{\Pi}_\Omega A_\Omega^{(\gamma_2)} A_\Omega^{-1} \tilde{\Pi}_\Omega \dots A_\Omega^{(\gamma_\varkappa)} Z. \tag{20}$$

Here $A_\Omega^{(\mu)}$ are the coefficients (10) of series (9); the summation is taken over all the collections $(\varkappa, \gamma_1, \dots, \gamma_\varkappa)$ of positive integers such that $j = \gamma_1 + \dots + \gamma_\varkappa$. The collections $(\varkappa_1, \gamma_1^1, \dots, \gamma_{\varkappa_1}^1)$ and $(\varkappa_2, \gamma_1^2, \dots, \gamma_{\varkappa_2}^2)$ differ if $\varkappa_1 \neq \varkappa_2$ or $\gamma_\mu^1 \neq \gamma_\mu^2$ for some $\mu = 1, \dots, \varkappa_1$. From (10) it follows that every expression $\langle \tilde{X}_j, \psi_p \rangle, p = 1, \dots, K$, is a linear combination of the derivatives $D_x^\alpha z_q, |\alpha| \leq j$, with coefficients smooth in $\bar{G}$. We define a $K \times K$-matrix operator $\tilde{S}$ in G by

$$\tilde{S}(\varepsilon, x, D_x) z = (\langle \tilde{X}, \psi_p \rangle)_{p=1}^K = \sum_{j=1}^{N} \varepsilon^j (\langle \tilde{X}_j, \psi_p \rangle)_{p=1}^K. \tag{21}$$

We shall make a similar construction for the vector-valued function R. However, in so doing it is necessary to take account of the discrepancy of $\tilde{Z}$ on $\partial \tilde{C}_\varepsilon \setminus \{\partial G \times \partial \Omega_\varepsilon\}$. To this end we expand expressions (19) into Taylor's series

$$\tilde{Z} \sim \sum_{\mu=0}^{\infty} (\mu!)^{-1} \tau^\mu \left((\partial/\partial\tau)^\mu \left(Z + \sum_{j=1}^{N} \varepsilon^j A_\Omega^{-1} \tilde{\Pi}_\Omega) \tilde{X}_j \right) \right) \Bigg|_{\tau=0},$$

$$\tilde{X} \sim \sum_{\mu=0}^{\infty} (\mu!)^{-1} \tau^\mu \left((\partial/\partial\tau)^\mu \sum_{j=1}^{N} \varepsilon^j (1 - \tilde{\Pi}_\Omega) \tilde{X}_j \right) \Bigg|_{\tau=0}.$$

Since $(\varepsilon^{\mathbf{s}} \tilde{L} \varepsilon^{\mathbf{t}}, \varepsilon^{\mathbf{r}} \partial \tilde{B} \varepsilon^{\mathbf{t}}) \tilde{Z} - \tilde{X} = o(\varepsilon^N)$ for $\tau > \varepsilon$, the coefficients of $Y^{(q)}$ by $\varepsilon^q, q = 0, 1, \dots, N$, vanish for $t > 1$ in the expression

$$\sum_{\mu=0}^{\infty} \varepsilon^\mu \Bigg\{ \left(A_\omega - \sum_{p=1}^{\infty} \varepsilon^p A_\omega^{(p)} \right) (\mu!)^{-1} t^\mu \left((\partial/\partial\tau)^\mu \left(Z + \sum_{j=1}^{N} \varepsilon^j A_\Omega^{-1} \tilde{\Pi}_\Omega) \tilde{X}_j \right) \right) \Bigg|_{\tau=0}$$

$$- (\mu!)^{-1} t^\mu \left((\partial/\partial\tau)^\mu \sum_{j=1}^{N} \varepsilon^j (1 - \tilde{\Pi}_\Omega) \tilde{X}_j \right) \Bigg|_{\tau=0} \Bigg\}.$$

The mentioned coefficients take the form

$$Y^{(q)}(s, \zeta) =$$

$$A_\omega \left\{ (q!)^{-1} t^q ((\partial/\partial\tau)^q Z) \Big|_{\tau=0} + \sum_{\mu=0}^{q-1} (\mu!)^{-1} t^\mu ((\partial/\partial\tau)^\mu A_\Omega^{-1} \tilde{\Pi}_\Omega \tilde{X}_{q-\mu}) \Big|_{\tau=0} \right\}$$

$$-\sum_{p=0}^{q-1} A_\omega^{(q-p)} \left\{ (p!)^{-1} t^p ((\partial/\partial\tau)^p Z)\Big|_{\tau=0} + \sum_{\mu=0}^{p-1} (\mu!)^{-1} t^\mu ((\partial/\partial\tau)^\mu A_\Omega^{-1} \tilde\Pi_\Omega \tilde X_{p-\mu})\Big|_{\tau=0} \right\}$$

$$-\sum_{\mu=0}^{q-1} (\mu!)^{-1} t^\mu ((\partial/\partial\tau)^\mu (1-\tilde\Pi_\Omega)\tilde X_{p-\mu})\Big|_{\tau=0} . \qquad (22)$$

By virtue of (12), (14), (15), and (20) the equality (22) can be written as follows

$$Y^{(q)}(s,\zeta) = \partial(\hat Y^{1,q}(\zeta,s,D_x), \quad \hat Y^{2,q}(\zeta,s,D_x))z, \quad q=0,1,\dots,N,$$

where ∂ is the restriction operator to the boundary ∂G while $\hat Y^{1,q}$ and $\hat Y^{2,q}$ are matrix operators of sizes $k\times K$ and $m\times K$ whose entries are differential operators of order q with coefficients in $\mathbf{C}^\infty(\bar\omega\times\partial G)$. Moreover,

$$\begin{aligned} &\hat Y^{1,q}(\zeta,s,D_x)z = 0, \quad (\zeta,s)\in\bar\omega\times\partial G, \quad t>1, \\ &\hat Y^{2,q}(\zeta,s,D_x)z = 0, \quad (\zeta,s)\in\partial\omega\times\partial G, \quad t>1. \end{aligned} \qquad (23)$$

We substitute the sum $\tilde Z$ and the vector-valued functions

$$\tilde R = R + \sum_{j=0}^{N} \varepsilon^j R_j \qquad (24)$$

(see (19) and (18)) into

$$(\varepsilon^{\mathbf{s}} \tilde L \varepsilon^{\mathbf{t}}, \varepsilon^{\mathbf{r}} \tilde B \varepsilon^{\mathbf{t}})(\tilde Z + \tilde R) = \tilde X + o(\varepsilon^N) \qquad (25)$$

and equate the coefficients by $\varepsilon^q, q=0,1,\dots,N$, to zero. Taking (19), (20), and (13) into account, we obtain the equalities

$$A_\omega R^{(q)} = \sum_{\mu=1}^{q} A_\omega^{(\mu)} R^{(q-\mu)} + A_\omega^{(q)} R - \partial(\hat Y^{1,q}, \hat Y^{2,q})z, \quad q=0,1,\dots,N. \qquad (26)$$

As before, we project the right-hand sides of (26) into $\operatorname{im} A_\omega$ and solve the equations obtained. As a result, we find

$$R^{(q)} = A_\omega^{-1} \tilde\Pi_\omega \hat W_q, \quad q=0,1,\dots,N, \qquad (27)$$

where

$$\hat W_q = \sum_{\gamma_1+\dots+\gamma_\varkappa=q} A_\omega^{(\gamma_1)} A_\omega^{-1} \tilde\Pi_\omega \dots A_\omega^{(\gamma_{\varkappa-1})} A_\omega^{-1} \tilde\Pi_\omega \{A_\omega^{(\gamma_\varkappa)} R - \partial(\hat Y^{1,\gamma_\varkappa}, \hat Y^{2,\gamma_\varkappa})z\}. \qquad (28)$$

In (28) $A_\omega^{(0)} = A_\omega$; the summation is taken over all collections $(\varkappa,\gamma_1,\dots,\gamma_\varkappa)$ such that $\varkappa,\gamma_1,\dots,\gamma_{\varkappa-1}$ are positive integers and $\gamma_\varkappa\in\mathbb{N}_0$; the collections $(\varkappa_j, \gamma_1^j, \dots, \gamma_{\varkappa_j}^j)$ $j=1,2$, differ if $\varkappa_1\neq\varkappa_2$ or $\gamma_\mu^1\neq\gamma_\mu^2$ for some $\mu=1,\dots,\varkappa_1$.

We substitute (24), (27), and (28) into (25) and see that the discrepancy is of the form $\hat W + o(\varepsilon^N)$, where

$$\hat W = \sum_{q=0}^{N} \varepsilon^q (1-\tilde\Pi_\omega)\hat W_q \qquad (29)$$

(cf. (27)–(29), (19), and (20)). From (13)–(15) and (28) it follows that every expression $\langle \hat W_q, \Psi_r\rangle$, $\Psi_1,\dots,\Psi_K$ being a basis in $\operatorname{coker} A_\omega$, is a linear combination of $T_j^{(q)}(s,D_s)r_j, j=1,\dots,K^k$, and $\partial\tilde T_p^{(q)}(s,D_x)z_p$, $p=1,\dots,K$. Here $T_j^{(q)}$ and $\tilde T_j^{(q)}$

are differential operators of order q on ∂G with smooth coefficients. We introduce a matrix differential operator $\hat{T}$ of sizes $K^c \times (K^k + K)$ on ∂G by the formula

$$\begin{aligned}\hat{T}(\varepsilon, s, D_x)(z, r) &= \partial\hat{T}_z(\varepsilon, s, D_x)z + \partial\hat{T}_r(\varepsilon, s, D_s)r = ((\hat{W}, \Psi_p))_{p=1}^{K^c} \\ &= \sum_{q=0}^{N} \varepsilon^q ((\hat{W}_q, \Psi_p))_{p=1}^{K^c},\end{aligned} \tag{30}$$

where $\hat{T}_z$ and $\hat{T}_r$ are $K^c \times K$- and $K^c \times K^k$-matrices. The system of differential equations

$$\begin{aligned}\tilde{S}(\varepsilon, x, D_x)z(x) &= E(x), \quad x \in G; \\ \tilde{T}(\varepsilon, x, D_x)(z, r)(s) &= \hat{C}(s), \quad x \in \partial G,\end{aligned} \tag{31}$$

with matrix operators (15) and (24) is called the third limit problem.

16.2 Asymptotics of Solutions

16.2.1 The Fredholm property of the original problem

In the Sobolev space $\mathbf{W}_2^l(\tilde{C}_\varepsilon)$ we introduce the equivalent norm

$$\sum_{j=0}^{l} \varepsilon^{j-l} ||v; \mathbf{W}_2^j(\tilde{C}_\varepsilon)||$$

depending on ε. The space $\mathbf{W}_2^l(\tilde{C}_\varepsilon)$ endowed with this norm will be denoted by $\mathbf{H}_\varepsilon^l(\tilde{C}_\varepsilon)$. As usual, we define the space of traces $\mathbf{H}_\varepsilon^{l-1/2}(\partial\tilde{C}_\varepsilon)$ of functions in $\mathbf{H}_\varepsilon^l(\tilde{C}_\varepsilon)$, $l \geq 1$, on $\partial\tilde{C}_\varepsilon$. Set

$$\mathbf{DH}_\varepsilon^l(\tilde{C}_\varepsilon) = \prod_{j=1}^{k} \mathbf{H}_\varepsilon^{l+t_j}(\tilde{C}_\varepsilon),$$

$$\mathbf{RH}_\varepsilon^l(\tilde{C}_\varepsilon) = \prod_{h=1}^{k} \mathbf{H}_\varepsilon^{l+s_h}(\tilde{C}_\varepsilon) \times \prod_{q=1}^{m} \mathbf{H}_\varepsilon^{l-r_q-1/2}(\partial\tilde{C}_\varepsilon).$$

Theorem 16.2.1. *Assume that the conditions formulated in Section* 16.1.3 *are satisfied. Then for sufficiently small* ε *the operator of problem* (1), (2), 16.1,

$$\tilde{A} = (\tilde{L}, \partial\tilde{B}) : \mathbf{DH}_\varepsilon^l(\tilde{C}_\varepsilon) \to \mathbf{RH}_\varepsilon^l(\tilde{C}_\varepsilon) \tag{1}$$

is Fredholm and

$$||u; \mathbf{DH}_\varepsilon^l(\tilde{C}_\varepsilon)|| \leq c\{||\tilde{A}u; \mathbf{RH}_\varepsilon^l(\tilde{C}_\varepsilon)|| + \varepsilon^{-1}||\varepsilon^{-\mathbf{t}}u; \mathbf{L}_2(\tilde{C}_\varepsilon)||\}, \tag{2}$$

with constant c *independent of* ε *and* u.

Proof. To prove the Fredholm property for the mapping (1) it suffices to verify (2) and construct a continuous operator

$$\tilde{R} : \mathbf{RW}_2^l(\tilde{C}_\varepsilon) \to \mathbf{DW}_2^l(\tilde{C}_\varepsilon)$$

such that the mapping

$$\tilde{A}\tilde{R} - 1 : \mathbf{RW}_2^l(\tilde{C}_\varepsilon) \to \mathbf{RW}_2^l(\tilde{C}_\varepsilon)$$

is compact.

Let $x^0 \in G \setminus \tilde{Q}_{2\varepsilon}$ and denote by $\hat{H}_\delta(x^0)$ the set $D^n_\delta(x^0) \times \Omega_\varepsilon$ with $\overline{D^n_\delta(x^0)} = \{x \in \mathbb{R}^n \colon |x - x^0| < \delta\}$. We also introduce the set $\hat{K}_\delta(x^0) = \partial\tilde{C}_\varepsilon \cap \hat{H}_\delta(x^0)$. After the change of variables $(x,\xi) \to (y,\eta) = \varepsilon^{-1}(x - x^0, \xi)$ the sets $\hat{H}_\varepsilon(x^0)$ and $\hat{K}_\varepsilon(x^0)$ convert into the standard sets $\hat{H}_1 = D^n_1(0) \times \Omega$ and $\hat{K}_1 = D^n_1(0) \times \partial\Omega$ while the system (1), (2), 16.1, takes the form

$$\begin{aligned}\tilde{L}(\varepsilon, x^0 + \varepsilon y, \varepsilon\eta, \varepsilon^{-1}D_y, \varepsilon^{-1}D_\eta)u(\varepsilon, x^0 + \varepsilon y, \varepsilon\eta) &= f(\varepsilon, x^0 + \varepsilon y, \varepsilon\eta), \quad (y,\eta) \in \hat{H}_1,\\ \partial\tilde{B}(\varepsilon, x^0 + \varepsilon y, \varepsilon\eta, \varepsilon^{-1}D_y, \varepsilon^{-1}D_\eta)u(\varepsilon, x^0 + \varepsilon y, \varepsilon\eta) &= g(\varepsilon, x^0 + \varepsilon y, \varepsilon\eta), \quad (y,\eta) \in \hat{K}_1.\end{aligned} \tag{3}$$

Using (3), 16.1, and (4), 16.1, we obtain the estimate

$$\begin{aligned}&\|(\varepsilon^{\mathbf{s}}\tilde{L}(\varepsilon, x^0 + \varepsilon y, \varepsilon\eta, \varepsilon^{-1}D_y, \varepsilon^{-1}D_\eta)\varepsilon^{\mathbf{t}} - L_0(x^0, \eta, D_y, D_\eta),\\ &\varepsilon^{\mathbf{r}}\partial\tilde{B}(\varepsilon, x^0 + \varepsilon y, \varepsilon\eta, \varepsilon^{-1}D_y, \varepsilon^{-1}D_\eta)\varepsilon^{\mathbf{t}} - \partial B_0(x^0, \eta, D_y, D_\eta)) \colon\\ &\mathbf{DW}^l_2(\hat{H}_1) \to \mathbf{RW}^l_2(\hat{H}_1, \hat{K}_1)\| \le c\varepsilon,\end{aligned} \tag{4}$$

where the constant c is independent of ε and $\varepsilon^{\mathbf{t}}v$ stands for the vector $(\varepsilon^{t_1}v_1, \dots, \varepsilon^{t_k}v_k)$. Since the operator $A_0(x^0) = \big(L_0(x^0, \eta, D_y, D_\eta), \partial B_0(x^0, \eta, D_y, D_\eta)\big)$ is elliptic, there holds the inequality

$$\|\varepsilon^{-\mathbf{t}}u; \mathbf{DW}^l_2(\hat{H}_{1/2})\| \le c\Big(\|\varepsilon^{-\mathbf{t}}u; \mathbf{L}_2(\hat{H}_1)\| + \|A_0(x^0)\varepsilon^{-\mathbf{t}}u; \mathbf{RW}^l_2(\hat{H}_1, \hat{K}_1)\|\Big) \tag{5}$$

(see Theorem 3.1.1) and there exists an operator $R(x^0)$ such that the mappings

$$\begin{aligned}&R(x^0) : \mathbf{RW}^{l,0}_2(\hat{H}_{1/2}, \hat{K}_{1/2}) \to \mathbf{DW}^{l,0}_2(\hat{H}_1),\\ &A_0(x^0)R(x^0) - E : \mathbf{RW}^{l,0}_2(\hat{H}_{1/2}, \hat{K}_{1/2}) \to \mathbf{RW}^{l+1,0}_2(\hat{H}_1, \hat{K}_1)\end{aligned} \tag{6}$$

are continuous.

Here $\mathbf{W}^{l+t_j,0}_2(\hat{H}_\delta)$ and $\mathbf{W}^{l-r_q-1/2,0}_2(\hat{K}_\delta)$ are the completions of $\mathbf{C}^\infty_0(\hat{H}_\delta \cup \hat{K}_\delta)$ and $\mathbf{C}^\infty_0(\hat{K}_\delta)$ with respect to the norms $\mathbf{W}^{l+t_j}_2(\hat{H}_\delta)$ and $\mathbf{W}^{l-r_q-1/2}_2(\hat{K}_\delta)$, respectively, and $E : \mathbf{W}^{q,0}_2(\Xi_1) \to \mathbf{W}^{q,0}_2(\Xi_2)$ is the operator that extends functions defined on Ξ_1 to Ξ_2 by zero.

Making the inverse change $(y,\eta) \to (x,\xi) = (x^0 + \varepsilon y, \varepsilon\eta)$ and taking (4) into account, we rewrite (5) in the form

$$\begin{aligned}\|u; \mathbf{DH}^l_\varepsilon(\hat{H}_{\varepsilon/2})(x^0))\| \le c\Big(&\|\tilde{A}u; \mathbf{RH}^l_\varepsilon(\hat{H}_\varepsilon(x^0), \hat{K}_\varepsilon(x^0))\| + \varepsilon\|u; \mathbf{DH}^l_\varepsilon(\hat{H}_\varepsilon(x^0))\|\\ &+ \varepsilon^{-l}\|\varepsilon^{-\mathbf{t}}u; \mathbf{L}_2\hat{H}_\varepsilon(x^0))\|\Big).\end{aligned} \tag{7}$$

The operator

$$R_\varepsilon(x^0) = \varepsilon^{-\mathbf{t}}R(x^o)(\varepsilon^{-\mathbf{s}}, \varepsilon^{-\mathbf{r}}) : \mathbf{RH}^{l,0}_\varepsilon(\hat{H}_\varepsilon(x^0), \hat{K}_\varepsilon(x^0)) \to \mathbf{DH}^{l,0}_\varepsilon(\hat{H}_\varepsilon(x^0))$$

is continuous uniformly with respect ε to and

$$\tilde{A}R_\varepsilon(x^0) = E + S_\varepsilon(x^0) + K_\varepsilon(x^0), \tag{8}$$

where S_ε has norm of order $O(\varepsilon)$ and K_ε is a compact mapping

$$K_\varepsilon : \mathbf{RH}^{l,0}_\varepsilon(\hat{H}_{\varepsilon/2}(x^0), \hat{K}_{\varepsilon/2}(x^0)) \to \mathbf{RH}^{l,0}_\varepsilon(\hat{H}_\varepsilon(x^0), \hat{K}_\varepsilon(x^0)).$$

Let $x^0 \in \partial G$ and let s be a local coordinate system in ∂G based on x^0. Denote by $\hat{H}_\delta(x^0)$ and $\hat{K}_\delta(x^0)$ the sets $\{(x,\xi) \in \tilde{C}_\varepsilon \colon |s| < 2\delta, |\tau| < 2\delta\}$ and $\partial\tilde{C}_\varepsilon \cap \overline{\hat{H}_\delta(x^0)}$. After the change of variables $(x,\xi) \to (\delta,\zeta) = \varepsilon^{-1}(s,\tau,\xi)$ the sets $\hat{H}_\varepsilon(x^0)$ and

$\hat{K}_\varepsilon(x^0)$ convert into $\hat{H}_1 = \{(\sigma,\zeta) \in \mathbb{R}^{n+\nu}\colon |\sigma| < 2, \zeta \in \omega, \zeta_1 = t \leq 2\}$, and $\hat{K}_1 = \{(\sigma,\zeta) \in \mathbb{R}^{n+\nu}\colon |\sigma| < 2, \zeta \in \partial\omega, t \leq 2\}$.

We apply the estimate (6), 3.1, to the problem obtained from (3) by means of that change and arrive at (5). Moreover, from the general theory of elliptic problems it follows that there exists an operator $R(x^0)$ for which the mappings (6) are uniformly continuous with respect to ε. Returning to the coordinates (x,ξ), from (5), 16.1, and (6), 16.1, we derive (7) and (8) for the operators $\tilde{P}$ and $\tilde{Q}$.

We introduce the covering $\bar{\tilde{C}}_\varepsilon \setminus \tilde{O}_\varepsilon$ consisting of the sets $\hat{H}_{\varepsilon/2}(x^{(i)}), x^{(i)} \in G \setminus \tilde{O}_\varepsilon, i = 1,\dots,I$, and the covering $\overline{\tilde{C}_\varepsilon \cap \tilde{O}_\varepsilon}$ consisting of $\hat{H}_{\varepsilon/2}(x^{(j)}), x^{(j)} \in \partial G, j = I+1,\dots,J$, such that for any point in $\bar{G}$ there exist at most $C(n)$ sets $\hat{H}_\varepsilon(x^{(p)}), p = 1,\dots,J$ containing the point; here $C(n)$ is a constant depending only on G and n. Adding up the estimates (7) for the sets $\hat{H}_\varepsilon(x^{(p)}), p = 1,\dots,I$, we obtain the inequality

$$||u; \mathbf{DH}^l_\varepsilon(\tilde{C}_\varepsilon)|| \leq cC(n)\Big(||\tilde{A}u; \mathbf{RH}^l_\varepsilon(\tilde{C}_\varepsilon)|| + \varepsilon||u; \mathbf{D}H^l_\varepsilon(\tilde{C}_\varepsilon)|| + \varepsilon^{-l}||\varepsilon^{-\mathbf{t}}u; \mathbf{L}_2(\tilde{C}_\varepsilon)||\Big),$$

which implies (2) for sufficiently small ε.

Let $\{\chi_p\}$ be a partition of unity in $\tilde{C}_\varepsilon$ subject to the covering $\{\hat{H}_{\varepsilon/2}(x^{(1)}),\dots, \hat{H}_{\varepsilon/2}(x^{(J)})\}$. We introduce the operator

$$R = \sum_{p=1}^{J} R_\varepsilon(x^{(p)})\,\chi_p.$$

From the representations (7) it follows that

$$\tilde{A}R = 1 + S + K,$$

where

$$S, K : \mathbf{RH}^l_\varepsilon(\tilde{C}_\varepsilon) \to \mathbf{RH}^l_\varepsilon(\tilde{C}_\varepsilon)$$

are continuous mappings, K is a compact operator, and the norm of S tends to zero as $\varepsilon \to 0$. Therefore, for small ε there exists the right regularizer $\tilde{R} = R(1+S)^{-1}$. □

Suppose that for any right-hand sides $(f,g) \in \mathbf{RH}^l_\varepsilon(\tilde{C}_\varepsilon)$ problem (1), (2) is uniquely solvable and its solution u satisfies

$$\varepsilon^{-l}||\varepsilon^{-\mathbf{t}}u; \mathbf{L}_2(\tilde{C}_\varepsilon)|| \leq c\varepsilon^M||(f,g); \mathbf{RH}^l_\varepsilon(\tilde{C}_\varepsilon)|| \tag{9}$$

with constant c independent of ε and f, g while M is a nonnegative number. Note that (2) together with Theorem 16.2.1 implies the estimate

$$||u; \mathbf{D}\hat{H}^l_\varepsilon(\tilde{C}_\varepsilon)|| \leq c\varepsilon^{-M}||(f,g); \mathbf{RH}^l_\varepsilon(\tilde{C}_\varepsilon)||. \tag{9'}$$

Lemma 16.2.2. *If $K^k + K > 0$ then $M > 0$.*

The proof is contained in Remark 16.2.5.

16.2.2 The case when limit problems are uniquely solvable

Assume that the right-hand sides f and g admit the expansions

$$(f,g)(\varepsilon,x,\xi) = \sum_{j=0}^{J} \varepsilon^{j-(\mathbf{s},\mathbf{r})}\Big((h^{(j)},k^{(j)})(x,\varepsilon^{-1}\xi) + \chi(x)(H^{(j)},K^{(j)})(s,\varepsilon^{-1}\tau,\varepsilon^{-1}\xi)\Big) + (\tilde{f}^{(J)},\tilde{g}^{(J)})(\varepsilon,x,\xi), \tag{10}$$

where $(h^{(j)}, k^{(j)}) \in \mathbf{C}^\infty(\bar{G} \times \bar{\Omega})$ and for every s the functions $(H^{(j)}, K^{(j)})$ belong to $\mathbf{RW}^l_{2,\beta}(\omega)$ with $\beta \in (0, \beta_0)$ and depend smoothly on $s \in \partial G$. The function χ is in $\mathbf{C}^\infty_0(\tilde{O}_d)$ and is equal to 1 in the neighborhood $\tilde{O}_{d/2}$ of ∂G. For the remainders in (10) there holds the estimate

$$||(\tilde{f}^{(J)}, \tilde{g}^{(J)}); \mathbf{RH}^l_\varepsilon(\tilde{C}_\varepsilon)|| \leq C_J \varepsilon^{J-l+1+\nu/2}, \tag{11}$$

where the constant C_J does not depend on $\varepsilon \in (0, 1]$.

We construct a formal asymptotic series

$$u(\varepsilon, x, \xi) \sim \sum_{p=0}^{\infty} \varepsilon^{p+\mathbf{t}} v^{(p)}(x, \varepsilon^{-1}\xi) + \chi(x) \sum_{p=0}^{\infty} \varepsilon^{p+\mathbf{t}} w^{(p)}(s, \varepsilon^{-1}\tau, \varepsilon^{-1}\xi), \tag{12}$$

for the solution to problem (1), (2), 16.1.

Substitute the representation (9), 16.1, of the operator $(\tilde{L}, \partial\tilde{B})$, the expansion (10) of the right-hand sides, and the first sum in (12) into problem (1), (2), 16.1. Then, equating the coefficients by the same degrees of ε written in the coordinates (x, η), we obtain

$$A_\Omega v^{(p)} - \sum_{\mu=1}^{p} A^\mu_\Omega v^{(p-\mu)} = (h^{(p)}, k^{(p)}), \quad p \in \mathbb{N}_0. \tag{13}$$

The vector-valued functions

$$v^{(p)} = \sum_{\gamma_1+\ldots+\gamma_k=p} A^{-1}_\Omega A^{(\gamma_1)}_\Omega A^{-1}_\Omega A^{(\gamma_2)}_\Omega \cdots A^{-1}_\Omega A^{(\gamma_{k-1})}_\Omega A^{-1}_\Omega (h^{(\gamma_k)}, k^{(\gamma_k)}) \tag{14}$$

are solutions to equation (13) (see, for instance, Theorem 4.4.2), where the notation is similar to that in (28), 16.1. Since the functions $v^{(p)}$ belong to $\mathbf{C}^\infty(\bar{G} \times \bar{\Omega})$, they admit Taylor's expansions

$$v^{(p)}(x, \eta) \sim \sum_{\mu=0}^{\infty} V^{(p)}_\mu(s, \eta) \tag{15}$$

with $V^{(p)}_\mu = ((\mu!)^{-1}(\partial/\partial\tau)^\mu h^{(p)})|_{\tau=0} \in \mathbf{C}^\infty(\partial G \times \bar{\Omega})$. Similarly,

$$(h^{(p)}, k^{(p)})(x, \eta) \sim \sum_{\mu=0}^{\infty} \tau^\mu (H^{(p)}_\mu, K^{(p)}_\mu)(s, \eta),$$

$$(H^{(p)}_\mu, K^{(p)}_\mu) = (\mu!)^{-1}\Big(((\partial/\partial\tau)^\mu h^{(p)})|_{\tau=0}, ((\partial/\partial\tau)^\mu k^{(p)})|_{\tau=0}\Big) \in \mathbf{C}^\infty(\partial G \times \bar{\Omega}). \tag{16}$$

Outside a $c\varepsilon$-neighborhood of $\partial G \times \{0\}$ the series $\sum \varepsilon^{p+\mathbf{t}} v^{(p)}$ formally satisfies (1), (2), 16.1, with (f, g) replaced by $\sum \varepsilon^p(\varepsilon^{-\mathbf{s}} h^{(p)}, \varepsilon^{-\mathbf{r}} k^{(p)})$. Therefore, for $t > 1$ (or for $\tau > \varepsilon$)

$$\sum_{p=0}^{\infty} \sum_{\mu=0}^{\infty} \varepsilon^{p+\mu} \Big(\Big(A_\omega - \sum_{q=1}^{\infty} \varepsilon^q A^{(q)}_\omega\Big) t^\mu V^{(p)}_\mu - t^\mu (H^{(p)}_\mu, K^{(p)}_\mu)\Big) \sim 0, \tag{17}$$

i.e., the coefficients

$$
\begin{aligned}
(\hat{H}^{(N)}, \hat{K}^{(N)}) = \sum_{\mu=0}^{N} \big(A_\omega(t^\mu V_\mu^{(N-\mu)} - t^\mu(H_\mu^{(N-\mu)}, K_\mu^{(N-\mu)})\big) \\
- \sum_{p=1}^{N} \sum_{\mu=0}^{N-p} A_\omega^{(p)}(t^\mu V_\mu^{(N-\mu-p)})
\end{aligned}
\tag{18}
$$

by $\varepsilon^N, N = 0, 1, \dots$, on the left in (17) satisfy the equalities

$$\hat{H}^{(N)}(s,t,\eta) = 0, \quad (s,t,\eta) \in \partial G \times \bar{\omega}, \quad \hat{K}^{(N)}(s,t,\eta) = 0, \quad (s,t,\eta) \in \partial G \times \partial\omega).$$

Moreover, $\hat{H}^{(N)} \in \mathbf{C}^\infty(\partial G \times \bar{\omega})$, $\hat{K}^{(N)} \in \mathbf{C}^\infty(\partial G \times \partial\omega)$.

Now we substitute the series (12) and representations (13), 16.1, and (10) into problem (1), (2), 16.1, and, taking (14)–(18) into account, equate the coefficients by the same degrees of ε. The equalities obtained are of the form

$$A_\omega w^{(p)} = \sum_{\mu=1}^{p} A_\omega^{(\mu)} w^{(p-\mu)} + (H^{(p)}, K^{(p)}) - (\hat{H}^{(p)}, \hat{K}^{(p)}). \tag{19}$$

As before, it follows that

$$w^{(p)} = \sum_{\gamma_1+\dots+\gamma_k=\mu} A_\omega^{-1} A_\omega^{(\gamma_1)} \dots A_\omega^{-1} A_\omega^{(\gamma_{\varkappa-1})} A_\omega^{-1}\big((H^{(\gamma_\varkappa)}, K^{(\gamma_\varkappa)}) - (\hat{H}^{(\gamma_\varkappa)}, \hat{K}^{(\gamma_\varkappa)})\big), \tag{20}$$

with summation taken over the same collections $(\gamma_1, \dots, \gamma_\varkappa)$ as in (28), 16.1. Since the coefficients of $A_\omega^{(\gamma)}$ are of power growth at most as $t \to \infty$, every term on the right in (20) belongs to $\mathbf{RW}^l_{2,\beta}(\omega)$ for any $\beta \in (0, \beta_0)$.

Theorem 16.2.3. *Let the mappings* (8), 16.1, (17), 16.1, *and* (1) *perform an isomorphism and let* (8) *be valid. We also suppose that the right-hand sides in* (1), (2), 16.1, *admit the representations* (10), (11). *Then*

$$||u - U^{(J)}; \mathbf{DH}^l_\varepsilon(\tilde{C}_\varepsilon)|| \le c_J \varepsilon^{J+1-l-M+\nu/2}, \tag{21}$$

where $J \in \mathbb{N}_0$, the constant c_J is independent of ε, u is the solution to problem (1), (2), 16.1, *and $U^{(J)}$ is the partial sum of the series* (7):

$$U^{(J)}(\varepsilon, x, \xi) = \sum_{j=0}^{J} \varepsilon^{j+\mathbf{t}} \big(v^{(j)}(x, \varepsilon^{-1}\xi) + \chi(x) w^{(j)}(s, \varepsilon^{-1}\tau, \varepsilon^{-1}\xi)\big), \tag{22}$$

$v^{(j)}$ and $w^{(j)}$ being the functions (14) *and* (20).

Proof. Substituting (22) into (1), (2), 16.1, we obtain

$$
\begin{aligned}
(\tilde{L}, \partial\tilde{B})U^{(J)} - (f,g) = \sum_{j=0}^{J} \varepsilon^j \{(\tilde{L}, \partial\tilde{B})\varepsilon^{\mathbf{t}} v^{(j)} + \chi(\tilde{L}, \partial\tilde{B})\varepsilon^{\mathbf{t}} w^{(j)}\} \\
+ \sum_{j=0}^{J} \varepsilon^j [(\tilde{L}, \partial\tilde{B}), \chi] \varepsilon^{\mathbf{t}} w^{(j)} - (f,g).
\end{aligned}
$$

Because $w^{(j)}$ decays exponentially as $t \to \infty$,

$$||[(\tilde{L}, \partial\tilde{B}), \chi]\varepsilon^{\mathbf{t}} w^{(j)}; \mathbf{RH}^l_\varepsilon(\tilde{C}_\varepsilon)|| \le c_1 \exp(-\varepsilon^{-1} c_2) \tag{23}$$

with $c_1, c_2 > 2$. We consider

$$\hat{M}^{(J)} = \sum_{j=0}^{J} \varepsilon^j (\tilde{L}, \partial\tilde{B})\varepsilon^{\mathbf{t}} \upsilon^{(j)} - (\varepsilon^{-\mathbf{s}} h, \varepsilon^{-\mathbf{r}} k).$$

By (14), for $x \in G \setminus \tilde{O}_\varepsilon$ we obtain

$$\begin{aligned} \varepsilon^{(\mathbf{s},\mathbf{r})} \hat{M}^{(J)} = \sum_{j=0}^{J} \varepsilon^j \Big(\varepsilon^{\mathbf{s}} (\tilde{L}, \varepsilon^{\mathbf{r}} \partial\tilde{B})\varepsilon^{\mathbf{t}} - A_\Omega + \sum_{p=1}^{J-j+1} \varepsilon^p A_\Omega^{(p)} \Big) \upsilon^{(j)} \\ - \Big((h,k) - \sum_{j=0}^{J} \varepsilon^j, (h^{(j)}, k^{(j)}) \Big). \end{aligned} \tag{24}$$

Moreover, on the set $\tilde{C}_\varepsilon \cap (\tilde{O}_{2\varepsilon} \times \mathbb{R}^\nu)$ there holds the equality

$$\begin{aligned} \varepsilon^{(\mathbf{s},\mathbf{r})} \hat{M}^{(J)} &= \sum_{j=0}^{J} \varepsilon^j \Big(A_\omega \upsilon^{(j)} - \sum_{p=1}^{j} A_\omega^{(p)} \upsilon^{(j-p)} - (h^{(j)}, k^{(j)}) \Big) - (h,k) \\ &+ \sum_{j=0}^{J} \varepsilon^j (h^{(j)}, k^{(j)}) + \sum_{j=0}^{J} \varepsilon^j \Big(\varepsilon^{\mathbf{s}} \tilde{L}, \varepsilon^{\mathbf{r}} \partial\tilde{B})\varepsilon^{\mathbf{t}} - A_\omega + \sum_{p=1}^{J-j+1} \varepsilon^p A_\omega^{(p)} \Big) \upsilon^{(j)}. \end{aligned} \tag{25}$$

From (15)–(18) it follows that

$$\left| D_s^\alpha D_\zeta^\gamma \Big\{ \sum_{j=0}^{J} \varepsilon^j \Big(A_\omega \upsilon^{(j)} - \sum_{p=1}^{j} A_\omega^{(p)} \upsilon^{(j-p)} - (h^{(j)}, k^{(j)}) - (\hat{H}^{(j)}, \hat{K}^{(j)}) \Big) \Big\} \right| = O(\varepsilon^{J+1}).$$

Combining (24), (25), (11), and (4), 16.1, we find

$$\Big\| \hat{M}^{(J)} + \sum_{j=0}^{J} \varepsilon^j (\varepsilon^{-\mathbf{s}} \hat{H}^{(j)}, \varepsilon^{-\mathbf{r}} \hat{K}^{(j)}; \mathbf{RH}_\varepsilon^l (\tilde{C}_\varepsilon) \Big\| \le c\varepsilon^{J-l+1+\nu/2}. \tag{26}$$

In a similar manner,

$$\begin{aligned} \Big\| \chi \sum_{j=0}^{J} \varepsilon^j \Big((\tilde{L}, \partial\tilde{B})\varepsilon^{\mathbf{t}} w^{(j)} + (\varepsilon^{-\mathbf{s}} \hat{H}^{(j)}, \varepsilon^{-\mathbf{r}} \hat{K}^{(j)}) - (H,K) \Big); \mathbf{RH}^l (\tilde{C}_\varepsilon) \Big\| \\ \le c\varepsilon^{J-l+1+(\nu+1)/2}. \end{aligned} \tag{27}$$

Taking account of (24), (27), (23), and (9), we arrive at (21). □

16.2.3 Solutions to the third limit problem

We introduce the differential operators $\tilde{S}$ and $\hat{T}$ by (21), 16.1, and (30), 16.1, where the number N has been replaced by a positive integer M (see Lemma 16.2.2). The operators depend on a small positive integer ε. In this section and in the next one we assume that for any right-hand sides $E = (E_1, \dots, E_K) \in \mathbf{C}^\infty(\bar{G})$ and $\hat{C} = (\hat{C}_1, \dots, \hat{C}_{K^k}) \in \mathbf{C}^\infty(\partial G)$ satisfying

$$\sum_{p=0}^{q} \varepsilon^{p\delta} (\|E; \mathbf{W}_2^p(G)\| + \varepsilon^{1/2} \|\hat{C}; \mathbf{W}_2^p(\partial G)\|) \le c_q, \tag{28}$$

the boundary value problem (31), 16.1, has asymptotic solution (z_T, r_T) such that

$$\sum_{p=0}^{q} \varepsilon^{p\delta}(||\tilde{S}z_T - E; \mathbf{W}_2^p(G)|| + \varepsilon^{1/2}||\hat{T}(z_T, r_T) - \hat{C}; \mathbf{W}_2^p(\partial G)||) \leq c_{qT}\varepsilon^T;$$

$$||z_T; \mathbf{W}_2^M(\partial G)|| + \varepsilon^{1/2}||r_T; \mathbf{W}_2^M(\partial G)|| + \sum_{j=0}^{q} \varepsilon^{jq}(||z_T; \mathbf{W}_2^{M+j}(G)|| \tag{29}$$

$$+ \varepsilon^{1/2}||r_T; \mathbf{W}_2^{M+j}(\partial G)||)c_{qT}\varepsilon^{-M}.$$

In (28) and (29) q and T are arbitrary nonnegative integers, $\delta \in [0, 1)$, and the constants c_q and c_{qT} are independent of ε. For example, the regularly degenerating boundary value problems with small parameter by the higher derivatives considered in VISHIK and LYUSTERNIK [1], AGRANOVICH and VISHIK [1], ALVINO and TROMBETTI [1] and NAZAROV [3, 4, 8] possess such a property. It is clear that in case the operators $\varepsilon^{\mathbf{a}}\tilde{S}\varepsilon^{\mathbf{c}}, \varepsilon^{\mathbf{b}}\hat{T}_z\varepsilon^{\mathbf{c}}$, and $\varepsilon^{\mathbf{b}}\hat{T}_r S\varepsilon^{\mathbf{d}}$ with some multiindices $\mathbf{a} = (a_1, \dots, a_K)$, $\mathbf{b} = (b_1, \dots, b_{K^c}), \mathbf{c} = (c_1, \dots, c_K)$ and $\mathbf{d} = (d_1, \dots, d_{K^k})$ do not depend on ε and the boundary value problem with operator $(\tilde{S}, \partial\hat{T}_z + \hat{T}_r)$ is uniquely solvable, the inequalities (29) are valid. The next theorem states the "uniqueness" of the asymptotic solution to problem (31), 16.1.

Theorem 16.2.4. *Let* (28) *be valid and let*

$$\sum_{p=0}^{q} \varepsilon^{p\delta}(||E; \mathbf{W}_2^p(G)|| + \varepsilon^{1/2}||\hat{C}; \mathbf{W}_2^p(\partial G)||) \leq c_q\varepsilon^{1-\delta}. \tag{30}$$

Then for every asymptotic solution (z_T, r_T) *to problem* (31) *for* $T \geq 1$ *there holds the inequality*

$$||z_T; \mathbf{L}_2(G)|| + \varepsilon^{1/2}||r_T; \mathbf{L}_2(\partial G|| \leq C(\sigma)\varepsilon^{\sigma-M}, \tag{31}$$

with $\sigma = 1 - \delta$ *for* $\delta > 0$ *and* $\sigma < 1$ *for* $\delta = 0$. *The constant* $C(\sigma)$ *is independent of* ε.

Proof. We define the functions $\tilde{Z}$ and $\tilde{R}$ by (19), (20), 16.1, and (24), (26), and (28), 16.1, where $N = M$ and the roles of Z and R are played by

$$Z_0 = \sum_{\mu=1}^{K}(z_T)_\mu\varphi_\mu, \quad R^{(0)} = \sum_{\mu=1}^{K^k}(r_T)_\mu\Phi_\mu. \tag{32}$$

For the domain $\tilde{C}_\varepsilon \setminus (\tilde{O}_\varepsilon \times R^\nu)$ we have

$$(\tilde{L}, \partial\tilde{B})\varepsilon^{\mathbf{t}}\tilde{Z} = \Big(\varepsilon^{-(\mathbf{s},\mathbf{r})}\sum_{j=0}^{M}\varepsilon^j\Big(A_\Omega Z_j - \sum_{p=1}^{M-j} A_\Omega^{(p)}Z_{j-p}\Big)\Big)$$

$$+ \Big(\varepsilon^{-(\mathbf{s},\mathbf{r})}\sum_{j=0}^{M}\varepsilon^j\Big\{((\varepsilon^{\mathbf{s}}\tilde{L}, \varepsilon^{\mathbf{r}}\partial\tilde{B})\varepsilon^{\mathbf{t}} - A_\Omega)Z_j - \sum_{p=1}^{M-j}\varepsilon^p A_\Omega^{(p)}Z_{j-p}\Big\}\Big) = \tilde{Q}_1 + \tilde{Q}_2$$

with $Z_j = A_\Omega^{-1}\tilde{\Pi}_\Omega\tilde{X}_j$, $j = 1, \dots, M$ (see (19), (20), 16.1). As in the proof of Theorem 16.2.3, we obtain the estimate

$$||\tilde{Q}_2; \mathbf{RH}_\varepsilon^l(\tilde{C}_\varepsilon \setminus (\tilde{O}_\varepsilon \times R^\nu), \partial\tilde{C}_\varepsilon \setminus (\tilde{O}_\varepsilon \times R^\nu))|| \leq c\varepsilon^{1-l+\nu/2-\delta}.$$

Moreover, by the definitions of $Z_j, \tilde{X}_j$, and the operator $\tilde{S}$

$$\tilde{Q}_1 = -\sum_{j=1}^{M} \varepsilon^{j-(\mathbf{s},\mathbf{r})}(1-\tilde{\Pi}_\Omega)\tilde{X}_j = -\sum_{p=1}^{K}(\tilde{S}z_T)_p\varepsilon^{-(\mathbf{s},\mathbf{r})}\chi^{(p)}. \tag{33}$$

Here for every $x \in \bar{G}$ the vectors $\chi^{(1)}, \dots, \chi^{(K)}$ belong to the subspace $(1-\tilde{\Pi}_\Omega)\mathbf{RW}_2^l(\Omega)$ and satisfy $\langle \chi^{(p)}, \psi_\mu \rangle = \delta_{p\mu}$. Recall that the projection $\tilde{\Pi}_\Omega$ is defined so that $\chi^{(p)} \in \Pi\mathbf{W}_2^{l-s_h}(\Omega) \times \{0\}$. Hence, one can consider that the function $\tilde{Q}_1$ is defined by (33) everywhere in $\tilde{C}_\varepsilon \times \partial\tilde{C}_\varepsilon$ and belongs to $\mathbf{RW}_2^l(\tilde{C}_\varepsilon)$.

Let us turn to the expression

$$(\tilde{L}, \partial\tilde{B})\varepsilon^{\mathbf{t}}(\tilde{Z}+\chi\tilde{R}) = \chi\{(\tilde{L}, \partial\tilde{B})\varepsilon^{\mathbf{t}}\tilde{Z} - \tilde{Q}_1 + (\tilde{L}, \partial\tilde{B})\varepsilon^{\mathbf{t}}\tilde{R}\} + [(\tilde{L}, \partial\tilde{B}), \chi]\varepsilon^{\mathbf{t}}\tilde{R}. \tag{34}$$

The norm of the last term in (34) is exponentially small. As before, the expression in the braces in (34) can be represented in the form $\tilde{Q}_3 + \tilde{Q}_4$, where

$$||\chi\tilde{Q}_4; \mathbf{RH}_\varepsilon^l(\tilde{C}_\varepsilon)|| \le c\varepsilon^{1-l-\delta+\nu/2},$$

$$\tilde{Q}_3 = -\sum_{\mu=1}^{K^c}(\partial\hat{T}_z z_T + \hat{T}_r r_T)_\mu \varepsilon^{-(\mathbf{s},\mathbf{r})}X^{(\mu)}.$$

For every $s \in \partial G$ the vectors $X^{(1)}, \dots, X^{(K^c)}$ belong to the subspace

$$(1-\tilde{\Pi}_\Omega)\mathbf{RW}_{2,\beta}^l(\omega), \quad \beta \in (0, \beta_0),$$

and there hold the equalities

$$\langle X^{(\mu)}, \Psi_p \rangle = \delta_{\mu p}.$$

Thus, the function

$$\tilde{u}(\varepsilon, x, \xi) = \varepsilon^t(\tilde{Z}(x, \varepsilon^{-1}\xi) + \chi(x)\tilde{R}(s, \varepsilon^{-1}\tau, \varepsilon^{-1}\xi))$$

satisfies the boundary value problem $(\tilde{L}, \partial\tilde{B})\tilde{u} = \tilde{N}_1 - \tilde{N}_2$, where

$$||\tilde{N}_1; \mathbf{RH}_\varepsilon^l(\tilde{C}_\varepsilon)|| \le c\varepsilon^{1-\delta-l+\nu/2}, \tag{35}$$

$$\tilde{N}_2 = \sum_{p=1}^{K}(\tilde{S}z_T)_p\varepsilon^{-(\mathbf{s},\mathbf{r})}\chi^{(p)} + \chi\sum_{p=1}^{K^c}\left(\hat{T}(z_T, r_T)\right)_p\varepsilon^{-(\mathbf{s},\mathbf{r})}X^{(p)} \tag{36}$$

(in (36) χ is the cut-off function introduced in 16.2.2, see (10)). From (36), (25), (30), and (4) it follows that

$$||\varepsilon^{-l-\mathbf{t}}\tilde{u}; \mathbf{L}^2(\tilde{C}_\varepsilon)|| \le ||\tilde{u}; \mathbf{DH}_\varepsilon^l(\tilde{C}_\varepsilon)|| \le c\varepsilon^{1-\delta-M-l+\nu/2}. \tag{37}$$

Since the components of $\tilde{R}(s,t,\zeta)$ decay exponentially as $t \to \infty$, from (37) we derive the inequality

$$||\tilde{Z}; \mathbf{L}_2(\tilde{C}_\varepsilon \setminus (\tilde{O}_{\varepsilon^\sigma} \times R^\nu))|| \le c\varepsilon^{\sigma-M+\nu/2},$$

where $\sigma = 1-\delta$ for $\delta > 0$ and σ is any number in $(0,1)$ for $\delta = 0$. Using the definition of the vector $\tilde{Z}$ and the linear independence of $\varphi_1, \dots, \varphi_M$, we find that

$$\begin{aligned}\varepsilon^{\nu/2}||z_T; \mathbf{L}_2(G \setminus (\tilde{O}_{\varepsilon^\sigma})|| &\le c_1||Z_0; \mathbf{L}_2(\tilde{C}_\varepsilon \setminus (\tilde{O}_{\varepsilon^\sigma} \times R^\nu))|| \\ &\le c_2||\tilde{Z}; \mathbf{L}_2(\tilde{C}_\varepsilon \setminus (\tilde{O}_{\varepsilon^\sigma} \times R^\nu))|| \\ &\le c\varepsilon^{\sigma-M+\nu/2}.\end{aligned} \tag{38}$$

The known inequality

$$C\|z_T;\mathbf{L}_2(G\cap\tilde{O}_{\varepsilon^\sigma})\|\le\varepsilon^\sigma\|Z_T;\mathbf{W}_2^l(G\cap\tilde{O}_{2\varepsilon^\sigma})\|+\|Z_T;\mathbf{L}_2(G\cap(\tilde{O}_{2\varepsilon^\sigma}\setminus\tilde{O}_{\varepsilon^\sigma}))\|$$

together with (38) and (29), where $M\ge 1$ (see Lemma 16.2.2), leads to the estimate

$$\|z_T;\mathbf{L}_2(G)\|\le c\varepsilon^{\sigma-M}. \tag{39}$$

Taking account of the inequality

$$\|\chi\tilde{R};\mathbf{L}_2(\tilde{C}_\varepsilon)\|\le\|\varepsilon^{-\mathbf{t}}\tilde{u};\mathbf{L}_2(\tilde{C}_\varepsilon)\|+\|\tilde{Z};\mathbf{L}_2(\tilde{C}_\varepsilon)\|,$$

the form of $\tilde{R}$, (37), (38), and the linear independence of $\Phi_1,\dots,\Phi_{K^k}$, we obtain

$$\varepsilon^{(\nu+1)/2}\|r_T;\mathbf{L}_2(\partial G)\|\le c_1\|\chi\tilde{R};\mathbf{L}_2(C_\varepsilon)\|\le c_2\varepsilon^{\sigma-M+\nu/2},$$

which together with (34) proves the theorem. □

Remark 16.2.5. The assertion of Lemma 16.2.2 follows from the proof of Theorem 16.2.4. Indeed, $\|(\tilde{L},\partial\tilde{B})\tilde{u};\mathbf{RH}_\varepsilon^l(\tilde{C}_\varepsilon)\|\le c\varepsilon$ for $\varphi\in\ker A_\Omega$, and $\tilde{u}(\varepsilon,x,\xi)=(1-\chi(x))\varepsilon^{\mathbf{t}}\varphi(x,\varepsilon^{-1}\xi)$. Because $\|\tilde{u};\mathbf{DH}_\varepsilon^l(\tilde{C}_\varepsilon)\|=O(1)$ the inequality (9) with $M=0$ is impossible. In the case $\dim\ker A_\omega\ge 1$ it suffices to set $\tilde{u}(\varepsilon,x,\xi)=\chi(x)\varepsilon^{\mathbf{t}}\Phi(s,\varepsilon^{-1}\tau,\varepsilon^{-1}\xi)$ and argue in a similar way.

16.2.4 Asymptotics in the case when $K+K^k>0$

If there arises the third limit problem we seek an asymptotic expansion for the solution to problem (1), (2), 16.1, in the form

$$\begin{aligned}u(\varepsilon,x,\xi)\sim&\sum_{p=0}^{\infty}\varepsilon^{p+\mathbf{t}}\big(v^{(p)}(x,\varepsilon^{-1}\xi)+\chi(x)w^{(p)}(s,\varepsilon^{-1}\tau,\varepsilon^{-1}\xi)\big)\\&+\sum_{p=0}^{\infty}\sum_{q=0}^{\infty}\varepsilon^{p+(1-\delta)q+\mathbf{t}}\big(\tilde{Z}_{pq}(\varepsilon,x,\varepsilon^{-1}\xi)+\chi(x)\hat{R}_{pq}(\varepsilon,s,\varepsilon^{-1}\tau,\varepsilon^{-1}\xi)\big).\end{aligned} \tag{40}$$

Here $v^{(p)},w^{(p)}$ are some solutions to problems (7), (16), 16.1,

$$\begin{aligned}\hat{Z}_{pq}&=Z_{pq}+\sum_{j=1}^{\infty}\varepsilon^jA_\Omega^{-1}\tilde{\Pi}_\Omega\tilde{X}_j^{(p,q)},\\\hat{R}_{pq}&=R_{pq}+\sum_{j=1}^{\infty}\varepsilon^jA_\omega^{-1}\tilde{\Pi}_\omega\tilde{W}_j^{(p,q)},\end{aligned} \tag{41}$$

the vectors $\tilde{X}_j^{(p,q)}$ and $\tilde{W}_j^{(p,q)}$ are defined by (20), 16.1, and (28), 16.1, with Z and R replaced by

$$Z_{pq}=\sum_{\mu=0}^{K}(z_T^{(p,q)})_\mu\varphi_\mu,\quad R_{pq}=\sum_{\mu=0}^{K^k}(r_T^{(p,q)})_\mu\Phi_\mu, \tag{42}$$

$z_T^{(p,q)}$ and $r_T^{(p,q)}$ being some asymptotic solutions to problem (31), 16.1.

We substitute the expansions (40) for u, (3), (4), 16.1, for $(\tilde{L},\partial\tilde{B})$, and (10) for (f,g) into problem (1), (2), 16.1, and separate out the terms written in the coordinates $(x,\eta) = (x,\varepsilon^{-1}\xi)$. The expressions obtained can be represented as formal series in degrees of ε:

$$\sum_{p=0}^{\infty}\varepsilon^p\Big(A_\Omega v^{(p)} - \sum_{\mu=0}^{p-1}A_\Omega^{(p-\mu)}v^{(\mu)}\Big) + \sum_{p=0}^{\infty}\sum_{q=0}^{\infty}\varepsilon^{p+\delta q}\Big(\sum_{j=1}^{M}\varepsilon^j(\tilde\Pi_\Omega\tilde X_j^{(p,q)} - A_\Omega^{(j)}Z_{pq})$$
$$-\sum_{j=1}^{M}\sum_{\mu=1}^{j-1}\varepsilon^j A_\Omega^{(j-\mu)}A_\Omega^{-1}\tilde\Pi_\Omega\tilde X_\mu^{(p,q)} - \sum_{k=0}^{p-1}\varepsilon^M\Big(A_\Omega^{(M+p-k)}Z_{kq}$$
$$+\sum_{\mu=1}^{M+p-k-1}A_\Omega^{M+p-k-\mu}A_\Omega^{-1}\tilde\Pi_\Omega\tilde X_\mu^{(k,q)} - \tilde\Pi_\Omega\tilde X_{M+p-k}^{(k,q)}\Big)\Big) = \sum_{j=0}^{\infty}\varepsilon^j(h^{(j)},k^{(j)}). \tag{43}$$

From (41) and (19), 16.1, it follows that

$$\tilde X_k^{(p,q)} = A_\Omega^{(k)}Z_{pq} + \sum_{\mu=1}^{k-1}A_\Omega^{(k-\mu)}A_\Omega^{-1}\tilde\Pi_\Omega\tilde X_\mu^{(p,q)} \tag{44}$$

(relations of such kind were established, in particular, in the proof of Theorem 16.2.4). We project (43) into A_Ω and equate the coefficients by ε^p:

$$A_\Omega v^{(p)} - \sum_{\mu=0}^{p-1}\tilde\Pi_\Omega A_\Omega^{(p-\mu)}v^{(\mu)} = \tilde\Pi_\Omega(h^{(p)},k^{(p)}),\quad p\in\mathbb{N}_0.$$

The vectors

$$v^{(p)} = \sum_{\gamma_1+\ldots+\gamma_\varkappa=p}A_\Omega^{-1}\tilde\Pi_\Omega A_\Omega^{(\gamma_1)}\ldots A_\Omega^{-1}\tilde\Pi_\Omega A_\Omega^{(\gamma_{\varkappa-1})}A_\Omega^{-1}\tilde\Pi_\Omega(h^{(\gamma_\varkappa)},k^{(\gamma_\varkappa)}), \tag{45}$$

where the summation is taken over the same collections of indices as in (28), 16.1, satisfy those equations (see (13), (14)). Taking (44) into account, we rewrite the coefficient by $\varepsilon^{p+\delta q}$ in the form

$$\sum_{j=1}^{M}\varepsilon^j(\tilde\Pi_\Omega - 1)\tilde X_j^{(p,q)} + \varepsilon^M\sum_{\mu=0}^{p-1}(\tilde\Pi_\Omega - 1)\tilde X_{M+p-\mu}^{(\mu,q)}. \tag{46}$$

The first term in (46) can be represented in the form

$$-\sum_{j=1}^{K}(\tilde S z^{(p,q)})_j\chi^{(j)} \tag{47}$$

(cf. (33)). We assume that for $x\in\bar G$

$$\begin{aligned}&\|\tilde X_h^{(\mu,q)};\mathbf{RW}_2^l(\Omega)\| = O(\varepsilon^{-M}),\quad h=1,\ldots,M,\\&\|\tilde X_{M+j}^{(\mu,q)};\mathbf{RW}_2^l(\Omega)\| = O(\varepsilon^{-M-\delta j}),\quad j\in\mathbb{N}_0,\end{aligned} \tag{48}$$

(in this connection see Lemma 16.2.6), project (43) into the linear hull $L(\chi^{(1)},\ldots,\chi^{(K)})$ and equate the projections of vectors having the same order in ε. In view of

(45)–(47),

$$-\sum_{j=1}^{K}(\tilde{S}z^{(p,q)})_j\chi^{(j)} = \sum_{j=1}^{K}\Big(\delta_{q0}\Big\langle (h^{(p)},k^{(p)}) + \sum_{\mu=0}^{p-1} A_\Omega^{(p-\mu)}v^{(\mu)}, \psi_j\Big\rangle + \sum_{\mu=0}^{q-1}\varepsilon^{M+\delta(q-\mu)}\Big\langle \tilde{X}^{(p,\mu)}_{M+q-\mu}, \psi_j\Big\rangle\Big)\chi^{(j)}. \tag{49}$$

Since the vectors $\chi^{(1)},\dots,\chi^{(K)}$ are linearly independent, the relation (49) is equivalent to the system

$$\tilde{S}z^{(p,q)} = -\Big(\Big\langle \delta_{q0}(h^{(p)},k^{(p)}) + \delta_{q0}\sum_{\mu=0}^{p-1} A_\Omega^{(p-\mu)}v^{(\mu)} + \sum_{\mu=0}^{q-1}\varepsilon^{M+\delta(q-\mu)}\tilde{X}^{(p,\mu)}_{M+q-\mu}, \psi_j\Big\rangle\Big)_{j=1}^{K}. \tag{50}$$

Equations for the coefficients $w^{(p)}$ and $\hat{R}_{pq}$ in (40) can be derived in a similar manner. In doing so, account must be taken of the discrepancy on the right in (1), (2), 16.1, originating from the constructed functions $v^{(p)}$ and $\hat{Z}_{pq}$. The additional terms in the right-hand sides of equations of the type (16), 16.1, are defined as in Section 16.1.4 with the help of Taylor's expansions of $v^{(p)}, (h^{(p)},k^{(p)})$, and $\hat{Z}_{pq}$. Namely, we set

$$\begin{aligned} V^{(p,\mu)} &= (\mu!)^{-1}\tau^\mu(\partial/\partial\tau)^\mu v^{(p)}|_{\tau=0}, \\ (H^{(p,\mu)},K^{(p,\mu)}) &= (\mu!)^{-1}\tau^\mu(\partial/\partial\tau)^\mu(\tilde{\Pi}_\Omega(h^{(p)},k^{(p)}))|_{\tau=0}, \\ Z^{(0,\mu)}_{pq} &= (\mu!)^{-1}\tau^\mu(\partial/\partial\tau)^\mu(Z_{pq})|_{\tau=0}, \\ Z^{(h,\mu)}_{pq} &= (\mu!)^{-1}\tau^\mu(\partial/\partial\tau)^\mu(A_\Omega^{-1}\tilde{\Pi}_\Omega\tilde{X}^{(p,q)}_h)|_{\tau=0}, \quad h\in\mathbb{N}. \end{aligned} \tag{51}$$

We introduce the vectors $Y^{(j)}_{pq}$ whose definition is similar to (22), 16.1, and the vectors

$$(\hat{H}^{(j)},\hat{K}^{(j)}) = \sum_{p=0}^{j}\Big(A_\omega V^{(p,j-p)} - \sum_{\mu=1}^{p} A_\omega^{(\mu)}V^{(p-\mu,j-p)} - (H^{(p,j-p)}-,K^{(p,j-p)})\Big).$$

By virtue of (45) (cf. (23), 16.1),

$$\begin{aligned} \hat{H}^{(j)}(s,\zeta) &= 0, \quad (s,\zeta)\in\partial G\times\bar{\omega} \quad \text{for } t>1, \\ \hat{K}^{(j)}(s,\zeta) &= 0, \quad (s,\zeta)\in\partial G\times\partial\omega \quad \text{for } t>1. \end{aligned} \tag{52}$$

Arguing as in the construction of $v^{(p)}$, we obtain that the vectors $w^{(p)}$ satisfy the equations

$$A_\omega w^{(p)} = \tilde{\Pi}_\omega\Big(\sum_{\mu=0}^{p-1} A_\omega^{(p-\mu)}w^{(\mu)} + (\tilde{H}^{(p)}-\hat{H}^{(p)},\tilde{K}^{(p)}-\hat{K}^{(p)})\Big).$$

Therefore,

$$w^{(p)} = \sum_{\gamma_1+\dots+\gamma_\varkappa=p} A_\omega^{-1}\tilde{\Pi}_\omega A_\omega^{(\gamma_1)}\dots A_\omega^{-1}\tilde{\Pi}_\Omega(\tilde{H}^{(\gamma_\varkappa)}-\hat{H}^{(\gamma_\varkappa)},\tilde{K}^{(\gamma_\varkappa)}-\hat{K}^{(\gamma_\varkappa)}), \tag{53}$$

the notation being similar to that in (28), 16.1. The inclusions $(\tilde{H}^{(j)},\tilde{K}^{(j)})\in \mathbf{RW}^l_{2,\beta}(\omega)$, $\beta\in(0,\beta_0)$ and formula (52) imply that the functions (53) belong to $\mathbf{DW}^l_{2,\beta}(\omega)$ for any β in $(0,\beta_0)$.

It remains to find equations for $r_T^{(p,q)}$ in (42). By the same reasoning as in the derivation of (50), we obtain the system of differential equations on ∂G:

$$\hat{T}(\varepsilon,s,D_x)(z^{(p,q)},r^{(p,q)}) = -\Big(\Big\langle \delta_{q0}(\tilde{H}^{(p)} - \bar{H}^{(p)},\tilde{K}^{(p)} - \bar{K}^{(p)}) + \delta_{q0}\sum_{\mu=0}^{p-1} A_\omega^{(p-\mu)} w^{(\mu)}$$
$$+\sum_{\mu=0}^{q-1} \varepsilon^{M+\delta(q-\mu)} \hat{W}^{(p,\mu)}_{M+q-\mu}, \Psi_j\Big\rangle\Big)_{j=1}^{K^c}, \tag{54}$$

where $\hat{W}_s^{(j,h)}$ are defined by (28), 16.1, with R and z replaced by R_{jh} and $z^{(j,h)}$ (see (42)).

The systems (50) and (54) form the third limit problem to find $z^{(p,q)}$ and $r^{(p,q)}$.

Lemma 16.2.6. *There exist asymptotic solutions to problems* (50) *and* (54).

Proof. In view of the assumptions formulated in Section 16.2.3, it suffices to verify (28) for the right-hand sides of the equations (50) and (54). To this end, we show by induction that there hold the relations

$$\begin{aligned}
&\varepsilon^{\delta\mu}\|\tilde{X}^{(p,q)}_{M+\mu}; \mathbf{W}_2^j(G\to \mathbf{RW}_2^l(\Omega))\| + \|z_T^{(p,q)}; \mathbf{W}_2^{j+M}(G)\| \\
&\quad + \varepsilon^{\delta\mu+1/2}\|\hat{W}^{(p,q)}_{M+\mu}; \mathbf{W}_2^j(\partial G\to \mathbf{RW}^l_{2,\beta}(\omega))\| + \varepsilon^{1/2}\|r_T^{(p,q)}; \mathbf{W}_2^{j+M}(\partial G)\| \\
&\le c^{(p,q)}_{\mu j}\varepsilon^{-M-\delta j}, \quad \mu\in\mathbb{N}_0; \\
&\varepsilon^{\delta j}\|\tilde{X}^{(p,q)}_{h}; \mathbf{W}_2^j(G\to \mathbf{RW}_2^l(\Omega))\| + \|z_T^{(p,q)}; \mathbf{W}_2^{h}(G)\| \\
&\quad + \varepsilon^{\delta j+1/2}\|\hat{W}^{(p,q)}_{h}; \mathbf{W}_2^j(\partial G\to \mathbf{RW}^l_{2,\beta}(\omega))\| + \varepsilon^{1/2}\|r_T^{(p,q)}; \mathbf{W}_2^{h}(\partial G)\| \\
&\le c^{(p,q)}_{hj}\varepsilon^{-M}, \quad h=0,\dots,M.
\end{aligned} \tag{55}$$

For $p=q=0$ the right-hand sides $E^{(p,q)}$ and $\hat{C}^{(p,q)}$ in (50) and (54) take the form

$$E^{(0,0)} = -(\langle (h^{(0)},k^{(0)}),\psi_j\rangle)_{j=1}^K,$$
$$\hat{C}^{(0,0)} = (\langle (\tilde{H}^{(0)} - \hat{H}^{(0)}, \tilde{K}^{(0)} - \hat{K}^{(0)}), \Psi_j\rangle)_{j=1}^{K^c}.$$

Therefore, there exists an asymptotic solution (50), (54) with $p=q=0$ subject to (29). By virtue of (20), 16.1, and (10), 16.1, $\tilde{X}_k^{(0,0)}$ is a linear combination of the derivatives of the components of $z_T^{(0,0)}$ of order no greater than k with coefficients in $\mathbf{RW}_2^l(\Omega)$. Similarly, in view of (28), (23), 16.1, and (14), (15), 16.1, $\hat{W}_k^{(0,0)}$ is a linear combination of the expressions $D_x^\alpha z_T^{(0,0)}$ and $D_s^\beta r_T^{(0,0)}$ with $|\alpha|,|\beta|\le k$. Therefore, the estimates (29) imply (55) for $p=q=0$.

Suppose that there have been found asymptotic solutions $z_T^{(p,q)}$ and $t_T^{(p,q)}$ to problems (50), (54) for (p,q) subject to the condition $p+(1-\delta)q < P+(1-\delta)Q$ while (55) holds for the solutions. We prove the existence of an asymptotic solution to problem (50), (54) and the estimate (55) for $p=P$ and $q=Q$. From the aforesaid, it follows that it suffices to verify (28) for the right-hand sides $E^{(P,Q)}, \hat{C}^{(P,Q)}$. In its turn, the estimate (28) follows inductively from (50) in view of the smoothness of the functions $(h^{(p)},k^{(p)}), v^{(p)}, (\tilde{H}^{(p)},\tilde{K}^{(p)}), (\hat{H}^{(p)},\hat{K}^{(p)}), w^{(p)}, \psi_j$, and Ψ_j, and of the coefficients of operators in (8), 1.16, and (12), 16.1. □

Retracing the proofs of Theorems 16.2.3 and 16.2.4 and using Lemma 16.2.6, we obtain the following assertion.

Theorem 16.2.7. *Let the conditions formulated in Sections* 16.1.1, 16.2.1, *and* 16.2.3 *be fulfilled and let the right-hand sides* (f, g) *of problem* (1), (2), 16.1, *admit the expansions* (10) *and* (11). *Then*

$$||u - U^{(I)}; \mathbf{DH}^l_\varepsilon(\tilde{C}_\varepsilon)|| = o(\varepsilon^{I-l-M+\nu/2}), \quad \varepsilon \to 0,$$

where u *is the solution to problem* (1), (2), 16.1, $U^{(I)}$ *the partial sum of the series* (35)*:*

$$U^{(I)}(\varepsilon, x, \xi) = \sum_{p=0}^{I} \varepsilon^{p+\mathbf{t}}\left(v^{(p)}(x, \varepsilon^{-1}\xi) + \chi(x) w^{(p)}(s, \varepsilon^{-1}\tau, \varepsilon^{-1}\xi)\right)$$

$$+ \sum_{p=0}^{I}\sum_{q=0}^{J(p)} \varepsilon^{p+(1-\delta)q+\mathbf{t}}\left(\hat{Z}^{(I)}_{pq}(\varepsilon, x, \varepsilon^{-1}\xi) + \chi(x)\hat{R}^{(I)}_{pq}(\varepsilon, s, \varepsilon^{-1}\tau, \varepsilon^{-1}\xi)\right),$$

$J(p) = [(I-p)(1-\delta)^{-1}]$, $v^{(p)}, w^{(p)}$ *stand for the functions* (45), (53),

$$\hat{Z}^{(I)}_{pq} = Z_{pq} + \sum_{j=1}^{J(p)-q} \varepsilon^j A_\Omega^{-1}\tilde{\Pi}_\Omega \tilde{X}_j^{(p,q)},$$

$$\hat{R}_{pq}(I) = R_{pq} + \sum_{j=0}^{J(p)-q} \varepsilon^j A_\omega^{-1}\tilde{\Pi}_\omega \hat{W}_j^{(p,q)}.$$

The vectors $\tilde{X}_j^{(p,q)}$ *and* $\hat{W}_j^{(p,q)}$ *are defined by* (20), 16.1, *and* (28), 16.1, *where* Z *and* R *are replaced by the vectors* (42). *Finally,* $z_T^{(p,q)}$ *and* $r_T^{(p,q)}$ *stand for the asymptotic solutions to problems* (50) *and* (54) *with* $T > (1-\delta)^{-1}(I-p) - q$.

16.3 Examples

In this section we consider some particular boundary value problems that differ by the type of third limit problem. The role of the thin domain is played by a three-dimensional cylinder of small height.

Example 16.3.1. (An algebraic or differential equation on the boundary of a domain.) Let G be an annular domain $D^2_a(O) \setminus D^2_b(O) = \{x \in \mathbb{R}^2 \colon \rho = |x| \in (b, a)\}$ with $a > b > 0$ and $\Omega = (-1, 1)$. We consider the boundary value problem in the cylinder $\tilde{C}_\varepsilon$:

$$\left(\Delta_x + (\partial/\partial\xi)^2 - (\varepsilon|x|)^{-1}A - B\right) u(\varepsilon, x, \xi) = f(x, \varepsilon^{-1}\xi), \quad (x, \xi) \in \tilde{C}_\varepsilon; \quad (1)$$

$$\begin{aligned} &u(\varepsilon, x, \pm\varepsilon) = 0, \quad x \in G; \quad u(\varepsilon, x, \xi) = 0, \quad |x| = b, \quad |\xi| < 1; \\ &-(\partial/\partial\rho)\, u(\varepsilon, x, \xi) + (2\varepsilon)^{-1}\pi u(\varepsilon, x, \xi) = 0, \quad |x| = a, |\xi| < 1, \end{aligned} \quad (2)$$

where $\rho = |x|$ and A, B are nonnegative constants.

Lemma 16.3.2. *Let* $A \geq \pi/2$ *and* $A + B > \pi/2$. *Then for any right-hand side* $f \in \mathbf{L}_2(\tilde{C}_\varepsilon)$ *there exists a unique generalized solution* $u \in \mathbf{W}^l_{2,0}(\tilde{C}_\varepsilon) = \{U \in \mathbf{W}^l_2(\tilde{C}_\varepsilon) \colon U = 0 \text{ for } \xi = \pm\varepsilon \text{ or } |x| = b\}$ *to problem* (1), (2).

Proof. Since problem (1), (2) is self-adjoint, it suffices to check only the absence of nontrivial solutions to the homogeneous equations (1), (2). We multiply (1) with $f=0$ by $\bar{u}$ and, integrating over $\tilde{C}_\varepsilon$, obtain

$$\int\limits_{\tilde{C}_\varepsilon}\Big(|\nabla_x u|^2+|(\partial/\partial\xi)u|^2-\big((\varepsilon\rho)^{-1}A+B\big)|u|^2\Big)dx\,d\xi-(2\varepsilon)^{-1}\pi a\int\limits_{-\varepsilon}^{\varepsilon}\int\limits_{|x|=a}|u|^2 ds\,d\xi=0.$$

This relation and the inequalities

$$4\int\limits_{-\varepsilon}^{\varepsilon}|(\partial/\partial\xi)\varkappa|^2 d\xi\ge(\pi/\varepsilon)^2\int\limits_{-\varepsilon}^{\varepsilon}|\varkappa|^2 d\xi,$$

$$(2\varepsilon)^{-1}\pi a\chi(a)^2\le\int\limits_b^a\rho|(\partial/\partial\rho)\chi|^2\,d\rho+\int\limits_b^a\big((\pi/2\varepsilon)^2\rho+(2\varepsilon)^{-1}\pi\big)|\chi|^2\,d\rho$$

(with $\varkappa(\pm\varepsilon)=0$ and $\chi(b)=0$) imply that

$$\int\limits_{\tilde{C}_\varepsilon}\big((\varepsilon\rho)^{-1}(A-\pi/2)+B\big)|u|^2\,dx\,d\xi\le 0.$$ □

In the case under consideration the first limit problem

$$(\partial/\partial\eta)^2\,v(\eta)=h(\eta),\quad \eta\in[-1,1];\quad v(\pm1)=k^{\pm} \tag{3}$$

is uniquely solvable. The second limit problem can be represented as two problems (see Remark 16.1.3): the problem corresponding to the outer boundary of the annulus

$$\begin{aligned}&\big((\partial/\partial t)^2+(\partial/\partial\eta)^2\big)w_a(t,\eta)=\tilde{H}_a(t,\eta),\quad (t,\eta)\in\omega=(0,\infty)\times(-1,1);\\&w_a(t,\pm1)=\tilde{K}_1^{\pm}(t),\quad t\in(0,\infty);\\&(\partial/\partial t)w(0,\eta)+2^{-1}\pi w(0,\eta)=\tilde{K}_2(\eta),\quad \eta\in(-1,1),\end{aligned} \tag{4}$$

and that corresponding to the inner boundary

$$\begin{aligned}\big((\partial/\partial t)^2+(\partial/\partial\eta)^2\big)w_b(t,\eta)r=\tilde{H}_b(t,\eta),(t,\eta)\quad&\in\omega;\\ w_b(t,\eta)=\tilde{K}(t,\eta),\quad (t,\eta)&\in\partial\omega.\end{aligned} \tag{5}$$

Problem (5) is uniquely solvable, while problem (4) has the one-dimensional kernel spanned by the function $\Phi(t,\eta)=\exp(-\pi t/2)\cos(\pi\eta/2)$ and a one-dimensional cokernel. The orthogonality to the elements in the cokernel can be written in the form

$$\int\limits_\omega\tilde{H}_a(t,\eta)\Phi(t,\eta)dt\,d\eta=2^{-1}\pi\int\limits_0^\infty\exp(-\pi t/2)\big(\tilde{K}_1^+(t)+\tilde{K}_1^-(t)\big)dt$$
$$-\int\limits_{-1}^{1}\cos(\pi\eta/2)\tilde{K}_2(\eta)d\eta.$$

In addition, to the change of variables $(x,\xi)\to(\vartheta,t,\eta)=(\vartheta,\varepsilon^{-1}(a-\rho),\varepsilon^{-1}\xi)$, where (ρ,ϑ) are polar coordinates, there corresponds the following expansion of the operator $\tilde{L}$ in equation (1):

$$\begin{aligned}\tilde{L} \sim & \varepsilon^{-2}\big((\partial/\partial t)^2 + (\partial/\partial\eta)^2\big) - (\varepsilon a)^{-1}(\partial/\partial t + A) \\ & - a^{-2}\big(t(\partial/\partial t + A) + a^2 B - (\partial/\partial\vartheta)^2\big) - \cdots .\end{aligned}$$

If $A > \pi/2$, then $(A + \partial/\partial t)\Phi = (A - \pi/2)\Phi$ and the third limit problem is the algebraic equation

$$a^{-1}(A - \pi/2)r(\vartheta) = \hat{C}(\vartheta), \quad \vartheta \in S^1. \tag{6}$$

In the case $A = \pi/2$ and $B > 0$, the third limit problem is the ordinary differential equation on the circle S^1:

$$Br(\vartheta) - a^2(\partial/\partial\vartheta)^2 r(\vartheta) = \hat{C}(\vartheta), \vartheta \in S^1. \tag{7}$$

Note that equations (6) and (7) are uniquely solvable.

Finally, we give the asymptotic representation of the solution to problem (1), (2) that follows from Theorem 16.2.7:

$$\begin{aligned} u(\varepsilon, x, \xi) \sim & \sum_{j=0}^{\infty} \varepsilon^j \big(v^{(j)}(x, \varepsilon^{-1}\xi) + w^{(j)}(\vartheta, \varepsilon^{-1}(a-\rho), \varepsilon^{-1}\xi) \\ & + w_b^{(j)}(\vartheta, \varepsilon^{-1}(\rho - b), \varepsilon^{-1}\xi)\big) + \varepsilon^k r^{(j)}(\vartheta) \exp\Big(- (2\varepsilon)^{-1}\pi(a-\rho)\Big) \cos(\pi\xi/2\varepsilon)\Big). \end{aligned}$$

Here $v^{(j)}$ are the solutions to the ordinary differential equations (3) with some right-hand sides; $w_a^{(j)}$ and $w_b^{(j)}$ are the exponentially decaying solutions to problems (1) and (2); if $A > \pi/2$, then $k = -1$, and $A = \pi/2$ in the case $k = -2$; finally, $r^{(j)}$ are the solutions to the equations (6) or (7) with some right-hand sides.

Example 16.3.3. (A boundary value problem with small parameter by the higher derivatives.) We consider the boundary value problem

$$\begin{aligned} & \big((\partial/\partial x_1)^4 + 2(\partial/\partial x_1)^2(\partial/\partial x_2)^2 + (\partial/\partial x_2)^4 + (\partial/\partial\xi)^4 + \varepsilon^{-1}M(x, \partial/\partial x)\big)u(\varepsilon, x, \xi) \\ & = f(x, \varepsilon^{-1}\xi), \quad (x, \xi) \in \tilde{C}_e; \\ & (\partial/\partial\rho)u(\varepsilon, x, \xi) = u(\varepsilon, x, \xi) = 0, \quad \text{for } |x| = a \quad \text{or for } |x| = b, \quad \xi \in (-1, 1); \\ & (\partial/\partial\xi)u(\varepsilon, x, \pm\varepsilon) = (\partial/\partial\xi)^3 u(\varepsilon, x, \pm\varepsilon) = 0, \quad x \in G. \end{aligned} \tag{8}$$

Here G and $\tilde{C}_\varepsilon$ are the same domains as in Example 16.3.1 and

$$M(x, \partial/\partial x) = -A\Delta_x + B\partial/\partial\rho + C; \quad A, C \geq 0, \quad A + C \neq 0.$$

The first limit problem

$$\begin{aligned} & (\partial/\partial\eta)^4 = h(\eta), \quad \eta \in [-1, 1]; \\ & (\partial/\partial\eta)v(\pm 1) = k_1^{\pm}, \quad (\partial/\partial\eta)^3\, v(\pm 1) = k^{\pm}, \end{aligned} \tag{9}$$

has a solution only in the case

$$\int_{-1}^{1} h(\eta)d\eta = k^+ - k^-, \tag{10}$$

and any solution to the homogeneous problem (9) is constant. The kernel of the second limit problem

$$\begin{aligned} &((\partial/\partial t)^4 + (\partial/\partial\eta)^4)w(t,\eta) = \tilde{H}(t,\eta), \quad (t,\eta)\in\omega;\\ &w(0,\eta) = \tilde{K}_1(\eta), \quad (\partial/\partial t)\, w(0,\eta) = \tilde{K}_2(\eta), \quad |\eta|<1; \\ &(\partial/\partial\eta)w(t,\pm1) = \tilde{K}_3^{\pm}(t), \quad (\partial/\partial\eta)^3(t,\pm1) = \tilde{K}_4^{\pm}(t), \quad t\in(0,\infty), \end{aligned} \tag{11}$$

is trivial. The compatibility conditions of problem (11) in the class of functions vanishing at infinity take the form

$$\int_\omega t^2\tilde{H}(t,\eta)dt\,d\eta = \int_0^\infty t^2\Big(\tilde{K}_4^+(t) - \tilde{K}_4^-(t)\Big)dt - 2\int_{-1}^{1}\tilde{K}_2(\eta)d\eta,$$

$$\int_\omega t^3\tilde{H}(t,\eta)dt\,d\eta = \int_0^\infty t^3\Big(\tilde{K}_4^+(t) - \tilde{K}_4^-(t)\Big)dt + 6\int_{-1}^{1}\tilde{K}_1(\eta)d\eta.$$

We construct the third limit problem using the scheme given in Section 16.1 and obtain

$$\varepsilon^2\Delta_x^2\, z(\varepsilon,x) + M(x,\partial/\partial x)\, z(\varepsilon,x) = E(x), \quad x\in G; \tag{12}$$

$$z(\varepsilon,x) = \hat{C}_1(x), \quad (\partial/\partial\rho)\, z(\varepsilon,x) = \hat{C}_2(x), \quad x\in\partial G. \tag{13}$$

Thus, the third limit problem for (8) turns out to be the Dirichlet problem for a differential equation with small parameter by the higher derivatives. The asymptotics of solutions to such problems were studied in VISHIK and LYUSTERNIK [1]. For $A\neq 0$ the equation (12) is regularly degenerated into an elliptic one (see VISHIK and LYUSTERNIK [1], NAZAROV [3]), while for $A=0, B\neq 0$ and $\varepsilon\to 0$ into an equation with simple characteristic (see VISHIK and LYUSTERNIK [1]). In the case $A=B=0$ the problem (12) is elliptic in the sense of Section 3.1.5. If $E\in \mathbf{C}^\infty(\bar{G})$ and $\hat{C}_j\in \mathbf{C}^\infty(\partial G), j=1,2$, the asymptotics of the solution to problem (12), (13) is of the form

$$z(\varepsilon,x)\sim\sum_{j=0}^{\infty}\varepsilon^{j\gamma}\Big(Z^{(j)}(x) + \tilde{Z}_a^{(j)}(\vartheta,\varepsilon^{-\gamma}(a-\rho)) + \tilde{Z}_b^{(j)}(\vartheta,\varepsilon^{-\gamma}(\rho-b))\Big)$$

(see VISHIK and LYUSTERNIK [1]). Here $Z^{(j)}\in\mathbf{C}^\infty(\bar{G})$ and $\tilde{Z}_a^{(j)},\tilde{Z}_b^{(j)}\in\mathbf{C}^\infty(S^1\times[0,\infty))$. The functions $\tilde{Z}_a^{(j)}(\vartheta,t)$ and $\tilde{Z}_b^{(j)}(\vartheta,t)$ tend to zero as $t\to\infty$. Besides, $A\neq 0$ for $\gamma=1/2$; $A=0, B\neq 0$ for $\gamma=1/3$; $A=B=0$ for $\gamma=1/4$. The functions $\tilde{Z}_a^{(0)}$ and $\tilde{Z}_b^{(0)}$ vanish for $A\neq 0$, the function $\tilde{Z}_b^{(0)}$ for $A=0, B>0$, and $\tilde{Z}_a^{(0)}$ for $A=0, B<0$.

Note that in the mentioned three cases the width of the boundary layer in problem (12), (13) is $O(\varepsilon^{1/2})$, $O(\varepsilon^{1/3})$, and $O(\varepsilon^{1/4})$, respectively. The boundary layer near the lateral area of the cylinder $\tilde{C}_\varepsilon$ described by the solutions of the second limit problem is of width $O(\varepsilon)$.

Example 16.3.3. (A regularly perturbed boundary value problem.) Let $G\subset\mathbb{R}^2$ be a domain in $\mathbf{C}^\infty$ with smooth boundary and let $\Omega=(-1,1)$. We consider the uniquely solvable boundary value problem

$$\begin{aligned} &\Big(\Delta_x + (\partial/\partial\xi)^2 - \varepsilon\Big) u(\varepsilon, x, \xi) = f(x, \varepsilon^{-1}\xi), \quad (x,\xi) \in \tilde{C}_\varepsilon; \\ &(\partial/\partial\xi)\, u(\varepsilon, x, \pm\varepsilon) = 0, \quad x \in G; \quad (\partial/\partial\tau)\, u(\varepsilon, x, \xi) = 0, \quad x \in \partial G, \quad |\xi| < \varepsilon, \end{aligned} \tag{14}$$

where τ is the inward normal on ∂G. The first limit problem

$$(\partial/\partial\eta)^2\, v(\eta) = h(\eta), \quad \eta \in [-1,1]; \quad (\partial/\partial\eta)\, v(\pm 1) = k^{\pm} \tag{15}$$

is solvable only if the equality (10) holds. Any constant satisfies the homogeneous problem (15). The second limit problem

$$\begin{aligned} &\big((\partial/\partial t)^2 + (\partial/\partial\eta)^2\big)\, w(t,\eta) = \tilde{H}(t,\eta), \quad (t,\eta) \in \omega = (0,\infty) \times (-1,1); \\ &(\partial/\partial t)\, w(0,\eta) = \tilde{K}_1(\eta), \quad |\eta| < 1; \\ &(\partial/\partial\eta)\, w(t,\pm 1) = \tilde{K}_2^{\pm}(t), \quad t \in (0,\infty), \end{aligned} \tag{16}$$

has unique solution in the class of functions vanishing at infinity for all exponentially decaying right-hand sides subject to the condition

$$\int_\omega \tilde{H}(t,\eta)\,dt\,d\eta = \int_0^\infty \Big(\tilde{K}_2^+(t) - \tilde{K}_2^-(t)\Big)\,dt + \int_{-1}^1 \tilde{K}_1(\eta)\,d\eta.$$

Thus, according to Section 16.1, the third limit problem for (14) is of the form

$$(\Delta_x - \varepsilon)\, z(\varepsilon, x) = E(x), \quad x \in G; \quad (\partial/\partial\tau)\, z(\varepsilon, x) = \hat{C}(x), \quad x \in \partial G. \tag{17}$$

For $\varepsilon = 0$ problem (17) has one-dimensional kernel and cokernel and for $\varepsilon > 0$ it is uniquely solvable. If $E \in \mathbf{C}^\infty(\bar{G})$ and $\hat{C} \in \mathbf{C}^\infty(\partial G)$, then in view of the results of Chapter 9 in VAINBERG and TRENOGIN [1] the asymptotic expansion of the solution to problem (17) can be written in the form

$$z(\varepsilon, x) \sim \varepsilon^{-1} C + \sum_{j=0}^\infty \varepsilon^j Z^{(j)}(x), \tag{18}$$

where $Z^{(j)}$ are in $\mathbf{C}^\infty(\bar{G})$, have zero means over G, and satisfy

$$\begin{aligned} &\Delta Z^{(0)}(x) = E(x) + C, \quad x \in G; \quad (\partial/\partial\tau) Z^{(0)}(x) = \hat{C}(x), \quad x \in \partial G; \\ &\Delta Z^{(j)}(x) = Z^{(j-1)}(x), \quad x \in G; \quad (\partial/\partial\tau) Z^{(j)}(x) = 0, \quad x \in \partial G, \quad j \in \mathbb{N}. \end{aligned}$$

The constant C in (18) is equal to

$$\operatorname{mes}_2(G)^{-1}\Big(\int_{\partial G} \hat{C}(x)\,ds - \int_G E(x)\,dx\Big).$$

In particular, (18) and the results in Section 16.2 imply that the asymptotics of the solution to problem (14) is of the form

$$u(\varepsilon, x, \xi) \sim \sum_{j=0}^\infty \varepsilon^{j+2}\Big(v^{(j)}(x, \varepsilon^{-1}\xi) + \chi(\tau)\, w^{(j)}(s, \varepsilon^{-1}\tau, \varepsilon^{-1}\xi) + \varepsilon^{-3}\tilde{Z}^{(j)}(x)\Big)$$

while $v^{(j)}$ and $w^{(j)}$ satisfy the limit problems (15) and (16) with some right-hand sides and $Z^{(0)}(x) = C$.

16.4 Bending of a Thin Plate

16.4.1 Statement of the problem

Let G be a domain in $\in \mathbb{R}^2$ with smooth boundary and compact closure $\bar{G}$. We consider a plate of the form $\tilde{C}_\varepsilon = G \times (-\varepsilon/2, \varepsilon/2)$. Suppose that mass forces F are applied to the plate and external loading $G_\pm = \{x \in \mathbb{R}^3 \colon x' = (x_1, x_2) \in G, x_3 = \pm\varepsilon/2\}$ to $P^\pm$. The lateral surface $S_\varepsilon = \{x \in \mathbb{R}^3 \colon x' \in \partial G, |x_3| < \varepsilon/2\}$ is rigidly fixed. Then the displacement vector $u = (u_1, u_2, u_3)$ satisfies the boundary value problem

$$\begin{aligned}
&\big(2(1+\nu)\big)^{-1} E \left(\Delta u_j(\varepsilon, x) + (1-2\nu)^{-1} (\partial/\partial x_j) \operatorname{div} u(\varepsilon, x)\right) - F_j(\varepsilon, x) = 0, \\
&x \in \tilde{C}_\varepsilon, \quad j = 1,2,3; && (1) \\
&\sigma_{3j}(u; \varepsilon, x) = P_j^\pm(\varepsilon, x'), \quad x \in G_\pm, \quad j = 1,2,3; && (2) \\
&u(\varepsilon, x) = 0, \quad x \in S_\varepsilon. && (3)
\end{aligned}$$

Here ν is the Poisson coefficient, E the Young modulus, and $\sigma_{jk}(u) = \big(2(1+\nu)\big)^{-1} \times E\left((\partial/\partial x_k) u_j + (\partial/\partial x_j) u_k + \delta_{jk}(1-2\nu)^{-1} 2\nu \operatorname{div} u\right)$ are the components of the stress tensor. Besides, we put

$$\begin{aligned}
&x = (x', \xi), \quad \Omega_\varepsilon = (-\varepsilon/2, \varepsilon/2), \quad F_j(\varepsilon, x) = -\varepsilon^{-2} f_j(\varepsilon, x', \varepsilon^{-1}\xi), \\
&P_j^\pm(\varepsilon, x') = \varepsilon^{-1} p_j^\pm(\varepsilon, x').
\end{aligned}$$

16.4.2 The first two limit problems

We make use of the iterative procedures given in Sections 16.1 and 16.2 and obtain the asymptotic expansions of the solution to problem (1)–(3) in degrees of ε.

The first limit problem for (1)–(4) consists of the three ordinary equations

$$\begin{aligned}
a_j (\partial/\partial\eta)^2 v_j(x', \eta) &= h_j(x', \eta), \qquad \eta = (-1/2, 1/2), \\
a_j (\partial/\partial\eta) v_j(x', \pm 1/2) &= k_j^\pm(x'), \qquad j = 1,2,3,
\end{aligned} \tag{4}$$

with $\eta = \varepsilon^{-1}\xi$, $a_1 = a_2 = \big(2(1+\nu)\big)^{-1} E$, and $a_3 = (1+2\nu)^{-1}(1-\nu)E$. Problem (4) has three-dimensional kernel spanned by the vectors $\varphi^{(1)} = (1,0,0)$, $\varphi^{(2)} = (0,1,0)$, and $\varphi^{(3)} = (0,0,1)$. The problem is solvable provided

$$\int_{-1/2}^{1/2} h_j(x', \eta)\, d\eta = k_j^+(x') - k_j^-(x'), \quad j = 1,2,3. \tag{5}$$

The series (9), 16.1, for problem (1)–(3) takes the form $A_\Omega^{(0)}(x) - \varepsilon A_\Omega^{(1)}(x) - \varepsilon^2 A_\Omega^{(2)}(x)$, where $A_\Omega^{(q)}(x) = \left(L^{(q)}(\partial/\partial x', \partial/\partial\eta), B^{(q)}(\partial/\partial x', \partial/\partial\eta)\right)$. Note that

$$\begin{aligned}
L^{(0)}(\partial/\partial\eta) &= \operatorname{diag}\left(a_1(\partial/\partial\eta)^2, a_2(\partial/\partial\eta)^2, a_3(\partial/\partial\eta)^2\right), \\
B^{(0)}(\partial/\partial\eta) &= \operatorname{diag}\left(a_1\partial/\partial\eta, a_2\partial/\partial\eta, a_3\partial/\partial\eta\right),
\end{aligned}$$

$$L^{(1)}(\partial/\partial x', \partial/\partial \eta) = -b_1\, a_1 \begin{pmatrix} 0 & 0 & (\partial/\partial x_1)(\partial/\partial \eta) \\ 0 & 0 & (\partial/\partial x_2)(\partial/\partial \eta) \\ (\partial/\partial \eta)(\partial/\partial x_1) & (\partial/\partial \eta)(\partial/\partial x_2) & 0 \end{pmatrix},$$

$$L^{(2)}(\partial/\partial x', \partial/\partial \eta) = -a_1 \begin{pmatrix} \Delta_{x'} + b_1(\partial/\partial x_1)^2 & b_1(\partial/\partial x_1)(\partial/\partial x_2) & 0 \\ b_1(\partial/\partial x_1)(\partial/\partial x_2) & \Delta_{x'} + b_1(\partial/\partial x_2)^2 & 0 \\ 0 & 0 & \Delta_{x'} \end{pmatrix},$$

$$B^{(1)}(\partial/\partial x', \partial/\partial \eta) = -a_1 \begin{pmatrix} 0 & 0 & \partial/\partial x_1 \\ 0 & 0 & \partial/\partial x_2 \\ b_2\partial/\partial x_1 & b_2\partial/\partial x_2 & 0 \end{pmatrix},$$

$$B^{(2)}(\partial/\partial x', \partial/\partial \eta) = 0.$$

Here $b_1 = (1-2\nu)^{-1}, b_2 = 2\nu(1-2\nu)^{-1}$.

We introduce the projector $\tilde{\Pi}_\Omega$ (see 16.1.4) by

$$\tilde{\Pi}_\Omega(h, k^\pm) = \Big(h(x', \eta) - \int_{-1/2}^{1/2} h(x', \zeta)d\zeta + k^+(x') - k^-(x');\, k^\pm\Big)$$

and the right inverse A_Ω^{-1} for $A_\Omega = A_\Omega^{(0)}$ by

$$A_\Omega^{-1}(h, k^\pm) = \Big(a_j^{-1} \int_{-1/2}^{\eta} (\eta - \zeta) h_j(x', \zeta)d\zeta + \eta a_j^{-1} k_j^\pm(x')$$

$$- a_j^{-1} \int_{-1/2}^{1/2} (1/8 - \zeta/2 + \zeta^2/2)\, h_j(x', \zeta)d\zeta\Big)_{j=1}^{3}.$$

Then

$$A_\Omega A_\Omega^{-1}\tilde{\Pi}_\Omega(h, k^\pm) - (h, k^\pm) = (1 - \tilde{\Pi}_\Omega)(h, k^\pm),$$

and the mean over $[-1/2, 1/2]$ of each component of the vector $A_\Omega^{-1}\tilde{\Pi}_\Omega(h, k^\pm)$ is equal to zero.

16.4.3 The complementary limit problem

We construct the third limit problem using the algorithm in 16.1.4. Introduce $Z^{(q)}(x') = z_q(x')\varphi^{(q)}, q = 1, 2, 3$, and first consider $Z^{(1)}$ and $Z^{(2)}$. As in (19), (20), 16.1, for $q = 1, 2$ we have

$$\tilde{X}_1^{(q)} = -\big(2(1+\nu)\big)^{-1} E\Big((0,0,0), (0,0,(1-2\nu)^{-1} 2\nu(\partial/\partial x_q)z_q)\Big),$$

$$(1 - \tilde{\Pi}_\Omega)\tilde{X}_1^{(q)} = 0, \quad A_\Omega^{-1}\tilde{X}_1^{(q)} = -(0, 0, (1-\nu)^{-1}\nu\eta(\partial/\partial x_q)z_q);$$

$$\tilde{X}_2^{(q)} = -\big(2(1+\nu)\big)^{-1} E\Big(\big(\delta_{q1}\Delta_{x'}z_q + (1-\nu)^{-1}(\partial/\partial x_1)(\partial/\partial x_q)z_q, \delta_{q2}\Delta_{x'}z_q$$

$$+ (1-\nu)^{-1}(\partial/\partial x_2)(\partial/\partial x_q)z_q, 0\big), \big(\pm\big(2(1-\nu)\big)^{-1}\nu(\partial/\partial x_1)(\partial/\partial x_q)z_q,$$

$$\pm\big(2(1-\nu)\big)^{-1}\nu(\partial/\partial x_2)(\partial/\partial x_q)z_q, 0\big)\Big),$$

$$(1-\tilde\Pi_\Omega)\tilde X_2^{(q)} = -\big(2(1+\nu)\big)^{-1}E\big(\delta_{q1}\Delta_{x'}z_q + (1-\nu)^{-1}(1+\nu)$$
$$\times (\partial/\partial x_1)(\partial/\partial x_q)z_q, \delta_{q2}\Delta_{x'}z_q$$
$$+(1-\nu)^{-1}(1+\nu)(\partial/\partial x_2)(\partial/\partial x_q)z_q, 0\big),$$
$$A_\Omega^{-1}\tilde\Pi_\Omega\tilde X_2^{(q)} = (1-\nu)^{-1}\nu(\eta^2/2-1/24)\big((\partial/\partial x_1)(\partial/\partial x_q)z_q, (\partial/\partial x_2)(\partial/\partial x_q)z_q, 0\big). \tag{6}$$

For $Z^{(3)}$ we obtain the following relations:

$$\tilde X_1^{(3)} = -\big(2(1+\nu)\big)^{-1}E\Big((0,0,0),\big((\partial/\partial x_1)z_3,(\partial/\partial x_2)z_3,0\big)\Big);$$
$$(1-\tilde\Pi_\Omega)\tilde X_1^{(3)} = 0,\quad A_\Omega^{-1}\tilde X_1^{(3)} = -\eta\big((\partial/\partial x_1)z_3,(\partial/\partial x_2)z_3,0\big);$$
$$\tilde X_2^{(3)} = -\big(2(1+\nu)\big)^{-1}E\Big(\big(0,0(1-2\nu)^{-1}2\nu\Delta_{x'}z_3\big),\big(0,0,\pm(1-2\nu)^{-1}\nu\Delta_{x'}z_3\big)\Big),$$
$$(1-\tilde\Pi_\Omega)\tilde X_2^{(3)} = 0,\quad A_\Omega^{-1}\tilde X_2^{(3)} = (1-\nu)^{-1}\nu(\eta^2/2-1/24)(0,0,\Delta_{x'}z_3);$$
$$\tilde X_3^{(3)} = -\big(2(1-\nu)\big)^{-1}E$$
$$\times\big((1-\nu)^{-1}(2-\nu)\eta\Delta_{x'}(\partial/\partial x_1)z_3,(1-\nu)^{-1}(2-\nu)\eta\Delta_{x'}(\partial/\partial x_2)z_3,0\big),$$
$$\Big(\big(2(1-\nu)\big)^{-1}\Delta_{x'}(\partial/\partial x_1)z_3,\big(2(1-\nu)\big)^{-1}\Delta_{x'}(\partial/\partial x_2)z_3,0\Big),$$
$$(1-\tilde\Pi_\Omega)\tilde X_3^{(3)} = 0;$$
$$A_\Omega^{-1}\tilde X_3^{(3)} = \big((1-\nu)^{-1}(2-\nu)\eta^3/6-(1-\nu)^{-1}\eta/4\big)\big(\Delta_{x'}(\partial/\partial x_1)z_3,\Delta_{x'}(\partial/\partial x_2)z_3,0\big);$$
$$\tilde X_4^{(3)} = -\big(2(1-\nu)\big)^{-1}E\bigg(\Big(0,0,\big((1-2\nu)^{-1}(1+\nu)\eta^2+\big(24(1-\nu)(1-2\nu)\big)^{-1}$$
$$\times(4\eta^2-\nu-6)\big)\Delta_{x'}^2z_3\Big),\Big(0,0,\pm\big(12(1-2\nu)\big)^{-1}\nu(\nu-1)^{-1}(\nu+1)\Delta_{x'}^2z_3\Big)\bigg),$$
$$(1-\tilde\Pi_\Omega)\tilde X_4^{(3)} = -\Big(0,0,\big(12(1-\nu^2)\big)^{-1}E\Delta_{x'}^2z_3\Big). \tag{7}$$

Thus, the system of differential equations in G corresponding to the third limiting problem is of the form

$$-\big(2(1+\nu)\big)^{-1}E\Big\{\Delta_{x'}z_1(x') + (1-\nu)^{-1}(1+\nu)(\partial/\partial x_1)\big((\partial/\partial x_1)z_1(x')$$
$$+(\partial/\partial x_2)z_2(x')\big)\Big\} = E_1(x'), \tag{8}$$
$$-\big(2(1+\nu)\big)^{-1}E\Big\{\Delta_{x'}z_2(x') + (1-\nu)^{-1}(1+\nu)(\partial/\partial x_2)\big((\partial/\partial x_1)z_1(x')$$
$$+(\partial/\partial x_2)z_2(x')\big)\Big\} = E_2(x'), \tag{9}$$
$$\big(12(1-\nu^2)\big)^{-1}E\,\Delta_{x'}^2z_3(x') = E_3(x'),\quad x'\in G. \tag{10}$$

Remark 16.4.1. System (8)–(10) decomposes into the following two: the system (8), (9) of plane stress state and equation (10).

16.4.4 The boundary layer

We turn to the construction of boundary conditions for equations (8)–(10). It necessitates analyzing the second limit problem. To this end in a small neighborhood $\tilde O$ of the boundary ∂G we rewrite the original problem into coordinates (s,τ,ξ), where $\xi = x_3$, τ is the distance to ∂G along the inward normal, and s is a coordinate on ∂G. After the change of variables $(s,\tau,\xi) \to (s,t,\eta) = (s,\varepsilon^{-1}\tau,\varepsilon^{-1}\xi)$ the matrix operator $\tilde{L}(s,\tau,\partial/\partial s,\partial/\partial\tau,\partial/\partial\xi)$ of system (1) and the matrix operator $\tilde{B}(s,\tau,\partial/\partial s,\partial/\partial\tau,\partial/\partial\xi)$ of boundary conditions (2) can be expanded into the series in degrees of ε

$$\begin{aligned}\tilde{L}(s,\tau,\partial/\partial s,\partial/\partial\tau,\partial/\partial\xi) &\sim \varepsilon^{-2}\sum_{k=0}^{\infty}\varepsilon^k P^{(k)}(s,t,\partial/\partial s,\partial/\partial t,\partial/\partial\eta),\\ \tilde{B}(s,\tau,\partial/\partial s,\partial/\partial\tau,\partial/\partial\xi) &\sim \varepsilon^{-1}\sum_{k=0}^{\infty}\varepsilon^k Q^{(k)}(s,t,\partial/\partial s,\partial/\partial t,\partial/\partial\eta).\end{aligned}\tag{11}$$

Here $P^{(k)}$ and $Q^{(k)}$ are 3×3-matrix differential operators of orders no greater than 2 and 1, respectively. Their coefficients depend smoothly on $s\in\partial G$ and polynomially on t. There hold the equalities

$$\begin{aligned}&P^{(0)}(\partial/\partial t,\partial/\partial\eta)w\\ &= \big(2(1+\nu)\big)^{-1}E\begin{pmatrix}\Delta_{t\eta}\,w_s\\ \Delta_{t\eta}\,w_\tau + (1-2\nu)^{-1}(\partial/\partial t)\big((\partial/\partial\tau)\,w_\tau + (\partial/\partial\eta)\,w_\xi\big)\\ \Delta_{t\eta}\,w_\xi + (1-2\nu)^{-1}(\partial/\partial\eta)\big((\partial/\partial\tau)\,w_\tau + (\partial/\partial\eta)\,w_\xi\big)\end{pmatrix},\\ &Q^{(0)}(\partial/\partial t,\partial/\partial\eta)w\\ &= \big(2(1+\nu)\big)^{-1}E\begin{pmatrix}(\partial/\partial\eta)\,w_s\\ (\partial/\partial\eta)\,w_\tau + (\partial/\partial t)\,w_\xi\\ 2\partial/\partial\eta)\,w_\xi + (1-2\nu)^{-1}\big((\partial/\partial t)\,w_\tau + (\partial/\partial\eta)\,w_\xi\big)\end{pmatrix},\end{aligned}\tag{12}$$

where w_s, w_τ, and w_ξ are the components of the vector w written in the coordinates (s,τ,ξ).

Remark 16.4.2. (i) If a part Γ of ∂G is an interval, then for $s\in\Gamma$ the operators $P^{(k)}$ and $Q^{(k)}$ with $k\geq 1$ in (29), 16.3, take the form

$$P^{(1)}(\partial/\partial s,\partial/\partial t,\partial/\partial\eta) = \big(2(1+\nu)(1-2\nu)\big)^{-1}E$$
$$\times\begin{pmatrix}0 & (\partial/\partial s)(\partial/\partial t) & (\partial/\partial s)(\partial/\partial\eta)\\ (\partial/\partial t)(\partial/\partial s) & 0 & 0\\ (\partial/\partial\eta)(\partial/\partial s) & 0 & 0\end{pmatrix},$$

$$P^{(2)}(\partial/\partial s) = \big(2(1+\nu)\big)^{-1}E\begin{pmatrix}2\nu(1-2\nu)^{-1} & 0 & 0\\ 0 & 1 & 0\\ 0 & 0 & 1\end{pmatrix}(\partial/\partial s)^2,$$

$$Q^{(1)}(\partial/\partial s) = \big(2(1+\nu)\big)^{-1}E\begin{pmatrix}0 & 0 & 1\\ 0 & 0 & 0\\ 2\nu(1-2\nu)^{-1} & 0 & 0\end{pmatrix}(\partial/\partial s),$$

$$Q^{(k)} = P^{(k+1)} = 0,\quad k=2,3,4.$$

(ii) Let a part Γ of ∂G coincide with a circle of radius R. If $s=\vartheta$, then for $s\in\Gamma$ the components of $P^{(k)}$ with $k\geq 1$ are given by

$$\begin{aligned}P^{(k)}_{ss} &= \big(2(1+\nu)\big)^{-1}E\\ &\quad\times\{(k-1)\,R^{-k}\,t^{k-2}\,(2(1-2\nu)^{-1}(1-\nu)(\partial/\partial s)^2 - 1) - R^{-k}\,t^{k-1}(\partial/\partial t)\},\end{aligned}$$

$$P^{(k)}_{s\tau} = \big(2(1+\nu)\big)^{-1} E\{(1-2\nu)^{-1}(3-4\nu)(k-1)R^{-k}t^{k-2}(\partial/\partial s) - (1-2\nu)^{-1}R^{-k}t^{k-1}(\partial/\partial t)(\partial/\partial s)\},$$

$$P^{(k)}_{s\xi} = \big(2(1+\nu)\big)^{-1}(1-2\nu)^{-1}E\,R^{-k}t^{k-1}(\partial/\partial\xi)(\partial/\partial s)\},\quad P^{(k)}_{\tau\xi} = 0,$$

$$P^{(k)}_{\tau s} = \big(2(1+\nu)\big)^{-1} E\{(1-2\nu)^{-1}R^{-k}t^{k-1}(\partial/\partial t)(\partial/\partial s) + (1-2\nu)^{-1}(3-4\nu)(k-1)R^{-k}t^{k-2}(\partial/\partial s)\},$$

$$P^{(k)}_{\tau\tau} = \big(2(1+\nu)\big)^{-1} E\{2(1-2\nu)^{-1}(1-\nu)R^{-k}t^{k-1}(\partial/\partial t) + (k-1)R^{-k}t^{k-2}\big(2(1-2\nu)^{-1}(1-\nu) - (\partial/\partial s)^2\big)\},$$

$$P^{(k)}_{\xi\xi} = 0,\quad P^{(k)}_{\xi s} = -\big(2(1+\nu)\big)^{-1}E(1-2\nu)^{-1}(3-4\nu)(k-1)R^{-k}t^{k-2}(\partial/\partial s),$$

$$P^{(k)}_{\xi\tau} = \big(2(1+\nu)\big)^{-1} E\{2(1-2\nu)^{-1}(1-\nu)R^{-k}t^{k-1}(\partial/\partial t) + (k-1)R^{-k}t^{k-2}\big(2(1-2\nu)^{-1}(1-\nu) - (\partial/\partial s)^2\big)\}.$$

The operators $Q^{(k)}$ for $k \geq 1$ and $s \in \Gamma$ are of the form

$$Q^{(k)} = \big(2(1+\nu)\big)^{-1}ER^{-k}t^{k-1}\begin{pmatrix} 0 & 0 & \partial/\partial s \\ 0 & 0 & 0 \\ \partial/\partial s & -1 & 0 \end{pmatrix}.$$

We return to the construction of the boundary layer near the lateral surface S_ε of the plate $\tilde{C}_\varepsilon$. From the expansions (11) of the operators $\tilde{L}$ and $\tilde{B}$ in (1) and (2) it follows that the boundary layer is described by means of the boundary value problem

$$P^{(0)}(\partial/\partial t, \partial/\partial\eta)w(t,\eta) = \tilde{H}(t,\eta),\quad (t,\eta)\in\omega; \tag{13}$$

$$Q^{(0)}(\partial/\partial t, \partial/\partial\eta)w(t,\pm 1/2) = \tilde{K}^{\pm}(t),\quad t\in(0,\infty); \tag{14}$$

$$w(0,\eta) = \tilde{K}(\eta),\quad |\eta| < 1/2, \tag{15}$$

in the half-strip $\omega = (0,\infty)\times(-1/2,1/2)$. (In view of (12), problem (13)–(15) decomposes into the following two: the problem of antiplanar deformation (torsion) for w_s and the two-dimensional problem in elasticity theory for w_τ, w_ξ.)

We study the compatibility conditions for problem (13)–(15) in the class of exponentially decaying functions.

Lemma 16.4.3. *The line* $\operatorname{Im}\lambda = 0$ *contains only one eigenvalue* $\lambda = 0$ *of the boundary value problem with complex parameter*

$$P^{(0)}(i\lambda, \partial/\partial\eta)W(\eta) = 0,\quad \eta\in(-1/2,1/2);\quad Q^{(0)}(i\lambda,\partial/\partial\eta)W(\pm 1/2) = 0. \tag{16}$$

To the eigenvalue there correspond three eigenvectors $\Phi^{(j,0)}(\eta) = \phi^{(j)}, j = 1,2,3.$ *Associated vectors are of the form*

$$\begin{aligned} &\Phi^{(1,1)}(\eta) = (0,0,0),\quad \Phi^{(2,1)}(\eta) = \big(0,0,-(1-\nu)^{-1}i\nu\eta\big),\\ &\Phi^{(3,1)}(\eta) = (0,-i\eta,0),\quad \Phi^{(3,2)}(\eta) = \Big(0,0,-\big(2(1-\nu)\big)^{-1}\nu\eta^2\Big),\\ &\Phi^{(3,3)}(\eta) = \Big(0,-i\big(6(1-\nu)\big)^{-1}(2-\nu)\eta^3 + \big(4(1-\nu)\big)^{-1}i\eta, 0\Big). \end{aligned} \tag{17}$$

Proof. Let $W^{(0)}$ satisfy the homogeneous problem (16). We multiply the system of equations (16) by $W^{(0)}$ (in the sense of scalar product) and integrate by parts over $(-1/2, 1/2)$. For real λ

$$0 = \int_{-1/2}^{1/2} \overline{W^{(0)}(\eta)}\, P^{(0)}(i\lambda, \partial/\partial\eta) W^{(0)}(\eta) d\eta$$

$$= \big(2(1+\nu)\big)^{-1} E \int_{-1/2}^{1/2} \big(|(\partial/\partial\eta)W_s^{(0)}(\eta)|^2 + \lambda^2 |W_s^{(0)}(\eta)|^2 + |(\partial/\partial\eta)W_\xi^{(0)}(\eta)|^2$$
$$+ \lambda^2 |W_\tau^{(0)}(\eta)|^2 + |(\partial/\partial\eta)W_\tau^{(0)}(\eta) + i\lambda W_\xi^{(0)}(\eta)|^2$$
$$+ 2(1-2\nu)^{-1}\nu |(\partial/\partial\eta)W_\xi^{(0)}(\eta) + i\lambda W_\tau^{(0)}(\eta)|^2 \, d\eta.$$

Therefore, $W^{(0)} = 0$ for $\lambda \neq 0$ and $W^{(0)} = (c_1, c_2, c_3)$ in the case $\lambda = 0$.

Thus, $\lambda = 0$ is the only eigenvalue of the operator $\hat{A}(\lambda) = \big(P^{(0)}(i\lambda, \partial/\partial\eta), Q^{(0)}(i\lambda, \partial/\partial\eta)\big)$ located on the line $\operatorname{Im}\lambda = 0$ and to the eigenvalue there correspond the eigenvectors

$$\Phi^{(1,0)}(\eta) = (1,0,0), \quad \Phi^{(2,0)}(\eta) = (0,1,0), \quad \Phi^{(3,0)}(\eta) = (0,0,1).$$

In order to check (17) one should directly solve the boundary value problems

$$\hat{A}(0)\Phi^{(j,p)} = -(\partial/\partial\lambda)\hat{A}(0)\Phi^{(j,p-1)} - 2^{-1}(\partial/\partial\lambda)^2 \hat{A}(0)\Phi^{(j,p-2)}, \tag{18}$$

with $p \in \mathbb{N}, j = 1,2,3$ and $\Phi^{(j,-1)} = 0$. Problem (18) for $j = 1,2, p = 1$ or for $j = 3, p = 1,2,3$ has the solution (17) while for $j = 1,2, p = 2$ and for $j = 3, p = 4$ is not solvable. □

Remark 16.4.4. In the strip $\Pi = \mathbb{R} \times (-1/2, 1/2)$ we consider the boundary value problem

$$\begin{aligned} P^{(0)}(\partial/\partial t, \partial/\partial\eta)\tilde{\Phi}(t,\eta) &= \tilde{F}(t,\eta), \quad (t,\eta) \in \Pi; \\ Q^{(0)}(\partial/\partial t, \partial/\partial\eta)\tilde{\Phi}(t, \pm 1/2) &= \tilde{G}^{\pm}(t), \qquad t \in \mathbb{R}. \end{aligned} \tag{19}$$

Theorem 3.2.4 and Lemma 16.4.2 imply that any solution to the homogeneous problem (19) having a polynomial growth as $t \to \pm\infty$ turns out to be a linear combination of the vector-valued functions

$$\begin{aligned} &V^{(1,0)}(t,\eta) = (1,0,0), \quad V^{(1,1)}(t,\eta) = (t,0,0), \quad V^{(2,0)}(t,\eta) = (0,1,0), \\ &V^{(2,1)}(t,\eta) = (0, t, -(1-\nu)^{-1}\nu\eta), \quad V^{(3,0)}(t,\eta) = (0,0,1), \\ &V^{(3,1)}(t,\eta) = (0, -\eta, t), \quad V^{(3,2)}(t,\eta) = (0, -t\eta, t^2/2 + (1-\nu)^{-1}\nu\eta^2/2), \\ &V^{(3,3)}(t,\eta) \\ &= (0, -t^2\eta/2 + (1-\nu)^{-1}(2-\nu)\eta^3/6 - (1-\nu)^{-1}\eta/4, t^3/6 + (1-\nu)^{-1}\nu t\eta^2/2). \end{aligned} \tag{20}$$

Proposition 16.4.5. *Let the right-hand sides of problem* (13)–(15) *satisfy the conditions*

$$\tilde K \in \mathbf{W}_2^{l/2}(-1/2,1/2), \quad \tilde K^{\pm} \in \mathbf{L}_2(0,\infty), \quad t\tilde H \in \mathbf{L}_2(\omega),$$
$$\tilde K^{\pm} \in \mathbf{C}^{\infty}[\delta,\infty), \quad \tilde H \in \mathbf{C}^{\infty}([\delta,\infty)\times[-1/2,1/2]),$$
$$(\partial/\partial t)^{q_1}\tilde K^{\pm}(t) = O\big(\exp(-\beta t)\big),$$
$$(\partial/\partial t)^{q_1}(\partial/\partial\eta)^{q_2}\tilde H(t,\eta) = O\big(\exp(-\beta t)\big), \quad t>\delta.$$

Then there exists a unique solution w *to problem* (13)–(15) *possessing the finite energy*

$$\tilde E(w) = \big(2(1+\nu)\big)^{-1} E \int_{\omega} \big(|\partial w_s/\partial t|^2 + |\partial w_s/\partial\eta|^2 + |\partial w_\tau/\partial t|^2 + |\partial w_\xi/\partial\eta|^2 + 2(1-2\nu)^{-1}\nu\,|\partial w_\tau/\partial t + \partial w_\xi/\partial\eta|^2 + |\partial w_\tau/\partial\eta + \partial w_\xi/\partial t|^2\big)\,dt\,d\eta. \tag{21}$$

For the solution the asymptotic representation

$$w(t,\eta) = (C_s, C_t - \tilde C_\eta, C_\eta + \tilde C t) + O\big(\exp(-\beta_0 t)\big), \quad t>\delta, \tag{22}$$

holds with some constants $C_s, C_t, C_\eta, \tilde C$ *and* $0<\beta_0\le\beta$.

Proof. It is not hard to see that any field with zero energy (21) is a linear combination of the vectors $(1,0,0)$, $(0,1,0)$, $(0,0,1)$, and $(0,-\eta,t)$, which is a rigid displacement. These vectors do not vanish for $t=0$. Therefore, the quadratic form (21) corresponding to problem (13)–(15) is positive definite. This provides the existence of the desired solution.

To verify (22) we introduce $\tilde\Phi = (1-\chi)w$, where χ is a function in $\mathbf{C}^{\infty}(\mathbb{R})$ such that $\chi(t)=1$ for $t\le 1/2$ and $\chi(t)=0$ for $t\ge 1$. Using the local estimates on solutions (see Theorem 3.1.2), we obtain $\tilde\Phi \in \mathbf{C}^{\infty}(\bar\Pi)$. Since $\tilde E((1-\chi)w)<\infty$, we find that $\tilde\Phi \in \mathbf{W}^l_{2,\gamma}(\Pi)$ for any $\gamma<0$ and $l\in\mathbb{N}_0$. Lemma 16.4.3 and Theorem 3.2.4 imply that

$$(1-\chi)w - \sum_{j=1}^{3}\sum_{k=0}^{K_j} c_{jk}V^{(j,k)} \in \mathbf{W}^l_{2,\gamma_1}(\Pi), \quad K_1=K_2=1, \quad K_3=3, \tag{23}$$

where $V^{(j,k)}$ are the functions (20), c_{jk} some constants, and γ_1 is a positive number. Because only $V^{(j,k)}$ with $j=1,2,3$, $k=0$ or $j=3$, $k=1$ possess finite energy (21) we have $c_{11}=c_{21}=c_{32}=c_{33}=0$ in (23). Putting $C_s=c_{10}, C_t=c_{20}, C_\eta=c_{30}$, and $\tilde C = c_{31}$, from (23) and the embedding theorem we derive (22). □

We introduce the solutions to the homogeneous boundary value problem (13)–(15) corresponding to the vector-valued functions (20):

$$\begin{aligned} &W^{(s)}(t,\eta) = V^{(1,1)}(t,\eta), \quad W^{(t)}(t,\eta) = V^{(2,1)}(t,\eta) + \Xi^{(t)}(t,\eta),\\ &W^{(\eta)}(t,\eta) = V^{(3,3)}(t,\eta) + \Xi^{(\eta)}(t,\eta),\\ &V(t,\eta) = V^{(3,2)}(t,\eta) + \Xi(t,\eta), \end{aligned} \tag{24}$$

where $\Xi^{(t)}$ and $\Xi^{(\eta)}$ are the functions with finite integrals (21) satisfying (13)–(15) for $\tilde{H} = 0, \tilde{K}^{\pm} = 0$, and the right-hand sides in the boundary condition are given by

$$\tilde{K} = \big(0, 0, -(1-\nu)^{-1}\nu\eta\big)$$
$$\tilde{K} = \Big(0, (1-\nu)^{-1}(2-\nu)\,\eta^3/6 - (1-\nu)^{-1}\eta/4, 0\Big),$$
$$\tilde{K} = \big(0, 0, (1-\nu)^{-1}\nu\eta^2/2\big).$$

We define the constants in the asymptotics (22) with the help of Theorem 3.2.5. To this end we need Green's formula in the rectangle $\Pi_R = [0, R] \times [-1/2, 1/2]$:

$$\begin{aligned}
&\int\limits_{\Pi_R} \Big(w(t,\eta)\cdot P^{(0)}(\partial/\partial t,\partial/\partial\eta)W(t,\eta) - W(t,\eta)\cdot P^{(0)}(\partial/\partial t,\partial/\partial\eta)w(t,\eta)\Big)\,dt\,d\eta \\
&= \int\limits_0^R \Big(w(t,\eta)\cdot Q^{(0)}(\partial/\partial t,\partial/\partial\eta)W(t,\eta) - W(t,\eta)\cdot Q^{(0)}(\partial/\partial t,\partial/\partial\eta)w(t,\eta)\Big)\Big|_{\eta=-1/2}^{\eta=1/2} dt \\
&+ \int\limits_{-1/2}^{1/2} \Big(W(0,\eta)\cdot\tilde{Q}(\partial/\partial t,\partial/\partial\eta)w(0,\eta) - w(0,\eta)\cdot\tilde{Q}(\partial/\partial t,\partial/\partial\eta)W(0,\eta)\Big)\,d\eta \\
&+ \int\limits_{-1/2}^{1/2} \Big(w(R,\eta)\cdot\tilde{Q}(\partial/\partial t,\partial/\partial\eta)W(R,\eta) - W(R,\eta)\cdot\tilde{Q}(\partial/\partial t,\partial/\partial\eta)w(R,\eta)\Big)\,d\eta
\end{aligned} \tag{25}$$

with

$$\tilde{Q}(\partial/\partial t, \partial/\partial\eta) = \big(2(1+\nu)\big)^{-1}E$$
$$\times \begin{pmatrix} \partial/\partial t & 0 & 0 \\ 0 & 2(1-2\nu)^{-1}(1-\nu)\partial/\partial t & 2(1-2\nu)^{-1}\nu\partial/\partial\eta \\ 0 & \partial/\partial\eta & \partial/\partial t \end{pmatrix}.$$

Proposition 16.4.6. *Let the assumptions of Proposition* 16.4.5 *be fulfilled. Then the constants* $C_s, C_t, C_\eta, \tilde{C}$ *in the asymptotics* (22) *of the solution* w *are defined by*

$$C_s = \int\limits_{-1/2}^{1/2} \tilde{K}_s(\eta)\,d\eta + 2(1+\nu)E^{-1}$$
$$\times \sum_{\pm} \pm \int\limits_0^{+\infty} \tilde{K}_s^{\pm}(t)\,ds - 2(1+\nu)E^{-1}\int\limits_\omega t\tilde{H}_s(t,\eta)\,dt\,d\eta, \tag{26}$$

$$\begin{aligned}
C_t &= (1-\nu^2)E^{-1}\Lambda(W^{(t)}; \tilde{H}, \tilde{K}^{\pm}, \tilde{K}),\\
C_\eta &= 12(1-\nu^2)E^{-1}\Lambda(W^{(\eta)}; \tilde{H}, \tilde{K}^{\pm}, \tilde{K}),\\
\tilde{C} &= 12(1-\nu^2)E^{-1}\Lambda(\tilde{V}; \tilde{H}, \tilde{K}^{\pm}, \tilde{K}),
\end{aligned} \tag{27}$$

where

$$\Lambda(W; \tilde{H}, \tilde{K}^{\pm}, \tilde{K}) = \sum_{\pm} \pm \int\limits_0^{+\infty} \Big(W_\tau(t, \pm 1/2)\tilde{K}_\tau^{\pm}(t) + W_\xi(t, \pm 1/2)\tilde{K}_\xi^{\pm}(t)\Big)\,dt$$

$$+\left(2(1+\nu)\right)^{-1}E\int_{-1/2}^{1/2}\Big\{\Big(2(\partial/\partial t)W_\tau(0,\eta)$$

$$+2(1-2\nu)^{-1}\nu\big((\partial/\partial t)W_\tau(0,\eta)+(\partial/\partial\eta)W_\xi(0,\eta)\big)\Big)\tilde{K}_\tau(\eta)$$

$$+\big((\partial/\partial\eta)W_\tau(0,\eta)+(\partial/\partial t)W_\xi(0,\eta)\big)\tilde{K}_\xi(\eta)\Big\}\,d\eta$$

$$+\int_\omega(W_\tau(t,\eta)\tilde{H}_\tau(t,\eta)+W_\xi(t,\eta)\tilde{H}_\xi(t,\eta)dt\,d\eta.$$

Proof. Denote by $I_1(R,W)$ the sum of the first three integrals in (25) and by $I_2(R,W)$ the last integral. We verify (26). Since $W^{(s)}$ satisfies the homogeneous equations (13)–(15), the integral on the left in (25) vanishes and the limit of $I_1(R,W^{(s)})$ as $R\to\infty$ is equal to the product of $\left(2(1+\nu)\right)^{-1}E$ and the right-hand side in (26). It remains to note that by virtue of (22) and the definition of $W^{(s)}$

$$I_2(R,W^{(s)})=\left(2(1+\nu)\right)^{-1}E\int_{-1/2}^{1/2}(C_s+o(1))d\eta=\left(2(1+\nu)\right)^{-1}EC_s+o(1)$$

and to let $R\to\infty$.

Now we consider relations (28). They arise as a result of substituting $W^{(t)}$, $W^{(\tau)}$, and $\tilde{V}$ into (25) and letting $R\to\infty$. Indeed, the aforementioned vector-valued functions satisfy the homogeneous problem (13)–(15). Therefore, the right-hand sides in (26) with W replaced by $W^{(t)}, W^{(\eta)}$ or $\tilde{V}$ are equal to zero. As $R\to\infty$,

$$I_1(R,W^{(t)})\to\Lambda(W^{(t)};\tilde{H},\tilde{K}^\pm,\tilde{K}),$$
$$I_1(R,W^{(\eta)})\to\Lambda(W^{(\eta)};\tilde{H},\tilde{K}^\pm,\tilde{K}),$$
$$I_1(R,\tilde{V})\to\Lambda(\tilde{V};\tilde{H},\tilde{K}^\pm,\tilde{K}).$$

Besides, using the asymptotics (22) of the solution w to problem (13)–(15), we find

$$I_2(R,W^{(t)})=\left(2(1+\nu)\right)^{-1}E\int_{-1/2}^{1/2}(2(C_t-\tilde{C}\eta)(1-\nu)^{-1}+o(1))d\eta$$
$$=(1-\nu^2)^{-1}EC_t+o(1),$$

$$I_2(R,W^{(\eta)})=\left(2(1+\nu)\right)^{-1}E\int_{-1/2}^{1/2}((\tilde{C}\eta-C_t)(1-\nu)^{-1}2R\eta+(C_\eta+\tilde{C}R)$$
$$\times(1-\nu)^{-1}(4\eta^2-1)/4+o(1))d\eta=-(1-\nu^2)^{-1}EC_\eta/12+o(1),$$

$$I_2(R,\tilde{V})=\left(2(1+\nu)\right)^{-1}E\int_{-1/2}^{1/2}((\tilde{C}\eta-C_t)(1-\nu)^{-1}2\eta+o(1))d\eta$$
$$=(1-\nu^2)^{-1}E\tilde{C}/12+o(1).$$

□

16.4.5 Boundary conditions in the third limit problem

From Propositions 16.4.5 and 16.4.6 it follows that the solution to problem (13)–(15) decreases provided the right-hand sides in (26) and (27) vanish. We derive the system of equations on ∂G forming the third limit problem (31), 16.1.

The projector $\tilde{\Pi}_\omega$ (see Section 16.1.4) is defined by

$$\begin{aligned}
&\tilde{\Pi}_\omega(\tilde{H},\tilde{K}^\pm,\tilde{K}) = (\tilde{H},\tilde{K}^\pm,\tilde{K}-\hat{K}),\\
&\quad\hat{K} = (\Lambda^{(s)}(\tilde{H},\tilde{K}^\pm,\tilde{K}), (1-\nu^2)\,E^{-1}\Lambda(W^{(t)};\tilde{H},\tilde{K}^\pm,\tilde{K}) - 12(1-\nu^2)E^{-1}\\
&\quad\times\Lambda(\tilde{V};\tilde{H},\tilde{K}^\pm,\tilde{K}), -12(1-\nu^2)E^{-1}\Lambda(W^{(\eta)};\tilde{H},\tilde{K}^\pm,\tilde{K})),
\end{aligned}\tag{28}$$

where $\Lambda^{(s)}(\tilde{H},\tilde{K}^\pm,\tilde{K})$ stands for the right-hand side in (26). The operator A_ω^{-1} sends the right-hand side $(\tilde{H},\tilde{K}^\pm,\tilde{K})$ of problem (13)–(15) to the solution with finite integral (21). From (22) and (26)–(28) it follows that the function $A_\omega^{-1}\Pi_\omega(\tilde{H},\tilde{K}^\pm,\tilde{K})$ exponentially decreases as $t\to\infty$. The expansion (13), 16.1, of the operator of problem (1)–(3) is given in (11) (see also Remark 16.4.2).

Now we find the discrepancies given by the auxiliary vectors $\tilde{Z}_j^{(q)} = A_\omega^{-1}\tilde{X}_j^{(q)}$ (see (6) and (7)) in the boundary conditions (3) and define the new auxiliary vectors $R_j^{(q)}$ and $\hat{W}_j^{(q)}$ (see (27), 16.1, and (28), 16.1). We have

$$\begin{aligned}
\hat{W}_0^{(1)} &= -(0,0,(z_{1s}|_{\partial G}, z_{1\tau}|_{\partial G},0)), \quad \tilde{\Pi}_\omega\hat{W}_0^{(1)} = 0, \quad R_0^{(1)} = 0;\\
\hat{W}_0^{(2)} &= -(0,0,(z_{2s}|_{\partial G}, z_{2\tau}|_{\partial G},0)), \quad \tilde{\Pi}_\omega\hat{W}_0^{(2)} = 0, \quad R_0^{(2)} = 0;\\
\hat{W}_0^{(3)} &= (0,0,(0,0,z_3|_{\partial G})), \quad \tilde{\Pi}_\omega\hat{W}_0^{(3)} = 0, \quad R_0^{(3)} = 0;\\
\hat{W}_1^{(3)} &= \Big(0,0,(\eta(\partial/\partial s)z_3|_{\partial G}, \eta(\partial/\partial\tau)z_3|_{\partial G},0))\Big),\\
\tilde{\Pi}_\omega\hat{W}_0^{(1)} &= \Big(0,0,(\eta(\partial/\partial s)z_3|_{\partial G},0,0))\Big),\\
R_1^{(3)} &= (\Xi(t,\tau)(\partial/\partial s)z_3|_{\partial G},0,0,),
\end{aligned}$$

where $(z_{1s},z_{1\tau})$ and $(z_{2s},z_{2\tau})$ are the vectors $(z_1,0)$ and $(0,z_2)$ written in the coordinates (s,τ), Ξ a function harmonic in ω such that $(\partial/\partial\eta)\Xi(t,\pm1/2) = 0$ for $t\in(0,\infty)$ and $\Xi(0,\eta)=\eta$ for $|\eta|<1/2$. Thus,

$$\begin{aligned}
(1-\tilde{\Pi}_\omega)\hat{W}_0^{(q)} &= -(0,0,e^{(q)})z_q|_{\partial G},\\
(1-\tilde{\Pi}_\omega)\hat{W}_1^{(3)} &= -(0,0,e^{(2)})(\partial/\partial\tau)z_3|_{\partial G},
\end{aligned}$$

$e^{(q)}$ being the basis vectors for $\mathbb{R}^3$ in the coordinates (s,τ,ξ). By (28), the vectors $(0,0,e^{(q)}), q=1,2,3$, and $(0,0,\eta e^{(2)})$ form a basis in the subspace coker A_ω. Hence, the boundary conditions for z_1,z_2,z_3 satisfying (8)–(10) are of the form

$$z_1(x') = \hat{C}_1(x'), \qquad z_2(x') = \hat{C}_2(x'), \quad x'\in\partial G;\tag{29}$$

$$z_3(x') = \hat{C}_3(x'), \quad (\partial/\partial\tau)z_3(x') = \hat{C}_4(x'), \quad x'\in\partial G.\tag{30}$$

16.4.6 Asymptotics of the solution

We have constructed the third limit problem. Let us formulate the assertion on the asymptotic expansion of the solution to problem (1)–(3) that follows from Theorem 16.2.5.

Theorem 16.4.7. *Let $F_j(\varepsilon,x) = -\varepsilon^{-2} f_j(\varepsilon,x',\varepsilon^{-1}\xi)$ and $P_j^{\pm}(\varepsilon,x') = \varepsilon^{-1} p_j^{\pm}(\varepsilon,x')$ while f_j and $p_j^{\pm}, j=1,2,3$, belong to $\mathbf{C}^\infty([0,1]\times\bar{G}\times[-1/2,1/2])$ and $\mathbf{C}^\infty([0,1]\times\bar{G}])$, respectively, $j=1,2,3$. Then for the solution to problem (1)–(3) there holds the inequality*

$$||u - U^{(N)}; \mathbf{W}_2^l(\tilde{C}_3)|| \le c_N \varepsilon^{N+1/2}.$$

Here $n \in \mathbb{N}$ is an arbitrary positive integer, the constant c_N independent of ε,

$$\begin{aligned} U^{(N)}(\varepsilon,x') = {} & \sum_{n=0}^{N} \varepsilon^n \big(v^{(n)}(x',\varepsilon^{-1}\xi) + \chi(x') w^{(n)}(s,\varepsilon^{-1}\tau,\varepsilon^{-1}\xi)\big) \\ & + \sum_{n=0}^{N+4} \Big\{ \varepsilon^{n-4}(0,0,z_3^{(n)}(x')) + \sum_{j=1}^{3} \varepsilon^{n-j} (\tilde{Z}_j^{(3,n)}(x',\varepsilon^{-1}\xi) \\ & + \chi(x') R_j^{(3,n)}(s,\varepsilon^{-1}\tau,\varepsilon^{-1}\xi)) \Big\} \\ & + \sum_{n=0}^{N+2} \Big\{ \varepsilon^{n-2}(z_1^{(n)}(x'), z_2^{(n)}(x'),0) + \sum_{q=1}^{2} \varepsilon^{n-1} (\tilde{Z}_1^{(q,n)}(x',\varepsilon^{-1}\xi) \\ & + \chi(x') R_1^{(q,n)}(s,\varepsilon^{-1}\tau,\varepsilon^{-1}\xi)) \Big\}, \end{aligned}$$

$v^{(n)}$ are some solutions to the first limit problem (4) with zero means over $(-1/2,1/2)$. The functions $w^{(n)}$ satisfy the second limit problem (13)–(15) and exponentially decrease. The vectors $z_1^{(n)}, z_2^{(n)}, z_3^{(n)}$ are solutions to the third limit problem (8)–(10), (29), (30) with some right-hand sides. The auxiliary vectors $\tilde{Z}_j^{(q,n)} = A_\Omega^{-1} \tilde{X}_j^{(q,n)}$ are defined for $z_1^{(n)}, z_2^{(n)}, z_3^{(n)}$ with the help of (6) and (7). The exponentially decaying solutions $R_j^{(q,N)}$ to problem (13)–(15) compensate the discrepancies in the boundary condition (3). Finally, χ is a cut-off function that equals 1 near ∂G and 0 outside the neighborhood $\tilde{O}$ with coordinates (s,τ).

Part VII

Elliptic Boundary Value Problems with Oscillating Coefficients or Boundary of Domain

Chapter 17

Elliptic Boundary Value Problems with Rapidly Oscillating Coefficients

Up to now, we have studied elliptic boundary value problems with perturbations located near isolated points or submanifolds of positive dimension. The last part of the book is devoted to rapidly oscillating perturbations extended over the whole domain or its boundary. At first glance the problems collected here essentially differ from those studied in the preceding chapters. Nevertheless, one can follow an analogy in the iterative procedures for the asymptotic expansions. It suffices to mention that as before the coefficients in the asymptotic series are found with the help of limit problems. Here the role of limit problems is played by a problem in the periodicity cell; a problem on a semicylinder with periodic boundary conditions on the lateral area that describes a boundary layer; a homogenized boundary value problem that arises to provide the compatibility conditions for the above limit problems.

The chapter contains the known method of homogenization of boundary value problems applied to the Dirichlet and Neumann problems for a second order elliptic equation with rapidly oscillating coefficients. In Section 17.1 we homogenize the differential equation without taking boundary conditions into account. The boundary layers are constructed in Sections 17.2 and 17.3. The asymptotics expansions are justified in Section 17.4. The final Section 17.5 contains auxiliary assertions on the solvability of elliptic boundary value problems with periodic coefficients on cylinders and on the asymptotics of their solutions.

17.1 Homogenization of Differential Equation

17.1.1 Statement of the problem

In the square $Q = (0,1) \times (0,1)$ we consider an elliptic equation of the form

$$\tilde{L}(\varepsilon, x, \partial/\partial x)u(\varepsilon, x) = (\partial/\partial x_j) l_{jk}(\varepsilon, x)(\partial/\partial x_k) u(\varepsilon, x) = f(\varepsilon, x), \quad x \in Q, \tag{1}$$

with small positive parameter $\varepsilon = N^{-1}$, N being a large positive integer. Everywhere in the chapter the indices written by Latin letters take values 1 and 2; we sum over repetitive indices. Suppose that the coefficients l_{jk} in (1) depend both on the slow variables x_1 and x_2 and on the fast variables $y_1 = \varepsilon^{-1}x_1$ and $y_2 = \varepsilon^{-1}x_2$, that is $l_{jk}(\varepsilon, x) = a_{jk}(x, \varepsilon^{-1}x)$. We also admit that the real functions a_{jk} are periodic with period 1 in x_1, y_1, y_2 and smooth for $x \in \bar{Q}$ and $y \in S = [0,1] \times [0,1]$. The matrix $A(x,y) = \big(a_{jk}(x,y)\big)_{j,k=1}^{2}$ is self-adjoint and uniformly positive definite:

$$c(\xi_1^2 + \xi_2^2) \le \xi_j a_{jk}(x,y)\xi_k \le C(\xi_1^2 + \xi_2^2), \quad \xi \in \mathbb{R}^2, \quad x \in \bar{Q}, \quad y \in S, \tag{2}$$

c and C being positive constants.

17.1.2 The limit problem in the cell

We construct a formal asymptotic solution to equation (1) with smooth right-hand side. Assume that the principal part u_0 of the asymptotics of the solution to problem (1) depends only on the slow variables x. We substitute u_0 into equation (1) and obtain

$$\begin{aligned}\tilde{L}(\varepsilon, x, \partial/\partial x)u_0(x) &= a_{jk}(x, \varepsilon^{-1}x)(\partial/\partial x_j)(\partial/\partial x_k)u_0(x)\\ &\quad + (\partial a_{jk}/\partial x_j)(x, \varepsilon^{-1}x)(\partial/\partial x_k)u_0(x)\\ &\quad + \varepsilon^{-1}(\partial/\partial y_j)a_{jk}(x, \varepsilon^{-1}x)(\partial/\partial x_k)u_0(x).\end{aligned} \tag{3}$$

The last term in (3) is of impermissible order $O(\varepsilon^{-1})$. Let us try to compensate it with a correction $\varepsilon u_1(x, \varepsilon^{-1}x)$ depending both on the slow and fast variables.

Lemma 17.1.1. *Let $\varphi(x, y)$ be a smooth function periodic in y. Then*

$$\begin{aligned}\tilde{L}(\varepsilon,x,\partial/\partial x)\varphi(x,\varepsilon^{-1}x) = \big(&L_0(x,y,\partial/\partial x)\varphi(x,y) + \varepsilon^{-1}L_1(x,y,\partial/\partial x,\partial/\partial y)\varphi(x,y)\\ &+\varepsilon^{-2}L_2(x,y,\partial/\partial y)\varphi(x,y)\big)\big|_{y=\varepsilon^{-1}x},\end{aligned} \tag{4}$$

where

$$\begin{aligned}L_0(x, y, \partial/\partial x) &= (\partial/\partial x_j)\, a_{jk}(x, y)\partial/\partial x_k,\\ L_1(x, y, \partial/\partial x, \partial/\partial y) &= (\partial/\partial x_j)\, a_{jk}(x, y)\partial/\partial y_k + (\partial/\partial y_j)\, a_{jk}(x, y)\partial/\partial x_k,\\ L_2(x, y, \partial/\partial y) &= (\partial/\partial y_j)\, a_{jk}(x, y)\partial/\partial y_k.\end{aligned} \tag{5}$$

The proof is a straightforward calculation.

Using (4), we find

$$\varepsilon\tilde{L}(\varepsilon x, x, \partial/\partial x)\, u_1(x, \varepsilon^{-1}x) = \varepsilon^{-1}L_2(x, y, \partial/\partial y)\, u_1(x, y)|_{y=\varepsilon^{-1}x} + O(1). \tag{6}$$

We compare the terms of order $O(\varepsilon^{-1})$ in (3) and (6) and obtain that u_1 must satisfy the equation

$$L_2(x, y, \partial/\partial y)\, u_1(x, y) = -(\partial/\partial y_j)\, a_{jk}(x, y)(\partial/\partial x_k)\, u_0(x), \quad y \in S. \tag{7}$$

We seek a solution to equation (7) in the space $\widehat{\mathbf{W}}_2^l(S)$ consisting of functions $U \in \mathbf{W}_2^l(S)$ that are subject to the conditions $(\partial/\partial y_2)^\mu U(y_1, 0) = (\partial/\partial y_2)^\mu U(y_1, 1)$, $(\partial/\partial y_1)^\mu U(0, y_2) = (\partial/\partial y_1)^\mu U(1, y_2)$, $y_j \in [0, 1], \mu = 0\dots, l-1$. The equation

$$L_2(x, y, \partial/\partial x)U(y) = F(y), \quad y \in S, \tag{8}$$

together with the conditions on ∂S is called the limit problem on the periodicity cell S.

Proposition 17.1.2. *Let $F \in \mathbf{L}_2(S)$ and $x \in \bar{Q}$. The equation (8) is solvable in the space $\widehat{\mathbf{W}}_2^1(S)$ if and only if*

$$\langle F\rangle :\, = \int_S F(y)dy = 0. \tag{9}$$

Under the complementary condition $F \in \widehat{\mathbf{W}}_2^l(S)$ we have $U \in \widehat{\mathbf{W}}_2^{l+2}(S)$.

Proof. The existence of the generalized solution $U \in \widehat{\mathbf{W}}_2^l(S)$ under condition (9) can be obtained by a traditional argument (see, for instance, LADYZHENSKAYA [1], LIONS and MAGENES [1]). Since equation (8) is self-adjoint, its compatibility condition consists in the orthogonality of the right-hand side F to the solutions U^0 of the homogeneous equation (8). From the periodicity conditions for U^0 we obtain

$$\int\limits_S a_{jk}(x,y)(\partial U^0/\partial y_j)(y)(\partial U^0/\partial y_k)(y)dy = 0.$$

In view of (2) it follows that $U^0 = \text{const}$. The verification of the inclusion $U \in \hat{\mathbf{W}}_2^{l+2}(S)$ for the right-hand side F in $\hat{\mathbf{W}}_2^l(S)$ is contained in Theorem 2.3.3 in BERS, JOHN and SCHECHTER [1]. □

Remark 17.1.3. A solution to equation (8) is determined up to an additive constant. From now on by a solution to equation (8), we mean the solution with mean value $\langle U \rangle$ in S equal to zero.

We return to equation (7). Since

$$\int\limits_S (\partial/\partial y_j)\, a_{jk}(x,y)dy = \int\limits_0^1 a_{jk}(x,y)|_{y_j=0}^{y_j=1} dy_p = 0,$$

there exists a solution of the form

$$p_1(x,y)(\partial/\partial x_1)\, u_0(x) + p_2(x,y)(\partial/\partial x_2)\, u_0(x)$$

while p_k has zero mean value and satisfies the equation

$$L_2(x,y,\partial/\partial y)\, p_k(x,y) = -(\partial/\partial y_j)\, a_{jk}(x,y), \quad y \in S. \tag{10}$$

Thus, the function u_1 is of the form

$$u_1(x,y) = v_1(x) + p_s(x,y)(\partial/\partial x_s)\, u_0(x). \tag{11}$$

17.1.3 The homogenized equation

We substitute the sum $u_0(x) + \varepsilon u_1(x,y) + \varepsilon^2 u_2(x,y)$ into equation (1). Taking (4) into account and separating out the terms of order $O(1)$, we find

$$L_0(x,y,\partial/\partial x)u_0(x) + L_1(x,y,\partial/\partial x,\partial/\partial y)u_1(x,y) + L_2(x,y,\partial/\partial y)u_2(x,y) = f_0(x,y).$$

In so doing we suppose that $f(\varepsilon,x) = f_0(x,\varepsilon^{-1}x) + O(\varepsilon)$, the function f_0 smoothly depends on $x \in \bar{Q}$ and is periodic in y and x_1.

We set

$$u_2(x,y) = v_2(x) + p_s(x,y)(\partial/\partial x_s)\, v_1(x) + w_2(x,y), \tag{12}$$

where $\langle w_2 \rangle = 0$. Then (5) and (11) imply

$$\begin{aligned} &L_0(x,y,\partial/\partial x)\, u_0(x) + (\partial/\partial y_j)\, a_{jk}(x,y)(\partial/\partial x_k)p_s(x,y)(\partial/\partial x_s)\, u_0(x) \\ &\quad + (\partial/\partial x_j)\, a_{jk}(x,y)(\partial/\partial y_k)p_s(x,y)(\partial/\partial x_s)\, u_0(x) + L_2(x,y,\partial/\partial y)\, w_2(x,y) \\ &= f_0(x,y). \end{aligned} \tag{13}$$

We consider (13) as a differential equation in y on S for the function w_2 depending on the parameter $x \in \bar{Q}$. According to Proposition 17.1.2, the compatibility condition for such an equation is the relation

$$\int_S \big(L_0(x,y,\partial/\partial x)\, u_0(x) + (\partial/\partial x_j)\, a_{jk}(x,y)(\partial/\partial y_k) p_s(x,y)(\partial/\partial x_k)\, u_0(x)\big)\, dy$$
$$= \int_S f_0(x,y) dy,$$

which can be rewritten in the form

$$\mathbf{L}(x,\partial/\partial x)\, u_0(x) = \tilde{F}_0(x), \quad x \in \bar{Q}. \tag{14}$$

Here $\tilde{F}_0(x) = \langle f_0(x,y)\rangle$ and

$$\mathbf{L}(x,\partial/\partial x) = (\partial/\partial x_j)\langle a_{jk}(x,y) + a_{jq}(x,y)(\partial/\partial y_q) p_k(x,y)\rangle \partial/\partial x_k. \tag{15}$$

Lemma 17.1.4. *The operator* $\mathbf{L}$ *is formally self-adjoint and elliptic.*

Proof. Let A_{jk} be the coefficients of the operator (15). The equality $A_{jk} = A_{kj}$ is evident. We check the positive definiteness of the form $\xi_j A_{jk}(x)\xi_k$. By virtue of (10),

$$\int_S a_{jq}(x,y)(\partial/\partial y_q)\, p_k(x,y) dy$$
$$= \int_S \big(2a_{jq}(x,y)(\partial/\partial y_q)\, p_k(x,y) + p_k(x,y)(\partial/\partial y_q)\, a_{jq}(x,y)\big) dy$$
$$= \int_S \big(2a_{jq}(x,y)(\partial/\partial y_q)\, p_k(x,y) - p_k(x,y) L_2(x,y,\partial/\partial y)\, p_j(x,y)\big) dy$$
$$= \int_S \big(2a_{jq}(x,y)(\partial/\partial y_q)\, p_k(x,y) + (\partial p_k/\partial y_q)(x,y)\, a_{qn}(x,y)(\partial/\partial y_n) p_j(x,y)\big) dy.$$

Therefore, using (2), we find

$$\xi_j A_{jk}(x)\xi_k = \int_S \xi_j \big(a_{jk}(x,y) + a_{jq}(x,y)(\partial/\partial x_q)\, p_k(x,y)\big)\, \xi_k dy$$
$$= \int_S \xi_j \big(a_{jk}(x,y) + 2a_{jq}(x,y)(\partial/\partial y_q)\, p_k(x,y)$$
$$+ (\partial p_k/\partial y_q)(x,y)\, a_{qn}(x,y)\, (\partial/\partial y_k)\, p_j(x,y)\big)\xi_k dy$$
$$= \int_S a_{qn}(x,y)\big(\xi_j(\partial/\partial y_q)\, p_j(x,y) + \xi_q\big)\, \big(\xi_k(\partial/\partial y_n)\, p_k(x,y) + \xi_n\big) dy$$
$$\geq c \int_S \{\big(\xi_j(\partial/\partial y_1)\, p_j(x,y) + \xi_1\big)^2 + \big(\xi_k(\partial/\partial y_2)\, p_k(x,y) + \xi_2\big)^2\} dy \geq 0.$$

The last integral vanishes only if the relations

$$(\partial/\partial y_1)\, \big(\xi_n p_n(x,y)\big) = -\xi_1, \quad (\partial/\partial y_2)\, \big(\xi_n p_n(x,y)\big) = -\xi_2 \tag{16}$$

hold for all y. From (16) it follows that $\xi_n p_n(x,y) = -\xi_1 y_1 - \xi_2 y_2 + c$. The right-hand side is periodic only for $\xi_1 = \xi_2 = 0$. Therefore, the expression $\xi_j A_{jk}(x)\xi_k$ is nonnegative and vanishes for $|\xi| = 0$ only. □

We see that the principal term u_0 of the asymptotics of the solution u to equation (1) must be subject to (14). Then the equation (13) for w_2 can be rewritten in the form

$$\begin{aligned}
&L_2(x,y,\partial/\partial y)\, w_2(x,y)\\
&= f_0(x,y) - \langle f_0(x,y)\rangle - (\partial/\partial y_j)a_{jk}(x,y)(\partial/\partial x_k)\, p_s(x,y)\,(\partial/\partial x_s)\, u_0(x)\\
&- (\partial/\partial x_j)\big(a_{jk}(x,y)\,(\partial/\partial y_k)\, p_s(x,y) - \langle a_{jk}(x,y)\,(\partial/\partial y_k)\, p_s(x,y)\rangle\big)(\partial/\partial x_s)\, u_0(x)\\
&- (\partial/\partial x_j)\big(a_{jk}(x,y) - \langle a_{jk}(x,y)\rangle\big)(\partial/\partial x_k)\, u_0(x), \quad y \in S.
\end{aligned} \tag{17}$$

The right-hand side in (17) satisfies the compatibility condition in the class of periodic functions for all $x \in \bar{Q}$. Thus, there exists a solution $w_2(x,y)$ with zero mean value periodically depending on S.

17.1.4 Asymptotic series

We construct an asymptotic series for the function satisfying (1). Suppose that the right-hand side f in the equation admits the representation

$$f(\varepsilon,x) \sim \sum_{\mu=0}^{\infty} \varepsilon^{\mu} f_{\mu}(x,\varepsilon^{-1}x), \tag{18}$$

where f_μ smoothly depend on $x \in \bar{Q}$ and $y \in S$ and have unit period in y and x_1.

We seek formal asymptotics for u in the form

$$u(\varepsilon,x) \sim \sum_{\mu=0}^{\infty} \varepsilon^{\mu} u_{\mu}(x,\varepsilon^{-1}x) \tag{19}$$

with $u_0(x,y) = u_0(x)$. Substitute the series (18) and (19) into the equation $Lu = f$ and, using (4), separate out the coefficients by the same degrees ε^ν with $\nu = -1,0,1,\dots$. Let u_1 and u_2 be defined by (11) and (12). In view of Sections 17.1.2 and 17.1.3, the sum of the coefficients by ε^{-1} and ε^0 is equal to zero. All the remaining terms vanish under the condition

$$\begin{aligned}
&L_0(x,y,\partial/\partial x)\, u_\nu(x,y)\\
&+L_1(x,y,\partial/\partial x,\partial/\partial y)\, u_{\nu+1}(x,y) + L_2(x,y,\partial/\partial y)\, u_{\nu+2}(x,y) = f_\nu(x,y),
\end{aligned} \tag{20}$$

where $\nu \in \mathbb{N}$. By analogy with (12) we represent u_λ in the form

$$u_\lambda(x,y) = -v_\lambda(x) + p_s(x,y)\,(\partial/\partial x_s)\, v_{\lambda-1}(x) + w_\lambda(x,y), \quad \lambda = 2,3\dots \tag{21}$$

with $\langle w_\lambda\rangle = 0$. Then (20) can be written in the form

$$\begin{aligned}
&\{L_0(x,y,\partial/\partial x)\, v_\nu(x,y) + (\partial/\partial x_j)\, a_{jk}(x,y)\,(\partial/\partial y_k)p_s(x,y)\,(\partial/\partial x_s)\, v_\nu(x)\}\\
&+\Big\{(\partial/\partial y_j)\, a_{jk}(x,y)\,(\partial/\partial x_k)\big(p_s(x,y)\,(\partial/\partial x_s)v_\nu(x) + w_{\nu+1}(x,y)\big)\Big\}\\
&+\Big\{(\partial/\partial x_j)\, a_{jk}(x,y)\Big((\partial/\partial y_k)\, w_{\nu+1}(x,y)\\
&+ (\partial/\partial x_k)\big(p_s(x,y)\,(\partial/\partial x_s)\, v_{\nu-1}(x) + w_\nu(x,y)\big)\Big)\Big\}\\
&+ L_2(x,y,\partial/\partial y)\, w_{\nu+2}(x,y) = f_\nu(x,y).
\end{aligned} \tag{22}$$

Denote the sums in the braces in (22) by $I_1(x,y)$, $I_2(x,y)$ and $I_3(x,y)$, respectively. By the definition (15) of the operator $\mathbf{L}$, from the equalities $\langle p_s\rangle = \langle w_{\nu+1}\rangle = 0$ it follows that

$$\begin{aligned}
&\langle I_1(x,y)\rangle = \mathbf{L}(x,\partial/\partial x)\,v_\nu(x), \quad \langle I_2(x,y)\rangle = 0,\\
&\langle I_3(x,y)\rangle = (\partial/\partial x_j)\langle a_{jk}(x,y)(\partial/\partial x_k)\,p_s(x,y)\rangle(\partial/\partial x_s)\,v_{\nu-1}(x)\\
&\qquad + (\partial/\partial x_j)\langle a_{jk}(x,y)\Big((\partial/\partial y_k)\,w_{\nu+1}(x,y) + (\partial/\partial x_k)\,w_\nu(x,y)\Big)\rangle.
\end{aligned}$$

Therefore, the compatibility condition of the equation (22) with respect to $w_{\nu+2}(x,y)$ in the class of periodic functions takes the form

$$\mathbf{L}(x,\partial/\partial x)\,v_\nu(x) = \tilde{F}_\nu(x), \quad x\in Q, \quad \nu\in\mathbb{N}, \tag{23}$$

with $\tilde{F}_\nu(x) = \langle f_\nu(x,y)\rangle - \langle I_3(x,y)\rangle$. Finally, the equation (22) can be written in the following manner:

$$\begin{aligned}
L_2(x,y,\partial/\partial y)\,w_{\nu+2}(x,y) &= f_\nu(x,y) - \langle f_\nu(x,y)\rangle - I_2(x,y) - I_1(x,y)\\
&\quad + \langle I_1(x,y)\rangle - I_3(x,y) + \langle I_3(x,y)\rangle, \quad y\in S.
\end{aligned} \tag{24}$$

The construction of the coefficients in (19) is completed.

17.2 Boundary Layer for the Dirichlet Problem

17.2.1 The boundary value problem for the boundary layer

We consider the boundary problem for equation (1), 17.1, with Dirichlet data on the sides $\Gamma_\beta = \{x\in\mathbb{R}^2\colon x_2=\beta,\ x_1\in(0,1)\}, \beta = 0,1$, of the square Q and with periodicity conditions on the sides $\tilde{\Gamma}_\beta = \{x\in\mathbb{R}^2\colon x_1=\beta,\ x_2\in(0,1)\}, \beta=0,1$:

$$u(\varepsilon,x_1,0) = g_0(\varepsilon,x_1), \quad u(\varepsilon,x_1,1) = g_1(\varepsilon,x_1), \quad x_1\in(0,1); \tag{1}$$

$$u(\varepsilon,0,x_2) = u(\varepsilon,1,x_2),\ (\partial/\partial x_1)u(\varepsilon,0,x_2) = (\partial/\partial x_1)u(\varepsilon,1,x_2),\ x_2\in(0,1). \tag{2}$$

Suppose that g_0 and g_1 can be expanded into the asymptotic series

$$g_\beta(\varepsilon,x_1) \sim \sum_{\mu=0}^{\infty}\varepsilon^\mu g_{\beta\mu}(x_1,\varepsilon^{-1}x_1), \quad \beta=0,1, \tag{3}$$

where $g_{\beta\mu}$ are smooth periodic functions in $x_1, y_1\in[0,1]$.

Generally, the series (19), 17.1, does not satisfy the boundary conditions (1). Therefore, as usual, near the sides Γ_0 and Γ_1 of Q there arises a boundary layer. We shall describe the boundary layer only near Γ_0. The layer near Γ_1 can be constructed in the same manner.

We introduce the coordinates $t_1 = \varepsilon^{-1}x_1$ and $t_2 = \varepsilon^{-1}x_2$ (in the case of Γ_1 the corresponding coordinates are $t_1 = \varepsilon^{-1}x_1$ and $t_2 = \varepsilon^{-1}(1-x_2)$). Passing to the variables t_1, t_2 and expanding the coefficients a_{jk} into MacLaurin's series in x_2, we obtain the following expansion of the operator $\tilde{L}$ into powers of ε:

$$\tilde{L}(\varepsilon,x,\partial/\partial x) \sim \varepsilon^{-2}\sum_{y=0}^{\infty}\tilde{M}_{0\gamma}(\varepsilon,t,\partial/\partial t), \tag{4}$$

$$\tilde{M}_{0\gamma}(\varepsilon,t,\partial/\partial t) = (\partial/\partial t_j)(\gamma!)^{-1}t_2^\gamma(\partial/\partial x_2)^\gamma\, a_{jk}(t,\varepsilon t_1,0)\,(\partial/\partial t_k). \tag{5}$$

We shall seek a boundary layer of the form

$$z_0(\varepsilon, x) \sim \sum_{\mu=0}^{\infty} \varepsilon^{\mu} z_{0\mu}(x_1, \varepsilon^{-1}x) \tag{6}$$

with functions $z_0(x_1, t)$ smooth in $x_1 \in (0,1)$ and $t \in \tilde{\Pi} = (0,1) \times (0,\infty)$, periodic in t_1 and x_1, and decreasing as $t_2 \to \infty$.

Lemma 17.2.1. *Let $\psi(x,t)$ be a smooth periodic function in t_1 and x_1. Then*

$$\begin{aligned}\tilde{M}_{0\gamma}(\varepsilon, t, \partial/\partial t)\psi(\varepsilon t_1, t) = \{&\varepsilon^2 M_{0\gamma}^{(0)}(x_1, t, \partial/\partial x_1)\psi(x_1, t)\\ &+ \varepsilon M_{0\gamma}^{(1)}(x_1, t, \partial/\partial x_1, \partial/\partial t)\psi(x_1, t)\\ &+ M_{0\gamma}^{(2)}(x_1, t, \partial/\partial t)\psi(x_1, t)\}|_{x_1=\varepsilon t_1},\end{aligned} \tag{7}$$

where

$$\begin{aligned} M_{0\gamma}^{(0)}(x_1, t, \partial/\partial x_1) &= (\partial/\partial x_j)(\gamma!)^{-1} t_2^{\gamma} (\partial/\partial x_2)^{\gamma}\, a_{1\gamma}(x_1, 0, t)(\partial/\partial x_1);\\ M_{0\gamma}^{(1)}(x_1, t, \partial/\partial x_1, \partial/\partial t) &= (\partial/\partial x_1)(\gamma!)^{-1} t_2^{\gamma} (\partial/\partial x_2)^{\gamma}\, a_{1k}(x_1, 0, t)(\partial/\partial t_k)\\ &\quad + (\partial/\partial t_j)(\gamma!)^{-1} t_2^{\gamma} (\partial/\partial x_2)^{\gamma}\, a_{j1}(x_1, 0, t)(\partial/\partial x_1);\\ M_{0\gamma}^{(2)}(x_1, t, \partial/\partial t) &= (\partial/\partial t_j)(\gamma!)^{-1} t_2^{\gamma} (\partial/\partial x_2)^{\gamma}\, a_{jk}(x_1, 0, t)\,(\partial/\partial t_k). \end{aligned} \tag{8}$$

To prove the lemma it suffices to apply the operator (5) to ψ.

We substitute the sum of series (19), 17.1, and (6) into the boundary condition on Γ_0. Taking (7) into account and equating the coefficients by the same degrees of ε, we obtain

$$z_{0\mu}(x_1, t_1, 0) = g_{0\mu}(x_1, t_1) - u_{\mu}(x_1, 0, t_1, 0). \tag{9}$$

Moreover, according to Lemma 2.1 and the expansion (4),

$$\begin{aligned}\tilde{L}(\varepsilon, x, \partial/\partial x)\, z_0(\varepsilon, x) \sim \varepsilon^{-2} \sum_{\mu=0}^{\infty} \varepsilon^{\mu} \Big\{ &M_{00}^{(2)}(x_1, t, \partial/\partial t)\, z_{0\mu}(x_1, t)\\ + \sum_{\gamma=1}^{\mu} M_{0\gamma}^{(2)}(x_1, t, \partial/\partial t)\, z_{0,\mu-\gamma}(x_1, t) &+ \sum_{\gamma=0}^{\mu-1} M_{0\gamma}^{(1)}(x_1, t, \partial/\partial x_1, \partial/\partial t)\\ \times z_{0,\mu-\gamma-1}(x_1, t) &+ \sum_{\gamma=0}^{\mu-2} M_{0\gamma}^{(0)}(x_1, t, \partial/\partial t)\, z_{0,\mu-\gamma-2}(x_1, t)\Big\}.\end{aligned}$$

Therefore, the functions $z_{0\mu}$ must be subject to the conditions

$$M_{00}^{(2)}(x_1, t, \partial/\partial t) z_{0\mu}(x_1, t) = h_{0\mu}(x_1, t), \quad t \in \tilde{\Pi}, \tag{10}$$

where the right-hand sides $h_{0\mu}$ are expressed in terms of the preceding coefficients $z_{00}, \dots, z_{0,\mu-1}$ in (6) by the formulas

$$\begin{aligned} h_{0\mu}(x_1, t) = &- \sum_{\gamma-1}^{\mu} M_{0\gamma}^{(2)}(x_1, t, \partial/\partial t)\, z_{0,\mu-\gamma}(x_1, t)\\ &- \sum_{q=1}^{2} \sum_{\gamma=0}^{\mu-q} M_{0\gamma}^{(2-q)}(x_1, t, \partial/\partial x_1, \partial/\partial t)\, z_{0,\mu-\gamma-q}(x_1, t).\end{aligned} \tag{11}$$

We consider equations (11) and (9) as boundary value problem on the domain $\tilde{\Pi} = \{t \in \mathbb{R}^2 \colon t_2 \in (0,\infty), t_1 \in (0,1)\}$ with parameter $x_1 \in [0,1]$ and periodicity conditions in t_1.

17.2.2 Conditions on the boundary layer

We study the solvability of problem (9), (10) and the behavior of its solution as $t_2 \to \infty$.

Lemma 17.2.2. *The equation*

$$\hat{M}(\lambda, t, \partial/\partial t)\tilde{U}(t) := (\partial/\partial t_j + \delta_{j2}\lambda)a_{jk}(x_1, 0, t)(\partial/\partial t_k + \delta_{k2}\lambda)\tilde{U}(t) = \tilde{F}(t), \quad t \in S, \tag{12}$$

with complex parameter λ is uniquely solvable in the class of functions periodic in t_1 and t_2 provided λ is pure imaginary and $\lambda \neq 2\pi ni$, where $n \in \mathbb{Z}$. The numbers $\lambda = 2\pi ni$ are eigenvalues for the operator $\hat{M}(\lambda, t, \partial/\partial t)$ and to every of them there corresponds the only eigenfunction $\Phi_0^{(n)}$ and the only associated function $\Phi_1^{(n)}$.

Proof. In the case of $\operatorname{Re}\lambda = 0$ equation (12) is formally self-adjoint. Therefore, it suffices to show the absence of nontrivial solutions $\tilde{U}_0$ to the corresponding homogeneous equation. For such a solution we have

$$\begin{aligned} 0 &= -\operatorname{Re}\int_S \overline{\tilde{U}_0(t)}\hat{M}(\lambda, t, \partial/\partial t)\tilde{U}_0(t)dt \\ &= \operatorname{Re}\int_S \overline{\Big((\partial/\partial t_j)\tilde{U}_0(t) + \delta_{j2}\lambda\tilde{U}_0(t)\Big)} a_{jk}(x_1, 0, t)\Big((\partial/\partial t_k)\tilde{U}_0(t) + \delta_{k2}\lambda\tilde{U}_0(t)\Big)dt \\ &\geq c\int_S \Big(|(\partial/\partial t_1)\tilde{U}_0(t)|^2 + |(\partial/\partial t_2)\tilde{U}_0(t) + \lambda\tilde{U}_0(t)|^2\Big)dt, \end{aligned}$$

c being the constant in (2), 17.1. Thus,

$$(\partial/\partial t_1)\tilde{U}_0(t) = 0, \quad (\partial/\partial t_2)\tilde{U}_0(t) + \lambda\tilde{U}_0(t) = 0 \quad \text{for} \quad t \in S. \tag{13}$$

Hence, $\tilde{U}_0(t) = c_1 \exp(-\lambda t_2)$. However, for $c_1 \neq 0$ and $\lambda \neq 2\pi ni$ the function is not periodic. This implies that $c_1 = 0$ and the solution $\tilde{U}_0$ is trivial.

Let $\lambda = 0$. From (13) it follows that any solution to the homogeneous equation (12) is constant and consequently $\Phi_0^{(0)}(t) = 1$. By definition, the first associated function $\Phi_1^{(0)}$ of the operator $\hat{M}(\lambda, t, \partial/\partial t)$ corresponding to $\lambda = 0$ is a solution to the equation

$$\hat{M}(0, t, \partial/\partial t)\,\Phi_1^{(0)}(t) = -(\partial/\partial\lambda)\,\hat{M}(0, t, \partial/\partial t)\,\Phi_0^{(0)}(t), \quad t \in S,$$

which in the case under consideration takes the form

$$(\partial/\partial t_j)\,a_{jk}(x_1, 0, t)(\partial/\partial t_k)\,\Phi_1^{(0)}(t) = -(\partial/\partial t_j)\,a_{j2}(t), \quad t \in S. \tag{14}$$

Hence, $\Phi_1^{(0)}(t) = p_2(x_1, 0, t)$ (see (10), 17.1).

We check the absence of the second associated function $\Phi_2^{(0)}$ which has to satisfy the equality

$$\begin{aligned} \hat{M}(0, t, \partial/\partial t)\,\Phi_2^{(0)} &= -(\partial/\partial\lambda)\,\hat{M}(0, t, \partial/\partial t)\,\Phi_1(0)(t) \\ &\quad - 2^{-1}(\partial/\partial\lambda)^2\,\hat{M}(0, t, \partial/\partial t)\,\Phi_0(0)(t), \quad t \in S. \end{aligned} \tag{15}$$

Since

$$(\partial/\partial\lambda)\,\hat{M}(0,t,\partial/\partial t) = a_{2k}(x_1,0,t)(\partial/\partial t_k) + (\partial/\partial t_j)\,a_{j2}(x_1,0,t)$$

and

$$(\partial/\partial\lambda)^2\,\hat{M}(0,t,\partial/\partial t) = 2a_{22}(x_1,0,t),$$

equation (15) can be rewritten in the form

$$\begin{aligned}\hat{M}(0,t,\partial/\partial t)\,\Phi_2^{(0)}(t) &= -a_{2k}(x_1,0,t)\,(\partial/\partial t_k)\,\Phi_1^{(0)}(t)\\ &\quad - (\partial/\partial t_j)a_{j2}(x_1,0,t)\,\Phi_1^{(0)}(t) - a_{22}(x_1,0,t), \quad t \in S.\end{aligned} \tag{16}$$

The equation is not solvable. To prove this it suffices to show that the mean value of the right-hand side $\hat{C}$ in (16) differs from zero (see Proposition 17.1.2). Integrating by parts and using (2), 17.1, we obtain

$$\begin{aligned}\int_S \hat{C}(t)dt &= -\int_S \Big(\Phi_1^{(0)}(t)(\partial/\partial t_k)\,a_{2k}(x_1,0,t) + a_{2k}(x_1,0,t)(\partial/\partial t_k)\,\Phi_1^{(0)}(t)\\ &\quad + a_{22}(x_1,0,t)\Big)dt\\ &= \int_S \Big(\Phi_1^{(0)}(t)\hat{M}(0,t,\partial/\partial t)\,\Phi_1^{(0)}(t) - 2a_{2k}(x_1,0,t)(\partial/\partial t_k)\,\Phi_1^{(0)}(t)\\ &\quad - a_{22}(x_1,0,t)\Big)dt\\ &= -\int_S \Big(\big((\partial/\partial t_j)\,\Phi_1^{(0)}(t)\big)\,a_{jk}(x_1,0,t)(\partial/\partial t_k)\,\Phi_1^{(0)}(t)\\ &\quad + 2a_{2k}(x_1,0,t)(\partial/\partial t_k)\,\Phi_1^{(0)}(t) + a_{22}(x_1,0,t)\Big)dt\\ &= -\int_S \big((\partial/\partial t_j)\,\Phi_1^{(0)}(t) + \delta_{j2}\big)\,a_{jk}(x_1,0,t)\big((\partial/\partial t_k)\,\Phi_1^{(0)}(t) + \delta_{k2}\big)dt\\ &\le -c\int_S \Big(|(\partial/\partial t_1)\,\Phi_1^{(0)}(t)|^2 + |(\partial/\partial t_2)\,\Phi_1^{(0)}(t) + 1|^2\Big)\,dt.\end{aligned} \tag{17}$$

Since $\Phi_1^{(0)}(t) \neq -t_2 + \text{const}$, the last integral in (17) is positive and, consequently, there is no second associated function.

It remains to note that the eigenfunctions and associated functions $\Phi_j^{(m)}$ and $\Phi_j^{(n)}$ corresponding to $\lambda = 2\pi mi$ and $\lambda = 2\pi ni$ are connected by the equalities

$$\Phi_j^{(n)}(t) = \exp\Big(2\pi(m-n)it_2\Big)\,\Phi_j^{(m)}(t). \qquad \square$$

The next theorem states the existence of a bounded solution to the problem on the boundary layer.

Theorem 17.2.3. *Let $G(t_1)$ and $H(t)$ be smooth functions periodic in t_1 and let $H(t) = O\big(\exp(-\beta t_2)\big)$ with $\beta > 0$. Then the problem*

$$\begin{aligned}&(\partial/\partial t_j)\,a_{jk}(x_1,0,t)(\partial/\partial t_k)\,U(t) = H(t), \quad t \in \tilde{\Pi};\\ &U(t_1,0) = G(t_1), \quad t_1 \in (0,1),\end{aligned} \tag{18}$$

has a periodic in t_1 solution with finite Dirichlet integral. For the solution there holds the asymptotic formula

$$U(t) = C(H,G) + O\big(\exp(-\gamma t_2)\big), \quad t_2 \to \infty. \tag{19}$$

Here $C(H,G)$ is a constant depending on H and G; γ an arbitrary positive number less than $\min\{\beta, \beta_0\}$ with $\beta_0 = -\mathrm{Re}\lambda_0$, λ_0 being the first eigenvalue of the operator $\hat{M}(\lambda, t, \partial/\partial t)$ in the upper half-plane.

Proof. The existence of a solution v to problem (18) with finite energy

$$\int\limits_{\tilde{\Pi}} \Big((\partial/\partial t_j)\, U(t)\Big)\, a_{jk}(x_1, 0, t)(\partial/\partial t_k)\, U(t)dt \tag{20}$$

is obvious. From (2), 17.1, it follows that such a solution has finite Dirichlet integral. The smoothness of the solution and the asymptotic expansion (19) can be derived from Lemma 17.2.2 and Theorem 17.5.9 in the same manner as Proposition 1.4.5 from Proposition 1.2.4 and Lemma 1.4.1. □

Corollary 17.2.4. *There exists a solution ζ to the homogeneous problem (18) with infinite Dirichlet integral and with polynomial growth as $t_2 \to \infty$. The solution is of the form*

$$\zeta(x,t) = t_2 + p_2(x_1, 0, t) + \zeta_1(x,t), \tag{21}$$

where ζ_1 is a bounded solution to problem (18) for $H=0$ and $G(t_1)=-p_2(x_1,0,t_1,0)$.

Let us find the explicit dependence of the constant $C(H,G)$ in (19) on the right-hand sides of problem (18).

Theorem 17.2.5. *Let the assumptions of Theorem 17.2.3 be fulfilled. Then the constant $C(H,G)$ in (19) determined by the formula*

$$C(H,G)=\sigma^{-1}\left(\int\limits_0^1 G(t_1)a_{j2}(x_1,0,t_1,0)(\partial/\partial t_j)\zeta(x_1,t_1,0)dt_1 - \int\limits_{\tilde{\Pi}} H(t)\zeta_1(x_1,t)dt\right), \tag{22}$$

where ζ is the solution (21) to the homogeneous problem (18) and

$$\sigma = \int\limits_S \Big(a_{2j}(x_1,0,t)(\partial/\partial t_j)\, p_2(t) + a_{22}(x_1,0,t)\Big)dt > 0.$$

Proof. We substitute the functions ζ and U into Green's formula for the rectangle $\tilde{\Pi}_R = (0,1) \times (0,R)$ and obtain

$$\begin{aligned}\int\limits_{\tilde{\Pi}_R} \zeta(t)(\partial/\partial t_j)\, a_{jk}(t)(\partial/\partial t_k)\, U(t)dt &= \int\limits_{\tilde{\Pi}_R} U(t)(\partial/\partial t_k)\, a_{jk}(t)(\partial/\partial t_j)\, \zeta(t)dt \\ &+ \int\limits_0^1 \Big(U(t_1,0)\, a_{j2}(t_1,0)(\partial/\partial t_j)\, \zeta(t_1,0) - \zeta(t_1,0)\, a_{2k}(t_1,0)(\partial/\partial t_k)\, U(t_1,0)\Big)dt_1 \\ &+ \int\limits_0^1 \Big(\zeta(t_1,R)\, a_{2k}(t_1,R)(\partial/\partial t_k)\, U(t_1,R) - U(t_1,R)\, a_{j2}(t_1,R)(\partial/\partial t_j)\, \zeta(t_1,R)\Big)dt_1.\end{aligned} \tag{23}$$

(For ease of notation, we do not indicate the dependence on x_1.) Denote by $I_1(R)$ the integral on the left in (23) and by $I_2(R)$, I_3, and $I_4(R)$ the integrals on the right. The assumptions of the theorem, Corollary 17.2.4, and formula (18) imply that

$$\lim_{R\to\infty} I_1(R) = \int_{\tilde{\Pi}} H(t)\,\zeta(t)dt,$$
$$I_3 = \int_0^1 G(t)\,a_{j2}(t_1,0)(\partial/\partial t_j)\,\zeta(t_1,0)dt_1, \quad I_2(R) = 0. \tag{24}$$

Moreover, combining the representation (19) for U, the similar formula for ζ_1, and the definition (21) for ζ, we arrive at

$$I_4(R) = C(H,G)\int_0^1 \{a_{21}(t_1,R)(\partial/\partial t_1)\,p_2(t_1,R)$$
$$+\, a_{22}(t_1,R)\big(1+(\partial/\partial t_2)\,p_2(t_1,R)\big)\}dt_1 + o(1), \quad R\to\infty. \tag{25}$$

We rewrite the integral I in (25) in the form

$$I = \int_{R-1}^{R}\int_0^1 (\partial/\partial t_2)\Big\{t_2\Big(a_{21}(t)(\partial/\partial t_1)\,p_2(t) + a_{22}(t)\big(1+(\partial/\partial t_2)\,p_2(t)\big)\Big)\Big\}dt$$
$$= \int_S \Big(a_{21}(t)(\partial/\partial t_1)\,p_2(t) + a_{22}(t)\,(\partial/\partial t_2)p_2(t) + a_{22}(t)\Big)\,dt$$
$$+ \int_{R-1}^{R}\int_0^1 t_2\Big((\partial/\partial t_2)\,a_{21}(t)(\partial/\partial t_1)\,p_2(t)$$
$$+\,(\partial/\partial t_2)\,a_{22}(t)(\partial/\partial t_2)\,p_2(t) + (\partial/\partial t_2)\,a_{22}(t)\Big)\,dt. \tag{26}$$

According to the definition of p_2 (see (10), 17.1) and the periodicity of a_{jk} and p_2, the last integral in (26) vanishes. Indeed, the integral is equal to

$$\int_{R-1}^{R}\int_0^1 t_2\Big(-(\partial/\partial t_1)a_{11}(t)(\partial/\partial t_1)p_2(t)$$
$$-\,(\partial/\partial t_1)a_{12}(t)(\partial/\partial t_2)p_2(t) - (\partial/\partial t_1)a_{12}(t)\Big)dt \tag{27}$$
$$= -\int_{R-1}^{R} t_2\Big(a_{11}(t)(\partial/\partial t_1)p_2(t) + a_{12}(t)(\partial/\partial t_2)p_2(t) + a_{12}(t)\Big)\Big|_{t_1=0}^{t_1=1} dt_2 = 0.$$

Passing to the limit in (23), from (13)–(27) we derive (22). It remains to note that we checked the positiveness of σ in Lemma 17.1.4. □

Corollary 17.2.6. *Since for $H = 0$ and $G = 1$ the solution to problem* (18) *equals* 1, *we have* $C(0,1) = 1$.

From Theorems 17.2.3 and 17.2.5 it follows that the boundary layer, i.e., the solution to problem (18), is exponentially decaying provided the right-hand side in (22) vanishes. This enables us to obtain the boundary values of the functions v_μ constructed in Section 17.1. Indeed, the solutions z_μ to problems (9), (10) are exponentially decaying under

$$C(h_{0\mu}, g_{0\mu} - u_\mu) = 0, \tag{28}$$

$h_{0\mu}$ being the functions (11). According to Corollary 17.2.6 and the representations (11), (12), (21), 17.1, of the coefficients u_λ in (19), 17.1, the formula (28) can be rewritten in the form

$$\begin{aligned} u_0(x_1,0) &= C\big(0, g_{00}(x_1,t_1)\big) \\ v_1(x_1,0) &= C\Big(h_{01}(x_1,t), g_{01}(x_1,t_1) - p_s(x_1,0,t_1,0)(\partial/\partial x_s)v_0(x_1,0)\Big), \\ v_\lambda(x_1,0) &= C\Big(h_{0\lambda}(x_1,t), g_{0\lambda}(x_1,t_1) - p_s(x_1,0,t_1,0)(\partial/\partial x_s)v_{\lambda-1}(x_1,0) \\ &\qquad - w_\lambda(x_1,0,t_1,0)\Big), \lambda = 2,3,\dots . \end{aligned} \tag{29}$$

The equalities (29) are boundary conditions on Γ_0 for the solutions to equations (14), (23), 17.1, satisfying the periodicity conditions on $\tilde\Gamma_0$ and $\tilde\Gamma_1$,

$$\begin{aligned} &v_\nu(0,x_2) = v_\nu(1,x_2), \\ &(\partial/\partial x_1)v_\nu(0,x_2) = (\partial/\partial x_1)v_\nu(1,x_2), \quad x_2 \in (0,1). \end{aligned} \tag{30}$$

Note that by Lemma 17.1.4 the Dirichlet problem for the operator (15), 17.1, is uniquely solvable. Moreover, Theorems 17.2.3 and 17.2.5 imply that the relations (29) provide the existence of solutions to problem (9), (10) vanishing at infinity. Thus, we have found the boundary layer near Γ_0 which is exponentially decaying. The boundary layer near Γ_1 can be found similarly. We have constructed the formal asymptotics of the solution to the Dirichlet problem (1), 17.1, (1), (2).

17.3 Boundary Layer for the Neumann Problem

In this section we show how to homogenize the Neumann boundary conditions. To this end we consider the boundary value problem for equation (1), 17.1, with periodicity conditions (2), 17.2, on the sides $\tilde\Gamma_0$ and $\tilde\Gamma_1$ of the square Q and with Dirichlet and Neumann data on the sides Γ_0 and Γ_1, respectively,

$$u(\varepsilon,x_1,0) = g_0(\varepsilon,x_1), \quad x_1 \in (0,1); \tag{1}$$

$$a_{2k}(x,1,\varepsilon^{-1}x_2,\varepsilon^{-1})(\partial/\partial x_k)u(\varepsilon,x_1,1) = g_1(\varepsilon,x_1), \quad x_1 \in (0,1). \tag{2}$$

As in Section 17.2, we suppose that the functions g_β in (1) and (2) admit the expansions (3), 17.2.

The boundary layer near Γ_0 was constructed in Section 17.2. To describe the boundary layer corresponding to the side Γ_1 we introduce the coordinates $t_1 = \varepsilon^{-1}x_1$, $t_2 = \varepsilon^{-1}(1-x_2)$. In the coordinates t the operator $\tilde L$ decomposes into the series (4), 17.2, with coefficients

$$\tilde M_{1\gamma}(\varepsilon,t,\partial/\partial t) = (-1)^{\delta_{jk}}(\partial/\partial t_j)(\gamma!)^{-1}(-t_2)^\gamma(\partial/\partial x_2)^\gamma a_{jk}(\varepsilon t_1,1,t)\partial/\partial t_k \tag{3}$$

(cf.(5),17.2). Making similar changes in (8), 17.2 (t_2 has to be replaced by $-t_2$ and the point $x_2 = 0$ by $x_2 = 1$), we introduce the operators $M_{1\gamma}^{(q)}$. Besides, we need the differential operators

$$\begin{aligned} N_0^{(0)}(x_1, y_1, \partial/\partial x) &= a_{2k}(x_1, 1, y_1, 0)\partial/\partial x_k, \\ N_0^{(1)}(x_1, y_1, \partial/\partial y) &= a_{2k}(x_1, 1, y_1, 0)\partial/\partial y_k, \\ N_1^{(0)}(x_1, t_1, \partial/\partial x_1) &= a_{21}(x_1, 1, t_1, 0)\partial/\partial x_1, \\ N_1^{(1)}(x_1, t_1, \partial/\partial t) &= (-1)^{\delta_{2k}} a_{2k}(x_1, 1, t_1, 0)\partial/\partial t_k. \end{aligned} \tag{4}$$

The next assertion is similar to Lemmas 17.1.1 and 17.2.1.

Lemma 17.3.1. *Let φ and ψ be the functions in Lemmas* 17.1.1 *and* 17.2.1. *There hold the inequalities*

$$\begin{aligned} &\tilde{M}_{1\gamma}(\varepsilon, t, \partial/\partial t)\psi(\varepsilon t_1, t) \\ &= \Big(\varepsilon^2 M_{1\gamma}^{(0)}(x_1, t, \partial/\partial x_1)\psi(x_1, t) + \varepsilon M_{1\gamma}^{(1)}(x_1, t, \partial/\partial x_1, \partial/\partial t)\psi(x_1, t) \\ &+ M_{1\gamma}^{(2)}(x_1, t, \partial/\partial t)\psi(x_1, t)\Big)|_{x_1=\varepsilon t_1}, \\ &a_{2k}(\varepsilon t_1, 1, t_1, 0)(\partial/\partial t_k)\psi(\varepsilon t_1, t) \\ &= \Big(\varepsilon N_1^{(0)}(x_1, t_1, \partial/\partial x_1)\psi(x_1, t) + N_1^{(1)}(x_1, t_1, \partial/\partial t)\Big)\psi(x_1, t)|_{x_1=\varepsilon t_1}, \\ &a_{2k}(\varepsilon t_1, 1, t_1, 0)(\partial/\partial t_k)\varphi(\varepsilon t_1, 1, t) \\ &= \Big(\varepsilon N_0^{(0)}(x_1, t_1, \partial/\partial x)\varphi(x_1, 1, t) + N_0^{(1)}(x_1, t_1, \partial/\partial t)\varphi(x_1, 1, t)\Big)|_{x_1=\varepsilon t_1}. \end{aligned} \tag{5}$$

As in Section 17.2, we seek a boundary layer near Γ_1 in the form

$$z_1(\varepsilon, x) \sim \varepsilon \sum_{\mu=0}^{\infty} \varepsilon^\mu z_{1\mu}\Big(x_1, \varepsilon^{-1}x_1, \varepsilon^{-1}(1 - x_2)\Big), \tag{6}$$

where $z_{1\mu}(x_1, t)$ are smooth functions periodic in x_1 and t_1 and exponentially decaying as $t_2 \to \infty$. (The factor ε is introduced to match the orders on the right and on the left in (2) when substituting the series (6).)

We substitute the series (6) into the homogeneous equation (1), 17.1, take (4), 17.1, (3), and (5) into account, and equate the coefficients by the same degrees of ε. As a result, we obtain

$$M_{10}^{(2)}(x_1, t, \partial/\partial t)z_{1\mu}(x_1, t) = h_{1\mu}(x_1, t), \quad t \in \tilde{\Pi}, \tag{7}$$

where $h_{1\mu}$ are expressed with the coefficients $z_{10}, \dots, z_{1,\mu-1}$ of the series (6) by means of (11), 17.2.

We replace u in the boundary condition (2) by the sum of series (19), 17.1, and (6). From Lemma 17.3.1 and the expansion (3), 17.2, of the right-hand side g_1 we derive the following equations:

$$\begin{aligned} &N_1^{(1)}(x_1, t_1, \partial/\partial t)z_{1\mu}(x_1, t_1, 0) = g_{1\mu}(x_1, t_1) \\ &- N_1^{(0)}(x_1, t_1, \partial/\partial x_1)z_{1,\mu-1}(x_1, t_1, 0) - N_0^{(1)}(x_1, t_1, \partial/\partial t_1)u_{\mu+1}(x_1, 1, t_1, 0) \\ &- N_0^{(0)}(x_1, t, \partial/\partial x)u_\mu(x_1, 1, t_1, 0), \quad t_1 \in (0, 1). \end{aligned} \tag{8}$$

As usual, we consider (7) and (8) as a boundary value problem in $\tilde{\Pi}$ with parameter x_1.

Theorem 17.3.2. *Let G and H be the same functions as in Theorem 17.2.3. The boundary value problem*

$$\begin{aligned}(-1)^{\delta_{jk}}(\partial/\partial t_j)a_{jk}(x_1,1,t)(\partial/\partial t_k)U(t) &= H(t), \quad t\in\tilde{\Pi};\\ (-1)^{\delta_{2k}}a_{2k}(x_1,1,t_1,0)(\partial/\partial t_k)U(t_1,0) &= G(t_1), \quad t\in(0,1),\end{aligned} \tag{9}$$

is solvable in the class of functions periodic in t_1 with finite Dirichlet integral if and only if

$$\int_{\tilde{\Pi}} H(t)dt + \int_0^1 G(t_1)dt_1 = 0. \tag{10}$$

A solution to problem (9) is determined up to an additive constant. The unique solution satisfies the relation

$$U(t) = O\Big(\exp(-\gamma t_2)\Big), \quad t_2\to\infty,$$

γ being the same number as in Theorem 17.2.3.

Proof. Problem (9) is self-adjoint in the class of functions with finite energy integral (20), 17.2. Therefore, its compatibility condition consists in the orthogonality of the right-hand sides H and G to the solutions of the corresponding homogeneous problem. By Lemma 17.2.1 and Corollary 17.2.6 every solution to such a problem is a constant. Equality (10) coincides with the orthogonality H and G to 1 with respect to Green's formula

$$\int_{\tilde{\Pi}} \Big(V(t)M_{10}^{(2)}(x_1,t,\partial/\partial t)U(t) - U(t)M_{10}^{(2)}(x_1,t,\partial/\partial t)V(t)\Big)dt$$

$$= -\int_0^1 \Big(V(t_1,0)N_1^{(1)}(x_1,t_1,\partial/\partial t)U(t_1,0) - U(t_1,0)N_1^{(1)}(x_1,t_1,\partial/\partial t)V(t_1,0)\Big)dt_1.$$

As in Theorem 17.2.3, we obtain that any solution to problem (9) satisfies

$$U(t) = C_U + O(\exp(-\gamma t_2), \quad C_U = \text{const}.$$

Thus, $U - C_U$ is the desired exponentially decaying solution because the constant is a solution to the homogeneous problem. □

Let us write the compatibility conditions for problems (7) and (8):

$$\int_0^1 \Big(N_0^{(0)}(x_1,t_1,\partial/\partial x)v_\mu(x_1,1)$$

$$+ N_0^{(1)}(x_1,t_1,\partial/\partial t)p_s(x_1,1,t_1,0)(\partial/\partial x_s)v_\mu(x_1,1)\Big)dt_1$$

$$= \int_{\tilde{\Pi}} h_{1\mu}(x_1,t)\,dt + \int_0^1 \{g_{1\mu}(x_1,t_1) - N_1^{(0)}(x_1,t_1,\partial/\partial x_1)z_{1,\mu-1}(x_1,t_1,0)$$
$$- N_0^{(1)}(x_1,t_1,\partial/\partial t)w_{\mu+1}(x_1,1,t_1,0) \tag{11}$$
$$- N_0^{(0)}(x_1,t_1,\partial/\partial x)\Big(p_s(\partial/\partial x_s)v_{\mu-1}(x_1,1,t_1,0) + w_\mu(x_1,1,t_1,0)\Big)\}dt_1.$$

Here $\mu \in \mathbb{N}_0$, $v_0(x) = u_o(x)$, $v_{-1}(x) = 0$, $z_{1,-1}(x_1,t) = 0$. Each equality in (11) can be rewritten in the form

$$\hat{A}_k(x_1)(\partial/\partial x_k)v_\mu(x_1,1) = \hat{C}_\mu(x_1), \quad x_1 \in (0,1), \tag{12}$$

where $\hat{C}_\mu$ is the right-hand side in (11). In view of definition (4) of the operators $N_0^{(q)}$, the functions $\hat{A}_k$ are given by

$$\hat{A}_k(x_1) = \int_0^1 \Big(a_{2k}(x_1,1,y_1,0) + a_{2q}(x_1,1,y_1,0)(\partial/\partial y_q)p_k(x_1,1,y_1,0)\Big)dy_1. \tag{13}$$

Lemma 17.3.3. *There hold the equalities*

$$\hat{A}_k(x_1) = A_{2k}(x_1,1), \quad k = 1,2,$$

$A_{jk}(x)$ *being the coefficients of the operator* (15), 17.1.

Proof. By virtue of the periodicity of a_{2s} and p_k, we have

$$\hat{A}_k(x_1) = \int_S (\partial/\partial y_2)\Big(y_2 a_{2k}(x_1,1,y) + y_2 a_{2q}(x_1,1,y)(\partial/\partial y_q)p_k(x_1,1,y)\Big)dy$$
$$= \int_S \Big(a_{2k}(x_1,1,y) + a_{2k}(x_1,1,y)(\partial/\partial y_q)p_k(x_1,1,y)\Big)dy$$
$$+ \int_S y_2\Big((\partial/\partial y_2)a_{2k}(x_1,1,y) + (\partial/\partial y_2)a_{2q}(x_1,1,y)(\partial/\partial y_q)p_k(x_1,1,y)\Big)dy. \tag{14}$$

The first integral on the right in (14) is equal to $A_{2k}(x_1,1)$. According to (10), 17.1, the second one takes the form

$$-\int_S y_2\Big((\partial/\partial y_1)a_{1q}(x_1,1,y)(\partial/\partial y_q)p_k(x_1,1,y) + (\partial/\partial y_1)a_{1k}(x_1,1,y)\Big)\,dy$$
$$= -\int_0^1 \Big(y_2 a_{1q}(x_1,1,y)(\partial/\partial y_q)p_k(x_1,1,y) + y_2 a_{1k}(x_1,1,y)\Big)\Big|_{y_1=0}^{y_1=1} dy_2 = 0. \quad \square$$

We have constructed the coefficients of the formal asymptotic expansions (19), (30), 17.1, and (6). Indeed, the coefficients u_μ in (19), 17.1, are defined by (11), (12), (21), 17.1, where w_λ is a periodic solution to equation (24), 17.1, and u_0 and v_λ are solutions to problems (14), (23), 17.1, (30), (29), 17.2, and (12). (Note that by Lemma 17.3.3 the boundary conditions (12) are intrinsic and, consequently, the mentioned boundary value problems are uniquely solvable.) The coefficients $z_{0\mu}$ and $z_{1\mu}$ in (6), 17.2, and (6) are exponentially decaying solutions to problems (10), (9), 17.2, and (7), (8) on the semistrip $\tilde{\Pi}$.

17.4 Justification of Asymptotic Expansions

Three preceding sections of this chapter were devoted to the formal asymptotic expansions. In the present section the expansions will be justified. To this end we introduce partial sums of series (19), (20), 17.1, and (6):

$$U_D^{(M)}(\varepsilon,x)=\sum_{\mu=0}^{M}\varepsilon^{\mu}\Big(u_\mu(x,\varepsilon^{-1}x)+z_{0\mu}(x_1,\varepsilon^{-1}x)+z_{1\mu}(x_1,\varepsilon^{-1}x_1,\varepsilon^{-1}(1-x_2)\Big)\quad(1)$$

and

$$U_N^{(M)}(\varepsilon,x)=\sum_{\mu=0}^{M}\varepsilon^{\mu}\Big(u_\mu(x,\varepsilon^{-1}x)+z_{0\mu}(x_1,\varepsilon^{-1}x)+\varepsilon z_{1\mu}(x_1,\varepsilon^{-1}x_1,\varepsilon^{-1}(1-x_2))\Big)\quad(2)$$

corresponding to the Dirichlet problem (1), 17.1, (1), 17.2, and to the mixed problem (1), 17.1, (2), 17.2, (1), (2), 17.3. (The terms $z_{1\mu}$ in (2) are defined in Section 17.3 and the coefficients $z_{1\mu}$ of the series for the boundary layer corresponding to the Dirichlet condition on Γ_1 are given in (1). The function $z_{1,\mu}$ is defined as well as $z_{0,\mu}$.)

Now we suppose that the expansions (18), 17.1, and (3), 17.2, of the right-hand sides f and g_β are subject to the conditions

$$\left\|f-\sum_{\mu=0}^{J}\varepsilon^{\mu}f_\mu;\mathbf{L}_2(Q)\right\|\le c_J\varepsilon^{J+1},$$

$$\varepsilon\left\|g_\beta-\sum_{\mu=0}^{J}\varepsilon^{\mu}g_{\beta\mu};\mathbf{W}_2^1(0,1)\right\|+\left\|g_\beta-\sum_{\mu=0}^{J}\varepsilon^{\mu}g_{\beta\mu};\mathbf{L}_2(0,1)\right\|\le c_J\varepsilon^{J+1}$$

with arbitrary positive integer $J\in\mathbb{N}$ and constant c_J independent of ε.

Let $\tilde F_D^{(M)}$, $\tilde G_{\beta D}^{(M)}$ and $\tilde F_N^{(M)}$, $\tilde G_{\beta N}^{(M)}$ $(\beta=0,1)$ be the discrepancies corresponding to the sums (1) and (2) in equation (1), 17.1, and in the original boundary conditions:

$$\begin{aligned}
\tilde F_D^{(M)}(\varepsilon,x)&=\tilde L(\varepsilon,x,\partial/\partial x)U_D^{(M)}(\varepsilon,x)-f(\varepsilon,x),\\
\tilde F_N^{(M)}(\varepsilon,x)&=\tilde L(\varepsilon,x,\partial/\partial x)U_N^{(M)}(\varepsilon,x)-f(\varepsilon,x),\\
\tilde G_{0D}^{(M)}(\varepsilon,x_1)&=U_D^{(M)}(\varepsilon,x_1,0)-g_0(\varepsilon,x_1),\\
\tilde G_{1D}^{(M)}(\varepsilon,x_1)&=U_D^{(M)}(\varepsilon,x_1,1)-g_1(\varepsilon,x_1),\\
\tilde G_{0N}^{(M)}(\varepsilon,x_1)&=U_N^{(M)}(\varepsilon,x_1,0)-g_0(\varepsilon,x_1),\\
\tilde G_{1N}^{(M)}(\varepsilon,x_1)&=a_{2k}(x_1,1,\varepsilon^{-1}x_1,0)(\partial/\partial x_k)U_N^{(M)}(\varepsilon,x_1,1)-g_1(\varepsilon,x_1).
\end{aligned}\quad(3)$$

Without dwelling on a simple but unwieldy analysis of the coefficients in series (19), 17.1, (6), 17.2, and (6), we state the properties of the functions listed in (3).

(i) There hold the representations

$$\tilde F_T^{(M)}(\varepsilon,x)=\varepsilon^{M-1}\hat F_T^{(M)}(x,\varepsilon^{-1}x)+\check F_T^{(M)}(\varepsilon,x)\quad(4)$$

with T equal to D or N, while

$$\begin{aligned}&\langle \hat{F}_T^{(M)}(x,y)\rangle = 0, \quad ||\check{F}_T^{(M)}; \mathbf{L}_2(Q)|| = O(\varepsilon^{M-1/2}),\\ &\max_{x\in\bar{Q}, y\in S} |(\partial/\partial x_j)\hat{F}_T^{(M)}(x,y)| \leq \text{const}.\end{aligned} \tag{5}$$

(ii) There is the relation

$$||\tilde{G}_{1N}^{(M)}; \mathbf{L}_2(0,1)|| = O(\varepsilon^M). \tag{6}$$

(iii) There exist functions $\tilde{U}_{0T}^{(M)}$ with $T = D, N$ and $\tilde{U}_{1D}^{(M)}$ periodic in x_1 such that

$$\begin{aligned}&\tilde{U}_{0T}^{(M)}(\varepsilon, x_1, 0) = \tilde{G}_{0T}^{(M)}(\varepsilon, x_1), \quad \tilde{U}_{1D}^{(M)}(\varepsilon, x_1, 1) = \tilde{G}_{1D}^{(M)}(\varepsilon, x_1),\\ &||\tilde{U}_{0T}^{(M)}; \mathbf{W}_2^1(Q)|| + ||\tilde{U}_{1D}^{(M)}; \mathbf{W}_2^1(Q)|| = O(\varepsilon^M).\end{aligned} \tag{7}$$

In what follows we prove estimates on the difference $\tilde{R}_N^{(M)} = u^{(N)} - U_N^{(M)}$, where $u^{(N)}$ is the solution to problem (1), 17.1, (1), (2), 17.3, (2), 17.2, and $U_N^{(M)}$ is the function (2). The same is true for the difference $\tilde{R}_D^{(M)} = u^{(D)} - U_D^{(M)}$. To verify that, it suffices to repeat the argument used in the proof of the following assertion.

Theorem 17.4.1. *For the solution $u^{(N)}$ to problem* (1), 17.1, (1), (2), 17.3, (2), 17.2, *there holds the inequality*

$$||u^{(N)} - U_N^{(M)}; \mathbf{W}_2^1(Q)|| \leq c\varepsilon^{M-1/2} \tag{8}$$

with constant c independent of ε.

Proof. The function $\check{R}_N^{(M)} = \tilde{R}_N^{(M)} - \chi(x_2)\tilde{U}_{0N}^{(M)}$ satisfies the boundary value problem

$$\begin{aligned}&\tilde{L}(\varepsilon, x, \partial/\partial x)\check{R}_N^{(M)}(\varepsilon, x) = -\tilde{F}_N^{(M)}(\varepsilon, x)\\ &\quad +(\partial/\partial x_j)a_{jk}(x, \varepsilon^{-1}x)(\partial/\partial x_k)\Big(\chi(x_2)\tilde{U}_{0N}^{(M)}(\varepsilon, x)\Big), \quad x\in Q;\end{aligned} \tag{9}$$

$$\begin{aligned}&\check{R}_N^{(M)}(\varepsilon, 0, x_2) = \check{R}_N^{(M)}(\varepsilon, 1, x_2),\\ &(\partial/\partial x_1)\check{R}_N^{(M)}(\varepsilon, 0, x_2) = (\partial/\partial x_1)\check{R}_N^{(M)}(\varepsilon, 1, x_2), \quad x\in[0,1];\\ &\check{R}_N^{(M)}(\varepsilon, x_1, 0) = 0,\\ &\quad a_{2k}(x_1, 1, \varepsilon^{-1}x, 0)(\partial/\partial x_k)\check{R}_N^{(M)}(\varepsilon, x_1, 0) = -\tilde{G}_{1N}^{(M)}(\varepsilon, x_1), \quad x_1\in[0,1].\end{aligned} \tag{10}$$

Here $\chi \in \mathbf{C}_0^\infty[0,\infty)$ is a cut-off function such that $\chi(t) = 1$ for $t \leq 1/3$ and $\chi(t) = 0$ for $t \geq 2/3$. Since $\langle \hat{F}_N^{(M)}(x,y)\rangle = 0$ (see (5)), there exists a solution to the equation

$$\Delta_y \tilde{U}_N^{(M)}(x,y) = \hat{F}_N^{(M)}(x,y), \quad y\in S,$$

with zero mean value and periodic in y (see Proposition 17.1.2). Set $\hat{P}_k^{(M)}(\varepsilon, x) = \Big((\partial/\partial y_k)\hat{U}_N^{(M)}(x,y)\Big)|_{y=\varepsilon^{-1}x}$. Then

$$\varepsilon(\partial/\partial x_k)\hat{P}_k^{(M)}(\varepsilon, x) = \hat{F}_N^{(M)}(x, \varepsilon^{-1}x) + \varepsilon\Big((\partial/\partial x_k)(\partial/\partial y_k)\hat{U}_N^{(M)}(x,y)\Big)|_{y=\varepsilon^{-1}x}. \tag{11}$$

Denote by $\varepsilon\hat{P}_0^{(M)}(\varepsilon, x)$ the last term on the right in (11). From (5) it follows that

$$||\hat{P}_\lambda^{(M)}; \mathbf{L}_2(Q)|| \leq c, \quad \lambda = 0, 1, 2. \tag{12}$$

We multiply equation (9) by $\check{R}_N^{(M)}$ and integrate over Q by parts using boundary conditions (10). We have

$$\begin{aligned}
&\int_Q (\partial \check{R}_N^{(M)}/\partial x_j)(\varepsilon,x)a_{jk}(x,\varepsilon^{-1}x)(\partial/\partial x_k)\check{R}_N^{(M)}(\varepsilon,x)dx \\
&= \int_0^1 \check{R}_N^{(M)}(\varepsilon,x_1,1)\Big(\tilde{G}_{1N}^{(M)}(\varepsilon,x_1) - \hat{P}_2^{(M)}(\varepsilon,x_1,1)\Big)dx_1 \\
&+ \int_Q \{\Big(\check{F}_N^{(M)}(\varepsilon,x) - \hat{P}_0^{(M)}(\varepsilon,x)\Big)\check{R}_N^{(M)}(\varepsilon,x) \\
&- \hat{P}_k^{(M)}(\varepsilon,x)(\partial/\partial x_k)\check{R}_N^{(M)}(\varepsilon,x) \\
&+ (\partial \check{R}_N^{(M)}/\partial x_j)(\varepsilon,x)a_{jk}(x,\varepsilon^{-1}x)(\partial/\partial x_k)\Big(\chi(x_2)\tilde{U}_{0N}^{(M)}(\varepsilon,x)\Big)\}dx.
\end{aligned} \tag{13}$$

Using inequality (2), 17.1, and the fact that $\check{R}_N^{(M)}$ vanishes for $x_1 = 0$, we obtain that $c_1||\check{R}_N^{(M)}; \mathbf{W}_2^1(Q)||$ is no greater than the left-hand side in (13). Besides, from (4), (12), (7), and (6) we derive that for the integrals $I(Q)$ over Q and $I(0,1)$ over $[0,1]$ on the right in (13) there hold the inequalities

$$|I(Q)| \le c_2\varepsilon^{2M-1} + 2^{-1}c_1||\check{R}_N^{(M)}; \mathbf{W}_2^1(Q)||^2$$

and

$$|I(0,1)| \le c_3\varepsilon^{2M-1} + \varepsilon||\check{R}_N^{(M)}(\varepsilon,\cdot,1); \mathbf{L}_2(0,1)||.$$

Combining the estimate

$$||\check{R}_N^{(M)}(\varepsilon,\cdot,1); \mathbf{L}_2(0,1)|| \le c_4||\check{R}_N^{(M)}; \mathbf{W}_2^1(Q)||$$

and the above inequalities, we arrive at

$$||\check{R}_N^{(M)}; \mathbf{W}_2^1(Q)|| = O(\varepsilon^{M-1/2}), \quad \varepsilon \to 0.$$

From this, (7) and the relation

$$||\tilde{R}_N^{(M)}; \mathbf{W}_2^1(Q)|| \le ||\check{R}_N^{(M)}; \mathbf{W}_2^1(Q)|| + ||\tilde{U}_{1N}^{(M)}; \mathbf{W}_2^1(Q)||$$

we obtain (8). □

Remark 17.4.2. The estimate (8) given by Theorem 17.4.1 is exact in the sense of asymptotics. Indeed, the difference $U_N^{(M+1)} - U_N^{(M)}$ contains the term

$$\varepsilon^M z_{0M}(x_1,\varepsilon^{-1}x).$$

Since z_{0M} is exponentially decaying as $t_2 \to \infty$, we have

$$||\varepsilon^M z_{0M}(x_1,\varepsilon^{-1}x); \mathbf{W}_2^1(Q)||^2 = O\Big(\varepsilon^{2M-2}\int_Q \exp(-\gamma\varepsilon^{-1}x_2)dx_2\Big) = O(\varepsilon^{2M-1}).$$

Therefore, the norm in $\mathbf{W}_2^1$ of the boundary layer dropped from (2) is of the same order as the right-hand side in (8).

Corollary 17.4.3. *There is the inequality*

$$||u(\varepsilon,x) - u_0(x); \mathbf{L}_2(Q)|| \le c\varepsilon^{1/2}. \tag{14}$$

In particular, the solution u of the original problem (1), 17.1, (1), (2), 17.3, (2), 17.2, with rapidly oscillating coefficients converges in $\mathbf{L}_2(Q)$ to the solution u_0 of problem (23), 17.1, (29), 17.2, (12), 17.3, (30), 17.2.

Indeed, from Theorem 17.4.1 for $M = 1$ it follows that

$$||u - u_0 - z_{00} - \varepsilon u_1 - \varepsilon z_{10} - \varepsilon z_{01} - \varepsilon^2 z_{11}; \mathbf{L}_2(Q)|| = O(\varepsilon^{1/2}).$$

Combining this inequality with the evident relations $||u_1; \mathbf{L}_2(Q)|| = O(1)$ and $||z_{\alpha\beta}; \mathbf{L}_2(Q)|| = O(\varepsilon^{1/2})$ with $\alpha, \beta = 0, 1$, we obtain (14).

17.5 Elliptic Boundary Value Problems with Periodic Coefficients in a Cylinder

Here we briefly describe auxiliary results used in Sections 17.2 and 17.3.

17.5.1 Model problem in a cylinder

Let $\tilde{C}$ be the cylinder $\Omega \times \mathbb{R}$, where Ω is a compact n-dimensional manifold with boundary $\partial\Omega$ or without boundary. We consider the boundary value problem

$$\mathbf{L}(x, t, D_x, D_t)u(x, t) = f(x, t), \quad (x, t) \in \tilde{C}; \tag{1}$$

$$\partial\mathbf{B}_j(x, t, D_x, D_t)u(x, t) = g_j(x, t), \quad (x, t) \in \partial\tilde{C}, \quad j = 1, \dots, m; \tag{2}$$

Here

$$\mathbf{L}(x, t, D_x, D_t) = \sum_{k=0}^{m} \mathbf{L}^{(k)}(x, t, D_x)D_t^k$$

$$\mathbf{B}_j(x, t, D_x, D_t) = \sum_{k=0}^{m} \mathbf{B}_j^{(k)}(x, t, D_x)D_t^k, \quad j = 1, \dots, m,$$

$\mathbf{L}^{(k)}(x, t, D_x)$ and $\mathbf{B}_j^{(k)}(x, t, D_x)$ being differential operators of order $m - k$ and $m_j - k$, respectively, with smooth coefficients that are periodic in t (the period equals 1); $m \in \mathbb{N}$ and $m_j \in \mathbb{N}_0$, $j = 1, \dots, m$. We suppose that $\mathbf{L}$ is a uniformly elliptic operator in $S = \{(x, t) \in \overline{\tilde{C}} : t \in [0, t]\}$ and the system $\{\mathbf{B}_1, \dots, \mathbf{B}_m\}$ is subject to the ellipticity conditions on $\Gamma = \{(x, t) \in \partial\tilde{C} : t \in [0, t]\}$.

Remark 17.5.1. If Ω is a manifold without boundary, the boundary conditions (5) are absent. Let Ω be the unit cube with opposite faces identified. Then equation (1) is considered as a boundary value problem in the cylinder $Q \times \mathbb{R}$ with periodicity conditions on the opposite faces (see Sections 17.2 and 17.3).

As in 3.2.1, we denote by $\mathbf{W}_{2,\beta}^l(\tilde{C})$ the space of functions in $\tilde{C}$ endowed with the norm

$$||v; \mathbf{W}_{2,\beta}^l(\tilde{C})|| = \Big(\sum_{j=0}^{l} \exp(\beta t)\, ||v; \mathbf{W}_2^j(\tilde{C})||^2\Big)^{1/2},$$

where $\beta \in \mathbb{R}$, $l \in \mathbb{N}_0$ a real number. Let $\tilde{\mathbf{H}}^l(S)$ be the closure in $\mathbf{C}^\infty(\overline{\tilde{C}})$ of the space of periodic functions in $\mathbf{W}_2^l(S)$. As usual we define the spaces $\mathbf{W}_{2,\beta}^{l-1/2}(\partial\tilde{C})$,

$\tilde{\mathbf{H}}^{l-1/2}(\Gamma)$ of traces and set

$$\mathbf{RW}^l_{2,\beta}(\tilde{C}) = \mathbf{W}^l_{2,\beta}(\tilde{C}) \times \prod_{j=1}^{m} \mathbf{W}^{l+2m-m_j-1/2}_{2,\beta}(\partial\tilde{C}),$$

$$\mathbf{R}\tilde{\mathbf{H}}^l(S) = \tilde{\mathbf{H}}^l(S) \times \prod_{j=1}^{m} \tilde{\mathbf{H}}^{l+2m-m_j-1/2}(\Gamma).$$

The next assertion is obvious.

Lemma 17.5.2. *The operator*

$$\mathbf{A} = (\mathbf{L}, \partial\mathbf{B}_1, \dots, \partial\mathbf{B}_m)\colon \mathbf{W}^{l+2m}_{2,\beta}(\tilde{C}) \to \mathbf{RW}^l_{2,\beta}(\tilde{C}) \tag{3}$$

of problem (1), (2) *is continuous for any* β *and* $l > \max\{r_1, \dots, r_m\} - 2m$.

17.5.2 Problem with a complex parameter

We introduce the operator

$$\tilde{A}(\lambda) = \Big(\mathbf{L}(x, \tau, D_x, -i\lambda + D_\tau),\ \partial\mathbf{B}_1(x, \tau, D_x, -i\lambda + D_\tau), \dots,$$
$$\partial\mathbf{B}_m(x, \tau, D_x, -i\lambda + D_\tau)\Big)$$

in S. Suppose that for some λ the operator

$$\tilde{A}(\lambda)\colon \tilde{\mathbf{H}}^{l+2m}(S) \to \mathbf{R}\tilde{\mathbf{H}}^l(S) \tag{4}$$

is an isomorphism.

The next assertion follows from Theorem 5.1. in Gohberg and Krein [1].

Theorem 17.5.3. *The operator* (4) *implements an isomorphism* (4) *for all* $\lambda \in \mathbb{C}$ *with the exception of some isolated points* λ_j.

Let us describe in more detail the set of eigenvalues λ_j and the form of the inverse operator $\tilde{R}(\lambda)$ for $\tilde{A}(\lambda)$. We multiply the equation $\tilde{A}(\lambda)V(\lambda) = \Phi(\lambda)$ by $\exp(2\pi k i\tau)$ with $k \in \mathbb{Z}$. Commuting $\tilde{A}(\lambda)$ and the exponential, we obtain

$$\tilde{A}(\lambda - 2\pi k i)\Big(\exp(2\pi k i\tau)V(\lambda)\Big) = \exp(2\pi k i\tau)\Phi(\lambda).$$

Thus, the numbers λ and $\lambda + 2\pi k i$ with $k \in \mathbb{Z}$ are eigenvalues simultaneously and the operator-valued function $\tilde{R}(\lambda)$ satisfies

$$\tilde{R}(\lambda + 2\pi k i) = \exp(-2\pi k i\tau)\tilde{R}(\lambda)\exp(2\pi k i\tau). \tag{5}$$

Let λ_j be an eigenvalue of $\tilde{A}(\lambda)$ and $I_j = \dim\ker\tilde{A}(\lambda_j)$. We denote by $\varphi_j^{(p,0)}, \dots, \varphi_j^{(p,\varkappa_{jp}-1)}$, $p = 1, \dots, I_j$, a canonical system of Jordan chains corresponding to λ_j. The next assertion is similar to Theorem 3.1.10.

Theorem 17.5.4. *The operator-valued function* $\tilde{R}(\lambda)$ *is holomorphic everywhere except for the poles, which are the eigenvalues of the operator pencil* $\tilde{A}(\lambda)$. *The order of a pole* λ_j *is equal to the greatest of the ranks* $\varkappa_{jp}, p = 1, \dots, I_j$, *of the eigenvectors* $\varphi_j^{(p,0)}$ *of* $\tilde{A}(\lambda)$. *In a neighborhood* $\tilde{V}$ *of* $\lambda_j \in \mathbb{C}$ *there holds the expansion*

$$\tilde{R}(\lambda) = (\lambda - \lambda_j)^{-\varkappa_j}\tilde{P}_{\varkappa_j} + (\lambda - \lambda_j)^{-\varkappa_j+1}\tilde{P}_{\varkappa_j-1} + \dots + (\lambda - \lambda_j)^{-1}\tilde{P}_1 + \Lambda(\lambda), \tag{6}$$

where $\varkappa_j = \max\{\varkappa_{j1}, \dots, \varkappa_{jI_j}\}$, $\tilde{P}_s$ *are finite-dimensional operators independent of* λ. *The operator* $\tilde{P}_{\varkappa_j}$ *acts into the subspace* $\ker\tilde{A}(\lambda_j)$ *and the operators* $\tilde{P}_{\varkappa_j-1}$,

..., $\tilde{P}_1$ *into the subspaces of the corresponding associated vectors. Finally,* Λ *is an operator-valued function holomorphic in* $\tilde{V}$*.*

We introduce the functions

$$\Phi_j^{(p,s)}(x,t) = \exp(\lambda_j t) \sum_{q=0}^{s} (q!)^{-1} t^q \varphi_j^{(p,s-q)}(x, t - [t]), \tag{7}$$

with $p = 1, \dots, I_j$, $s = 0, \dots, \varkappa_{jp} - 1$. Obviously, $\Phi_j^{(p,s)} \in \mathbf{W}_{2,\mathrm{loc}}^{l+2m}(\bar{\tilde{C}})$.

17.5.3 Analog of the Fourier transform

Let $\hat{H}$ be a Hilbert space. We denote by $\mathbf{L}_2([0, 2\pi] \to \hat{H})$ the space of functions $z\colon [0, 2\pi] \to \hat{H}$ with norm

$$|z; \hat{H}| = \left(\int_0^{2\pi} ||z(\xi); \hat{H}||^2 d\xi \right)^{1/2}.$$

Suppose that $v \in \mathbf{W}_{2,\beta}^l(\tilde{C})$ and set

$$(\tilde{G}_\beta v)(\xi, x, \tau) = \sum_{j=-\infty}^{\infty} \exp\Big((\beta - i\xi)(j + \tau)\Big)\, v(x, j + \tau) \tag{8}$$

for $\xi \in [0, 2\pi]$ and $(x, \tau) \in S$.

Lemma 17.5.5. *The transform* (8) *is an isomorphism between*

$$\mathbf{W}_{2,\beta}^l(\tilde{C}) \qquad \textit{and} \qquad \mathbf{L}_2\Big([0, 2\pi] \to \tilde{\mathbf{H}}^l(S)\Big).$$

The inverse transform is of the form

$$(G_\beta^{-1} z)(x, t) = (2\pi)^{-1} \int_0^{2\pi} \exp\Big((i\xi - \beta)\, t\Big)\, z(\xi, x, t - [t])\, d\xi. \tag{9}$$

Proof. In view of the definition of $\mathbf{W}_{2,\beta}^l(\tilde{C})$ it suffices to check the lemma for $\beta = 0$. For $v \in \mathbf{C}_0^\infty(\bar{\tilde{C}})$ we have

$$\begin{aligned}((\partial/\partial t)^s (\tilde{G}_0 v))(\xi, x, 1) &= \sum_{p=0}^{s} \big(p!(s-p)!\big)^{-1} s!(i\xi)^{s-p} \\ &\quad \times \sum_{j=-\infty}^{\infty} \exp\big(-2\pi i(j+1)\big)(\partial/\partial\tau)^p\, v(x, j+1) \\ &= \big((\partial/\partial\tau)^s\, (\tilde{G}_0 v)\big)\, (\xi, x, 0).\end{aligned}$$

Therefore, for any $\xi \in [0, 2\pi]$ the function $\tilde{G}_0 v$ is in $\tilde{\mathbf{H}}^l(s)$. Moreover,

$$\begin{aligned}|G_0 v; \mathbf{L}_2(S)|^2 &= \int_0^{2\pi} \left\| \sum_{j=-\infty}^{\infty} \exp\Big(-i\xi(j+\tau)\Big) v(x, j+\tau); \mathbf{L}_2(S) \right\|^2 d\xi \\ &= \int_0^{2\pi} \sum_{j,q=-\infty}^{\infty} \exp\Big(i\xi(q-j)\Big) \langle v(x,j+\tau), v(x,q+\tau)\rangle_{\mathbf{L}_2(S)}\, d\xi \\ &= 2\pi \sum_{j=-\infty}^{\infty} \|v(x,j+\tau); \mathbf{L}_2(S)\|^2 = 2\pi \|v; \mathbf{W}^0_{2,0}(\tilde{C})\|^2.\end{aligned} \tag{10}$$

From (10) and the obvious relation

$$D_x^\alpha(-\xi + D_\tau)\,(\tilde{G}_0 v)(\xi, x, \tau) = \Big(\tilde{G}_0(D_x^\alpha D_t v)\Big)(\xi, x, \tau)$$

we derive

$$\begin{aligned}&|G_0 v; \tilde{\mathbf{H}}^l(S)|^2 \\ &\le c \sum_{q=0}^{l} \sum_{|\alpha| \le l-q} \int_0^{2\pi} \left\| \sum_{j=-\infty}^{\infty} \exp\Big(-i\xi(j+\tau)\Big)\,(D_x^\alpha D_t^j v)(x, j+\tau); \mathbf{L}_2(S) \right\|^2 d\xi \\ &\le 2\pi c \sum_{q=0}^{l} \sum_{|\alpha|\le l-q} \sum_{j=-\infty}^{\infty} \|(D_x^\alpha D_t^j v)(x,j+\tau); \mathbf{L}_2(S)\|^2 \le C\|v; \mathbf{W}^l_{2,0}(\tilde{C})\|^2.\end{aligned} \tag{11}$$

Thus, $\tilde{G}_0 v \in \mathbf{L}_2\Big([0, 2\pi] \to \tilde{\mathbf{H}}^l(S)\Big)$. Since the space $\mathbf{C}_0^\infty(\overline{\tilde{C}})$ is dense in $\mathbf{W}^l_{2,0}(\tilde{C})$, there exists a unique extension of $\tilde{G}_0$ from $\mathbf{C}_0^\infty(\overline{\tilde{C}})$ to $\mathbf{W}^l_{2,0}(\tilde{C})$ while (11) remains true.

Let $z \in \mathbf{L}_2\Big([0, 2\pi] \to \tilde{\mathbf{H}}^l(S)\Big)$. We set

$$e_p(z; x, \tau) = \int_0^{2\pi} \exp(i\xi p)\, z(\xi, x, \tau) d\xi, \quad p \in \mathbb{Z}.$$

The convergence of the series $\sum_{p=-\infty}^{\infty} \|e_p(z;\cdot); \tilde{\mathbf{H}}^l(S)\|^2$ follows from the obvious inequality

$$\sum_{p=-M}^{N} \|e_p(z;\cdot); \tilde{\mathbf{H}}^l(S)\|^2 \le 2\pi \|z; \tilde{\mathbf{H}}^l(S)\|^2.$$

If $e_p(z; x, \tau) = 0$ for all $p \in \mathbb{Z}$, then

$$\int_0^{2\pi} \exp(i\xi p) \langle z, h\rangle_{\tilde{\mathbf{H}}^l(S)}\, d\xi = 0$$

for any $h \in \tilde{\mathbf{H}}^l(S)$. Therefore, $\langle z, h\rangle_{\tilde{\mathbf{H}}^l(S)} = 0$ and $z = 0$ almost everywhere in $[0, 2\pi]$. We obtain Parseval's equality

$$2\pi\|z; \tilde{\mathbf{H}}^l(S)\|^2 = \sum_{p=-\infty}^{\infty} \|e_p(z;\cdot); \tilde{\mathbf{H}}^l(S)\|^2. \tag{12}$$

The transform (9) with $\beta = 0$ defines the function

$$v(x,t) = (2\pi)^{-1} e_{[t]} \left(\exp(i\xi\tau)\, z;\; x, t - [t]\right).$$

Since

$$\left(D_x^\alpha D_t^j e_p(i\xi\tau)\, z; \cdot\right)(x,0) = \left(D_x^\alpha D_\tau^j e_{p-1}\left(\exp(i\xi\tau)\, z; \cdot\right)\right)(x,1),$$

there is the inclusion $v \in \mathbf{W}^l_{2,\mathrm{loc}}(\overline{\tilde{C}})$. Besides, from (12) it follows that

$$\|v; \mathbf{W}^l_{2,0}(\tilde{C})\| \le c\|z; \tilde{\mathbf{H}}^l(S)\|.$$

It remains to check the equality $\tilde{G}_0^{-1}\tilde{G}_0 v = v$. We have

$$e_p\left(\exp(i\xi\tau)\tilde{G}_0 v; x, \tau\right) = \int_0^{2\pi} \exp\left(i\xi(p+\tau)\right) \sum_{j=-\infty}^{\infty} \exp\left(-i\xi(j+\tau)\right) v(x, j+\tau)\, d\xi$$

$$= \sum_{j=-\infty}^{\infty} v(x, j+\tau) \int_0^{2\pi} \exp\left(i\xi(p-j)\right) d\xi = 2\pi v(x, p+\tau).$$

□

Lemma 17.5.6. *Let $v \in \mathbf{W}^l_{2,\beta}(\tilde{C})$ and let P be one of the operators $\mathbf{L}, \mathbf{B}_1, \dots, \mathbf{B}_m$. Then*

$$\tilde{G}_\beta\left(P(x,t,D_x,D_t)v\right) = P(x,\tau,D_x,\xi + i\beta + D_\tau)\tilde{G}_\beta v.$$

The proof immediately follows from (8).

17.5.4 Unique solvability of the model problem

We return to problem (1), (2).

Theorem 17.5.7. *Let the operator* (4) *with $\lambda = \lambda_0$ be an isomorphism. Then the mapping* (3) *is an isomorphism if and only if the set $\Delta_\beta = \{\lambda \in \mathbb{C}\colon \operatorname{Re}\lambda = \beta, -\operatorname{Im}\lambda \in [0, 2\pi)]\}$ contains no eigenvalues of the pencil $\tilde{A}(\lambda)$.*

Proof. Sufficiency. Let $F = (f, g_1, \dots, g_m) \in \mathbf{RW}^l_{2,\beta}(\tilde{C})$. From Lemma 17.5.6 it follows that $\tilde{G}_\beta F \in \mathbf{L}_2([0,2\pi] \to \mathbf{R}\tilde{\mathbf{H}}^l(S)$. Since the set Δ_β is free from eigenvalues of $\tilde{A}(\lambda)$, for any $\xi \in [0, 2\pi]$ there is a continuous operator

$$\tilde{R}(\beta - i\xi) : \mathbf{R}\tilde{\mathbf{H}}^l(S) \to \tilde{\mathbf{H}}^{l+2m}(S)$$

such that

$$\tilde{A}(\beta - i\xi)U(\xi, x, \tau) = \tilde{G}_\beta F(\xi, x, \tau)$$

with $U = \tilde{R}(\beta - i\xi)\tilde{G}_\beta F$ (see Theorems 17.5.3. and 17.5.4). Because the norms of the operators $\tilde{R}(\beta - i\xi)$ with $\xi \in [0, 2\pi]$ are uniformly bounded,

$$\|U; \tilde{\mathbf{H}}^{l+2m}(S)\| \le c\|\tilde{G}_\beta F; \mathbf{R}\tilde{\mathbf{H}}^l(S)\| \le C\|F; \mathbf{RW}^l_{2,\beta}(\tilde{C})\|. \tag{13}$$

We set $u = \tilde{G}_\beta^{-1} U$. By Lemma 17.5.5, $u \in \mathbf{W}^{l+2m}_{2,\beta}(\tilde{C})$. Moreover, from Lemma 17.5.6 it follows that u satisfies problem (1), (2). Thus, the mapping (3) is surjective. Lemma 17.5.5 and the inequality (13) imply the estimate

$$\|u; \mathbf{W}^{l+2m}_{2,\beta}(\tilde{C})\| \le c\|F; \mathbf{RW}^l_{2,\beta}(\tilde{C})\|, \tag{14}$$

with constant c independent of $F = (f, g_1, \dots, g_m)$.

If u is a solution to the homogeneous problem (1), (2), then $(\tilde{G}_\beta u)(\xi,\cdot)$ is in the kernel of $\tilde{A}(\beta - i\xi)$. Since the subspaces $\ker \tilde{A}(\beta - i\xi)$ for $\xi \in [0, 2\pi)$ are trivial, we have $\tilde{G}_\beta u = 0$ and $u = 0$.

Necessity. Let the set Δ_β contain an eigenvalue λ_0. We denote by Φ_0 a nonzero solution to the equation $\tilde{A}(\lambda_0)\Phi_0 = 0$. Let

$$u_N(x,t) = \exp(\lambda_0 t)\, \Phi_0(x, t - [t])\chi(t-N)\chi(N-t),$$

where $N \in \mathbb{N}$, χ a function in $\mathbf{C}_0^\infty(\mathbb{R})$ such that $\chi(t) = 1$ for $t \le -1$ and $\chi(t) = 0$ for $t > 0$. Clearly, $u_N \in \mathbf{W}^{l+2m}_{2,\beta}(\tilde{C})$ and

$$\|\exp(\beta t)u_N; \mathbf{L}_2(\tilde{C})\| \ge 2(N-1)\|\Phi_0; \mathbf{L}_2(S)\|.$$

The function u_N satisfies problem (1), (2) with right-hand side $F_N = \exp(\lambda t) \times [\mathbf{A}(x,t,D_x,-i\lambda + D_t), \chi(t-N)\chi(N-t)]\Phi_0$, whose norm in $\mathbf{RW}^l_{2,\beta}(\tilde{C})$ is majorized by a constant. Therefore, the inequality (14) with constant c independent of F_N cannot be true (cf. Proposition 1.3.11). □

Remark 17.5.8. From the proof of Theorem 17.5.7 it follows that problem (1), (2) may have closed range only if the operator $\tilde{A}(\lambda)$ in (4) is an isomorphism for some λ.

17.5.5 Asymptotics of solutions

Suppose that

$$F = (f, g_1, \dots, g_m) \in \mathbf{RW}^l_{2,\beta_1}(\tilde{C}) \cap \mathbf{RW}^l_{2,\beta_2}(\tilde{C}) \tag{15}$$

with $\beta_1 < \beta_2$. For $\gamma \in [\beta_1, \beta_2]$ we have the inclusion $F \in \mathbf{RW}^l_{2,\gamma}(\tilde{C})$ and the inequality

$$\|F; \mathbf{RW}^l_{2,\gamma}(\tilde{C})\| \le c(\|F; \mathbf{RW}^l_{2,\beta_1}(\tilde{C})\| + \|F; \mathbf{RW}^l_{2,\beta_2}(\tilde{C})\|),$$

where the constant c does not depend on γ. The function

$$\hat{F}(\lambda) = (\tilde{G}_{\mathrm{Re}\lambda}F)(-\mathrm{Im}\lambda, x, \tau)$$

is holomorphic in the rectangle $\{\lambda \in \mathbb{C} \colon \mathrm{Re}\lambda \in (\beta_1, \beta_2), -\mathrm{Im}\lambda \in (-\delta, 2\pi + \delta)\}$ with $\delta > 0$ and takes values in the space $\mathbf{R\tilde{H}}^l(S)$.

Theorem 17.5.9. *Let* (15) *be fulfilled and let* Δ_{β_1} *and* Δ_{β_2} *contain no eigenvalues of* $\tilde{A}(\lambda)$. *We also assume that* u_q *solves problem* (1), (2) *in* $\mathbf{W}^{l+2m}_{2,\beta_q}(\tilde{C})$, $q = 1, 2$. *Then there holds the representations*

$$u_1(x,t) = \sum_{j=1}^{J} \sum_{p=1}^{I_j} \sum_{q=0}^{\varkappa_{jp}-1} c_j^{p,q} \Phi_j^{(p,q)}(x,t) + u_2(x,t), \tag{16}$$

where $c_j^{p,q}$ *are some constants,* $\Phi_j^{(p,q)}$ *the functions* (7), *and* $\lambda_1, \dots, \lambda_J$ *the eigenvalues of* $\tilde{A}(\lambda)$ *located in the rectangle* $\{\lambda \in \mathbb{C} \colon \mathrm{Re}\lambda \in (\beta_1, \beta_2), -\mathrm{Im}\lambda \in [0, 2\pi)\}$.

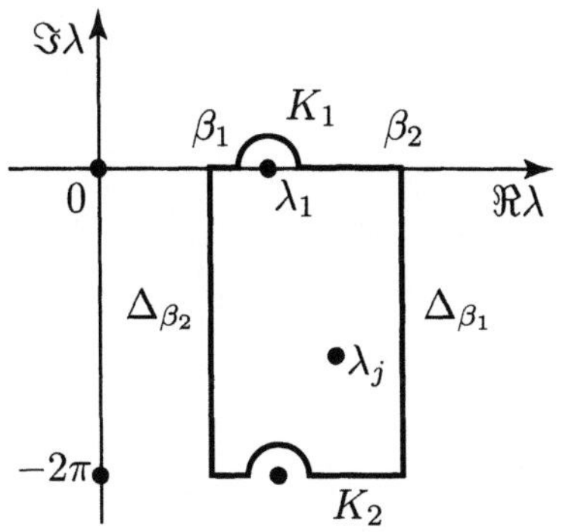

Fig. 17.1

Proof. Let K be a simple closed contour consisting of the intervals Δ_{β_1} and Δ_{β_2} and arcs K_1 and K_2 connecting the endpoints of Δ_{β_1} and Δ_{β_1} (see Fig. 17.1). We consider that the arcs K_1 and K_2 are free from the poles of $\tilde{R}(\lambda)$ and differ by the shift by 2π along the imaginary axis.

Denote the poles located within the contour K by $\lambda_1, \dots, \lambda_J$. From (5) and (8) we obtain

$$\int\limits_{K_1} \exp(\lambda t)\tilde{R}(\lambda)\hat{F}(\lambda)d\lambda = \int\limits_{K_2} \exp(\lambda t)\tilde{R}(\lambda)\hat{F}(\lambda)d\lambda. \tag{17}$$

By virtue of Theorem 17.5.7,

$$u_k = \int\limits_{\Delta_{\beta_k}} \exp(\lambda t)\tilde{R}(\lambda)\hat{F}(\lambda)d\lambda. \tag{18}$$

Substituting (17) and (18) into Cauchy's formula

$$\oint_K \exp(\lambda t)\tilde{R}(\lambda)\hat{F}(\lambda)d\lambda = 2\pi i \sum_{j=1}^{J} \operatorname{Res}\{\exp(\lambda_j t)\tilde{R}(\lambda_j)\hat{F}(\lambda_j)\},$$

we see that

$$u_2 - u_1 = 2\pi i \sum_{j=1}^{J} \operatorname{Res}\{\exp(\lambda_j t)\tilde{R}(\lambda_j)\hat{F}(\lambda_j)\}. \tag{19}$$

Finally, we substitute (19) into the expansion (6) of $\tilde{R}$ and find the residues. □

Chapter 18

Paradoxes of Limit Passage in Solutions of Boundary Value Problems When Smooth Domains Are Approximated by Polygons

In the present chapter we approximate smooth domains by polygonal ones and analyze passing to the limit in solutions of boundary value problems. For this purpose we use an asymptotic approach. Though the techniques have a broader field of applications, we restrict ourselves to studying some particular boundary value problems in the theory of thin plates. More precisely, we consider a number of questions related to the known paradox that goes back to SAPONDZHYAN [2] and BABUŠKA [2]. The case in point is the following example of instability: if one makes approximations of a thin circular plate by regular polygons with freely supported boundaries, one obtains the limiting solution that does not satisfy the freely supporting condition. We explain the asymptotic genesis of the Sapondzhyan-Babuška paradox, point a way to eliminate it and reveal some new phenomena of instability.

In Section 18.1, we consider approximations of a smooth convex plate by a sequence of freely supported polygons whose boundaries γ_N consist of N segments. The boundary condition $\Delta w_N = 0$ remains valid after passing to the limit as $N \to \infty$. The freely supporting condition for the limiting contour γ is of the form

$$\Delta w + (1-\sigma)\rho^{-1}\partial u/\partial \nu = 0,$$

σ being the Poisson coefficient, ρ the curvature radius, and ν the inward normal. It is shown that assigning some point moments at the vertices of the polygons, one can obtain any distribution of moments on the limiting contour. In particular, choosing the mentioned point moments in a special way, one obtains the freely supporting condition and eliminates the Sapondzhyan-Babuška paradox.

These results are based on the asymptotic formula for the solutions w_N as $N \to \infty$. Far from the contour γ_N the principal term of the asymptotics is a solution w_N to the Sophie Germain equation. Near γ_N there arises a boundary layer. Outside a small neighborhood of the vertices the layer is defined as a solution to the biharmonic equation in a half-strip Π. We set the periodicity conditions on the sides of Π and the solution and its Laplacian on the base of Π. The decaying condition of the solution at infinity defines boundary conditions for w. The large discrepancy given by the boundary layer near a corner point is compensated by a special solution to a boundary value problem for the biharmonic equation in the angular domain.

The construction of the boundary layer is typical for the method of homogenization of boundary value problems with rapidly oscillating coefficients or boundary of the domain (see BENSOUSSAN, LIONS and PAPANICOLAOU [1], BAKHVALOV and PANASENKO [1] and Sections 17.2 and 17.3). The only innovation here is the correction of the layer near corner points.

In Section 18.2, by the same techniques we study the problem on a plate freely supported along the smooth boundary of a convex hole approximated by polygons. Such a situation has been considered for the first time and differs from that in Section 18.1 in a qualitative sense. In the case of a convex plate the boundary condition $\Delta w = 0$ remains valid after passing to the limit. However, when the hole is approximated, there arises the new condition $\Delta w + 4\rho^{-1}\partial w/\partial\nu = 0$ on the limiting contour. True, as before in Section 18.1, assigning point moments at the corner points, one can obtain any given value for Δw on the boundary of the hole.

It is shown in Section 18.3, that using other approximations to the smooth contour by polygons, we have the rigidly fixing conditions as a limiting case of the freely supporting ones. This fact has been established with the help of some integral estimates on the restrictions to γ_N of the gradients of functions in the intersections of the spaces $\mathbf{W}_2^2$ and $\overset{\circ}{\mathbf{W}}{}_2^1$ in the domain bounded by the contour γ_N. Therefore, the results in Section 18.3 for solutions to the Sophie Germain equations pertain equally to the nonlinear bending equations and to the dynamic problems in plate theory.

18.1 Approximation to a Freely Supported Convex Plate

18.1.1 Statement of the problem and description of the results

We consider the plate ω bounded by a smooth convex closed contour γ of unit length. The bending u of the plate under the freely supporting conditions on γ is a solution to the boundary value problem

$$D\Delta^2 u(x) = q(x), \quad x \in \omega; \tag{1}$$

$$u(x) = \Delta u(x) + (1-\sigma)k(x)(\partial/\partial\nu)\,u(x) = 0, \quad x \in \gamma, \tag{2}$$

where D is the cylindrical stiffness of the plate, k the curvature of γ, $q \in \mathbf{C}^\infty(\bar\omega)$ a normal loading, σ the Poisson coefficient, and ν the inward normal to γ.

Let us describe an approximation to the plate ω by polygonal ones. We denote by $P^{(N)} = \big(p_j^{(N)}\big)_{j=1}^N$ a collection of points in γ dividing the contour into N arcs equal in length. Let ω be a polygonal line with vertices at the points $p_j^{(N)}$. The plate bounded by the contour γ_N is denoted by ω_N (see Fig. 18.1). The bend w_N of the plate with freely supported boundary satisfies the boundary value problem

$$D\Delta^2 w_N(x) = q(x), \quad x \in \omega; \tag{3}$$

$$w_N(x) = \Delta w_N(x) = 0, \quad x \in \gamma_N \setminus P^{(N)}. \tag{4}$$

Since the plate is convex, the integrals

$$\int_{\omega_N} |\nabla w_N(x)|^2\,dx, \quad \int_{\omega_N} |\nabla \Delta w_N(\chi)|^2\,dx$$

are uniformly bounded with respect to N. Therefore, the limiting solution ω and its Laplacian Δw are weak limits of sequences in the Sobolev space $\overset{\circ}{\mathbf{W}}{}_2^1(\omega)$. Hence, w satisfies the boundary conditions

$$w(x) = \Delta w(x) = 0, \quad x \in \gamma, \tag{5}$$

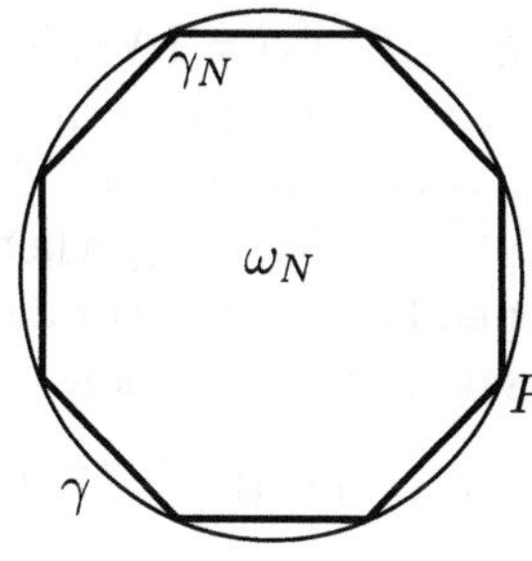

Fig. 18.1

which differ from the freely supporting conditions (2) of the contour γ. The Sapondzhyan-Babuška paradox consists of such an instability.

18.1.2 Formal asymptotics

We construct the principal term in the asymptotics of the solution to problem (3), (4) as $N \to \infty$. According to what was said on the limit passage from w_N to w, the solution to the Sophie Germain equation

$$D\Delta^2 w_N(x) = q(x), \quad x \in \omega, \tag{6}$$

with boundary conditions (5) can be taken as basic approximation to w_N far from γ_N.

We construct a boundary layer compensating the discrepancy of w_N in (4). To this end we introduce coordinates (ν, τ) in a d-neighborhood U_d of γ, where ν is the distance to γ along the inward normal and τ is the length of arc on γ. The Laplace operator takes the form

$$(\partial/\partial\nu)^2 - \big(1 - k(\tau)\big)^{-1} k(\tau)(\partial/\partial\nu) + \Big\{\big(1 - \nu k(\tau)\big)^{-1}\partial/\partial\tau\Big\}^2.$$

Therefore,

$$\Delta^2 = \Big((\partial/\partial\nu)^2 + (\partial/\partial\tau)^2\Big)^2 + \nu\Lambda_4(\nu,\tau,\partial/\partial\nu,\partial/\partial\tau) + \Lambda_3(\nu,\tau,\partial/\partial\nu,\partial/\partial\tau), \tag{7}$$

Λ_j being a differential operator of order j with smooth coefficients. We set

$$w(x) = W(\nu,\tau), \quad \Delta w(x) = V(\nu,\tau), \quad (\partial/\partial\nu)w(x) = Y(\nu,\tau). \tag{8}$$

By virtue of (5),

$$W(\nu,\tau) = \nu Y(0,\tau) + O(\nu^2), \quad V(\nu,\tau) = O(\nu), \quad \nu \to 0. \tag{9}$$

We choose a vertex of the polygonal line γ_N as origin of the coordinates (ν, τ). Let us find the equation of γ_N for $|\tau| < N^{-1}$. If near the origin the line γ coincides with the circle of radius $R = k(0)^{-1}$, then $(\nu, \tau) \in \gamma_N$ if and only if

$$\nu = R - R\Big\{\cos\Big(\big(|\tau| - 1/2N\big)/R\Big)\Big\}^{-1}\cos(1/2N).$$

Hence, for any smooth contour the equation of the two segments of γ_N outgoing from the origin is of the form

$$\nu = k(0)^{-1}\Big\{1 - \big\{\cos\big(k(0)(|\tau| - 1/2N)\big)\big\}^{-1}\cos\big(k(0)/2N\big)\Big\} + O(N^{-3})$$

or, equivalently,

$$\nu = -(2N^2)^{-1}k(0)\Big(\big(N|\tau| - 1/2\big)^2 - 1/4\Big) + O(N^{-3}). \tag{10}$$

We introduce the variables $\xi = (\xi_1, \xi_2) = (N\tau, N\nu)$, as usual in the homogenization method. In these variables the principal term of the discrepancy of the solution w to problem (6), (5) on γ_N coincides with $N^{-1}\xi_2 Y(0,\tau)$. Therefore, we seek the boundary layer in the form $N^{-2}z(\tau, N\tau, N\nu)$, where $(\tau,\xi_1,\xi_2) \to z(\tau,\xi_1,\xi_2)$ is a bounded function smoothly depending on τ, periodic in ξ_1 with unit period. In view of (7), the function z must satisfy the equation

$$\Delta_\xi^2 z(\tau,\xi) = 0, \quad \xi = (\xi_1,\xi_2) \in \Pi = (-1/2, 1/2) \times (0,\infty). \tag{11}$$

On the vertical sides of Π there are given the periodicity conditions

$$(\partial/\partial\xi_1)^k z(\tau, -1/2, \xi_2) = (\partial/\partial\xi_1)^k z(\tau, 1/2, \xi_2), \quad \xi_2 \in (0,\infty), \quad k = 0,1,2,3. \tag{12}$$

According to (10), in the coordinates ξ the equation of γ_N is of the form

$$\xi_2 = (2N)^{-1} k(\tau)((1/4 - (|\xi_1| - 1/2)^2) + O(N^{-2}). \tag{13}$$

From (9) it follows that z is subject to the boundary conditions

$$z(\tau,\xi_1,0) = k(\tau)\big(1/4 - (|\xi_1| - 1/2)^2\big)\, Y(0,\tau)/2, \quad |\xi_1| < 1/2; \tag{14}$$

$$\Delta_\xi z(\tau,\xi_1,0) = 0, \qquad |\xi_1| < 1/2, \quad \xi_1 \neq 0. \tag{15}$$

We require $\xi_1 \neq 0$ because of the discontinuity of the derivative with respect to ξ_1 of the right-hand side in (14).

The next assertion contains all the information on problem (11), (12), (14), and (15) that will be used in what follows.

Lemma 18.1.1. *There exists a unique solution Z to the problem*

$$\begin{aligned}
&\Delta^2 Z(\xi) = 0, \quad \xi \in \Pi;\\
&Z(\xi_1, 0) = A_1\big(1/4 - (|\xi_1| - 1/2)^2\big),\\
&\Delta_\xi Z(\xi_1,0) = A_2 + A_3\delta(\xi_1), \quad |\xi_1| < 1/2;\\
&(\partial/\partial\xi_1)^k Z(-1/2,\xi_2) = (\partial/\partial\xi_1)^k Z(1/2,\xi_2), \quad \xi_2 > 0, \quad k = 0,1,2,3,
\end{aligned} \tag{16}$$

which admits the asymptotic representations

$$\begin{aligned}
Z(\xi) = (2\pi)^{-1}\Big((A_3 - 4A_1)\xi_2 \log\rho - 4A_1\vartheta\xi_1\Big)&\\
+ C_z\xi_2 + O(\rho^2|\log\rho|^2) \qquad &\text{as} \quad \rho \to 0;
\end{aligned} \tag{17}$$

$$Z(\xi) = \Big(A_1 + 3(A_2 + A_3)\xi_2^2\Big)/6 + o\Big(\exp\big(-(\pi - \beta)\xi_2\big)\Big) \quad \text{as} \quad \xi_2 \to \infty, \tag{18}$$

δ being Dirac's measure, (ρ,ϑ) the polar coordinates based on

$$\xi = 0 \quad (\vartheta \in [-\pi/2, \pi/2]),$$

with C_z, A_j some constants, and β any positive number.

Proof. We first note that the function $\varphi(\xi) = 2\pi^{-1}\vartheta\rho\sin\vartheta$ is biharmonic on the half-plane $\mathbb{R}^2_+ = \{\xi \colon \xi_2 > 0\}$ and $\varphi(\xi_1, 0) = \rho$.

Moreover, $\Delta_\xi\varphi(\xi) = 4\pi^{-1}\rho^{-1}\cos\vartheta$, $\Delta_\xi\varphi(\xi_1,0) = 4\delta(\xi_1)$. So we seek a solution to problem (16) in the form

$$Z(\xi) = \chi(\rho)(B\rho\log\rho\cos\vartheta + 2\pi^{-1}\, A_1\vartheta\rho\cos\vartheta) + R(\xi), \tag{19}$$

where χ is a cut-off function in $\mathbf{C}_0^\infty(0,1/2)$, $\chi(\rho) = 1$ for $\rho \le 1/4$ and $B = (2\pi)^{-1}(A_3 - 4A_1)$. The remainder R in (19) satisfies the boundary value problem

$$\begin{aligned} &\Delta^2 R(\xi) = F(\xi), \quad \xi \in \Pi; \\ &R(\xi_1, 0) = \Phi_1(\xi_1), \quad \Delta_\xi R(\xi_1,0) = \Phi_2(\xi_1), \quad |\xi_1| < 1/2; \\ &(\partial/\partial\xi_1)^k R(-1/2,\xi_2) = (\partial/\partial\xi_1)^k R(1/2,\xi_2), \quad \xi_2 > 0, \quad k = 0,1,2,3, \end{aligned} \tag{20}$$

with smooth functions Φ_1, Φ_2, and F that satisfy the periodicity conditions at $\xi_2 = \pm 1/2$ while $F(\xi) = 0$ for $\xi_2 > 1/2$.

Since problem (30) decomposes into two boundary value problems for the Poisson equation, it is easy to check that there exists a solution in $\mathbf{C}^\infty(\bar{\Pi})$ admitting the asymptotic representation

$$R(\xi) = C_0 + C_1\xi_2 + C_2\xi_2^2 + C_3\xi_2^3 + o\Big(\exp\big((\beta - \pi)\xi_2\big)\Big), \tag{21}$$

as $\xi_2 \to \infty$. The functions ξ_2 and ξ_2^3 satisfy the homogeneous problem (20) so in (21) we can set $C_1 = 0$ and $C_3 = 0$.

Thus the solution Z to problem (16) subject to (17) and

$$Z(\xi) = C_0 + C_2\xi_2^2 + o\Big(\exp\big((\beta - \pi)\xi_2\big)\Big) \tag{22}$$

can be found with the help of (19). It remains to calculate the constants C_0 and C_2 in (22). To this end we shall follow the method given in MAZ'YA and PLAMENEVSKI [4].

Denote by Π_{hH} the rectangle $(-1/2, 1/2) \times (0, H)$ which the half-disk $\{\xi\colon \rho \le h, |\vartheta| \le \pi/2\}$ is cut from. By Green's formula

$$\begin{aligned} &\int_{\Pi_{hH}} \Big(u(\xi)\Delta_\xi^2 v(\xi) - v(\xi)\Delta_\xi^2 u(\xi)\Big)d\xi \\ &= - \int_{\partial\Pi_{hH}} \Big\{u(\xi)(\partial/\partial\nu_\xi)\Delta_\xi v(\xi) - ((\partial/\partial\nu_\xi)u(\xi))\Delta_\xi v(\xi) + (\Delta_\xi u(\xi))(\partial/\partial\nu_\xi)v(\xi) \\ &\quad - v(\xi)(\partial/\partial\nu_\xi)\Delta_\xi u(\xi)\Big\}ds_\xi \end{aligned}$$

for the functions ξ_2 and Z, we obtain

$$\begin{aligned} 0 &= - \int_{\partial\Pi_{hH}} \Big\{\xi_2(\partial/\partial\nu_\xi)\Delta_\xi Z(\xi) - ((\partial/\partial\nu_\xi)\xi_2)\Delta_\xi Z(\xi)\Big\}ds_\xi \\ &= - \int_{-1/2}^{1/2} \Big(H(\partial/\partial\xi_2)\Delta_\xi Z(\xi_1,H) - \Delta_\xi Z(\xi_1,H)\Big)d\xi_1 + \int_{h<|\xi_1|<1/2} \Delta_\xi Z(\xi_1,0)d\xi_1 \\ &\quad - h\int_{-\pi/2}^{\pi/2} \Big\{\xi_2(\partial/\partial\rho)\Delta_\xi Z(\xi) - ((\partial/\partial\rho)\xi_2)\Delta_\xi Z(\xi)\Big\}\Big|_{\rho=h} d\vartheta. \end{aligned} \tag{23}$$

By (22), the first integral on the right in (23) is $-2C_2 + o(1)$ as $H \to \infty$. In view of (17) the last term on the right-hand side of (23) is equal to

$$\int\limits_{-\pi/2}^{\pi/2} \Big(2\pi^{-1} A_3 \cos^2 \vartheta + o(1)\Big)\, d\vartheta = A_3 + o(1) \quad \text{as} \quad h \to 0.$$

Finally, by virtue of the boundary conditions for $\Delta_\xi Z$ in (16) the limit of the integral

$$\int\limits_{h<|\xi_1|<1/2} \Delta_\xi Z(\xi_1, 0) d\xi_1$$

is equal to A_2. Therefore, $C_2 = (A_2 + A_3)/2$. Similarly,

$$\begin{aligned} 0 = -\int\limits_{\partial\Pi_{hH}} &\Big\{\xi_2^3(\partial/\partial\nu_\xi)\Delta_\xi Z(\xi) - \big((\partial/\partial\nu_\xi)\xi_2^3\big)\Delta_\xi Z(\xi)\Big\} \\ &+ 6\xi_2(\partial/\partial\nu_\xi)Z(\xi) - 6\big((\partial/\partial\nu_\xi)\xi_2\big)Z(\xi)\Big\} ds_\xi. \end{aligned} \tag{24}$$

The function Z being periodic, the sum of the integrals over the vertical sides of the rectangle $(-1/2, 1/2) \times (0, H)$ vanishes. Using (22), we find that the integral over the upper side is $6C_0 + o(1)$ as $H \to \infty$. From (17) it follows that the integral over the set $\{\xi\colon \xi_2 = 0, h < |\xi_1| < 1/2\}$ tends to

$$-6A_1 \int\limits_{-1/2}^{1/2} \Big(1/4 - (|\xi_1| - 1/2)^2\Big) = -A_1.$$

Thus $c_0 = A_1/6$ and (18) has been proved.

The uniqueness of the solution to problem (16)–(18) is obvious. □

As the boundary layer z we take the solution Z to problem (16) under the corresponding choice of A_j. To satisfy (14), (15), we set $A_1 = k(\tau)Y(0,\tau)/2$ and $A_2 = 0$. By virtue of (18) the boundedness condition for z leads to $A_3 = 0$. Therefore, according to (17), near the point $\xi = 0$

$$\begin{aligned} z(\tau, \xi) = -\pi^{-1}k(\tau)Y(0,\tau)\big(\rho\cos\vartheta\log\rho - \vartheta\rho\sin\vartheta + C_z\xi_2\big) \\ + O(\rho^2|\log\rho|) \quad \text{as} \quad \rho \to 0. \end{aligned} \tag{25}$$

Had the function z had square integrable derivatives of second order, the principal term in the asymptotics of the solution to the original problem (3), (4) would have been $v(x) + N^{-1}\chi(d^{-1}\nu)z(\tau, N\tau, N\nu)$. However, as (25) shows, z does not belong to the space $\mathbf{W}^2_{2,\text{loc}}(\bar\Pi)$. It means that we have to improve the boundary layer near the corner points. We shall construct the needed corrections.

Let (r, ϑ) be polar coordinates based on one of the vertices denoted by O. In a neighborhood of the point O the domain ω_N is given by the inequality $|\vartheta| < \alpha(N)/2$. The solution w_N to problem (3), (4) has the asymptotics

$$w_N(x) = Kr^{\pi/\alpha(N)} \cos\big(\pi\vartheta/\alpha(N)\big) + o(r^2) \quad \text{as} \quad r \to 0 \tag{26}$$

(see Lemma 5.8.1). The length of the part of γ between the two vertices adjacent to O is $2N^{-1}$. The curvature radius of γ at O is equal to $k(0)^{-1}$. Therefore,

$$\alpha(N) = \pi - k(0)N^{-1} + O(N^{-2}). \tag{27}$$

It follows that

$$w_N = Kr^{1+k(0)/\pi N+O(N^{-2})} \cos\left\{\vartheta\Big(1+(\pi N)^{-1}k(0)+O(N^{-2})\Big)\right\} + O(r^2)$$

or for $r = O(N^{-3/2})$

$$w_N(x) = K\{r\cos\vartheta + (\pi N)^{-1}k(0)(r\log r\cos\vartheta - r\vartheta\sin\vartheta)\} + O(N^{-2}). \tag{28}$$

By virtue of (25),

$$\begin{aligned} w(x) = N^{-2}z(\tau, N\tau, N\nu) &= (\partial/\partial\nu)w(0)\Big\{r\cos\vartheta + (\pi N)^{-1}k(0)r \\ &\times\Big(\cos\vartheta\log(rN) - \vartheta\sin\vartheta + C_z\cos\vartheta\Big)\Big\} + O(r^2|\log r|) \quad \text{as} \quad r\to 0. \end{aligned} \tag{29}$$

Comparing (28) with (29), we arrive at

$$K = (\partial/\partial\nu)w(0)\Big(1-(\pi N)^{-1}k(0)(\log N + C_z)\Big) + O(N^{-2}).$$

Thus, as the correction to the boundary layer near the point $p_j^{(N)}$ we have to take

$$\begin{aligned} \Psi_N^{(j)}(r_j,\vartheta_j) &= (\partial/\partial\nu)w(p_j^{(N)})\Big(1-(\pi N)^{-1}k(p_j^{(N)})(\log N + C_z)\Big) \\ &\times r_j^{\pi/\alpha_j(N)}\cos\big(\pi\vartheta_j/\alpha_j(N)\big), \end{aligned} \tag{30}$$

where (r_j, ϑ_j) are polar coordinates based on $p_j^{(N)}$ and $\alpha_j(N)$ is the opening of the angle of the polygon ω_N at $p_j^{(N)}$.

18.1.3 Justification of the asymptotics

The argument in 18.1.2 shows that in the zone $D_N^{(0)} = \{x\in\omega_N\colon \operatorname{dist}(x, P^{(N)}) > N^{-3/2}\}$ the asymptotics of $w_N(x)$ coincides with the difference between the solution $w(x)$ to problem (6), (5) and the boundary layer $N^{-2}z(\tau, N\tau, N\nu)$ while in the small neighborhoods $D_N^{(j)} = \{x\in\omega_N\colon \operatorname{dist}(x, p_j^{(N)}) < 2N^{-3/2}\}$ of the corner points it coincides with the functions (30). Therefore, we set

$$A_N(x) = X_N^{(0)}(x)\big(w(x) - N^{-2}\chi(d^{-1}\nu)z(\tau, N\tau, N\nu)\big) + \sum_{j=1}^{N} X_N^{(j)}(x)\,\Psi_N^{(j)}(r_j,\vartheta_j), \tag{31}$$

where $X_N^{(0)}, X_N^{(1)}, \dots, X_N^{(N)}$ is the partition of unity subject to the covering of the domain ω_N by the neighborhoods $D_N^{(0)}, D_N^{(1)}, \dots, D_N^{(N)}$:

$$X_N^{(0)}(x) = \sum_{k=1}^{N}\big(1-\chi(N^{3/2}r_k/4)\big), \quad X_N^{(j)}(x) = \chi(N^{3/2}r_j/4), \quad j = 1,\dots,N,$$

χ being the cut-off function in (19). The factor $\chi(d^{-1}\nu)$ is introduced in (31) because the coordinates (τ,ν) are defined only on the d-neighborhood of γ.

Let us estimate the discrepancy of the function A_N in (3). There holds the equality

$$\begin{aligned}\Delta^2 A_N(x) = {} & X_N^{(0)} D^{-1} q(x) + [\Delta^2, X_N^{(0)}]\{w(x) - N^{(-2)} z(\tau, N\tau, N\nu)\} \\ & - N^{-2}[\Delta^2, \chi(d^{-1}\nu)] z(\tau, N\tau, N\nu) \\ & - N^{-2} X_N^{(0)}(x)\chi(d^{-1}\nu)] z\{[\Delta^2 - (\partial/\partial\nu)^2 - (\partial/\partial\tau)^2\}(\tau, N\tau, N\nu) \\ & + \sum_{j=1}^{N} [\Delta^2, X_N^{(j)}(x)] \Psi_N^{(j)}(r_j, \vartheta_j).\end{aligned} \tag{32}$$

Denote the terms on the right in (32) by $F_N^{(1)}, \dots, F_N^{(5)}$. By virtue of (17), (18), and the representation (7) of the biharmonic operator,

$$\begin{aligned}&|F_N^{(3)}(x)| \le cN^{-2}, \\ &|F_N^{(4)}(x)| \le c\Big(N^{-2} + N(N\nu + 1)\exp\big(-(\pi - \beta)N\nu\big)\Big)\log N.\end{aligned} \tag{33}$$

Combining (29) and (30), we obtain the inequality

$$|F_N^{(2)}(x) + F_N^{(5)}(x)| \le cN^3 \log N. \tag{34}$$

Note that

$$\operatorname{supp} F_N^{(2)} \cup \operatorname{supp} F_N^{(5)} \subset D_N^{(1)} \cup \dots \cup D_N^{(N)}.$$

Now we estimate the discrepancy of A_N in the boundary conditions (4). According to (30),

$$X_N^{(j)}(x)\Psi_N^{(j)}(r_j, \vartheta_j) = \Delta\Big(X_N^{(j)}(x)\Psi_N^{(j)}(r_j, \vartheta_j)\Big) = 0$$

on γ_N and the restrictions G_N^1 and G_N^Δ of A_N and ΔA_N to γ_N vanish outside $\gamma_N \cap D_N^{(0)}$, i.e., for $r_j < N^{-3/2}$. Moreover, taking (9), (10) and the boundary conditions for z into account, we obtain

$$|r_j^p(\partial/\partial l)^p G_N^1(x)| + |r_j^{p+2}(\partial/\partial l)^p G_N^\Delta(x)| < cN^{-3}, \quad r_j < 2N^{-1}/3, \tag{35}$$

on every segment of the polygon line for $p \in \mathbb{N}_0$, where $\partial/\partial l$ is the derivative along the segment.

We construct a function B_N compensating the above discrepancies. Let $\eta_1^{(0)}$, $\dots$, $\eta_N^{(0)}$ and $\eta_1^{(1)}, \dots, \eta_N^{(1)}$ be functions in $\mathbf{C}_0^\infty[0,1)$ such that

$$\sum_{k=1}^{N} \eta_k^{(0)}(Nr_k) = 1 \quad \text{on} \quad \gamma_N, \quad \eta_j^{(1)} = 1, \quad t \in \operatorname{supp} \eta_j^{(0)}. \tag{36}$$

Denote by $B_N^{(j)}$ the solution to the boundary value problem

$$\begin{aligned}&\Delta_\xi^2 B_N^{(j)}(\xi) = 0, \quad \xi \in K_N^{(j)} = \{\xi \in \mathbb{R}^2 \colon \rho > 0, |\vartheta| < \alpha_j(N)/2)\}, \\ &B_N^{(j)}(\xi) = -\eta_j^{(0)}(|\xi|) G_N^1(N^{-1}\xi + p_j^{(N)}), \\ &\Delta_\xi B_N^{(j)}(\xi) = -\eta_j^{(0)}(|\xi|) G_N^\Delta(N^{-1}\xi + p_j^{(N)}), \quad \xi \in \partial K_N^{(j)}.\end{aligned} \tag{37}$$

From (35) it follows that the energy integral

$$\tilde{E}(B_N^{(j)}; K_N^{(j)}) = 2^{-1} D \int_{K_N^{(j)}} \Big\{ |\Delta_\xi B_N^{(j)}|^2 - 2(1-\sigma)\Big((\partial^2 B_N^{(j)}/\partial\xi_1^2)(\partial^2 B_N^{(j)}/\partial\xi_2^2) - (\partial^2 B_N^{(j)}/\partial\xi_1\partial\xi_2)^2\Big)\Big\} d\xi,$$

corresponding to the solution to problem (37) does not exceed const$\cdot N^{-6}$. Moreover, the function $\eta_j^{(1)} B_N^{(j)}$ satisfies the same boundary conditions as $B_N^{(j)}$ (see (36)) and

$$|\Delta_\xi^2 (\eta_j^{(1)}(|\xi|)\, B_N^{(j)}(\xi))| \le cN^{-3}.$$

We set

$$B_N(x) = \sum_{j=1}^{N} \eta_j^{(1)}(Nr_j)\, B_N^{(j)}(N(x - p_j^{(N)})). \tag{38}$$

In view of the properties (36) of the cut-off functions $\eta_j^{(0)}$ and $\eta_j^{(1)}$

$$B_N(x) = -G_N^1(x), \quad \Delta B_N(x) = -G_N^\Delta(x), \quad x \in \gamma_N.$$

From what we have said about the solutions to problem (37), it follows that

$$\tilde{E}(B_N; \omega_N) \le cN^{-3}, \quad |\Delta^2 B_N(x)| \le cN, \quad \operatorname{supp} \Delta^2 B_N \subset \{x \in \omega_N : \nu < 2N^{-1}\}. \tag{39}$$

The difference $R_N = w_N - A_N - B_N$ between the solution to problem (3), (4) and the constructed asymptotic representation belongs to $\mathbf{W}_2^2(\omega_N)$ and satisfies the boundary value problem

$$\Delta^2 R_N(x) = (1 - X_N^{(0)}(x)) D^{-1} q(x) - \Delta^2 B_N(x) - \sum_{h=2}^{5} F_N^{(h)}(x), \quad x \in \omega_N; \tag{40}$$

$$R_N(x) = \Delta R_N(x) = 0, \quad x \in \gamma_N. \tag{41}$$

To estimate the energy integral

$$\tilde{E}(R_N; \omega_N) = 2^{-1} D \int_{\omega_N} \Big\{ |\Delta R_N|^2 - 2(1-\sigma)\Big((\partial^2 R_N/\partial x_1^2)(\partial^2 R_N/\partial x_2^2) - \big((\partial/\partial x_1)(\partial/\partial x_2) R_N\big)^2\Big)\Big\} dx, \tag{42}$$

we need the following assertion.

Lemma 18.1.2. *The inequality*

$$\int_{\omega_N} d(x)^3 \big(1 + |\log d(x)|\big)^{-2} |R_N(x)|^2\, dx \le c_0^{-1} \tilde{E}(R_N; \omega_N) \tag{43}$$

holds with $c_0 > 0$ *and* $d(x) = \operatorname{dist}(x, \gamma)$.

To prove (43) it suffices to apply Hardy's inequality for functions of one variable two times (see Lemma 1.4.2).

Theorem 18.1.3. *The difference* $w_N - w$ *between the solutions to problems* (3), (4) *and* (6), (5) *satisfies the estimate*

$$\tilde{E}(w_N - w; \omega_N) \le \text{const}\, N^{-1/2} |\log N|^4, \tag{44}$$

$\tilde{E}$ *being the energy integral.*

Proof. Let us multiply (40) by R_N and integrate by parts taking account of the boundary conditions (41). We obtain

$$\tilde{E}(R_N;\omega_N) = 2^{-1}D\int\limits_{\omega_N}\tilde{F}_N(x)R_n(x)dx, \tag{45}$$

where $\tilde{F}_N$ stands for the right-hand side of (40). From (32), (33), (39) it follows that the integral on the right in (45) does not exceed

$$c\Bigg(\int\limits_{\omega_N} N^{-2}|R_N(x)|dx + \int\limits_{\{x\in\omega_N:\,\nu<d\}} N(N\nu+1)\log N\exp\big((\gamma-\pi)N\nu\big)\,|R_N(x)|dx$$
$$+\int\limits_{\{x\in\omega_N:\,\nu<2N^{-1}\}} N|R_N(x)|dx + \sum_{j=1}^{N}\int\limits_{D_N^{(j)}} N^3\log N\,|R_N(x)|dx\Bigg).$$

Hence, according to Lemma 18.1.1,

$$\tilde{E}(R_N;\omega_N) \le C\Bigg(\int\limits_{\omega_N} N^{-4}dx + \int\limits_{\{x\in\omega_N:\,\nu<d\}} N^2(N\nu+1)^2|\log N|^2\,\nu^3\,|\log\nu|^2$$
$$\exp\big(-2(\gamma-\pi)N\nu\big)\,dx$$
$$+\int\limits_{\{x\in\omega_N:\,\nu<2N^{-1}\}}\nu^3\,|\log\nu|^2N^2dx + \sum_{j=1}^{N}\int\limits_{D_N^{(j)}} N^6|\log N|^2\nu^3\,|\log\nu|^2\,dx\Bigg)$$
$$+2^{-1}c_0\int\limits_{\omega_N} d(x)^{-3}\big(1+|\log d(x)|\big)^{-2}\,|R_N(x)|^2\,dx$$
$$\le CN^{-1/2}|\log N|^4 + 2^{-1}\tilde{E}(R_N;\omega_N),$$

where c_0 is the constant in (43). Therefore,

$$\tilde{E}(R_N;\omega_N) \le 2CN^{-1/2}|\log N|^4. \tag{46}$$

Since

$$\tilde{E}(w-A_N;\omega_N) \le c\Bigg\{\int\limits_{\{x\in\omega_N:\,\nu<d\}} N^{-4}\Big(1+N^2\exp\big(-(\pi-\gamma)N\nu\big)\Big)^2dx$$
$$+\sum_{j=1}^{N}\int\limits_{D_N^{(j)}} r_j^{-1}\,dx \le CN^{-1/2},$$

the relations (46) and (39) imply (44). □

18.1.4 Concentrated moments at the vertices of the polygon

We impose the supplementary condition

$$\begin{aligned}M_r(x) &= -D\Big\{(\partial/\partial r_j)^2 w_N(x) + r_j^{-2}\sigma\big((\partial/\partial\vartheta_j)^2 w_N(x) + r_j(\partial/\partial r_j)w_N(x)\big)\Big\}\\ &= D(4\pi N)^{-1}\Big\{2-\alpha_j(N)^{-1}\pi + \sigma\big(2+\alpha_j(N)^{-1}\pi\big)\Big\}\\ &\quad\times m(p_j^{(N)})r_j^{-\pi/\alpha_j(N)}\cos\big(\pi\vartheta_j/\alpha_j(N)\big) + O(r_j^{-\pi/\alpha_j(N)}),\quad j=1,\dots,N\end{aligned} \tag{47}$$

on the function w_N, where m is a smooth function on γ. The relations (47) mean that the point moments $M_{r_j}(x)$ (or $M_{\vartheta_j}(x) = -M_{r_j}(x) + o(r_j^{\pi/\alpha_j(N)})$) of small intensity are given at the vertices of the polygon γ_N; see SAPONDZHYAN [2]. We shall show that choosing the function m, one can provide any limiting distribution of moments on γ. In particular, for $m = -(1-\sigma)k\partial u/\partial\nu$, u being the solution to problem (1), (2) on the bending of the freely supported plate ω, we obtain the convergence of the sequence $\{w_N\}$ to u. Formulating the problem in this way resolves the Sapondzhyan-Babuška paradox, albeit in a somewaht artificial way.

Problem (3), (4), (47) is uniquely solvable. However, the energy integral $\tilde{E}(w_N;\omega_N)$ for the solution turns out to be infinite. Indeed, let us seek w_N in the form

$$\begin{aligned}w_N(x) &= W_N(x) - (4\pi N)^{-1}\sum_{j=1}^{N} X_N^{(j)}(x)m(p_j^{(N)})\big(\alpha_j(N)^{-1}\pi - 1\big)^{-1}\\ &\quad\times r_j^{2-\pi/\alpha(N)}\cos\big(\pi\vartheta_j/\alpha_j(N)\big),\end{aligned} \tag{48}$$

where $X_N^{(j)}$ is the cut-off function given in 18.1.3. The function $r^{2-\pi/\alpha_j(N)}\cos\big(\pi\vartheta_j/\alpha_j(N)\big)$ is biharmonic in the angle $K_N^{(j)}$ and vanishes together with its Laplacian on $\partial K_N^{(j)}$. Substituting (48) into (3), (4), we obtain the following boundary value problem for W_N:

$$\begin{aligned}\Delta^2 W_N(x) &= D^{-1}q(x) + (4\pi N)^{-1}\sum_{j=1}^{N} m(p_j^{(N)})\big(\alpha_j(N)^{-1}\pi - 1\big)^{-1}\\ &\quad\times[\Delta^2, X_N^{(j)}(x)]\{r_j^{2-\pi/\alpha_j(N)}\cos\big(\pi\vartheta_j/\alpha_j(N)\big)\},\quad x\in\omega_N;\\ W_N(x) &= \Delta W_N(x) = 0,\quad x\in\gamma_N.\end{aligned} \tag{49}$$

The right-hand side of the equation in (49) is in $\mathbf{L}_2(\omega_N)$ so there exists a solution $W_N \in \mathbf{W}_2^2(\omega_N)$ to problem (49) with finite energy integral $\tilde{E}(W_N;\omega_N)$. Finally, (47) follows immediately from (48).

As the principal term in the asymptotics (as $N\to\infty$) of the solution w_N to problem (3), (4), (47), we take the function w satisfying the equation (6) and the boundary condition

$$w(x) = 0,\quad x\in\gamma. \tag{50}$$

The assumption on the boundedness of the boundary layer leads to the second boundary condition for z.

As in 18.1.2, the boundary value problem for the boundary layer w is of the form (16), where

$$A_1 = k(\tau)Y(0,\tau)/2,\quad A_2 = W(0,\tau) \tag{51}$$

(see the notation in (8)). To find the factor A_3 by $\delta(\xi_1)$ one should take account of the coincidence of the asymptotics of the functions $w(x) - N^{-2}z(\tau, N\tau, N\nu)$ and $w_N(x)$ as $r_j \to 0$. According to (18) and (51),

$$\begin{aligned} w(x) - N^{-2}z(\tau, N\tau, N\nu) &= (\partial/\partial\nu)w(p_j^{(N)})\Big\{r\cos\vartheta_j + (\pi N)^{-1}k(p_j^{(N)})r_j \\ &\quad\times\Big(\cos\vartheta_j\log(Nr_j) - \vartheta_j\sin\vartheta_j + c_z\cos\vartheta_j\Big)\Big\} \qquad (52) \\ &\quad - (2\pi N)^{-1}A_3 r_j\log(Nr_j)\cos\vartheta_j + O(r_j^2|\log r_j|). \end{aligned}$$

The solution w_N to problem (3), (4), (47) admits the asymptotics

$$\begin{aligned} w_N(x) &= \Big\{-(4\pi N)^{-1}m(p_j^{(N)})\big(\alpha_j(N)^{-1}\pi - 1\big)^{-1}r_j^{2-\pi/\alpha_j(N)} + K_j r_j^{\pi/\alpha_j(N)}\Big\} \\ &\quad\times\cos\big(\pi\vartheta_j/\alpha_j(N)\big) + O(r_j^2), \end{aligned}$$

with some number K_j, cf. (47), (48) and (26). By virtue of (27)

$$\begin{aligned} w_N(x) &= \Big\{K_j + \big(4k(p_j^{(N)})\big)^{-1}m(p_j^{(N)})\Big\}r_j\cos\vartheta_j + (\pi N)^{-1}\Big\{\big(k(p_j^{(N)})K_j \\ &\quad + 4^{-1}m(p_j^{(N)})\big)\cos\vartheta_j\log r_j - \big(k(p_j^{(N)})K_j - 4^{-1}m(p_j^{(N)})\big)\vartheta_j\sin\vartheta_j\Big\}r_j \\ &\quad + m(p_j^{(N)})C_w(N)r_j\cos\vartheta_j + O(N^{-3}|\log N|^2), \end{aligned} \qquad (53)$$

where $r_j = O(N^{-3/2})$ and $C_w(N) = O(N^{-1})$.

Comparing the coefficients of $r_j\cos\vartheta_j\log r_j$ and $r_j\vartheta_j\sin\vartheta_j$ in (52) and (51), we find

$$\begin{aligned} k(p_j^{(N)})K_j + m(p_j^{(N)})/4 &= (\partial/\partial\nu)w(p_j^{(N)})k(p_j^{(N)}) - A_3/2, \\ k(p_j^{(N)})K_j - m(p_j^{(N)})/4 &= (\partial/\partial\nu)w(p_j^{(N)})k(p_j^{(N)}). \end{aligned}$$

Hence,

$$A_3 = m(p_j^{(N)}). \qquad (54)$$

In view of (18), the boundary layer z is bounded only if $A_2 = -A_3$ or, equivalently,

$$\Delta w(x) = m(x) \quad \text{for} \quad x \in \gamma. \qquad (55)$$

The last equality is the missing boundary condition in the boundary value problem for w. Thus, the formal asymptotics of w_N has been constructed. We formulate the result obtained.

Theorem 18.1.4. *The inequality*

$$\tilde{E}(w_N - A_N; \omega_N) \le \text{const}\cdot N^{-1/2}|\log N|^4 \qquad (56)$$

holds, where $A_N(x)$ is the asymptotics of $w_N(x)$ defined by (31) *with the solution w to problem* (6), (50), (55), *z is the bounded solution to problem* (16) *while the constants $A_1, A_2,$ and A_3 are given by* (51) *and* (54), *and*

$$\begin{aligned} \Psi_N^{(j)}(r_j, \vartheta_j) &= \Big\{-(4\pi N)^{-1}m(p_j^{(N)})\big(\alpha_j(N)^{-1}\pi - 1\big)^{-1}r_j^{2-\pi/\alpha_j(N)} \\ &\quad + K_j^{(N)}r_j^{\pi/\alpha(N)}\Big\}\cos\big(\pi\vartheta_j/\alpha_j(N)\big) \end{aligned}$$

with

$$\begin{aligned} K_j^{(N)} &= (\partial/\partial\nu)w(p_j^{(N)})\big(1 - (\pi N)^{-1}k(p_j^{(N)})(\log N + C_z)\big) \\ &\quad - (4\pi N)^{-1}m(p_j^{(N)})\Big\{\big(1 - \alpha_j(N)^{-1}\pi\big)^{-1} - 2\log N\Big\}. \end{aligned}$$

To prove (56) we can proceed exactly as in 18.1.3.

Remark 18.1.5. In the boundary value problem without point moments considered in 18.1.3, the contribution of the boundary layer $N^{-2}z$ and corrections $X_N^{(j)}$ and $\Psi_N^{(j)}$) into the energy integral are small. That is why the inequality (46) for $w_N - A_N$ implies the estimate (44) on $w_N - w$.

For the problem studied in the present section this estimate is not true, if for no other reason than the difference $w_N - w$ between the solutions to problems (3), (4), (47) and (6), (50), (55) does not belong to $\mathbf{W}_2^2(\omega_N)$. However, the next theorem establishes the convergence of w_N to w in the norm of $\mathbf{W}_2^2(G)$, where G is any strictly interior subdomain of ω.

Theorem 18.1.6. *(i) Let G be a strictly interior subdomain of ω. Then for N sufficiently large there holds the inequality*

$$\tilde{E}(w_N - w; G) \le \text{const} \cdot N^{-1/2} |\log N|^4. \tag{57}$$

(ii) There holds the estimate

$$||w_N - w; \mathbf{W}_2^1(\omega_N)|| \le \text{const} \cdot N^{-1/4} |\log N|^2. \tag{58}$$

Proof. For sufficiently large N the restriction of the asymptotics A_N to G coincides with $w(x) - N^{-2}z(\tau, N\tau, N\nu)\chi(d^{-1}\nu)$. By virtue of (18),

$$\begin{aligned}
&\tilde{E}\big(N^{-2}\chi(d^{-1}\nu)z(\tau, N\tau, N\nu); G\big) \\
&\le c \int\limits_{\{x \in \omega_N :\, d_G < \nu < d\}} N^{-4}\Big(1 + N^2 \exp\big(-(\pi - \gamma)N\nu\big)\Big)^2 dx \\
&\le cN^{-1/2},
\end{aligned}$$

with $d_G = \operatorname{dist}(\partial G, \gamma)$. Therefore, (56) implies the inequality (57).

To verify (58) it suffices to make use of the inequalities

$$||w_N - A_N; \mathbf{W}_2^1(\omega_N)||^2 \le \text{const}\, \tilde{E}(w_N - A_N; \omega_N),$$

$$\begin{aligned}
||w_N - A_N; \mathbf{W}_2^1(\omega_N)|| \le c\Bigg\{ &\int\limits_{\{x \in \omega_N :\, \nu < d\}} N^{-4}\Big(1 + N \exp\big(-(\pi - \gamma)N\nu\big)\Big)^2 dx \\
&+ \sum_{j=1}^{N} \int\limits_{D_N^{(j)}} r_j^{2\left(1 - \pi/\alpha_j(N)\right)} dx \Bigg\} \le cN^{-1}.
\end{aligned}$$

□

18.2 Approximation of a Hole in a Freely Supported Plate

18.2.1 Statement of the problem

Let Γ and γ be smooth closed contours. Assume that Γ encloses γ and the length of γ equals 1. We consider the plate Ω bounded by Γ and γ. We suppose that the boundary γ of the strictly convex hole ω in the plate is freely supported while the contour Γ is rigidly fixed. The bend u of the plate Ω is a solution to the boundary value problem

$$\begin{aligned} &D\Delta^2 u(x) = q(x), \quad x \in \Omega; \\ &u(x) = (\partial/\partial\nu)\, u(x) = 0, \quad x \in \Gamma; \\ &u(x) = \Delta u(x) - (1-\sigma)k(x)(\partial/\partial\nu)\, u(x) = 0, \quad x \in \gamma, \end{aligned} \tag{1}$$

where D is a cylindrical rigidity of the plate, σ the Poisson coefficient, ν the inward (with respect to Ω) directed normal on γ, and k the curvature of γ. The orientation of γ is chosen so that the curvature is nonnegative.

Let ω_N be the same polygon as in 18.1.1 with boundary γ_N and vertices $p_j^{(N)}$. We consider the plate Ω_N bounded by Γ and γ_N (see Fig. 18.2). Its bend satisfies the boundary value problem

$$\begin{aligned} &D\Delta^2 w_N(x) = q(x), \quad x \in \Omega_N; \\ &w_N(x) = (\partial/\partial\nu)\, w_N(x) = 0, \quad x \in \Gamma; \\ &w_N(x) = \Delta w_N(x) = 0, \quad x \in \gamma_N. \end{aligned} \tag{2}$$

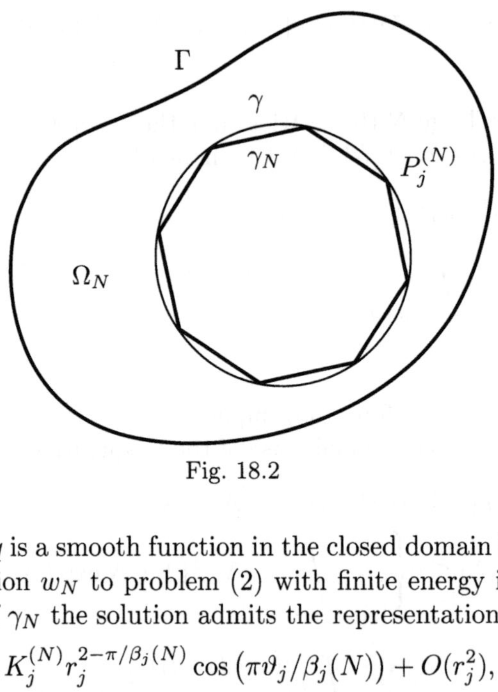

Fig. 18.2

Suppose that q is a smooth function in the closed domain bounded by Γ. Then there exists a solution w_N to problem (2) with finite energy integral $\tilde{E}(w_N;\Omega_N)$. Near the vertices of γ_N the solution admits the representation

$$w_N(x) = K_j^{(N)} r_j^{2-\pi/\beta_j(N)} \cos\left(\pi\vartheta_j/\beta_j(N)\right) + O(r_j^2), \quad r_j \to 0, \tag{3}$$

where (r_j, ϑ_j) are polar coordinates based on $p_j^{(N)}$ such that in a neighborhood of $p_j^{(N)}$ the domain Ω_N is given by $|\vartheta| < \beta_j(N)/2$, $\beta_j(N)$ is the opening of the exterior angle of the polygon ω_N, and $K_j^{(N)}$ are some constants.

18.2.2 Asymptotics of the solution for the plate with a polygonal hole

When constructing the asymptotics of the solution to problem (2), we can essentially proceed as in Sections 18.1.2 and 18.1.4. As the leading term in the asymptotics of the solution w_N we take the solution to the Sophie Germain equation

$$D\Delta^2 w(x) = q(x), \quad x \in \Omega, \tag{4}$$

with boundary conditions

$$u(x) = (\partial/\partial\nu)\, w(x) = 0, \quad x \in \Gamma; \tag{5}$$

$$w(x) = 0, \quad x \in \gamma. \tag{6}$$

As in 18.1.4, we derive the second boundary condition on γ from the assumption on the boundedness of the boundary layer.

In a d-neighborhood of the contour γ we introduce curvilinear coordinates (ν, τ), ν being the distance to γ along the inward normal to γ and τ the length of arc on γ. The functions w, Δw, and $(\partial/\partial\nu)w$ written in the coordinates (ν, τ) are denoted by W, V, and Y (see (8), 18.1). We assume that the functions are smoothly extended to negative ν. In the same way as in Section 18.1.2 we obtain the equation of the two segments of γ_N outgoing from the point $P_j^{(N)}$:

$$\nu = (2N^2)^{-1} k(p_j^{(N)})\big((N|\tau| - 1/2)^2 - 1/4\big) + O(N^{-3}). \tag{7}$$

The opening $\beta_j(N)$ of the greater angle between them satisfies

$$\beta_j(N) = \pi + k(p_j^{(N)})N^{-1} + O(N^{-2}). \tag{8}$$

Passing to the stretched-out coordinates $\xi = N^{-1}(\tau, \nu)$ and arguing as in Section 18.1.2, we obtain a boundary value problem for the boundary layer $-N^{-2} z(\tau, N\tau, N\nu)$. The problem is of the form (16), 18.1, with

$$A_1 = -k(\tau) Y(0, \tau)/2, \quad A_2 = W(0, \tau). \tag{9}$$

We find the unknown constant A_3 comparing the asymptotics of the functions $w_N(x)$ and $w(x) - N^{-2} z(\tau, N\tau, N\nu)$ in the zone $r_j = O(N^{-3/2})$.

From (3) and (8) we derive the relation

$$\begin{aligned} w_N(x) = K_j^{(N)} \Big(r_j \cos\vartheta_j + (\pi N)^{-1} k(p_j^{(N)}) r_j (\log r_j \cos\vartheta_j + \vartheta_j \sin\vartheta_j) \Big) \\ + O(N^{-3} |\log N|^2), \quad r_j = O(N^{-3/2}). \end{aligned} \tag{10}$$

Now we take account of the representation (17), 18.1, for the solution to problem (16), 18.1, return to the coordinates (ν, τ), and arrive at

$$\begin{aligned} w(x) - N^{-2} z(\tau, N\tau, N\nu) = (\partial/\partial\nu) w(p_j^{(N)}) \Big\{ r \cos\vartheta_j - (\pi N)^{-1} k(p_j^{(N)}) r_j \\ \times \Big(\cos\vartheta_j \log(N r_j) - \vartheta_j \sin\vartheta_j + C_z \cos\vartheta_j \Big) \Big\} \\ - (2\pi N)^{-1} A_3 r_j \log(N r_j) \cos\vartheta_j + O(r_j^2 |\log r_j|). \end{aligned} \tag{11}$$

Comparing the coefficients by $r_j \log r_j \cos\vartheta_j$ and $r_j \vartheta_j \sin\vartheta_j$ in (10) and (11), we find the equation for A_3:

$$(\pi N)^{-1} k(p_j^{(N)}) (\partial/\partial\nu) w(p_j^{(N)}) = -(\pi N)^{-1} k(p_j^{(N)}) (\partial/\partial\nu) w(p_j^{(N)}) - (2\pi N)^{-1} A_3.$$

It follows that

$$A_3 = -4k(\tau) Y(0, \tau). \tag{12}$$

In view of (18), 18.1, the function z is bounded only in the case $A_2 = -A_3$. Thus, from (9) and (12) we derive the complementary boundary condition for w on γ:

$$\Delta w(x) - 4k(x)(\partial/\partial\nu) w(x) = 0, \quad x \in \gamma. \tag{13}$$

Note that problem (4)–(6), (13) is uniquely solvable because the corresponding energy integral

$$\int_\Omega \Big\{|\Delta w|^2 - 8\Big((\partial^2 w/\partial x_1^2)(\partial^2 w/\partial x_2^2) - \big((\partial/\partial x_1)(/\partial/\partial x_2)w\big)^2\Big)\Big\}\,dx$$

is positive by virtue of the known identity (see BIRMAN [1])

$$2\int_\Omega \Big\{(\partial^2 w/\partial x_1^2)(\partial^2 w/\partial x_2^2) - \big((\partial/\partial x_1)\,(\partial/\partial x_2)w\big)^2\Big\}\,dx = \int_\gamma k(\partial w/\partial \nu)^2\,ds.$$

The formal asymptotics of the solution w_N to problem (2) has been constructed. Far from the contour γ the principal term of the asymptotics is the solution w to problem (4)–(6), (13). In the part of a d-neighborhood of γ_N located outside $O(N^{-3/2})$-neighborhoods of the vertices of γ, this term coincides with the difference between w and the boundary layer $N^{-2}z(\tau, N\tau, N\nu)$. For $r_j < 2N^{-3/2}$, $j = 1, \dots, N$, the principal term of the asymptotics of w_N is equal to

$$\begin{aligned}\Psi_N^{(j)}(r_j,\vartheta_j) &= (\partial/\partial\nu)w(p_j^{(N)})\Big(1 + (\pi N)^{-1}k(p_j^{(N)})(\log N - C_z)\Big)\\ &\quad\times r_j^{2-\pi/\beta_j(N)}\cos\big(\pi\vartheta_j/\beta_j(N)\big).\end{aligned} \tag{14}$$

To justify the asymptotics we can proceed as in the proof of Theorem 18.3.1. Therefore we restrict ourselves to the formulation of the result.

Theorem 18.2.1. *The difference $w_N - w$ between the solutions to problems* (2) *and* (4)–(6), (13) *satisfies the estimate*

$$\tilde E(w_N - w; \Omega_N) \le \text{const}\cdot N^{-1/2}|\log N|^4, \tag{15}$$

where $\tilde E$ is the energy integral (42), 18.1, *in the domain Ω_N.*

18.2.3 Point moments at the vertices of the polygon

We change the statement of the problem and impose on w_N the supplementary condition

$$\begin{aligned}M_{r_j}(x) &= -D\Big\{(\partial/\partial r_j)^2 w_N(x) + r_j^{-2}\sigma\big((\partial/\partial\vartheta_j)^2 w_N(x) + r_j(\partial/\partial r_j)w_N(x)\big)\Big\}\\ &= 4^{-1}D(1-\sigma)k(p_J^{(N)})^{-1}m(p_J^{(N)})(\pi^2\beta_j(N)^{-2} - 1)r_j^{-2+\pi/\beta_j(N)}\\ &\quad\times\cos\big(\pi\vartheta_j/\beta_j(N)\big) + O(r_j^{-\pi/\beta_j(N)}),\quad j = 1,\dots,N,\end{aligned} \tag{16}$$

where m is a smooth function on γ. As well as (47), 18.1, the formulas (16) mean that some small point moments are assigned to the vertices of the polygon ω_N.

Problem (2), (16) is uniquely solvable. However, the energy integral $\tilde E(w_N;\Omega_N)$ for the solution turns out to be infinite. The proof is the same as in Section 18.1.4 for problem (30, (4), (47), 18.1. Note that w_N admits the representations

$$\begin{aligned}w_N(x) &= \big(-k(p_J^{(N)})^{-1}m(p_J^{(N)})r^{\pi/\beta_j(N)}/4 + K_j(N)r^{2-\pi/\beta_j(N)}\big)\\ &\quad\times\cos\big(\pi\vartheta_j/\beta_j(N)\big) + O(r_j^2)\end{aligned} \tag{17}$$

with constants $K_j(N)$, $j = 1,\dots,N$.

We construct the asymptotics for the solution w_N to problem (2), (16) as $N \to \infty$. As the principal term we take the function w satisfying the equation (4) and boundary conditions (5), (6). As before, the second boundary condition for w on

γ we derive requiring the boundedness of the boundary layer $-N^{-2}z(\tau, N\tau, N\nu)$ as $\nu \to \infty$. The same reason as in Section 18.2.2 provides for the function z a boundary value problem of the form (16), 18.1. The A_1 and A_2 in (16), 18.1, are defined by (9). We shall seek A_3 taking account of the coincidence of the asymptotics of $w_N(x)$ and $w(x) - N^{-2}z(\tau, N\tau, N\nu)$ in the zone $r = O(N^{-3/2})$.

From (17) and (8) it follows that

$$\begin{aligned} w_N(x) = &\Big(- k(p_J^{(N)})^{-1} m(p_J^{(N)})/4 + K_j^{(N)}\Big) r_j \cos\vartheta_j \\ &+ (\pi N)^{-1} k(p_J^{(N)}) r_j \Big\{ \Big(k(p_J^{(N)})^{-1} m(p_J^{(N)})/4 + K_j^{(N)}\Big) \log r_j \cos\vartheta_j \\ &+ \Big(- k(p_J^{(N)})^{-1} m(p_J^{(N)})/4 + K_j^{(N)}\Big) \vartheta_j \sin\vartheta_j \Big\} + O(N^{-2}|\log N|^2), \\ &r_j = O(N^{-3/2}). \end{aligned} \tag{18}$$

The asymptotic formula (17), 18.1, for the solution to problem (16), 18.1, enables us to obtain the representation (11) for the difference $w(x) - N^{-2}z(\tau, N\tau, N\nu)$ in the coordinates (τ, ν). Comparing the coefficients of $r_j \log r_j \cos\vartheta_j$ and $r_j \vartheta_j \sin\vartheta_j$ in (18) and (11), we arrive at

$$\begin{aligned} k(p_j^{(N)})^{-1} m(p_j^{(N)})/4 + K_j^{(N)} &= -(\partial/\partial\nu) w(p_j^{(N)}) - k(p_j^{(N)})^{-1} A_3/2 \\ - k(p_j^{(N)})^{-1} m(p_j^{(N)})/4 + K_j^{(N)} &= (\partial/\partial\nu) w(p_j^{(N)}). \end{aligned}$$

Therefore,

$$A_3 = -4k(p_j^{(N)})(\partial/\partial\nu) w(p_j^{(N)}) - m(p_j^{(N)}). \tag{19}$$

In view of (18), 18.1, the solution z to problem (16), 18.1, is bounded provided $A_2 + A_3 = 0$. Thus, from (9) and (19) it follows that

$$\Delta w(x) - 4k(x)(\partial/\partial\nu) w(x) = m(x), \quad x \in \gamma, \tag{20}$$

which is the desired boundary condition to problem (4)–(6).

The formal asymptotics of the solution w_N to problem (2), (16) has been constructed. Far from the contour γ the principal term of the asymptotics is the solution w to problem (4)–(6), (20). In the part of a d-neighborhood of γ located outside of $O(N^{-3/2})$-neighborhoods of the vertices it coincides with the difference of the solution w and the boundary layer $N^{-2}z(\tau, N\tau, N\nu)$. For $r_j < 2N^{-3/2}$, $j = 1, \dots, N$, the principal terms are

$$\begin{aligned} \Psi_N^{(j)}(r_j, \vartheta_j) = &\big(- k(p_J^{(N)})^{-1} m(p_J^{(N)}) r_j^{\pi/\beta_j(N)}/4 + K_j^{(N)} r_j^{2-\pi/\beta_j(N)}\big) \\ &\times \cos\big(\pi\vartheta_j/\beta_j(N)\big) \end{aligned} \tag{21}$$

with

$$\begin{aligned} K_j^{(N)} = &\big((\partial/\partial\nu) w(p_j^{(N)}) + k(p_j^{(N)})^{-1} m(p_j^{(N)})/4\big)\big(1 + 2(\pi N)^{-1} k(p_j^{(N)}) \log N\big) \\ &- (\partial/\partial\nu) w(p_j^{(N)})(\pi N)^{-1} k(p_j^{(N)})(\log N + C_z). \end{aligned}$$

The justification of the asymptotics can be obtained exactly as in Theorems 18.1.3 and 18.1.6. We formulate the corresponding results on the convergence of w_N to w. As in Section 18.1.4, there is no inclusion $w_N \in \mathbf{W}_2^2(\Omega_N)$ because of the singularity of the second derivatives of w_N. Therefore we are dealing with the convergence on any compact $G \subset \Omega$.

Theorem 18.2.2. *(i) Let G be a strictly interior subdomain of Ω. There for sufficiently large N*

$$\tilde{E}(w_N - w; G) \leq \text{const} \cdot N^{-1/2} |\log N|^4.$$

(ii) There holds the inequality

$$||w_N - w; \mathbf{W}_2^1(\Omega_N)|| \leq \text{const} \cdot N^{-1/4} |\log N|^2.$$

18.3 Passage to Conditions of Rigid Suppoert

In the present section, a smooth plate G is approximated by freely supported polygonal plates G_N so that the limiting boundary conditions on the contour Γ coincide with the conditions of rigid support.

Let G be approximated by the sequence of domains G_N constructed in the following manner. Suppose that $\eta_1^{(N)}, \dots, \eta_N^{(N)}$ are some closed arcs on Γ having no common interior points. We denote by $\zeta_j^{(N)}, j = 1, \dots, N-1$, the chord between the nearest endpoints of the arcs $\eta_j^{(N)}$ and $\eta_{j+1}^{(N)}$ and by $\zeta_N^{(N)}$ the chord between the endpoints of $\eta_N^{(N)}$ and $\eta_1^{(N)}$. Generally speaking, $\eta_j^{(N)}$ and $\zeta_j^{(N)}$ may degenerate to a point.

Let $q_j^{(N)}$ be the point at a distance $h_j^{(N)}$ from Γ on the inward normal to Γ at the middle of $\eta_j^{(N)}$; $h_j^{(N)} = o(1)$ as $N \to \infty$.

We connect $q_j^{(N)}$ and the endpoints of $\eta_j^{(N)}$ with the straight lines $t_{j1}^{(N)}$ and $t_{j2}^{(N)}$ (if $\eta_j^{(N)}$ is a point, then $t_{j1}^{(N)} = t_{j2}^{(N)}$). Denote by Γ_N the polygon line consisting of $t_{jk}^{(N)}$ and $\zeta_j^{(N)}$, by $|\eta_j^{(N)}|$ and $|\zeta_j^{(N)}|$ the lengths of $\eta_j^{(N)}$ and $\zeta_j^{(N)}$. Finally, let G_N stand for the domain bounded by the contour Γ_N.

We assume that the plate G_N is freely supported and is subjected to a normal force $q \in \mathbf{C}^\infty(\bar{G})$. Then the bending w of G_N satisfies the boundary value problem

$$D\Delta^2 w_N(x) = q(x), \quad x \in G_N; \tag{1}$$

$$w_N(x) = \Delta w_N(x) = 0, \quad x \in \partial G_N. \tag{2}$$

Further we consider the polygon lines of two types. For example, to the first type we relate gears with small acute cuts, to the second one gears with large slanting cuts (see Figs. 18.3 and 18.4).

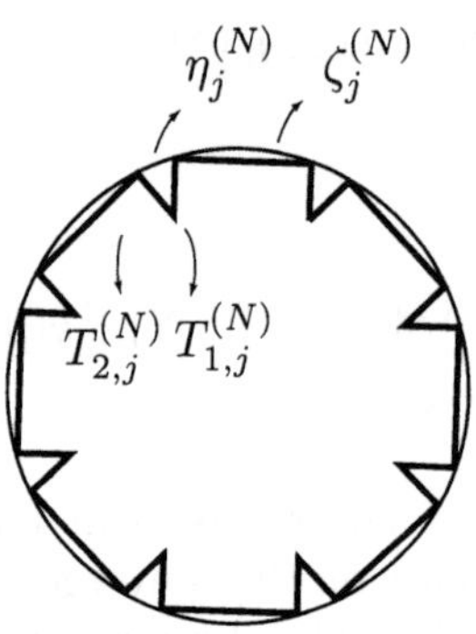

Fig. 18.3

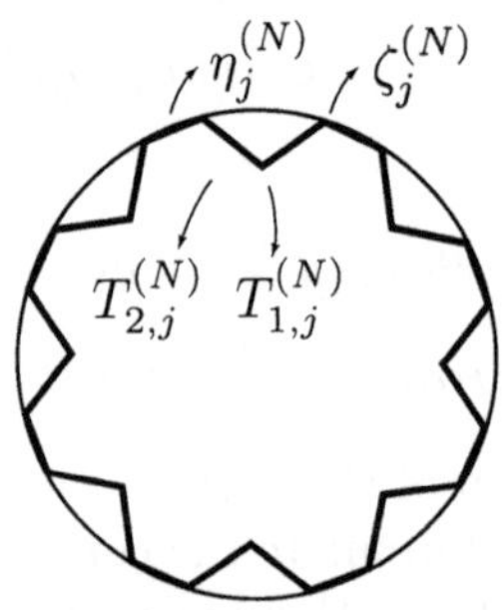

Fig. 18.4

Theorem 18.3.1. *Let*

$$\max_{1\le j\le N}|\zeta_j^{(N)}|/|\eta_j^{(N)}| = O(1), \quad \max_{1\le j\le N}(h_j^{(N)})^{-2/3}/|\eta_j^{(N)}| = o(1) \tag{3}$$

as $N\to\infty$. Then for any strictly interior subdomain g of G the sequence of the solutions $\{w_N\}$ of problems (1), (2) converges in $\mathbf{W}_2^2(g)$ to the solution $w\in\mathbf{W}_2^2(G)$ of the boundary value problem

$$D\Delta^2 w_N(x) = q(x), \quad x\in G; \tag{4}$$

$$w(x) = (\partial/\partial\nu)w(x) = 0, \quad x\in\Gamma. \tag{5}$$

Proof. As before, let (ν,τ) be curvilinear coordinates in a d-neighborhood U_d of Γ. We denote by $\xi_j^{(N)}$ the arc with span $\zeta_j^{(N)}$ and introduce the sets

$$P_j^{(N)}(\delta) = \{(\nu,\tau)\in G\colon 0\le\nu\le\delta, \quad \tau\in\eta_j^{(N)}\},$$

$$Q_j^{(N)}(\delta) = \{(\nu,\tau)\in G\colon 0\le\nu\le\delta, \quad \tau\in\eta_j^{(N)}\cup\xi_j^{(N)}\},$$

$$S_j^{(N)}(\delta) = \{(\nu,\tau)\in G\colon \nu=\delta, \quad \tau\in\eta_j^{(N)}\cup\xi_j^{(N)}\}.$$

The set $G_N\cap P_j^{(N)}(h_j^{(N)})$ is the union of the curvilinear right triangles $T_{j1}^{(N)}$ and $T_{j2}^{(N)}$ with hypotenuses $t_{j1}^{(N)}$ and $t_{j2}^{(N)}$. The legs of the triangles lying on the arc $S_j^{(N)}(h_j^{(N)})$ will be denoted by $s_{j1}^{(N)}$ and $s_{j2}^{(N)}$ and the adherent acute angles with vertex $q_j^{(N)}$ by $\alpha_{j1}^{(N)}$ and $\alpha_{j2}^{(N)}$.

We fix N and drop the indices for ease of notation. Introduce the Cartesian coordinates (y_1,y_2) based on the point q with axis directed along the outward normal to $S(h)$.

We set

$$\begin{aligned} v_1 &= \cos\alpha_1\,\partial w_N/\partial y_1 + \sin\alpha_1\,\partial w_N/\partial y_2,\\ v_2 &= -\cos\alpha_2\,\partial w_N/\partial y_1 + \sin\alpha_2\,\partial w_N/\partial y_2. \end{aligned} \tag{6}$$

Since $w_N=0$ on $t_1\cup t_2$, we have $v_k=0$ on t_k, $k=1,2$. Therefore,

$$\int_{s_k} v_k^2 d\tau \le ch\iint_{T_k}|\operatorname{grad} v_k|^2\,dx. \tag{7}$$

Moreover, in view of $|\zeta|=O(|\eta|)$, we obtain

$$\int_{S(h+|\eta|)} v_k^2 d\tau \le c\Big(\int_{s_k} v_k^2 d\tau + |\eta|\iint_{Q(h+|\eta|)\setminus Q(h)}|\operatorname{grad} v_k|^2\,dx\Big), \quad k=1,2. \tag{8}$$

Combining (7) and (8), we conclude that

$$\int_{S(h+|\eta|)} v_k^2 d\tau \le c(h+|\eta|)\iint_{Q(h+|\eta|)\cap G_N}|\operatorname{grad} v_k|^2\,dx, \quad k=1,2.$$

By virtue of (6),

$$\partial w_N/\partial y_1 = (\varphi_1\sin\alpha_2 - \varphi_2\sin\alpha_1)/\sin(\alpha_1+\alpha_2)$$

and

$$\partial w_N/\partial y_2 = (\varphi_1\cos\alpha_2 - \varphi_2\cos\alpha_1)/\sin(\alpha_1+\alpha_2).$$

Since

$$|\partial w_N/\partial\nu| \le |\partial w_N/\partial y_2| + |\eta|\,|\partial w_N/\partial y_1|$$

on $S(h+|\eta|)$, we arrive at

$$\int\limits_{S(h+|\eta|)} |\partial w_N/\partial\nu|^2 d\tau \leq c(h+|\eta|)\big(\sin(\alpha_1+\alpha_2)\big)^{-2}\big(\cos^2\alpha_1+\cos^2\alpha_2 + |\eta|^2(\sin^2\alpha_1+\sin^2\alpha_2)\big) \iint\limits_{Q(h+|\eta|)\cap G_N} |\operatorname{grad}_2 w_N|^2\, dx. \tag{9}$$

First we suppose that $h \leq |\eta|$. Taking the inequality $\sin(\alpha_1+\alpha_2) \geq ch|\eta|^{-1}$ into account, we majorize the constant in (9) by $c|\eta|^3 h^{-2}$. In the case $h > |\eta|$, we apply the obvious relations $\sin(\alpha_1+\alpha_2) \geq c|\eta|h^{-1}$ and $\cos\alpha_k \leq c|\eta|h^{-1}$ and obtain the estimate $c_1 h^3|\eta|^{-2}(|\eta|^2h^{-2}+|\eta|^2) \leq ch$ for the constant in (9). Thus,

$$\int\limits_{S_j^{(N)}(h_j^{(N)}+|\eta_j^{(N)}|)} |\partial w_N/\partial\nu|^2 d\tau \leq c(h+h^{-2}|\eta|^3) \iint\limits_{Qj^{(N)}(h_j^{(N)}+|\eta_j^{(N)}|)\cap G_N} |\operatorname{grad}_2 w_N|^2\, dx.$$

Let δ be a sufficiently small positive number. For large N there holds the inequality $\delta > |\eta_j^{(N)}| + h_j^{(N)}$. Clearly,

$$\int\limits_{S_j^{(N)}(\delta)} |\partial w_N/\partial\nu|^2 d\tau \leq c\Big(\int\limits_{S_j^{(N)}(h_j^{(N)}+|\eta_j^{(N)}|)} |\partial w_N/\partial\nu|^2 d\tau + \delta \iint\limits_{Q_j^{(N)}(\delta)\backslash Q_j^{(N)}(h_j^{(N)}+|\eta_j^{(N)}|)} |\operatorname{grad}\partial w_N/\partial\nu|^2\, dx\Big).$$

Summing over j, we have

$$\int\limits_{\nu=\delta} |\partial w_N/\partial\nu|^2 d\tau \leq c\Big(\delta + \max_{j=1,\dots,N} (h_j^{(N)})^{-2}|\eta_j^{(N)}|^3\Big)\, ||w_N; \mathbf{W}_2^2(U_\delta\cap G_N)||^2.$$

Therefore,

$$\overline{\lim_{N\to\infty}} \int\limits_{\nu=\delta} |\partial w_N/\partial\nu|^2 d\tau \leq c\delta||q; \mathbf{L}_2(G)||^2. \tag{10}$$

Similarly,

$$\overline{\lim_{N\to\infty}} \int\limits_{\nu=\delta} |w_N|^2 d\tau \leq c\delta||q; \mathbf{L}_2(G)||^2.$$

Let g be an arbitrary strictly interior subdomain of G. Because the norms $||w_N; \mathbf{W}_2^2(G)||$ are uniformly bounded, there exists the weak limit w in $\mathbf{W}_2^2(g)$ of the sequence $\{w_N\}$ independent of g. The norms $||w_N; \mathbf{W}_2^2(g)||$ are uniformly bounded with respect to g so $w \in \mathbf{W}_2^2(G)$.

The restriction operator $\mathbf{W}_1^2(G) \to \mathbf{L}_2\big(\{(\nu,\tau)\colon \nu=\delta\}\big)$ is compact, which together with (10) implies

$$\int\limits_{\nu=\delta} (|\partial w/\partial\nu|^2 + |w|^2) d\tau \leq c\delta||q; \mathbf{L}_2(G)||^2.$$

Now the inclusion $w \in \overset{\circ}{\mathbf{W}}{}_2^2(G)$ can be derived in a standard way. □

Theorem 18.3.2. *Suppose that for $N \to \infty$ there hold the relations*

$$\begin{aligned}&\max_{1\le j\le N} h_j^{(N)}/|\zeta_j^{(N)}| = O(1), \quad \max_{1\le j\le N} |\eta_j^{(N)}|/h_j^{(N)} = o(1),\\ &\max_{1\le j\le N} \big(|\log h_j^{(N)}|\,|\zeta_j^{(N)}|\big) = o(1).\end{aligned} \tag{11}$$

Then the sequence of solutions $\{w_N\}$ of problem (1), (2) *converges in $\mathbf{W}_2^2(g)$ to the solution $w \in \mathbf{W}_2^2(G)$ of problem* (4), (5), *where g is any strictly interior subdomain of G.*

Proof. Let $M_{j1}^{(N)}$ and $M_{j2}^{(N)}$ be the endpoints of the arc $S_j^{(N)}(h_j^{(N)} + |\zeta_j^{(N)}|)$, where $S_j^{(N)}(\delta)$ is the segment of the arc defined at the beginning of the proof of the preceding theorem. The notation $Q_j^{(N)}(\delta)$ was introduced in the same proof. Let $R_j^{(N)}$ stand for the pentagon bounded by the arc $S_j^{(N)}(h_j^{(N)})$, the normal to Γ with endpoint $M_{j1}^{(N)}$, the straight lines $\zeta_j^{(N)}$, $t_{1j}^{(N)}$, and the one connecting the points $q_j^{(N)}$ and $M_{j2}^{(N)}$. We denote by $B_j^{(N)}$ the ball of radius $C(h_j^{(N)} + |\zeta_j^{(N)}|)$ with center $q_j^{(N)}$, where C is a sufficiently large number such as and $B_j^{(N)} \supset \bar{R}_j^{(N)}$. Let us fix N and j and drop these indices for ease of notation. Finally, we denote by $\operatorname{cap}(t_1, B)$ the harmonic capacity of the segment t_1 with respect to the ball B. It is known (see, for example, Maz'ya [5]) that

$$c_1\big(\log C(h+|\zeta|)/h\big)^{-1} \le \operatorname{cap}(t_1, B) \le c_2\big(\log C(h+|\zeta|)/h\big)^{-1}.$$

Since the function v_1 defined in (7) is equal to zero on t_1, from Theorem 10.1.2 in Maz'ya [5] it follows that

$$\int\limits_{S(h+|\zeta|)} v_1^2 d\tau \le c(h+|\zeta|)\log\big(C(h+|\zeta|)/h\big) \iint\limits_R |\operatorname{grad} v_1|^2\, dx. \tag{12}$$

Let δ denote a sufficiently small positive number. For large N there holds the inequality $\delta > 2(h_j^{(N)} + |\zeta_j^{(N)}|)$. The angle between the axes $0y_1$ and the normal to $S(\delta)$ will be denoted by $\beta(\tau)$. We introduce the function

$$\begin{aligned}z(x) = (\partial/\partial y_1)w_N(x) + \big\{\big(\tan\beta(\tau) - \tan\alpha_1\big)(\delta - h - |\zeta|)^{-1}(\nu - \delta) + \tan\beta(\tau)\big\}\\ \times (\partial/\partial y_2)w_N(x),\end{aligned}$$

which coincides with $(\cos\alpha_1)^{-1}v_1$ on $S(h+|\zeta|)$ and with $\big(\cos\beta(t)\big)^{-1}\partial w_n/\partial\nu$ on $S(\delta)$. Obviously,

$$\int\limits_{S(\delta)} z^2 d\tau \le c\Big(\int\limits_{S(h+|\zeta|)} z^2 d\tau + \delta \iint\limits_{Q(\delta)\setminus Q(h+|\zeta|)} |\operatorname{grad} z|^2\, dx\Big).$$

It follows that

$$\begin{aligned}\int\limits_{S(\delta)} |\partial w_N/\partial\nu|^2 d\tau \le c \int\limits_{S(h+|\zeta|)} v_1^2 d\tau + \delta \iint\limits_{Q(\delta)\setminus Q(h+|\zeta|)} |\operatorname{grad}_2 w_N|^2\, dx\\ + \delta^{-1}\Big(\max_{\tau\in S(\delta)} |\tan\beta(\tau)|^2 + |\tan\alpha_1|^2 + \delta^2\Big) \iint\limits_{Q(\delta)\setminus Q(h+|\zeta|)} |\operatorname{grad} w_N|^2\, dx.\end{aligned} \tag{13}$$

Combining (12) and (13) and indicating the dependence on j and N, we obtain

$$\int_{S_j^{(N)}(\delta)} |\partial w_N/\partial\nu|^2 d\tau$$
$$\leq c\Big\{\delta + (\delta N)^{-1} + (\delta h_j^{(N)})^{-1}|\eta_j^{(N)}| + (h_j^{(N)} + |\zeta_j^{(N)}|)$$
$$\times \log\Big(C(h_j^{(N)} + |\zeta_j^{(N)}|)/h_j^{(N)}\Big)\Big\} \|w_N; \mathbf{W}_2^2(Q_j^{(N)}(\delta) \cap G_N)\|^2.$$

We sum over $j = 1, \dots, N$ and arrive at the inequality

$$\int_{\nu=\delta} |\partial w_N/\partial\nu|^2 d\tau \leq c\Big\{\delta + (\delta N)^{-1} + \max_{j=1,\dots,N} \Big((\delta h_j^{(N)})^{-1}|\eta_j^{(N)}|$$
$$+ (h_j^{(N)} + |\zeta_j^{(N)}|)\log\big(C(h_j^{(N)} + |\zeta_j^{(N)}|)/h_j^{(N)}\big)\Big)\Big\}\|w_N; \mathbf{W}_2^2(U_\delta \cap G_N)\|^2.$$

The expression in the braces tends to zero as $N \to \infty$ (see (11)) so

$$\overline{\lim_{N\to\infty}} \int_{\nu=\delta} |\partial w_N/\partial\nu|^2 d\tau \leq C\delta\|q; \mathbf{L}_2(G)\|^2.$$

It remains to repeat the conclusion of the proof of Theorem 18.3.1. □

Remark 18.3.3. In the proofs of Theorems 18.3.1 and 18.3.2, the fact that w_N satisfies the Sophie Germain equation (1) was used only to obtain the inequality

$$\|w_N; \mathbf{W}_2^2(G_N)\| \leq \|q; \mathbf{L}_2(G)\|.$$

Therefore, what was said in these theorems on the passage to the conditions of rigid support for the solution to problem (1), (2) applies equally to the nonlinear bending equations and to the dynamical problems in the theory of plates.

Chapter 19

Homogenization of a Differential Operator on a Fine Periodic Net of Curves

Here we deal with the homogenization of an ordinary differential operator on an ε-periodic net of curves in $\mathbb{R}^n$. The role of the operator homogenized is played by a partial differential operator acting on functions in n variables. As a model we take the problem of the propagation of heat in a fine wire-cloth. On its segments there are given some ordinary second order differential equations. At the nodes, the sum of the flows rates is equal to zero. Finally, we impose the homogeneous Dirichlet condition at the boundary points.

Section 19.1 contains the statement of the boundary value problem on an ε-periodic net S_ε. The principal terms of the asymptotics $u_0(x(l))+\varepsilon u_1(x(l),\varepsilon^{-1}l)$ of the solution and the estimate on the remainder in the energy norm on S_ε are given in Section 19.2. Here l is a parameter indicating the position of a point on S_ε, u_0 the solution to the homogenized partial differential equation, and u_1 is the solution to the ordinary differential equation on the net stretched-out. In the next section we describe the method of computing the coefficients of the homogenized operator that reduces to solving certain algebraic systems. It is shown that the homogenized operator is elliptic if and only if the system mentioned is of maximal rank. The complete asymptotics expansion of the solution to the Dirichlet problem on the net S_ε filling the half-space $\mathbb{R}^n_+$ as $\varepsilon \to 0$ is constructed in Sections 19.4 and 19.5. The techniques are equally applicable to the case of the nets formed by surfaces.

19.1 Statement of the Problem on a Net

19.1.1 The net S

Let $x = (x_1, x_2, \dots, x_n) \in \mathbb{R}^n$, let e_j be the unit vector of the axis Ox_j , and let Q be a half-open cube in $\mathbb{R}^n$. We suppose that Y is the union of finitely many smooth closed simple arcs located in the cube $\bar{Q}$. Introduce a partition of the space consisting of unit cubes each containing the same set Y. The collection of all such Y will be denoted by S and called the net.

We suppose that the partition of the space $\mathbb{R}^n$ is shifted with respect to the net so that ∂Q contains no endpoints of the arcs of the net S. Then the set Y has the same number of common points with opposite faces of Q, which are symmetrically located. It means that if $x^{(0)}$ is a common point of Y and a face of $\bar{Q}$ orthogonal to the axis x_m, then Y intersects ∂Q either at $x^{(0)} + e_m$ or at $x^{(0)} - e_m$.

The set $Y = Q \cap S$ is called the cell of the net, the curves forming Y are called the edges. We define the endpoints of the edges to be the nodes. Let T_Y be the set of nodes of Y and T the set of nodes of the net S. We denote by σ_j the *j-th* edge of Y, $j = 1, 2, \dots, J$. Further, let Λ_j be the length of the *j-th* edge and λ_j the coordinate of a point on σ_j, $0 \le \lambda_j \le \Lambda_j$.

We identify the symmetric common points of Y and opposite faces of ∂Q and obtain the set U. Suppose that any point of U can be connected with a pair of symmetric points of opposite faces of Q by a path located in U.

We introduce a parameter λ indicating the position of a point in S. A function u on S is called Y-periodic if

$$u(\lambda^+) = u(\lambda^-), \tag{1}$$

where λ^+ and λ^- are points of the set such that the differences of their corresponding Cartesian coordinates are integers.

19.1.2 The net S_ε

We consider the family of nets S_ε obtained from S by the transformation $x \to \varepsilon^{-1}x$ with small positive ε. Introduce the space $\mathbf{H}^1(S_\varepsilon)$ of functions on the net that are continuous at the nodes and absolutely continuous on the edges; the norm is defined by

$$||u; \mathbf{H}^1(S_\varepsilon)|| = \left\{\varepsilon^{n-1} \int\limits_{S_\varepsilon} \left(\left((d/dl)u(l)\right)^2 + \left(u(l)\right)^2\right) dl\right\}^{1/2}. \tag{2}$$

From now on $l = \varepsilon\lambda$ is the parameter indicating the position of a point in S_ε. The set of the nodes of S_ε will be denoted by T_ε.

Let $\mathbf{C}^k(\bar\Omega; W), k \in \mathbb{N}_0$, be the space of functions on $\bar\Omega$ taking values in a Hilbert space W. If v is in $\mathbf{C}^1(\bar\Omega; \mathbf{H}^1(Y \setminus T_Y))$, then

$$(d/dl)v(x\varepsilon^{-1}l) = w(\lambda) \cdot \nabla v(x, \lambda) + \varepsilon^{-1}(\partial/\partial\lambda)v(x, \lambda). \tag{3}$$

Here $\nabla = (\partial/\partial x_1, \dots, \partial/\partial x_n)$ and the vector w at the point $\lambda_j^{(0)}$ is equal to $\left(w^{(1)}(\lambda_j^{(0)}), \dots, w^{(n)}(\lambda_j^{(0)})\right)$, where $w_j^{(m)}(\lambda_j^{(0)})$ is the cosine of the angle between the j-*th* edge at $\lambda_j^{(0)}$ and the axis x_m.

Let Ω be a domain in $\mathbb{R}^n$ with compact closure and smooth boundary. We suppose that $\partial\Omega \cap T_\varepsilon = \emptyset$ and consider the boundary value problem

$$\begin{aligned} &(d/dl)\Big(a\big(x(l), \varepsilon^{-1}l\big)(d/dl)\, u(\varepsilon, l)\Big) - b\big(x(l), \varepsilon^{-1}l\big)\, u(\varepsilon, l) = f\big(x(l), \varepsilon^{-1}l\big), \\ &l \in S_\varepsilon \cap \Omega \setminus T_\varepsilon; \\ &C_\tau(a\, du/dl) = 0, \quad \tau \in T_\varepsilon \cap \Omega; \quad u(\varepsilon, l) = 0, \quad l \in S_\varepsilon \cap \partial\Omega, \end{aligned} \tag{4}$$

a, b, f being functions Y-periodic in the second argument, $a \in \mathbf{C}^3(\bar\Omega; \mathbf{H}^1(Y \setminus T_\gamma))$ and $b \in \mathbf{C}^1(\bar\Omega; \mathbf{H}^1(Y \setminus T_\gamma))$, $a > 0, b \geq 0$, and $f \in \mathbf{C}^3(\bar\Omega; \mathbf{L}^2(Y))$. From now on C_τ are the operators defined by

$$C_\tau(v) = \sum_{\{j:\, \tau \in \sigma_j\}} \pm \lim_{\lambda_j \to \Lambda_j(\tau)} v(\lambda_j),$$

$\Lambda_j^{(\tau)}$ equals λ_j at a point τ, the sign "+" corresponds to the case $\Lambda_j^{(\tau)} > 0$ and "−" to the case $\Lambda_j^{(\tau)} = 0$. The function u is supposed to be continuous at the nodes of S_ε.

Being definite negative, problem (4) is uniquely solvable in $\mathbf{H}^1(S_\varepsilon \cap \Omega)$. We shall seek the asymptotics of the solution to problem (4) in the form

$$u(\varepsilon, l) = u_0(x) + \sum_{k=1}^{\infty} \varepsilon^k u_k(x, \varepsilon^{-1}l), \tag{5}$$

with functions u_k periodic in the second argument.

19.2 The Principal Term of Asymptotics

19.2.1 The formal asymptotics

Let $u(x,\lambda)$ be a function Y-periodic in λ. Taking (3), 19.1, into account, we rewrite the first equality in (4), 19.1, in the form

$$\Big(\varepsilon^{-2}L_2(x,\lambda,\partial/\partial\lambda)+\varepsilon^{-1}L_1(x,\lambda,\partial/\partial x,\partial/\partial\lambda)+L_0(x,\lambda,\partial/\partial x)\Big)u(x,\lambda)=f(x,\lambda), \tag{1}$$

where

$$\begin{aligned}L_2(x,\lambda,\partial/\partial\lambda)&=(\partial/\partial\lambda)\big(a(x,\lambda)\,\partial/\partial\lambda\big),\\ L_1(x,\lambda,\partial/\partial x,\partial/\partial\lambda)&=(\partial/\partial\lambda)\big(a(x,\lambda)\,w(\lambda)\cdot\nabla\big)+w(\lambda)\cdot\nabla\big(a(x,\lambda)\,\partial/\partial\lambda\big),\\ L_0(x,\lambda,\partial/\partial x)&=w(\lambda)\cdot\nabla\big(a(x,\lambda)w(\lambda)\cdot\nabla\big)-b(x,\lambda).\end{aligned}$$

By virtue of (3), 19.1, the second equality in (4), 19.1, coincides with

$$\varepsilon^{-1}S_1^{(\tau)}(u)+S_0^{(\tau)}(u)=0, \tag{2}$$

where

$$S_1^{(\tau)}(u)=C_\tau(a\,\partial u/\partial\lambda),\quad S_0^{(\tau)}(u)=C_\tau(aw\cdot\nabla u).$$

Note that the operators $S_m^{(\tau)}$ depend on the parameter x.

We substitute series (5), 19.1, into (1) and (2), equate the coefficients at the same powers of ε, and obtain equations for the functions u_0 and u_1. The coefficient at ε^{-2} vanishes because u_0 does not depend on λ. The coefficient by ε^{-1} is equal to zero provided

$$\begin{aligned}&L_2(x,\lambda,\partial/\partial\lambda)u_1(x,\lambda)=-L_1(x,\lambda,\partial/\partial x,\partial/\partial\lambda)u_0(x),\\ &S_1^{(\tau)}(u_1)=S_0^{(\tau)}(u_0),\quad \tau\in T,\end{aligned} \tag{3}$$

which determines u_1. We represent u_1 as the scalar product $N(x,\lambda)\cdot u_0(x)$, N being the vector with components $N^{(m)}$ that are Y-periodic in the second argument. Then from (2) it follows that $N^{(m)}$ satisfies

$$\begin{aligned}&(\partial/\partial\lambda)\big(a(x,\lambda)(\partial/\partial\lambda)N^{(m)}(x,\lambda)=-(\partial/\partial\lambda)\big(a(x,\lambda)w^{(m)}(\lambda)\big),\\ &S_1^{(\tau)}(N^{(m)})=S_0^{(\tau)}(x_m),\quad \tau\in T_Y.\end{aligned} \tag{4}$$

We introduce the space $\mathbf{H}_Y^1$ of Y-periodic functions continuous at the nodes and absolutely continuous on the edges. The space is endowed with the norm

$$\int_Y\Big\{\big((\partial/\partial\lambda)v(\lambda)\big)^2+\big(v(\lambda)\big)^2\Big\}d\lambda.$$

The proof of the next assertion is standard.

Lemma 19.2.1. *Let $F(x,\lambda)$ be a function Y-periodic in λ which is in $\mathbf{C}^k(\bar\Omega;\mathbf{L}_2(Y))$. The problem*

$$(\partial/\partial\lambda)\big(a(x,\lambda)(\partial/\partial\lambda)u(x,\lambda)=F(x,\lambda),\quad S_1^{(\tau)}(u)=g_\tau(x),\quad \tau\in T_Y, \tag{5}$$

is solvable in $\mathbf{C}^k(\bar\Omega;\mathbf{H}_Y^1)$ if and only if

$$\int_{Y^{(q)}}F(x,\lambda)d\lambda=\sum_{\tau\in T_Y^{(q)}}g_\tau(x),\quad q=1,2,\dots,\varkappa. \tag{6}$$

On every connected component $U^{(q)}$ of the set U (see Section 19.1.1) a solution to problem (4) is determined up to an arbitrary constant.

We choose the constants so that

$$\int\limits_{U^{(q)}} u(x,\lambda)d\lambda = 0, \quad q = 1, 2, \dots, \varkappa. \tag{7}$$

From now on, by a solution to problem (5) we mean the function subject to (7).

Since the right-hand sides in (4) satisfy (6), problem (4) for $N^{(m)}$ is solvable and the function u_1 has been constructed. The coefficient u_2 in the series (5), 19.1, is defined by the relations

$$\begin{aligned} &L_2(x,\lambda,\partial/\partial\lambda)\,u_2(x,\lambda) \\ &= -L_1(x,\lambda,\partial/\partial x,\partial/\partial\lambda)\,u_1(x,\lambda) - L_0(x,\lambda,\partial/\partial x)\,u_0(x) + f(x,\lambda), \\ &\quad S_1^{(\tau)}(u_2) = -S_0^{(\tau)}(u_1), \quad \tau \in T. \end{aligned} \tag{8}$$

The compatibility condition (6) of problem (8) leads to the equation for u_0,

$$\sum_{j,k=1}^{n} (\partial/\partial x_j)\big(a_{jk}(x)(\partial/\partial x_k)\,u_0(x)\big) - b_1(x)\,u_0(x) = \int\limits_Y f(x,\lambda)d\lambda \tag{9}$$

with coefficients

$$\begin{aligned} a_{jk}(x) &= \int\limits_Y \big(w^{(j)}(\lambda)w^{(k)}(\lambda)\,a(x,\lambda) + w^{(j)}(\lambda)(\partial/\partial\lambda)N^{(k)}(x,\lambda)\,a(x,\lambda)\big)d\lambda, \\ b_1(x) &= \int\limits_Y b(x,\lambda)d\lambda. \end{aligned} \tag{10}$$

The function u_0 satisfies the boundary condition

$$u_0(x) = 0, \quad x \in \partial\Omega. \tag{11}$$

We shall prove Theorem 19.3.2 that gives the necessary and sufficient condition of the ellipticity of the operator

$$A(x,\partial/\partial x) = \sum_{m,k=1}^{n} (\partial/\partial x_m)\big(a_{mk}(x)\partial/\partial x_k\big) - b_1(x).$$

Assume that such a condition is fulfilled. Then, the coefficient u_0 in (5), 19.1, can be obtained from (9), (11) and $u_1 = N \cdot \nabla u_0$.

19.2.2 Justification of the asymptotics

To estimate the remainder in the space $\mathbf{H}^1(S_\varepsilon \cap \Omega)$ we prove the following assertions.

Proposition 19.2.2. *For a function $u \in \mathbf{H}^1(S_\varepsilon \cap \Omega)$ equal to zero on $S_\varepsilon \cap \partial\Omega$ there holds the inequality*

$$\int\limits_{S_\varepsilon \cap \Omega} \big(u(l)\big)^2\,dl \le c \int\limits_{S_\varepsilon \cap \Omega} \big((d/dl)u(l)\big)^2\,dl$$

with $c = O(1)$.

Proof. Let $y(Z_\varepsilon)$ be the shortest way in S_ε from the cell Z_ε to the boundary $\partial\Omega$. For a function v vanishing on $\partial\Omega$

$$c \sum_{\{Y_\varepsilon \subset S_\varepsilon :\, Y_\varepsilon \cap y(Z_\varepsilon) \neq \emptyset\}} \operatorname*{osc}_{Y_\varepsilon} |v| \geq \max_{Z_\varepsilon} |v|.$$

From this and the evident inequality

$$\operatorname*{osc}_{Y_\varepsilon} |v| \leq c \, ||dv/dl;\, \mathbf{L}_1(Y_\varepsilon)||$$

it follows that

$$c \sum_{\{Y_\varepsilon \subset S_\varepsilon :\, Y_\varepsilon \cap y(Z_\varepsilon) \neq \emptyset\}} \int_{Y_\varepsilon} |(d/dl) v(l)| dl \geq \max_{Z_\varepsilon} |v|.$$

Replace the right-hand side by $|Z_\varepsilon|^{-1} ||v; \mathbf{L}_1(Z_\varepsilon)||$, $|Z_\varepsilon|$ being the length of all the edges of Z_ε, and sum over the cells Z_ε in S_ε. Then

$$\int_{S_\varepsilon \cap \Omega} |v(l)| dl \leq c |Y_\varepsilon| \sum_{Z_\varepsilon \subset S_\varepsilon \cap \Omega} \sum_{\{Y_\varepsilon :\, Y_\varepsilon \cap y(Z_\varepsilon) \neq \emptyset\}} \int_{Y_\varepsilon} |(d/dl)\, v(l)| dl.$$

The sum on the right is majorized by the integral of $|dv/dl|$ over S_ε multiplied by a constant equal to $O(\varepsilon^{-1})$. Besides, $|Y_\varepsilon|$ is $O(\varepsilon)$. Therefore,

$$\int_{S_\varepsilon \cap \Omega} |v(l)| dl \leq c \int_{S_\varepsilon \cap \Omega} |(d/dl)\, v(l)| dl$$

with $c = O(1)$. It remains to substitute the function u^2 for v in the last relation and apply the Schwarz inequality. □

Proposition 19.2.3. *(i) If* $W \in \mathbf{C}^1(\bar{\Omega}; \mathbf{H}^1(Y \backslash T_Y))$, *then the map* $l \to W(x(l), \varepsilon^{-1} l)$ *belongs to* $\mathbf{L}_2(S_\varepsilon \cap \Omega)$ *and*

$$\varepsilon^{(n-1)/2} ||W; \mathbf{L}_2(S_\varepsilon \cap \Omega)|| \leq c_1 ||W; \mathbf{C}(\bar{\Omega}; \mathbf{L}_2(Y))|| + \varepsilon c_2 ||W; \mathbf{C}^1(\bar{\Omega}; \mathbf{H}^1(Y \setminus T_Y))|| \tag{12}$$

with constants c_1 *and* c_2 *which are* $O(1)$.

(ii) Under the supplementary condition that the support of W *is located in the* ε*-neighborhood of* $\partial\Omega$, *we have* $c_j = O(\varepsilon^{1/2}), j = 1, 2$.

Proof. From the triangle inequality it follows that

$$\begin{aligned} &\Big\{\varepsilon^{n-1} \int_{S_\varepsilon \cap \Omega} \big(W(x(l), \varepsilon^{-1} l)\big)^2 dl\Big\}^{1/2} \\ &\leq \Big\{\varepsilon^{n-1} \sum_{\{Y_\varepsilon \subset S_\varepsilon :\, Y_\varepsilon \cap \Omega \neq \emptyset\}} \int_{Y_\varepsilon \cap \Omega} \big(W(x(Y_\varepsilon), \varepsilon^{-1} l)\big)^2 dl\Big\}^{1/2} \\ &+ \Big\{\varepsilon^{n-1} \sum_{\{Y_\varepsilon \subset S_\varepsilon :\, Y_\varepsilon \cap \Omega \neq \emptyset\}} \int_{Y_\varepsilon \cap \Omega} |W(x(l), \varepsilon^{-1} l) - W(x(Y_\varepsilon), \varepsilon^{-1} l)|^2 dl\Big\}^{1/2}, \end{aligned} \tag{13}$$

where $x(Y_\varepsilon)$ is the coordinate of the center of the cell Y_ε in S_ε. Since the number of the cells in S_ε located in Ω is $O(\varepsilon^{-n})$, the first term on the right is no greater than $c\|W;\mathbf{C}(\bar\Omega;\mathbf{L}_2(Y))\|$ with constant c independent of ε. We estimate the second term on the right in (13). There hold the inequalities

$$\begin{aligned}
&|W\big(x(l),\varepsilon^{-1}l)\big)-W\big(x(Y_\varepsilon),\varepsilon^{-1}l\big)|\\
&\quad\le c\varepsilon\left|\int_0^1\nabla W\big(x(l)(1-t)+x(Y_\varepsilon)t,\varepsilon^{-1}l\big)\,dt\right|\\
&\quad\le c\varepsilon\bigg\{\int_0^1\int_Y\Big(\big|(\partial/\partial\mu)\big(\nabla W\big(x(l)(1-t)+x(Y_\varepsilon)t,\mu\big)\big)\big|\\
&\quad+|\nabla W\big(x(l)(1-t)+x(Y_\varepsilon)t,\mu\big)|\Big)\,d\mu\,dt\bigg\},\qquad l\in Y_\varepsilon.
\end{aligned}$$

Thus,

$$\begin{aligned}
&\int_{Y_\varepsilon}|W\big(x(l),\varepsilon^{-1}l\big)-W\big(x(Y_\varepsilon),\varepsilon^{-1}l\big)|^2dl\\
&\quad\le c\varepsilon^2\int_{Y_\varepsilon}\int_0^1\|\nabla W\big(x(l)(1-t)+x(Y_\varepsilon)t,\mu\big);\mathbf{H}^1(Y\setminus T_Y)\|^2dt\,dl\\
&\quad\le c\varepsilon^3\|W;\mathbf{C}^1(\bar Y_\varepsilon;\mathbf{H}^1(Y\setminus T_Y))\|^2.
\end{aligned}$$

Hence, the second term on the right in (13) does not exceed

$$\|W;\mathbf{C}^1(\bar\Omega;\mathbf{H}^1(Y\setminus T_Y))\|$$

because the number of cells participating in the sum is $O(\varepsilon^{-n})$. The assertion (i) is proven.

If W is supported by the ε-neighborhood of $\partial\Omega$, then the sum in the right-hand side of (13) is over the cells whose number is $O(\varepsilon^{1-n})$. Therefore, the constants c_1 and c_2 satisfy Proposition 19.2.3(ii). □

Let us turn to estimating the remainder

$$R(\varepsilon,l)=u(\varepsilon,l)-u_0\big(x(l)\big)-\varepsilon u_1\big(x(l),\varepsilon^{-1}l\big),\qquad l\in S_\varepsilon\cap\bar\Omega,$$

where u is the solution to problem (4), 19.1. Let $\eta_\varepsilon(x)$ be a function in $\mathbf{C}^\infty(\mathbb{R}^n)$ equal to 1 in the ε-neighborhood of $\partial\Omega$ such that $\operatorname{supp}\eta_\varepsilon\cap T_\varepsilon=\emptyset$. Then the function $\tilde R=R+\varepsilon\eta_\varepsilon u_1$ satisfies the boundary value problem

$$\begin{aligned}
&(d/dl)\Big(a\big(x(l),\varepsilon^{-1}l\big)(d/dl)\tilde R(\varepsilon,l)\Big)-b\big(x(l),\varepsilon^{-1}l\big)\tilde R(\varepsilon,l)\\
&=\tilde F^{(0)}\big(x(l),\varepsilon^{-1}l\big)+\varepsilon\tilde F^{(1)}\big(x(l),\varepsilon^{-1}l\big),\qquad l\in S_\varepsilon\cap\Omega;\\
&C_\tau(a\,d\tilde R/dl)=G_\tau,\quad \tau\in T_\varepsilon\cap\Omega;\quad \tilde R(\varepsilon,l)=0,\quad l\in S_\varepsilon\cap\partial\Omega
\end{aligned}\tag{14}$$

with $\tilde F^{(0)}=-L_1u_1-L_0u_0+f$, $\tilde F^{(1)}=-L_0u_1+(d/dl)\big(a(d/dl)(\eta_\varepsilon u_1)\big)$, and $G_\tau=-S_0^{(\tau)}(u_1)$. Since for $\tilde F^{(0)}$ and G_τ there holds (6), Lemma 19.2.1 provides the solvability of the problem

$$L_2(x,\lambda,\partial/\partial\lambda)\,v(x,\lambda)=\tilde F^{(0)}(x,\lambda),\quad \lambda\in S;\quad S_1^{(\tau)}(v)=-G_\tau,\quad \tau\in T.$$

We set $W_1(x,\lambda)=a(x,\lambda)(\partial/\partial\lambda)v(x,\lambda)$. From the formula (3), 19.1, for the function W_1 it follows that

$$\varepsilon(d/dl)\,W_1(x,\varepsilon^{-1}l)=\varepsilon w(\lambda)\cdot\nabla W_1(x,\lambda)+\tilde F^{(0)}(x,\lambda).$$

Therefore, (14) takes the form

$$
\begin{aligned}
&(d/dl)\Big(a(x,\varepsilon^{-1}l)(d/dl)\,\tilde{R}(\varepsilon,l)\Big) - b(x,\varepsilon^{-1}l)\tilde{R}(\varepsilon,l)\\
&= \varepsilon\Big\{(d/dl)\,W_1(x,\varepsilon^{-1}l) + (d/dl)\Big(a(x,\varepsilon^{-1}l)(d/dl)\big(\eta_\varepsilon(x)\,u_1(x,\varepsilon^{-1}l)\big)\Big)\\
&- b(x,\varepsilon^{-1}l)\eta_\varepsilon(x)\,u_1(x,\varepsilon^{-1}l) - w(\varepsilon^{-1}l)\cdot\nabla W_1(x,\varepsilon^{-1}l) - L_0(x,\varepsilon^{-1}l,\partial/\partial x)\\
&\times u_1(x,\varepsilon^{-1}l)\Big\}.
\end{aligned}
$$

We multiply the last equality by $\tilde{R}$ and integrate over $S_\varepsilon\cap\Omega$. Integrating by parts, we obtain

$$
\begin{aligned}
&\int\limits_{S_\varepsilon\cap\Omega}\Big\{a(x(l),\varepsilon^{-1}l)\big((d/dl)\,\tilde{R}(\varepsilon,l)\big)^2 + b(x(l),\varepsilon^{-1}l)\tilde{R}(\varepsilon,l)\big)^2\Big\}\,dl\\
&= \varepsilon\int\limits_{S_\varepsilon\cap\Omega}\Big\{\big((d/dl)\,\tilde{R}(\varepsilon,l)\,W_1(x(l),\varepsilon^{-1}l)\\
&+\Big((d/dl)\,\tilde{R}(\varepsilon,l)\,a(x(l),\varepsilon^{-1}l)(d/dl)\big(\eta_\varepsilon(x)\,u_1(x(l),\varepsilon^{-1}l)\big)\Big)\\
&+\Big(\tilde{R}(\varepsilon,l)\big(b(x(l),\varepsilon^{-1}l)\,\eta_\varepsilon(x)\,u_1\,(x(l),\varepsilon^{-1}l)\big)\\
&+w(\varepsilon^{-1}l)\cdot\nabla W_1(x(l),\varepsilon^{-1}l) + L_0(x(l),\varepsilon^{-1}l,\partial/\partial x)\,u_1(x(l),\varepsilon^{-1}l)\big)\Big)\Big\}\,dl\\
&-\int\limits_{\tau\in T_\varepsilon\cap\Omega}\big(C_\tau(\tilde{R}\,a\,d\tilde{R}/dl) - C_\tau(\tilde{R}\,a\,\partial v/\partial\lambda)\big).
\end{aligned}
\tag{15}
$$

The continuity of $\tilde{R}$ and the definition of v imply that the last sum in (15) vanishes. Introduce the notation

$$
aw\cdot\nabla(\eta_\varepsilon u_1) = W_2,\quad a(\partial/\partial\lambda)(\eta_\varepsilon u_1) = W_3,\quad b\eta_\varepsilon u_1 + w\cdot\nabla W_1 + L_0u_1 = W_4.
$$

We estimate the left-hand side of (15) with the help of Proposition 19.2.2 and apply the Schwarz inequality and Proposition 19.2.2 to the terms on the right in (15). Then

$$
\begin{aligned}
&\varepsilon^{(1-n)/2}||\tilde{R};\mathbf{H}^1(S_\varepsilon\cap\Omega)||^2\\
&\le c\Big\{\varepsilon\big[||\tilde{R};\mathbf{H}^1(S_\varepsilon\cap\Omega)||\,\big(||W_1;\mathbf{L}_2(S_\varepsilon\cap\Omega)||\\
&+||W_2;\mathbf{L}_2(S_\varepsilon\cap\Omega)||\big) + ||\tilde{R};\mathbf{L}_2(S_\varepsilon\cap\Omega||\,||W_4;\mathbf{L}_2(S_\varepsilon\cap\Omega)||\big]\\
&+||\tilde{R};\mathbf{H}^1(S_\varepsilon\cap\Omega)||\,||W_3;\mathbf{L}_2(S_\varepsilon\cap\Omega)||\Big\}.
\end{aligned}
\tag{16}
$$

By virtue of the assumptions on the coefficients and right-hand side of the equations (4), 19.1, the functions $W_j, j = 1,2,3,4$, are in $\mathbf{C}^1\big(\bar{\Omega};\mathbf{H}^1(Y\setminus T_Y)\big)$. We apply Proposition 19.2.3 (i) to the terms in braces on the right in (16) and Proposition 19.2.3 (ii) to the last term to obtain the estimate

$$
||\tilde{R};\mathbf{H}^1(S_\varepsilon\cap\Omega)|| \le c\varepsilon^{1/2}. \tag{17}
$$

Taking the definition (2), 19.1, and relation (3), 19.1, into account, we see that the norm of $\varepsilon\eta_\varepsilon u_1$ in $\mathbf{H}^1(S_\varepsilon\cap\Omega)$ is equal to

$$
\varepsilon^{(n-1)/2}||(\partial/\partial\lambda + \varepsilon w\cdot\nabla)\eta_\varepsilon u_1;\mathbf{L}_2(S_\varepsilon\cap\Omega)||.
$$

Proposition 19.2.3 applied to this quantity leads to the inequality

$$||\varepsilon\eta_\varepsilon u_1; \mathbf{H}^1(S_\varepsilon \cap \Omega)|| \leq \varepsilon^{1/2}. \tag{18}$$

The next assertion follows from (17) and (18).

Theorem 19.2.4. *There holds the inequality*

$$||R; \mathbf{H}^1(S_\varepsilon \cap \Omega)|| \leq c\varepsilon^{1/2}$$

with constant c independent of ε.

19.2.3 Asymptotics of solutions to nonstationary problems

The construction of the principal term of the asymptotics of solutions to nonstationary problems is similar to that of the solution to problem (4), 19.1, given in 19.2.1 (cf. BAKHVALOV and PANASENKO [1], Ch. 3, §1). For the solution u to the mixed boundary value problem for the heat equation there hold the relations

$$\begin{aligned}
&S\big(x(l), \varepsilon^{-1}l\big)(\partial/\partial t)\, u(\varepsilon, l, t) - L(\varepsilon, l, t, \partial/\partial l)\, u(\varepsilon, l, t)\\
&= f(l, \varepsilon^{-1}l, t), \quad l \in S_\varepsilon \cap \Omega \setminus T_\varepsilon;\\
&C_\tau(a\, \partial u/\partial l) = 0, \quad \tau \in T_\varepsilon \cap \Omega;\\
&u(\varepsilon, l, t) = 0, \quad l \in S_\varepsilon \cap \partial\Omega, \quad t \in [0, t_0]; \quad u(\varepsilon, l, t) = 0, \quad l \in S_\varepsilon \cap \Omega,
\end{aligned} \tag{19}$$

with $L = (d/dl)\Big(a\big(x(l), \varepsilon^{-1}l, t\big)\big)d/dl\Big) - b\big(x(l), \varepsilon^{-1}l, t\big)$, where the functions a and b smoothly depend on t and satisfy the assumptions on the coefficients of problem (4), 19.1. In the case of the mixed boundary value problem for the wave equation we have

$$\begin{aligned}
&P\big(x(l), \varepsilon^{-1}l\big)(\partial/\partial t)^2 u(\varepsilon, l, t) - L(\varepsilon, l, t, \partial/\partial l)\, u(\varepsilon, l, t)\\
&= f(l, \varepsilon^{-1}l, t), \quad l \in S_\varepsilon \cap \Omega \setminus T_\varepsilon;\\
&C_\tau(a\, \partial u/\partial l) = 0, \quad \tau \in T_\varepsilon; \quad u(\varepsilon, l, t) = 0, \quad l \in S_\varepsilon \cap \partial\Omega, \quad t \in [0, t_0];\\
&u(\varepsilon, l, t) = 0, \quad (\partial/\partial t)\, u(\varepsilon, l, t) = 0, \quad l \in S_\varepsilon \cap \Omega,
\end{aligned} \tag{20}$$

with positive Y-periodic functions S and P smooth on the edges of the net S_ε. We seek the asymptotics of the solutions to problems (18) and (19) in the form (5), 19.1, with coefficients depending on t (in contrast to the case of problem (4), 19.1). The construction of the principal terms is similar to that in Section 19.2.1. In the case of problem (18) the function u_0 satisfies the equation

$$S_1(x)(\partial/\partial t)u_0(x, t) - A(x, \partial/\partial x)u_0(x, t) = f_1(x). \tag{21}$$

In the case of problem (19) this function satisfies the equation

$$P_1(x)(\partial/\partial t)^2 u_0(x, t) - A(x, \partial/\partial x)u_0(x, t) = f_1(x). \tag{22}$$

Besides, there hold the homogeneous initial and boundary conditions. In (20) and (21)

$$S_1(x) = \int_Y S(x, \lambda)d\lambda; \quad P_1(x) = \int_Y P(x, \lambda)d\lambda; \quad f_1(x) = \int_Y f(x, \lambda)d\lambda.$$

The coefficient u_1 is equal to $N \cdot \nabla u_0$, N being the solution to problem (4). For the remainder $R = u - u_0 - \varepsilon u_1$ with u satisfying to (18) (respectively, (19)) there are the following inequalities that result from estimates on the solutions to problems (18) and (19).

Theorem 19.2.5. *In the case of problem* (19)

$$\Big\{\varepsilon^{n-1}\int\limits_0^{t_0}\int\limits_{S_\varepsilon\cap\Omega}\Big(\big(R(\varepsilon,l,t)\big)^2+\big((d/dl)R(\varepsilon,l,t)\big)^2+\big((d/dt)R(\varepsilon,l,t)\big)^2\Big)dl\,dt\Big\}^{1/2}\le c\varepsilon^{1/2}. \tag{23}$$

For problem (18) *there holds a similar inequality, but it does not contain the third term on the left.*

The proof is similar to that of Theorem 19.2.4.

19.3 Computation of Coefficients of the Homogenized Operator and Their Properties

19.3.1 Definition of coefficients of the homogenized operator

From (4), 19.2, it follows that on the *j-th* edge of the cell Y there hold the relations

$$a(x,\lambda_j)(\partial/\partial\lambda_j)\,N^{(m)}(x,\lambda_j)=-a(x,\lambda_j)\,w_j^{(m)}(\lambda_j)+c_j^{(m)}(x),\quad m=1,\dots,n, \tag{1}$$

with some functions $c_j^{(m)}$ independent of λ. Hence, by definition (10), 19.2, the coefficients a_{mk} of the operator A are given by

$$a_{mk}(x)=\int\limits_Y w^{(m)}(\lambda)c^{(k)}(x,\lambda)d\lambda. \tag{2}$$

Here $w^{(m)}$ coincides with $w_j^{(m)}$ on the edge σ_j and the function $c^{(k)}$ is equal to $c_j^{(k)}$ in (1) on σ_j.

The coefficients a_{mk} satisfy the equalities $a_{mk}=a_{km}$. Indeed, integrating by parts in the last term in the first equality (10), 19.2, and taking (4), 19.2, into account, we have

$$a_{mk}(x)=\int\limits_Y\Big\{a(x,\lambda)\big(w^{(m)}(\lambda)\,w^{(k)}(\lambda)+(\partial/\partial\lambda)\,N^{(m)}(x,\lambda)(\partial/\partial\lambda)\,N^{(k)}(x,\lambda)\big)\Big\}d\lambda.$$

The functions $c_j^{(m)}$ in (1) satisfy n systems of J linear algebraic equations. We describe the construction of the system. By virtue of (1), the second equality in (4), 19.2, is of the form

$$\sum_{\{j:\,\tau\in\sigma_j\}}\pm c_j^{(m)}=0,\quad \tau\in T_Y. \tag{3}$$

The Y-periodicity condition for $N^{(m)}$ leads to the relations

$$\pm c_{p^+}^{(m)}\pm c_{p^-}^{(m)}=0, \tag{4}$$

where $p^\pm$ are the indices corresponding to the edges σ_{p^+}, σ_{p^-} that contain the symmetric points γ^+, γ^- of opposite faces of the cube Q, the sign "+" by $c_q^{(m)}$ in (3) and (4) is related to the case $\Lambda_j^{(m)}>0$, the sign "−" to the case $\Lambda_j^{(m)}=0$ (the notation $\Lambda_j^{(m)}$ was introduced in Section 19.1.2 before the definition of the operators C_τ). Suppose that there are K relations (3) and P conditions (4). We identify the symmetric boundary points of Y located in opposite faces of Q and obtain the graph U containing $J-P$ edges. Let $\tilde{Y}$ be a basis in the space of cycles of the graph with the operation $y_1 + y_2 = (y_1 \cup y_2) \setminus (y_1 \cap y_2)$, y_1 and y_2 being the cycles of U. The dimension of $\tilde{Y}$ is equal to the cyclomatic number $J - P - K + \varkappa$ of U, where $\varkappa$ is the number of connected components of U (e.g., see ZYKOV [1]). From (1) it follows that on the j-*th* edge there hold the equalities

$$N^{(m)}(x, \Lambda_j) = N^{(m)}(x, 0) - \int_0^{\Lambda_j} w_j^{(m)}(\lambda_j) d\lambda_j + c_j^{(m)}(x) \int_0^{\Lambda_j} \big(a(x, \lambda_j)\big)^{-1} d\lambda_j, \tag{5}$$

$m = 1, \dots, n$. We add the relations (5) on the edges of the cycle y_r in the set $\tilde{Y}$, $r = 1, \dots, J - P - K + \varkappa$, and obtain

$$\sum_{\{j:\, l_j \in y_r\}} (-1)^{\nu_{jr}} c_j^{(m)}(x) \int_0^{\Lambda_j} \big(a(x, \lambda_j)\big)^{-1} d\lambda_j = \sum_{\{j:\, l_j \in y_r\}} (-1)^{\nu_{jr}} \int_0^{\Lambda_j} w_j^{(m)}(\lambda_j) d\lambda_j, \tag{6}$$

$m = 1, \dots, n$, where $\nu_{jr} = 1$ if the parameter λ_j increases in accordance with the orientation of the contour y_r and $\nu_{jr} = 2$ otherwise. Equation (6) and relations (3) and (4) form the system to find $c_j^{(m)}$ for $j = 1, \dots, J$. Note that $K + P$ equations (3) and (4) turn out to be linear dependent. Indeed, their sum vanishes on every connected component of Y because the left-hand sides of (3) and (4) contain each $c_j^{(m)}$ with opposite signs. Therefore, among $K + P$ relations (3) and (4), there are at least $\varkappa$ equations that depend linearly on the other ones. We exclude these $\varkappa$ relations and obtain the system consisting of $P + K - \varkappa$ equations (3) and (4) and $J - P - K + \varkappa$ equalities (6) to find J functions $c_j^{(m)}$ for every m.

We write down the system for $c_j^{(m)}$ in the form

$$\tilde{A} C = \tilde{W}, \tag{7}$$

where C is the $J \times n$ matrix with entries $c_j^{(m)}$, $\tilde{W}$ the $J \times n$ matrix formed by the right-hand sides of the systems for $c_j^{(m)}$: the first $P + K - \varkappa$ rows of the matrix consist of zeros, the entry of $\tilde{W}$ in the m-*th* column and $J - P - K + \varkappa + r$-*th* row is equal to the right-hand side of the r-*th* equation in (6). The entries of $\tilde{A}$ are the coefficients in the left-hand sides of the equations (3), (4), and (6).

We prove that the system for $c_j^{(m)}$ is uniquely solvable. The form of the right-hand sides in (4), 19.2, Lemma 19.2.1 and relations (1) provide the existence of $c_j^{(m)}$. Let us verify the uniqueness. Assume that $c_{j0}^{(m)}$ is a solution to the system (3), (4), and (6). We find the functions $N_0^{(m)}$ from the equalities

$$(\partial/\partial\lambda) N_0^{(m)}(x, \lambda) = -w_j^{(m)}(\lambda_j) + \big(a(x, \lambda_j)\big)^{-1} c_{j0}^{(m)}, \quad j = 1, \dots, J, \quad m = 1, \dots, n. \tag{8}$$

(Note that the equalities (6) are necessary and sufficient for the solvability of (8) relative to $N_0^{(m)}$.) In view of the uniqueness of the solution to problem (4), 19.2, $N_0^{(m)}$ and $N^{(m)}$ coincide. Therefore, from (1) it follows that $c_j^{(m)}$ and $c_{j0}^{(m)}$, too, coincide. Thus, the system (7) is uniquely solvable and $\tilde{A} \neq 0$.

Remark 19.3.1. The right-hand side of an equation in the system for $c_j^{(m)}$, $j = 1, \dots, J$, differs from zero only if the equation is the relation (6) for the cycle passing through the intersections of the cell Y and the opposite faces of Q orthogonal to the axis Ox_m. Indeed, the right-hand sides in (3) and (4) vanish while the module of that in (6) is equal to the length of projection of the cycle onto the axis Ox_m. In case the cycle contains no intersections of Y and the opposite faces of Q, the length equals zero.

19.3.2 Ellipticity of the homogenized operator

The ellipticity criterion of the homogenized operator is related to the structure of the matrix $\tilde{W}$ in the right-hand side in (7).

Theorem 19.3.2. *The operator A is elliptic if and only if $r(\tilde{W}) = n$, where $r(\tilde{W})$ is the rank of $\tilde{W}$.*

Proof. For Y-periodic functions u and v the integration by parts over Y takes the form

$$\int_Y v(\lambda)(\partial/\partial\lambda)u(\lambda)d\lambda = \sum_{\tau\in T_Y} C_\tau(v(\lambda)\,u(\lambda)) - \int_Y u(\lambda)(\partial/\partial\lambda)\,v(\lambda)d\lambda. \tag{9}$$

From (4), 19.2, and (9) it follows that

$$\int_Y a(x,\lambda)w^{(m)}(\lambda)(\partial/\partial\lambda)N^{(k)}(x,\lambda)d\lambda$$

$$= \int_Y \Big(2a(x,\lambda)w^{(m)}(\lambda)(\partial/\partial\lambda)N^{(k)}(x,\lambda) + N^{(k)}(x,\lambda)(\partial/\partial\lambda)\big(a(x,\lambda)w^{(m)}(\lambda)\big)\Big)d\lambda$$

$$- \sum_{\tau\in T_Y} C_\tau\big(N^{(k)}(x,\lambda)a(x,\lambda)w^{(m)}(\lambda)\big)$$

$$= \int_Y \Big(2a(x,\lambda)w^{(m)}(\lambda)(\partial/\partial\lambda)N^{(k)}(x,\lambda)$$

$$+ a(x,\lambda)(\partial/\partial\lambda)N^{(k)}(x,\lambda)(\partial/\partial\lambda)N^{(m)}(x,\lambda)\Big)d\lambda.$$

Thus,

$$a_{mk}(x) = \int_Y a(x,\lambda)\big(w^{(m)}(\lambda)w^{(k)}(\lambda) + 2w^{(m)}(\lambda)(\partial/\partial\lambda)N^{(k)}(x,\lambda)$$

$$+ (\partial/\partial\lambda)N^{(m)}(x,\lambda)(\partial/\partial\lambda)N^{(k)}(x,\lambda)\big)d\lambda.$$

Therefore,

$$\sum_{m,k=1}^{n} \xi_m a_{mk}\xi_k = \int_Y a(x,\lambda)\{\xi\cdot(w(\lambda) + (\partial/\partial\lambda)N(x,\lambda))\}^2 d\lambda \tag{10}$$

with $\xi = (\xi_1, \xi_2, \dots, \xi_n)$. By virtue of (1), the right-hand side of (10) is equal to

$$\sum_{j=1}^{J}\int_0^{\Lambda_j} \big(a(x,\lambda_j)\big)^{-1}\big(\xi\cdot c_j(x,\lambda_j)\big)^2 d\lambda_j, \tag{11}$$

c_j being the vector with components $c_j^{(m)}$.

Suppose that the assumption of Theorem 19.3.2 is fulfilled. Since $\det A \neq 0$, A being the matrix in (7), the equation (7) implies $r(C) = r(\tilde{W})$. Hence, there exist terms in the sum (11) that differ from zero. Thus, the expression (11) is positive for any $\xi \neq 0$, i.e., the operator A is elliptic.

In case the assumption of Theorem 19.3.2 is not fulfilled, the rank of $\tilde{W}$ (consequently, the rank of C, as well) is less than n. From (2) it follows that the matrix consisting of the coefficients a_{jk} of A is equal to the product $\tilde{V}C$, where $\tilde{V}$ is $n \times J$-matrix with entries

$$\int_0^{\Lambda_j} w_j^{(m)}(\lambda_j)d\lambda_j.$$

The operator A turns out to be nonelliptic because the rank of the product does not exceed the rank of any factor. □

Remark 19.3.3. If the cell Y does not intersect a pair of opposite faces of Q, then the operator A is not elliptic. Indeed, the column of $\tilde{W}$ corresponding to the pair of faces consists of zeros and, consequently, $r(\tilde{W}) < n$.

Remark 19.3.4. To provide the ellipticity for A it suffices to require for all n pairs of opposite faces of Q the existence of a connected component of Y intersecting both of the opposite faces. In such a case n linearly independent rows of $\tilde{W}$ are formed by the cycles each of which includes a curve connecting opposite faces, i.e., $r(\tilde{W}) = n$. By Theorem 19.3.2, the operator A is elliptic.

19.3.3 Examples of homogenized operators

Example 19.3.5. Let S_ε be the periodic net pictured in Figure 19.1. The parameter ρ runs the interval $0 \leq \rho < 1/2$. Suppose that the coefficients a and b of problem (4), 19.1, are equal to 1. We consider the cell Y of the net S_ε, see Figure 19.2. By virtue of (4),

$$c_5^{(m)} = c_2^{(m)}, \quad c_6^{(m)} = c_1^{(m)}, \quad c_7^{(m)} = c_8^{(m)} = c_3^{(m)}, \quad m = 1, 2.$$

Fig. 19.1

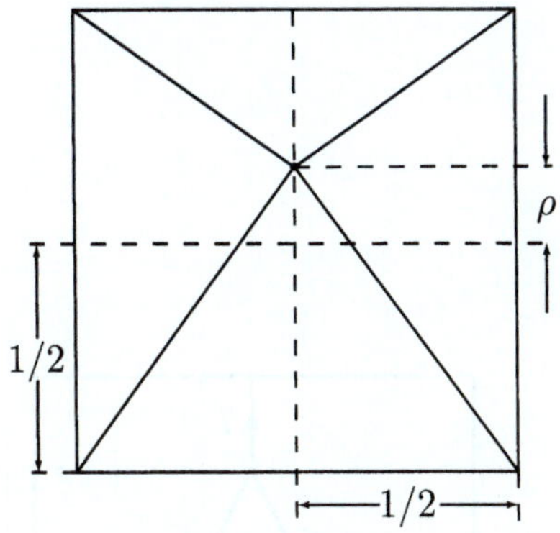

Fig. 19.2

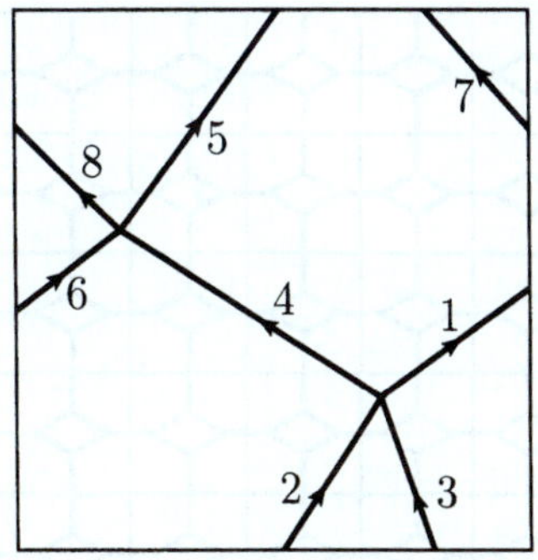

Fig. 19.3

Let us find $c_j^{(m)}$, $m = 1, 2; j = 1,2,3,4$. From (3) it follows that

$$c_1^{(m)} + c_4^{(m)} = c_2^{(m)} + c_3^{(m)}, \quad m = 1,2. \tag{12}$$

As the cycle y_1, we take the path consisting of the edges 8, 7, 3, 1, 6 (see Fig. 19.3). In view of (6) we obtain

$$R_1 c_1^{(m)} + R_2 c_3^{(m)} = \delta_{2m}, \quad m = 1,2, \tag{13}$$

with $R_1 = \big(1/4 + (1/2-\rho)^2\big)^{1/2}$ and $R_2 = \big(1/4 + (1/2+\rho)^2\big)^{1/2}$. Assume that the cycle y_2 consists of the edges 1, 6, 4. The equality (6) for y_2 is of the form

$$R_1 (c_1^{(m)} - c_4^{(m)}) = \delta_{1m}, \quad m = 1,2. \tag{14}$$

In the case of the cycle y_3 containing edges 8, 7, 3, 2, 5, formula (6) leads to the relations

$$R_2 (c_2^{(m)} - c_3^{(m)}) = \delta_{1m}, \quad m = 1,2. \tag{15}$$

From (12)–(15) we obtain

$$\begin{aligned} c_1^{(2)} &= c_2^{(2)} = c_3^{(2)} = c_4^{(2)} = \big\{\big(1/4 + (1/2+\rho)^2\big)^{1/2} + \big(1/4 + (1/2-\rho)^2\big)^{1/2}\big\}^{-1}, \\ c_1^{(1)} &= -c_4^{(1)} = (1/2)\big(1/4 + (1/2-\rho)^2\big)^{-1/2}, \\ c_2^{(1)} &= -c_3^{(1)} = (1/2)\big(1/4 + (1/2+\rho)^2\big)^{-1/2}. \end{aligned}$$

The cosines of the angles between the coordinate axes and the edges of the cell Y are equal to

$$\begin{aligned} &w_j^{(1)} = c_j^{(1)}, \quad j = 1,\ldots,8, \quad w_1^{(2)} = w_4^{(2)} = w_6^{(2)} = (1/2-\rho)\big((1/2-\rho)^2 + 1/4\big)^{-1/2}, \\ &w_2^{(2)} = w_3^{(2)} = w_5^{(2)} = w_7^{(2)} = w_8^{(2)} = (1/2+\rho)\big((1/2+\rho)^2 + 1/4\big)^{-1/2}. \end{aligned}$$

Substituting the expressions of $c_j^{(m)}$ and $w_j^{(m)}$ into (2), we calculate the coefficients of the operator A

$$\begin{aligned} &a_{11} = 2(2\rho^2+1)(4\rho^4+1)^{-1}, \quad a_{22} = 2(4\rho^4+1)^{-1/2}\big((4\rho^4+1)^{1/2} + 2\rho^2\big)^{-1/2}, \\ &a_{21} = a_{12} = 0, \quad b_1 = 2\big\{\big(1/4 + (1/2+\rho)^2\big)^{1/2} + \big((1/2-\rho)^2 + 1/4\big)^{-1/2}\big\}. \end{aligned}$$

Note that a_{11} takes the maximum $2+\sqrt{2}$ at $\rho = \pm\big((\sqrt{2}-1)/2\big)^{1/2}$ while a_{22} takes its maximum 2 at $\rho = 0$.

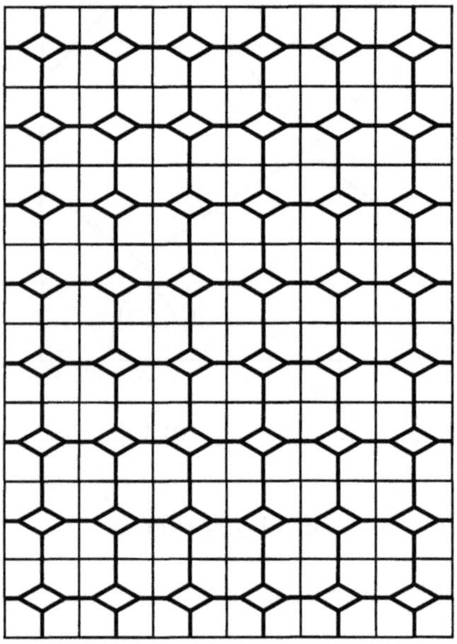

Fig. 19.4

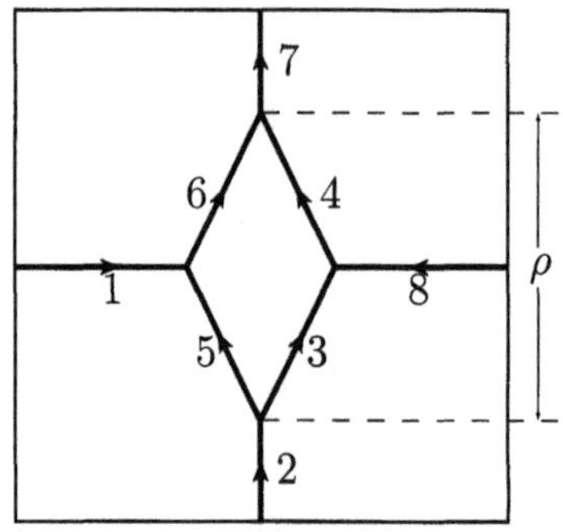

Fig. 19.5

Example 19.3.6. We consider the net shown in Figure 19.4. Suppose that a and b in (4), 19.1, are equal to 1. The cell of the net consists of 8 edges (see Fig. 19.5) and $0 < \rho n < 1$. Using (7), we find $c_j^{(m)}$. The Y-periodicity condition (4) leads to

$$c_1^{(m)} = c_8^{(m)}, \quad c_2^{(m)} = c_7^{(m)}, \quad m = 1, 2.$$

By virtue of (3), at the nodes of Y

$$c_1^{(m)} + c_5^{(m)} = c_6^{(m)}, \quad c_1^{(m)} + c_4^{(m)} = c_3^{(m)}, \quad c_3^{(m)} + c_5^{(m)} = c_2^{(m)}, \quad m = 1, 2. \tag{16}$$

If the cycle y_1 consists of the edges $3, 4, 6, 5$ (see Fig. 19.5), then (6) takes the form

$$c_3^{(m)} + c_4^{(m)} - c_5^{(m)} - c_6^{(m)} = 0, \quad m = 1, 2. \tag{17}$$

In case y_2 contains the edges $2, 3, 4, 7$, then by virtue of (6)

$$c_2^{(m)}(1 - \rho) + (c_3^{(m)} + c_4^{(m)})R_3 = \delta_{2m}, \quad m = 1, 2, \tag{18}$$

with $R_3 = (1+9\rho^2)^{1/2}/6$. Finally, for y_3 consisting of the edges $1, 6, 4, 8$ the formula (6) leads to the equation

$$2c_1^{(m)}/3 + (c_6^{(m)} - c_4^{(m)})R_3 = \delta_{1m}, \quad m = 1, 2. \tag{19}$$

The relations (16)–(19) form the system for $c_j^{(m)}$. From (16)–(19) we derive

$$c_1^{(1)} = c_8^{(1)} = 2c_3^{(1)} = 2c_6^{(1)} = -2c_4^{(1)} = 6\big(4 + (1 + 9\rho^2)^{1/2}\big)^{-1}, \quad c_2^{(1)} = c_1^{(2)} = 0,$$
$$c_2^{(2)} = c_7^{(2)} = 2c_3^{(2)} = 2c_4^{(2)} = 2c_5^{(2)} = 2c_6^{(2)} = 6\big(6 - 6\rho + (1 + 9\rho^2)^{1/2}\big)^{-1}.$$

We find $w_j^{(m)}, j = 1, 2, \ldots, 8$:

$$w_1^{(1)} = w_8^{(1)} = w_2^{(2)} = w_7^{(2)} = 1, \quad w_3^{(1)} = w_6^{(1)} = -w_4^{(1)} = -w_5^{(1)} = (1 + 9\rho^2)^{-1/2}$$
$$w_2^{(1)} = w_7^{(1)} = w_1^{(2)} = w_8^{(2)} = 0, \quad w_3^{(2)} = w_6^{(2)} = w_4^{(2)} = w_5^{(2)} = 3(1 + 9\rho^2)^{-1/2}.$$

Thus, according to (10), 19.2, the coefficients of the homogenized operator are of the form

$$a_{11} = 6\big(4 + (1+9\rho^2)^{1/2}\big)^{-1}, \quad a_{22} = 2(2+3\rho)\big(6 - 6\rho + (1+9\rho^2)^{1/2}\big)^{-1},$$
$$a_{12} = a_{21} = 0, \quad b_1 = \big(5 - \rho + 2(1+9\rho^2)^{1/2}\big)/3.$$

19.4 The Complete Asymptotic Expansion

19.4.1 Asymptotic solution outside a neighborhood of the boundary

In the interior of the domain Ω we seek the complete asymptotic expansion of the solution u to problem (4), 19.1, in the form

$$u(\varepsilon, l) \sim \sum_{q=0}^{\infty} \varepsilon^q u_q(x, \lambda), \tag{1}$$

where the series is formal whose coefficients u_q are smooth on the edges of the net S_ε and admit the representations

$$u_0(x,\lambda) = v_0(x), \quad u_q(x,\lambda) = v_q(x) + N(x,\lambda)\cdot \nabla v_{q-1}(x) + w_q(x,\lambda), q \in \mathbb{N}, \tag{2}$$

with functions w_q satisfying (7), 19.2 ($w_1 = 0$ in the formula (2) for u_1). We substitute series (1) into the two first equations of problem (4), 19.1, written in the form (1) and (2), 19.2, and equate the coefficients by the same degrees of ε. Then we obtain problems of the form (5), 19.2, for w_q. The compatibility conditions (6), 19.2, for such problems lead to equations for the coefficients v_q,

$$A(x, \partial/\partial x)\, v_q(x) = \tilde{F}_q(x), \quad x \in \Omega. \tag{3}$$

Here

$$\tilde{F}_q(x) = \int_Y \big\{\delta_{0q} f(x,\lambda) - (\partial/\partial l)\big(a(x,\lambda)(\partial/\partial\lambda) w_{q+1}(x,\lambda)\big)$$
$$- L_0(x,\lambda,\partial/\partial l)\big(w_q(x,\lambda) + N(x,\lambda)\cdot\nabla v_{q-1}(x)\big)\big\} d\lambda, \quad x \in S_\varepsilon \cap \Omega.$$

The functions v_q satisfy the boundary conditions

$$v_q(x) = h_q(x), \quad x \in \partial\Omega. \tag{4}$$

We shall define their right-hand sides, when considering the boundary layer near $\partial\Omega$ (see (14)).

The terms w_q are subject to

$$\begin{aligned} &- L_2(x,\lambda,\partial/\partial\lambda) w_q(x,\lambda) \\ &= (\partial/\partial\lambda)\big\{a(x,\lambda)(\partial/\partial l) w_{q-1}(x,\lambda) + N(x,\lambda)\cdot\nabla v_{q-2}(x)\big)\big\} \\ &\quad + (\partial/\partial l)\langle a(x,\lambda)\big\{(\partial/\partial\lambda)\big(w_{q-1}(x,\lambda) + N(x,\lambda)\cdot v_{q-2}(x)\big) \\ &\quad + (\partial/\partial l)\big(w_{q-2}(x,\lambda) + v_{q-2}(x) + N(x,\lambda)\cdot\nabla v_{q-3}(x)\big)\big\}\rangle - \delta_{0q}\langle f(x,\lambda)\rangle, \\ &S_l^{(\tau)}(w_q) = -S_0^{(\tau)}\big(w_{q-1}(x,\lambda) + N(x,\lambda)\cdot\nabla v_{q-2}(x)\big) \end{aligned} \tag{5}$$

with

$$\langle g(x,\lambda)\rangle = g(x,\lambda) - \int_Y g(x,\lambda) d\lambda.$$

Note that the compatibility condition (6), 19.2, is fulfilled for the right-hand side of problem (5).

The coefficients of series (1) do not satisfy the boundary condition of problem (4), 19.1. Therefore, near $\partial\Omega$ the asymptotic expansion (1) does not converge to the solution of problem (4), 19.1. To compensate the discrepancy in the boundary condition and obtain a complete asymptotic expansion in the whole domain Ω we add a boundary layer to the series (1).

19.4.2 Construction of the boundary layer

In what follows we suppose that the domain Ω is the half-space $\{(x_1,\dots,x_n)\colon x_n > 0\}$. The boundary layer arises near $\partial\Omega = \{(x_1,\dots,x_n)\colon x_n = 0\}$. Here our reasoning is similar to those in Sections 17.2 and 17.3. We introduce the coordinates $t = \varepsilon^{-1}x$. Expanding the coefficients a and b in (4), 19.1, into Taylor's series in x_n and passing to the variables t, we split the operators on the left in (4), 19.1, into the series with operator-valued coefficients

$$\begin{aligned}
&(\partial/\partial l)\Big(a\big(x(l),\varepsilon^{-1}l\big)\,\partial v/\partial l\Big) - b\big(x(l),\varepsilon^{-1}l\big)\,v \sim \varepsilon^{-2}\sum_{p=0}^{\infty}\varepsilon^p M_p(\varepsilon,t,\lambda,\partial/\partial\lambda)\,v,\\
&C_\tau(a\,\partial v/\partial l) \sim \varepsilon^{-1}\sum_{p=0}^{\infty}\varepsilon^p L_p^{(\tau)}(\varepsilon,t,\lambda,\partial v/\partial\lambda),
\end{aligned} \tag{6}$$

where

$$\begin{aligned}
M_p(\varepsilon,t,\lambda,\partial/\partial\lambda)v &= (\partial/\partial\lambda)\big((p!)^{-1}t_n^p(\partial/\partial x_n)^p\,a(\varepsilon t',0,\lambda)\,\partial v/\partial\lambda\big)\\
&\quad + \big((p-2)!\big)^{-1}t_n^{p-2}\,(\partial/\partial x_n)^{p-2}b(\varepsilon t',0,\lambda)v,\\
L_p^{(\tau)}(\varepsilon,t,\lambda,\partial v/\partial\lambda) &= C_\tau\big((p!)^{-1}t_n^p\,(\partial/\partial x_n)^p\,a(\varepsilon t',0,\lambda)\,\partial v/\partial\lambda\big)
\end{aligned} \tag{7}$$

with $t' = (t_1,t_2,\dots,t_{n-1})$. The point t in (6) depends on λ.

Let $h(x',t,\lambda)$ be a smooth Y-periodic function and let $x' = (x_1,x_2,\dots,x_{n-1})$. There hold the relations

$$\begin{aligned}
&M_p(\varepsilon,t,\lambda,\partial/\partial\lambda)\,h = \Big(\sum_{s=0}^{2}\varepsilon^s M_{ps}(\varepsilon,t,\lambda,\partial/\partial x',\partial/\partial\lambda)\Big)h,\\
&L_p^{(\tau)}(\varepsilon,t,\lambda,\partial h/\partial\lambda)(x',t,\lambda) = L_{p0}^{(\tau)}(\varepsilon,t,\lambda,\partial h/\partial\lambda) + \varepsilon L_{p1}^{(\tau)}(\varepsilon,t,\lambda,\partial h/\partial x').
\end{aligned} \tag{8}$$

Here

$$\begin{aligned}
&M_{p1}(\varepsilon,t,\lambda,\partial/\partial x',\partial/\partial\lambda)\,h\\
&\quad = \sum_{m=1}^{n-1}(\partial/\partial x_m)\big(w^{(m)}(\lambda)(p!)^{-1}t_n^p(\partial/\partial x_n)^p\,a(\varepsilon t',0,\lambda)\,\partial h/\partial\lambda\big)\\
&\qquad + (\partial/\partial\lambda)\Big(\sum_{m=1}^{n-1}w^{(m)}(\lambda)(p!)^{-1}t_n^p(\partial/\partial x_n)^p\,a(\varepsilon t',0,\lambda)\,\partial h/\partial x_m\Big),\\
&L_{p1}^{(\tau)}(\varepsilon,t,\lambda,\partial h/\partial x') = C_\tau\Big((p!)^{-1}t_n^p(\partial/\partial x_n)^p a(\varepsilon t',0,\lambda)\sum_{m=1}^{n-1}w^{(m)}(\lambda)\,\partial h/\partial x_m\Big),
\end{aligned}$$

$$M_{p2}(\varepsilon,t,\lambda,\partial/\partial x')h = \sum_{m,q=1}^{n-1} (\partial/\partial x_m)\big(w^{(m)}(\lambda)\, w^{(q)}(\lambda)(p!)^{-1} t_n^p (\partial/\partial x_n)^p\, a(\varepsilon t',0,\lambda)\,\partial h/\partial x_q\big) \tag{9}$$

while M_{p0} and $L_{p0}^{(\tau)}$ coincide with the operators M_p and $L_p^{(\tau)}$ in (7). It is supposed that in (7) and (9)

$$(\partial/\partial\lambda)\Psi(t) = \sum_{m=1}^{n} w^{(m)}(\lambda)(\partial/\partial t_m)\Psi(t).$$

In a neighborhood of $\partial\Omega$ we seek boundary layer in the form

$$h(\varepsilon,x) \sim \sum_{p=0}^{\infty} \varepsilon^p h_p(x',t,\lambda), \tag{10}$$

where the series is formal and h_p is a Y-periodic in λ function decreasing in t_n. We substitute the series (10) into (4), 19.1, and take (6) and (8) into account. Equating the coefficients at the same powers of ε, we obtain the equations for h_p,

$$\begin{aligned} &M_{00}(\varepsilon,t,\lambda,\partial/\partial\lambda)\,h_p(x',t,\lambda) = H_p(\varepsilon t',t,\lambda), \quad \lambda \in S\setminus T\cap\Omega,\\ &L_{00}^{(\tau)}(\varepsilon,t,\lambda,\,\partial h_p/\partial\lambda) = H_p^{(\tau)}(\varepsilon t',t,\lambda), \quad \lambda,\tau \in T\cap\Omega. \end{aligned} \tag{11}$$

The right-hand sides of (11) are defined by

$$H_p(\varepsilon t',t,\lambda) = \sum_{r=0}^{2}\ \sum_{q=1-\mathrm{sgn} r}^{p-r} M_{qr}(\varepsilon,t,\lambda,\partial/\partial\lambda,\partial/\partial x')h_{p-q-r}(x',t,\lambda),$$

$$\tilde H_p^{(\tau)}(\varepsilon t',t,\lambda) = \sum_{q=1}^{p} L_{q0}^{(\tau)}(\varepsilon,t,\lambda,\partial h_{p-q}/\partial\lambda) + \sum_{q=0}^{p-1} L_{q1}^{(\tau)}(\varepsilon,t,\lambda,\partial h_{p-q-1}/\partial x').$$

Substituting the sum of series (1) and (10) into the boundary condition on $\partial\Omega$ and equating the coefficients at the same powers of ε, we find the boundary conditions for h_p on $\partial\Omega$,

$$h(x',t',0,\lambda) = -u_p(x',0,\lambda), \quad \lambda\in\partial\Omega. \tag{12}$$

Thus, the functions h_p are defined by the equations (11) and (12). The next assertion on the solution to problem (11), (12) follows from Theorem 19.5.3 that will be proved in Section 19.5.

Theorem 19.4.1. *Let $\hat H_k(x',t,\lambda)$ and $g(x')$ be smooth Y-periodic in λ functions, $k = 0,1,\dots,K$, K being the number of nodes in Y. Suppose that $\hat H_k(x',t,\lambda) = o\big(\exp(-\gamma_k t_n)\big), \gamma_n > 0$. Then for $\lambda\in\Omega$ and $x'\in\partial\Omega$ there exists a solution $h(x',t,\lambda)$ to problem (11), (12) with right-hand sides $\hat H_k$ and g having the finite integral*

$$\int_{\Omega\cap S} \big\{a(\varepsilon t',0,\lambda)\big((\partial/\partial\lambda)h(\varepsilon t',t,\lambda)\big)^2\big\}d\lambda.$$

The solution is Y-periodic and satisfies the conditions (1), 19.1, in t on the faces of cells parallel to the axis x_m. Moreover, the solution admits the asymptotic representation

$$h(\varepsilon t',t,\lambda) = \sum_{q=1}^{\varkappa} \chi^{(q)} C^{(q)}(\hat H_k, g) + o\big(\exp(-\gamma t_n)\big), \quad t_n\to\infty. \tag{13}$$

Here $C^{(q)}$ are some constants depending on $\hat{H}_k$ and g; γ is a sufficiently small number; $\chi^{(q)}$ is the characteristic function of the q-th connected component of the set. (Clearly, the number of connected components of S coincides with the number of connected components of the graph U.)

Recall that t depends on the variable λ equal to λ_j on the edge σ_j. Since for $\hat{H}_k =; k = 0, 1, \dots, K$ and $g = 1$ the solution to problem (11), (12) is equal to 1, we have

$$\sum_{q=1}^{\varkappa} \chi^{(q)} C^{(q)}(0,1) = 1.$$

As will be shown in Section 19.4.3,

$$\sum_{q=1}^{\varkappa} \chi^{(q)} C^{(q)}$$

is a functional on $(\hat{H}_k, g)$. Therefore, the decay condition of the boundary layer (which means that the first sum on the right in (13) vanishes) leads to the boundary conditions for v_q in the representation (2) of the coefficients u_q,

$$\begin{aligned} &v_0(x',0) = 0, \quad v_m(x',0) \\ &= \sum_{q=1}^{\varkappa} \chi^{(q)} C^{(q)} \big(\hat{H}_k(x',t,\lambda) - w_m(x',0,\lambda) - N(x',0,\lambda)\cdot\nabla v_{m-1}(x',0)\big), \quad m \in \mathbb{N}. \end{aligned} \tag{14}$$

Hereafter we do not estimate the remainder in the asymptotics and restrict ourselves to the construction of formal expansions.

19.4.3 Definition of constants in the boundary layer

To compute the constants $C^{(q)}$ in (13) we need the next assertion that follows from Theorem 19.5.3.

Lemma 19.4.2. *There exist solutions $\Psi^{(q)}$ to the homogeneous problem (11), (12) admitting the asymptotics representations*

$$\Psi^{(q)}(x',t,\lambda) = \big(t_n + N^{(n)}(x',0,\lambda)\big)\chi^{(q)} + \Phi + o\big(\exp(-\gamma t_n)\big), \quad t_n \to \infty,$$

with function Φ constant on every connected component of S.

Theorem 19.4.3. *The constants $C^{(q)}$ in (13) are defined by the formula*

$$\begin{aligned} C^{(q)}(\hat{H}_k, q) = \Big(&- \int_{\Pi+} \hat{H}_0(x',t,\lambda)\,\Psi^{(q)}(x',t,\lambda)d\lambda \\ &- \sum_{\tau\in P_0} g(x')C_\tau\big(a(x',0,\lambda)(\partial/\partial\lambda)\,\Psi^{(q)}(x',t,\lambda)\big) \\ &+ \sum_{\tau\in T\cap\Pi+} C_\tau\big(\Psi^{(q)}(x',t,\lambda)\hat{H}_k(x',t,\lambda)\big)\Big)/\Xi^{(q)}, \end{aligned} \tag{15}$$

where $\Pi^+ = (\mathbb{R}_+ \times Q')\cap S$, Q' is the unit cube in $\mathbb{R}^{n-1}$, $P_y = \{(t', t_n) \colon t_n = y\}\cap S$,

$$\Xi^{(q)}(x') = \int_Y \chi^{(q)}\, w^{(n)}(\lambda)\, a(x',0,\lambda)\big((\partial/\partial\lambda)\, N^{(n)}(x',0,\lambda) + w^{(n)}(\lambda)\big)\, d\lambda > 0,$$

and $\Psi^{(q)}$ *are the solutions to problem* (11), (12) *given by Lemma* 19.4.2. *In the last sum on the right in* (15) *the index* k *is equal to the number of the node of* Y *corresponding to the node* τ *of* S.

Proof. Let $\Psi^{(q)}$ satisfy the homogeneous equations (11) and (12) (see Lemma 19.4.2) and let U be the solution to problem (11), (12) with right-hand sides g and $\hat{H}_k, k = 0, 1, \dots, K$. We substitute $\Psi^{(q)}$ and U into Green's formula for the equation (11) on the set $\Pi_R = \{Q' \times (0, R)\} \cap S$ and obtain

$$\int_{\Pi_R} \{\Psi^{(q)}(t,\lambda)\,(\partial/\partial\lambda)\big(a(\lambda)\,(\partial/\partial\lambda)\,U(t,\lambda)\big) - U(t,\lambda)\,(\partial/\partial\lambda)\big(a(\lambda)\,(\partial/\partial\lambda)\Psi^{(q)}(t,\lambda)\big)\}\,d\lambda = \sum_{\tau\in K_R} C_\tau\Big(a(\lambda)\big(\Psi^{(q)}(t,\lambda)\,(\partial/\partial\lambda)\,U(t,\lambda) - U(t,\lambda)\,(\partial/\partial\lambda)\,\Psi^{(q)}(t,\lambda)\big)\Big) \tag{16}$$

with $K_R = (T\cap\Pi_R)\cup P_0\cup P_R$. Hereafter we do not indicate the dependence of the functions on x'. Letting $R\to\infty$ in (16), we arrive at

$$\begin{aligned}\int_{\Pi+} \hat{H}_0(t,\lambda)\,\Psi^{(q)}(t,\lambda)\,d\lambda = &\sum_{\tau\in T\cap\Pi+} C_\tau\big(\Psi^{(q)}(t,\lambda)\,\hat{H}_k(t,\lambda)\big)\\ &- \sum_{\tau\in P_0} gC_\tau\,\big(a(\lambda)\,(\partial/\partial\lambda)\,\Psi^{(q)}(t,\lambda)\big)\\ &+ \lim_{R\to\infty}\sum_{\tau\in P_R} C_\tau\Big(a(\lambda)\big(\Psi^{(q)}(t,\lambda)\,(\partial/\partial\lambda)\,U(t,\lambda)\\ &- U(t,\lambda)\,(\partial/\partial\lambda)\,\Psi^{(q)}(t,\lambda)\big)\Big).\end{aligned} \tag{17}$$

By virtue of Lemma 19.4.2 and (13) the last limit $I^{(q)}$ in (17) is equal to

$$-C^{(q)}\lim_{R\to\infty}\sum_{\tau\in P_R} C_\tau\Big(a(\lambda)\big(w^{(n)}(\lambda) + (\partial/\partial\lambda)\,N^{(n)}(\lambda)\big)\chi^{(q)}\Big).$$

Let us write $I^{(q)}$ in the form

$$I^{(q)} = -\chi^{(q)}C^{(q)}\lim_{R\to\infty}\int_{\Pi_R\backslash\Pi_{R-1}} (\partial/\partial\lambda)\{t_n\,a(\lambda)\big(w^{(n)}(\lambda) + (\partial/\partial\lambda)\,N^{(n)}(\lambda)\big)\}\,d\lambda.$$

Differentiating with respect to λ, we obtain

$$I^{(q)} = C^{(q)}\Big(\Xi^{(q)} + \int_{\Pi_R\backslash\Pi_{R-1}} t_n\,(\partial/\partial\lambda)\{a(\lambda)\big(w^{(n)}(\lambda) + (\partial/\partial\lambda)\,N^{(n)}(\lambda)\big)\}\,d\lambda\Big). \tag{18}$$

The last integral on the right in (18) vanishes because the expression in the braces is constant on every edge of Y. It remains to show that $\Xi^{(q)} > 0$.

There holds the equality

$$\Xi^{(q)} = \chi^{(q)}\Big(\int_Y \Big\{2a(\lambda)\,w^{(n)}(\lambda)(\partial/\partial\lambda)\,N^{(n)}(\lambda) + N^{(n)}(\lambda)\,(\partial/\partial\lambda)\big(a(\lambda)w^{(n)}(\lambda)\big) + a(\lambda)\big(w^{(n)}(\lambda)\big)^2\Big\}\,d\lambda + \sum_{\tau\in T} C_\tau\big(a(\lambda)\,w^{(n)}(\lambda)\,N^{(n)}(\lambda)\big)\Big). \tag{19}$$

Using the definition of $N^{(n)}$ and integrating by parts over Y, we have

$$\int_Y N^{(n)}(\lambda)\,(\partial/\partial\lambda)\big(a(\lambda)\,w^{(n)}(\lambda)\big)\,d\lambda = \int_Y a(\lambda)\big((\partial/\partial\lambda)\,N^{(n)}(\lambda)\big)^2\,d\lambda$$
$$-\sum_{\tau\in T} C_\tau\big(N^{(n)}(\lambda)\,a(\lambda)(\partial/\partial\lambda)\,N^{(n)}(\lambda)\big). \tag{20}$$

Substituting (19) into (20), we find

$$\Xi^{(q)} = \chi^{(q)} \int_Y a(\lambda)\big((\partial/\partial\lambda)\,N^{(n)}(\lambda) + w^{(n)}(\lambda)\big)^2\,d\lambda. \tag{21}$$

The integral on the right in (21) vanishes if and only if

$$N^{(n)}(\lambda) = t_n(\lambda) + \text{const.}$$

However, the equality cannot be fulfilled in view of the Y- periodicity of the function $N^{(n)}$. Thus, $\Xi^{(q)}$ are positive. □

19.5 Asymptotics of the Solution to a Boundary Value Problem on a Net Located in a Cylinder

19.5.1 The boundary value problem in a cylinder

In this section we justify the asymptotics (13) for the boundary layer constructed in Section 19.4.

In $\Pi = \{(t', t_n) : t' \in Q',\ t_n \in \mathbb{R}\} \cap S$ we consider the boundary value problem

$$(\partial/\partial\lambda)\big(a(t)(\partial/\partial\lambda)\,u(t)\big) = \hat{H}_0(t), \quad t \in \Pi \setminus T, \tag{1}$$

$$C_\tau\,(a\,\partial u/\partial\lambda) = \hat{H}_k(t), \quad \tau, t \in T, \tag{2}$$

$$u(t'^{-}, t_n) = u(t'^{+}, t_n), \quad (\partial/\partial\lambda)$$
$$u(t'^{-}, t_n) = (\partial/\partial\lambda)\,u(t'^{+}, t_n), \quad t \in \partial\Pi, \tag{3}$$

where $k = 1, \ldots, K$ is the number of the node of Y corresponding to the node τ of S, and t'^{+}, t'^{-} are symmetric points of opposite faces of Q. (The function a is equal to $a(0,\lambda)$ in the definition (7), 19.4, of the operator of problem (11), 19.4.) We suppose that the right-hand sides of equations (1) and (2) are in the space $\mathbf{H}^1_\rho(\Pi)$ of functions f satisfying (3) and having the norm $||f; \mathbf{H}^1_\rho(\Pi)|| = ||\exp(\rho t_n) f; \mathbf{H}^1_\rho(\Pi)||$. It is easy to verify the next assertion.

Lemma 19.5.1. *The norm*

$$\Big(\sum_{m=-\infty}^{\infty} R^{2m} ||u(t', m + \{t_n\}); \mathbf{H}^1_\rho(Y)||^2\Big)^{1/2}, \quad R > 0, \tag{4}$$

is equivalent to the norm of u in $\mathbf{H}^1_{\log R}(\Pi)$. Here $\{y\} = y - [y]$ and $[y]$ is the integral part of y.

We associate to u in $\mathbf{H}^1_{\log R}(\Pi)$ the series

$$U(z,t) = \sum_{m=-\infty}^{\infty} u(t', m + \{t_n\}) z^m, \tag{5}$$

z being a complex number in the set $G_R := \{z \in \mathbb{C} \colon |z| = R\}$. From (4) it follows that U is an element of the space $\mathbf{L}_2(G_R; \mathbf{H}^1(Y))$ endowed with the norm

$$||U; \mathbf{L}_2(G_R; \mathbf{H}^1(Y))|| = \Big(\int_{|z|=R} ||U(z); \mathbf{H}^1(Y)||^2 dz \Big)^{1/2}.$$

If $v \in \mathbf{L}_2(G_R; \mathbf{H}^1(Y))$, then the function

$$V = (2\pi i)^{-1} \int_{|z|=R} z^{-[t_n]-1} v(t', \{t_n\}, z) dz \tag{6}$$

belongs to the space $\mathbf{H}^1_{\log R}(\Pi)$. From Lemma 19.5.1 and Parseval's equality it follows that

$$2\pi \int_{|z|=R} ||v(t,z); \mathbf{H}^1(Y)||^2 dz = \sum_{m=-\infty}^{\infty} \Big|\Big| \int_{|z|=R} z^{-m-1}\, v(t,z) dz; \mathbf{H}^1(Y) \Big|\Big|^2.$$

The transformation (4) is an isomorphism between $\mathbf{H}^1_{\log R}(\Pi)$ and $\mathbf{L}_2(G_R; \mathbf{H}^1(Y))$. The inverse transformation is given by (6).

19.5.2 The boundary value problem on a cell

Let u be a solution to the homogeneous problem (1)–(3). Then the function U in (5) satisfies the problem with complex parameter ζ:

$$\begin{aligned} &(\partial/\partial\lambda)\big(a(t)\,(\partial/\partial\lambda)\,U(t,\zeta)\big) = 0, \quad t \in Y \setminus T_Y; \quad C_\tau(a\,\partial U/\partial\lambda) = 0, \quad \tau \in T_Y;\\ &U(t'^-, t_n, \zeta) = U(t'^+, t_n, \zeta), \end{aligned} \tag{7}$$

$$\begin{aligned} &(\partial/\partial\lambda)\,U(t'^-, t_n, \zeta) = (\partial/\partial\lambda)\,U(t'^+, t_n, \zeta), \quad t \in (\partial Q' \times [0,1]) \cap S;\\ &U(t', 0, \zeta) = \zeta U(t', 1, \zeta),\\ &(\partial/\partial\lambda)\,U(t', 0, \zeta) = \zeta\,(\partial/\partial\lambda)\,U(t', 1, \zeta), \quad t' \in P_0. \end{aligned} \tag{8}$$

Lemma 19.5.2. *For $\zeta = 1$ there exist $\varkappa$ eigenfunctions $\Phi_0^{(q)}, q = 1, \dots, \varkappa$, of problem (7), (8), $\varkappa$ being the number of connected components of the set U (see 19.3.1). To every eigenfunction $\Phi_0^{(q)}$ there corresponds only one associated function $\Phi_1^{(q)}$. If $\zeta \in D = \{z \in \mathbb{C} \colon |z| = 1, z \neq 1\}$, then problem (7), (8) is uniquely solvable.*

Proof. A solution u of the problem adjoint to problem (7), (8) satisfies (7) and the equalities

$$\begin{aligned} &\bar\zeta\, u(t', 0, \zeta) = u(t', 1, \zeta),\\ &\bar\zeta(\partial/\partial\lambda)\, u(t', 0, \zeta) = (\partial/\partial\lambda)\, u(t', 1, \zeta), \quad t' \in P_0. \end{aligned} \tag{9}$$

For $|\zeta| = 1$ we represent the solution u to problem (7), (9) in the form $u(t,\zeta) = \bar\zeta^{1-t_n} u(t,\zeta)$. Then u satisfies the equations

$$(\partial/\partial\lambda - w^{(n)}(\lambda)\log\bar\zeta)\, a(t)(\partial/\partial\lambda - w^{(n)}(\lambda)\log\bar\zeta))\, u(t,\zeta) = 0, \tag{10}$$

$$C_\tau\Big(a(t)(\partial/\partial\lambda - w^{(n)}(\lambda)\log\bar\zeta))\, u(t,\zeta)\Big) = 0, \quad \tau \in T_Y, \tag{11}$$

and conditions (8). We multiply (10) by $\bar{u}$, integrate over Y and then integrate by parts. In view of (11) and (8) the terms outside the integral vanish and

$$\int_Y \left|\left(\partial/\partial\lambda - w^{(n)}(\lambda)\log\bar{\zeta}\right) u(\lambda,\zeta)\right|^2 d\lambda = 0.$$

Thus, $u(\lambda,\zeta) = \bar{\zeta}^{t_n}\,\Phi(\lambda)$, where Φ is a function constant on every connected component of U.

In view of ellipticity of the homogenized operator and Remark 19.3.3 the cell Y intersects the faces of ∂Q orthogonal to the axis Ox_n. Since the function u is Y-periodic, the number ζ in the representation of u is equal to 1. In case $\zeta \in D$ the problem adjoint to (7), (8) has no solution.

Now we turn to the Jordan chains corresponding to the eigenvalue $\zeta = 1$. Let us consider the functions $\Phi_0^{(q)}, q = 1, \dots, \varkappa$, that are equal to $C^{(q)}$ on the q-*th* connected component of U and vanish everywhere else. Such functions satisfy problem (7), (8). The associated functions are subject to

$$\begin{aligned}&\Phi_1^{(q)}(t',0) - \zeta\Phi_1^{(q)}(t',1) = -\Phi_0^{(q)}(t',1)\\&(\partial/\partial\lambda)\,\Phi_1^{(q)}(t',0) - \zeta(\partial/\partial\lambda)\,\Phi_1^{(q)}(t',1) = -(\partial/\partial\lambda)\,\Phi_0^{(q)}(t',1),\end{aligned} \tag{12}$$

$(t',1) \in P_1$, $q = 1,2,\dots,\varkappa$, and to the equalities (7). Hence, by virtue of (4), 19.2, $\Phi_1^{(q)} = \left(t_n + N^{(n)}(t',0,t_n)\right)\Phi_0^{(q)}$. The associated functions $\Phi_2^{(q)}$ satisfy (7) and (12) with right-hand sides $-\Phi_1^{(q)}(t',1)$ and $-(\partial/\partial\lambda)\Phi_1^{(q)}(t',1)$, respectively. The compatibility condition of problem (7), (12) with Ψ_1 and Ψ_2 on the right in (12) takes the form

$$\sum_{(t',1)\in P_1} a(t',1)\,\Psi_2(t') = 0. \tag{13}$$

We show there are no functions $\Phi_2^{(q)}$ because the corresponding condition (13) has not been fulfilled.

Indeed, the left-hand side of (13) with $\Psi_2(t') = (\partial/\partial\lambda)\Phi_1^{(q)}(t',1)$ can be rewritten in the form

$$\int_Y \chi^{(q)}(\partial/\partial\lambda)\{t_n a(t)\left(t_n + N^{(q)}(t',0,t)\right)\}d\lambda.$$

In view of (2), 19.3, the last integral equals $\Xi^{(q)}$ in (15), 19.4. The fact that $\Xi^{(q)}$ is positive was established in the proof of Theorem 19.4.3. Thus, to the eigenvalue $\zeta = 1$ there correspond q eigenfunctions and every of them possesses one associated function. □

19.5.3 Asymptotic expansion of solutions to the problem in a cylinder

Theorem 19.5.3. *Let u_1 be a solution to problem (1)–(3) of at most polynomial growth at infinity. Assume that the right-hand sides $\hat{H}_k, k = 0,1,\dots,K$, of (1) and (2) are in $\mathbf{H}^1_\rho(\Pi)$ with small positive ρ. Then u_1 admits the asymptotic representation*

$$u_1(t) = \Phi_1(t)\left(t_n + N^{(n)}(t',0,t)\right) + \Phi_0(t) + u_2(t),$$

where Φ_1 and Φ_0 are constant on the connected components of S, and $u_2 \in \mathbf{H}^1_\rho(\Pi)$.

Proof. Let $h_k, k = 0, 1, \dots, K$, be the functions obtained from $\hat{H}_k$ with the help of transform (4). By virtue of the properties of the transform, h_k are in $\mathbf{L}_2(G_{\exp(\rho)}; \mathbf{H}^1(Y))$.

We denote by $G_j, j = 1, 2$, the circle $\{z \in \mathbb{C} \colon |z| = \exp((-1)^j \rho)\}$. Since the sets G_j contain no eigenvalues of problem (7), (8), for all z in the circles we can define the operator $\tilde{K}$ such that the function $u = \tilde{K}(h_k)$ is a solution to problem (7), (8) with right-hand side h_k. Thus, by the definition of the transform (6), the functions

$$u_j(t) = (2\pi i)^{-1} \int_{z \in G_j} z^{-t_n} \tilde{K}(z, h_k)\, dz \tag{14}$$

are in $\mathbf{H}_{(-1)^j \rho}(\Pi)$ and satisfy the equations of problem (1)–(3).

By Lemma 19.5.2, the resolvent $\tilde{K}$ of problem (7), (8) admits the expansion

$$\tilde{K}(z) = (z-1)^{-2} \tilde{P}_1 + (z-1)^{-1} \tilde{P}_2 + \tilde{P}(z) \tag{15}$$

in a neighborhood $\tilde{G}$ of the eigenvalue $\zeta = 1$, where $\tilde{P}_1$ and $\tilde{P}_2$ are operators acting into the spaces spanned by $\Phi_0^{(q)}$ and $\Phi_1^{(q)}$, respectively, and $\tilde{P}$ is holomorphic in $\tilde{G}$ (e.g., see GOHBERG and KREIN [1]). Let Γ be a closed contour consisting of the circles G_1 and G_2 and a segment connected them. According to the Cauchy formula,

$$\int_{\Gamma} z^{-t_n} \tilde{K}(z, h_k) dz = 2\pi i \operatorname{Res}\big(\tilde{K}(1, h_k)\big).$$

Substituting the last relation into (14) and computing the residues with the help of (15), we obtain the assertion of the theorem. □

Chapter 20

Homogenization of Equations on a Fine Periodic Grid

Homogenization can be useful not only for studying differential equations but for difference ones, as well. Moreover, to difference equations it was applied even earlier, in the nineteenth century, to analyze the limit passages in discrete models of continua for obtaining continuous ones.

In the present chapter, it is shown how the techniques in Chapter 19 can be modified for constructing asymptotics of solutions to difference equations with rapidly oscillating coefficients on a fine periodic grid. In the first section, we describe the grid, the difference boundary value problem, and the formal procedure of homogenization. The second section is devoted to studying the coefficients of the homogenized system of partial differential equations. We find the ellipticity criterion for the homogenized system and indicate a way to compute its coefficients. In the final third section we apply the above mentioned results to describe a model of crystalline grid and present examples of homogenized operators in the case of hexagonal grids.

In contrast to the preceding chapter, here we do not justify the asymptotics. We do not consider the construction of the complete asymptotic expansions in the interior of the grid and the boundary layer near the rectilinear parts of the boundary. These questions can be studied as in Chapter 19.

20.1 Homogenization of Difference Equations

20.1.1 A grid in $\mathbb{R}^n$ and the interaction set of its points

Let $\{Y\}$ be the set of vectors admitting the representation $Y = z_1 d_1 + \cdots + z_n d_n$, where $z_1, \ldots, z_n$ are some integers and $d_1, \ldots, d_n$ form a basis in $\mathbb{R}^n$.

Definition 20.1.1. *A set $\tilde{G}$ is defined to be a grid if $\tilde{G}$ is invariant under the shifts by vectors $Y \in \{Y\}$.*

Definition 20.1.2. *The elementary cell G of a grid $\tilde{G}$ is defined to be the intersection of $\tilde{G}$ and the parallelepiped with edges $d_1, d_2, \ldots, d_n$.*

Let the shift of G by a vector $Y \in \{Y\}$ move points t, s to t', s'. For ease of notation we shall write ts instead of $(t, s) \in \tilde{G} \times \tilde{G}$.

Definition 20.1.3. *A set $\Gamma \subset \tilde{G} \times \tilde{G}$ is called the interaction set of the points of $\tilde{G}$ if for any element st the set Γ contains $s't'$ and $t's'$. The element $st \in \Gamma$ is called the interaction of s with t.*

Let $g(s,t)$ be the set of points $s, s_1, s_2, \ldots, s_p, t$ such that $ss_1, s_1s_2, \ldots, s_p, t \in \Gamma$ and $s_j \neq s_k$ for $j \neq k$. If for some s, t there are no points $s_1, s_2, \ldots, s_p$ with the mentioned property we put $g(s,t) = \emptyset$.

Definition 20.1.4. *A subset $\tilde{G}^{(j)}$ of a grid $\tilde{G}$ is called a subgrid if any s and t in $\tilde{G}^{(j)}$ satisfy $g(s,t) \neq \emptyset$.*

We assume that

$$\tilde{G} = \bigcup_{j=1}^{\varkappa} \tilde{G}^{(j)}.$$

Definition 20.1.5. *A function v on $\tilde{G}^{(j)} \times \Gamma$ is called G-periodic if for any s and t with interaction $st \in \Gamma$ and for any s' and t' there holds the equality $v(s,st) = v(s',s't')$.*

Let $\tilde{G}_\varepsilon$ stand for the grid $\tilde{G}$ obtained from $\tilde{G}$ by means of the change of variable $x \to \varepsilon^{-1}x$ with small positive ε. We denote by σ and τ the points of $\tilde{G}_\varepsilon$ corresponding to s and t in $\tilde{G}$ and by $x(\sigma)$ the Gartesian coordinates of $\sigma \in \tilde{G}_\varepsilon$. Suppose that the set $\gamma(\sigma)$ of points interacting with $\sigma \in \tilde{G}_\varepsilon$ consists of elements in $\tilde{G}_\varepsilon$ located at a distance not greater than $O(\varepsilon)$ from σ. The set of all interactions of the grid $\tilde{G}_\varepsilon$ will be denoted by γ. We introduce some operators on functions given on $\tilde{G}_\varepsilon$ or $\tilde{G}_\varepsilon \times \gamma$.

Definition 20.1.6. *Let u be a function given on $\tilde{G}_\varepsilon$. The difference operator ∂ is defined by the formula*

$$\partial u(\tau\sigma) = u(\tau) - u(\sigma).$$

Definition 20.1.7. *Let v be a function given on $\tilde{G}_\varepsilon \times \gamma$ with property $v(\sigma, \sigma\tau) = -v(\sigma, \tau\sigma)$. Then*

$$\partial^* v(\sigma) = \sum_{\tau \in \gamma(\sigma)} \big(v(\sigma, \sigma\tau) + v(\tau, \tau\sigma)\big). \tag{1}$$

Note that

$$-\langle v \partial u \rangle_{\tilde{G}_\varepsilon \cap \Omega} = \langle u \partial^* v \rangle_{G_\varepsilon \cap \Omega},$$

Ω being a domain in $\mathbb{R}^n$ with compact closure and smooth boundary $\partial\Omega$. Hereafter $\langle w \rangle_Q$ with $Q \subset \tilde{G}_\varepsilon$ stands for

$$|Q|^{-1} \sum_{\sigma \in Q} \sum_{\tau \in \gamma(\sigma)} w(\sigma, \tau\sigma),$$

where $|Q|$ is the number of elements in Q. The same notation $\langle w \rangle_Q$ we shall use for the mean of w over $Q \subset \tilde{G}$.

20.1.2 Statement of the problem

We introduce a space $\mathbf{S}(G)$ of periodic vector-valued functions v on $\tilde{G} \times \Gamma$ endowed with the norm

$$||v; \mathbf{S}(G)|| = \langle |v|^2 \rangle_G^{1/2}.$$

Here

$$|v| = \Big(\sum_{k=1}^{M} (v^{(k)})^2 \Big)^{1/2},$$

$v^{(k)}$ being the components of v. Let $\mathbf{C}^m(\bar{\Omega}; H), m \in \mathbb{N}_0$, be the space of functions on $\bar{\Omega}$ with values in a Hilbert space H which are continuously differentiable up to order m, and $C^0 = C$. We denote by $\tilde{A}$ and $\tilde{B}$ symmetric square matrices of order M with entries $a_{jk}(x, ts)$ and $b_{jk}(x, s)$ given on $\bar{\Omega} \times \Gamma$ and $\bar{\Omega} \times \tilde{G}$, respectively.

Suppose that the columns of $\tilde{A}$ and $\tilde{B}$ belong to $\mathbf{C}^1(\bar{\Omega};\mathbf{S}(G))$ and $\mathbf{C}(\bar{\Omega};\mathbf{S}(G))$ while the functions a_{jk} satisfy $a_{jk}(x,st)=a_{jk}(x,ts)$.

In the domain $\tilde{G}_\varepsilon\cap\Omega$ we consider the equation

$$\varepsilon^{-2}\partial^*(\tilde{A}\partial u)-\tilde{B}u=f \tag{2}$$

with $f\in\mathbf{C}\big(\bar{\Omega};\mathbf{S}(G)\big)$ which is a concise notation for the equality

$$\varepsilon^{-2}\sum_{\tau\in\gamma(\sigma)}\Big(\tilde{A}\big(x(\tau),ts\big)+\tilde{A}\big(x(\sigma),ts\big)\Big)\big(u(\tau)-u(\sigma)\big)-\tilde{B}\big(x(\sigma),s\big)\,u(\sigma)$$
$$=f\big(x(\sigma),s\big),\qquad \sigma\in\tilde{G}_\varepsilon\cap\Omega.$$

Let $\partial(\tilde{G}_\varepsilon\cap\Omega)$ be the set of points in $\tilde{G}_\varepsilon\cap\Omega$ interacting with points in $\tilde{G}_\varepsilon\setminus\Omega$. We assume that the solution u to equation (2) satisfies the boundary condition

$$u(\sigma)=0,\quad \sigma\in\partial(\tilde{G}_\varepsilon\cap\Omega). \tag{3}$$

In what follows we study the solvability of problem (2), (3) and find the principal term in the asymptotics of the solution as $\varepsilon\to 0$.

20.1.3 Solvability of the boundary value problem

We introduce a space $\tilde{\mathbf{H}}^0(\tilde{G}_\varepsilon\cap\Omega)$ of vector-valued functions v on the set $\tilde{G}_\varepsilon\cap\Omega$ endowed with the norm

$$||v;\tilde{\mathbf{H}}^0(\tilde{G}_\varepsilon\cap\Omega)||=\langle|v|^2\rangle^{1/2}_{\tilde{G}_\varepsilon\cap\Omega},$$

and a space $\tilde{\mathbf{H}}^1(\tilde{G}_\varepsilon\cap\Omega)$ with norm

$$||v;\tilde{\mathbf{H}}^1(\tilde{G}_\varepsilon\cap\Omega)||=\langle|v|^2+\varepsilon^{-2}|\partial v|^2\rangle^{1/2}_{\tilde{G}_\varepsilon\cap\Omega}.$$

The space $\tilde{\mathbf{H}}^1(\tilde{G}_\varepsilon\cap\Omega)$ is a Hilbert space with inner product

$$(v_1,v_2)=\langle\varepsilon^{-2}\partial v_1\cdot\partial v_2+v_1\cdot v_2\rangle_{\tilde{G}_\varepsilon\cap\Omega},$$

where the point denotes the inner product in $\mathbb{R}^M$.

Lemma 20.1.8. *There exists the unique solution $u\in\tilde{\mathbf{H}}^1(\tilde{G}_\varepsilon\cap\Omega)$ to equation* (2) *satisfying* (3) *and*

$$||u;\tilde{\mathbf{H}}^1(\tilde{G}_\varepsilon\cap\Omega)||\le c||f;\tilde{\mathbf{H}}^0(\tilde{G}_\varepsilon\cap\Omega)||.$$

Hereafter c stands for a constant independent of ε.

Proof. Any function u equal to zero on $\partial(\tilde{G}_\varepsilon\cap\Omega)$ satisfies the Friedrichs inequality

$$\langle u^2\rangle_{\tilde{G}_\varepsilon\cap\Omega}\le c\varepsilon^{-2}\langle(\partial u)^2\rangle_{\tilde{G}_\varepsilon\cap\Omega} \tag{4}$$

which can be proved similarly to Proposition 19.2.2.

We multiply (2) by $\eta\in\tilde{\mathbf{H}}^1(\tilde{G}_\varepsilon\cap\Omega)$ vanishing on $\partial(\tilde{G}_\varepsilon\cap\Omega)$. Summing over $\sigma\in G_\varepsilon\cap\Omega)$, we obtain

$$\varepsilon^{-2}\sum_{\sigma\in\tilde{G}_\varepsilon\cap\Omega}\sum_{\tau\in\gamma(\sigma)}\tilde{A}\big(x(\tau),\sigma\tau\big)\,\partial u(\tau\sigma)\cdot\eta(\sigma)$$
$$+\sum_{\sigma\in\tilde{G}_\varepsilon\cap\Omega}\Big(\varepsilon^{-2}\sum_{\tau\in\gamma(\sigma)}\tilde{A}\big(x(\sigma),\sigma\tau\big)\,\partial u(\tau\sigma)\cdot\eta(\sigma)-\tilde{B}\big(x(\sigma),\sigma\tau\big)\,u(\sigma)\cdot\eta(\sigma)\Big)$$
$$=\sum_{\sigma\in\tilde{G}_\varepsilon\cap\Omega}f\big(x(\sigma),s\big)\cdot\eta(\sigma),$$

which we rewrite in the form

$$-\langle \varepsilon^{-2}\tilde{A}\partial u \cdot \partial \eta + \tilde{B}u \cdot \eta \rangle_{\tilde{G}_\varepsilon \cap \Omega} = \langle f \cdot \eta \rangle_{\tilde{G}_\varepsilon \cap \Omega)}. \tag{5}$$

Taking account of (4) and the Riesz theorem on the general form of functionals in Hilbert space, we complete the proof. □

20.1.4 The leading terms in asymptotics

For functions $v(x,s)$ and $u(x,ts)$ in $\mathbf{C}^{k+1}(\bar{\Omega};\mathbf{S}(G))$ there hold the following asymptotic expansions

$$\begin{aligned}
v\big(x(\tau),t\big) - v\big(x(\sigma),s\big) &= Dv\big(x(\sigma),ts\big) + \sum_{m=1}^{k} (m!)^{-1}\big(l^{(j)}(\tau\sigma)\partial/\partial x_j\big)^m v\big(x(\sigma),t\big) \\
&\quad + (k!)^{-1}\int_0^1 \big(l^{(j)}(\tau\sigma)\partial/\partial x_j\big)^{k+1} v\big(x(\sigma)(1-y) \\
&\quad + x(\tau)\,y,t\big)(1-y)^k dy, \\
u\big(x(\tau),ts\big) + u\big(x(\sigma),ts\big) &= 2u\big(x(\sigma),ts\big) + \sum_{m=1}^{k} (m!)^{-1}\big(l^{(j)}(\tau\sigma)\partial/\partial x_j\big)^m u\big(x(\sigma),ts\big) \\
&\quad + (k!)^{-1}\int_0^1 \big(l^{(j)}(\tau\sigma)\partial/\partial x_j\big)^{k+1} u\big(x(\sigma)(1-y) \\
&\quad + x(\tau)\,y,ts\big)(1-y)^k dy, \quad k \in \mathbb{N}_0.
\end{aligned} \tag{6}$$

Here $l^{(k)}(\tau\sigma)$ are components of the vector $l(\tau\sigma)$ directed from σ to τ and $Dv(x,ts) = v(x,t) - v(x,s)$. In the sequel we sum over repeating indices from 1 to n.

We apply the operator on the left in (2) to the function v and take (6) into account. Then the result can be written as the formal series

$$\varepsilon^{-2}(A_0 v + \varepsilon A_1 v + \varepsilon^2 A_2 v + \cdots) \tag{7}$$

with coefficients

$$\begin{aligned}
A_0 v &= D^*(\tilde{A}\,Dv), \quad A_1 v = 2^{-1}D^*\big(l^{(j)}(\partial/\partial x_j)\,\tilde{A}\,Dv + 2\tilde{A}\,l^{(j)}\partial v/\partial x_j\big) \\
A_2 v &= 2^{-1}D^*\big\{(l^{(j)}\partial/\partial x_j)^2\tilde{A}D\,v + 2\tilde{A}\,(l^{(j)}\partial/\partial x_j)^2 v \\
&\quad + l^{(j)}(\partial/\partial x_j)\tilde{A}(l^{(m)}\partial v/\partial x_m)\big\} - \tilde{B}v,
\end{aligned}$$

where

$$D^* u(x,s) = 2\sum_{t\in\gamma(s)} v(x,ts)$$

and $\gamma(s)$ is the set of points interacting with $s \in \tilde{G}$.

One can seek the asymptotics of the solution to problem (2), (3) in the form

$$u_0 + \varepsilon u_1(x,s) + \varepsilon^2 u_2(x,s) + \cdots. \tag{8}$$

Coefficients of the series for $k \geq 1$ are vector-valued functions periodic in the second variable. Substituting (8) into (7) and equating the coefficients at the same powers of ε we obtain the equations for u_k. Since the coefficient at ε^{-1} vanishes, we have

$$A_0 u_1 = -A_1 u_0. \tag{9}$$

To find u_1 we introduce the matrices $N^{(k)}, k = 1, \dots, n$, that are periodic solutions to the equations

$$D^*(\tilde{A}DN^{(k)})(x,s) = -D^*(\tilde{A}l^{(k)})(x,s), \quad k = 1,\dots,n, \quad s \in G. \tag{10}$$

By virtue of (9),

$$u_1 = (N^{(k)}\,\partial/\partial x_k)\,u_0. \tag{11}$$

Let us consider the solvability of equations (10) in the class of functions with G-periodic components.

Lemma 20.1.9. *Let the columns of the matrix $\tilde{A}$ and a vector-valued function F exist in $\mathbf{C}^m\big(\bar{G};\mathbf{S}(G)\big)$ for some $m = 0, 1, \dots$. The equation*

$$D^*(\tilde{A}D\,u)(x,s) = F(x,s), \quad x \in \bar{\Omega}, \quad s \in G, \tag{12}$$

is solvable in $\mathbf{C}^m\big(\bar{G};\mathbf{S}(G)\big)$ if and only if

$$\langle F\rangle_G(x) = 0. \tag{13}$$

A solution u to problem (12) is determined up to a vector constant on every set $G^{(j)} = \tilde{G}^{(j)} \cap G$. The constants can be chosen so that

$$\langle u\rangle_{G^{(j)}} = 0, \quad j = 1,\dots,\varkappa.$$

The proof can be obtained from the Riesz theorem on a linear functional in Hilbert space and the Poincaré inequality

$$\langle u^2\rangle_G \le \langle u\rangle_G^2 + c\langle (D\,u)^2\rangle_G. \tag{14}$$

The estimate (14) follows from the evident relation

$$u(x,s_2) - u(x,s_1) = \sum_{s,t\in g(s_2,s_1)} \big(u(x,t) - u(x,s)\big),$$

valid for any two points $s_1, s_2 \in G$. We square the last equality, sum over $s_1, s_2 \in G$, apply Cauchy's inequality, and arrive at (14).

Let us turn to the asymptotics of the solution to problem (2), (3). In view of the G-periodicity of a_{jk} and the relations

$$a_{jk}(x,st) = a_{jk}(x,ts), \quad l(st) = -l(ts),$$

the right-hand sides of equations (10) satisfy (13). Hence, the equations (10) for $N^{(k)}$ are solvable.

The coefficient u_2 in (8) is defined by the equation

$$A_0u_2 = -A_2u_0 - A_1u_1 + f.$$

Its compatibility condition

$$\langle f - A_2u_0 - A_1u_1\rangle_G = 0$$

is equivalent to the following equation for u_0:

$$(\partial/\partial x_j)(\hat{A}^{(j,k)}(x)(\partial/\partial x_k)\,u_0(x)) - \langle\tilde{B}\rangle_G(x)u_0(x) = \langle f\rangle_G(x), \quad x \in \Omega. \tag{15}$$

Here $\hat{A}^{(j,k)}$ is the matrix with entries

$$\hat{A}^{(j,k)}_{pq}(x) = \langle l^{(j)}l^{(k)}\,a_{pq} + l^{(j)}\,a_{pm}D\,N^{(j)}_{mq}\rangle_G(x). \tag{16}$$

In Theorem 20.2.1 below there is a criterion for the strong ellipticity of system (15), which we assume to be fulfilled.

Thus, the coefficient u_1 in (5) is defined by (11) while u_0 satisfies (15) and the boundary condition

$$u_0(x) = 0, \quad x \in \partial\Omega. \tag{17}$$

(If (17) holds, then the function $u_0(x(\sigma))$, $\sigma \in \partial(\tilde{G}_\varepsilon \cap \Omega)$, is subject to $|u_0(x(\sigma))| \leq c\varepsilon$.) The proof of the next assertion is similar to that of Theorem 19.2.4.

Theorem 20.1.10. *Let the columns of the matrices $\tilde{A}$ and $\tilde{B}$ are in $\mathbf{C}^3(\bar{\Omega}; \mathbf{S}(G))$ and $\mathbf{C}^1(\bar{\Omega}; \mathbf{S}(G))$, respectively, while $f \in \mathbf{C}^3(\bar{\Omega}; \mathbf{S}(G))$. Then for the solution u to problem* (2), (3) *there holds the estimate*

$$||u - u_0 - \varepsilon u_1; \tilde{\mathbf{H}}^1(\tilde{G}_\varepsilon \cap \Omega)|| \leq c\varepsilon^{1/2},$$

where u_0 is the solution to problem (15), (17) *and u_1 is defined by* (11).

20.1.5 Asymptotics of the solution to the nonstationary problem

Let u satisfy the equation

$$\begin{aligned} &\tilde{R}(\partial/\partial T)^2 u(\sigma, T) - \varepsilon^{-2}\partial^*(\tilde{A}\partial u)(\sigma, T) + \tilde{B}u(\sigma, T) = f(s, x(\sigma), T), \\ &\sigma \in \tilde{G}_\varepsilon \cap \Omega, \quad T \in [0, T_0], \end{aligned} \tag{18}$$

the boundary condition (13) and the initial conditions

$$u(\sigma, 0) = 0, \quad (\partial/\partial T)\, u(\sigma, 0) = 0, \quad \sigma \in \tilde{G}_\varepsilon \cap \Omega. \tag{19}$$

Here $\tilde{R}$ is a positive definite matrix with entries $r_{jk}(x, s)$ that are G-periodic in the second variable and smooth in x. The entries a_{jk} and b_{jk} of $\tilde{A}$ and $\tilde{B}$ satisfy the assumptions on the coefficients of equation (2) and smoothly depend on T.

The construction of the principal term $u_0(x, T) + \varepsilon u_1(x, s, T)$ in the asymptotic expansion of the solution to problem (18), (3), and (19) is similar to that of the solution to stationary problem (2), (3) (see Section 20.1.3). The function u_0 satisfies the equation

$$\begin{aligned} &\langle\tilde{R}\rangle_G\, (\partial/\partial T)^2\, u_0(x, T) - (\partial/\partial x_j)\big(\hat{A}^{(j,k)}(x)(\partial/\partial x_k)\, u_0(x, T)\big) + \langle\tilde{B}\rangle_G\,(x)\, u_0(x, T) \\ &\quad = \langle f\rangle_G(x, T), \quad x \in \Omega, \quad T \in [0, T_0], \end{aligned}$$

and the homogeneous initial and boundary conditions. The coefficient u_1 is equal to $(N^{(k)}\partial/\partial x_k)\, u_0$, where $N^{(k)}$ is the solution to (10). The justification of the asymptotics follows from the a priori estimate for the solution to problem (18), (3), and (19) and can be carried out according to the scheme in Bakhvalov and Panasenko [1].

20.2 Calculation of Coefficients of the Homogenized Operator and Their Properties

Suppose that points of G participate in $2P$ interactions. To every interaction we associate a number $z(st) \in \{-1, 1\}$ such that $z(ts) = -z(st)$. Let $p(st) \in \{1, 2, \dots, P\}$ be the index of interaction pairs st and ts (i.e., $p(st) = p(ts)$ but $p(st) \neq p(s't')$ for $s't' \neq st$ and $s't' \neq ts$). From (10), 20.1, it follows that the matrices $N^{(k)}$ satisfy the equalities

$$\tilde{A}(x, ts)DN^{(k)}(x, ts) = -l^{(k)}(ts)\tilde{A}(x, ts) + z(ts)C^{(k, p(ts))}(x), \quad s \in G, \quad t \in \gamma(s), \tag{1}$$

with matrices $C^{(k,p)}$ subject to

$$\sum_{t\in\gamma(s)} z(ts)C^{\left(k,p(ts)\right)}(x) = 0, \quad s \in G, \quad k = 1,\dots,n. \tag{2}$$

Using (1), we write coefficients of the homogenized operator in the form

$$\hat{A}^{(j,k)}_{mq}(x) = |G|^{-1} \sum_{\sigma\in G} \sum_{t\in\gamma(s)} z(ts) l^{(j)}(ts) C^{\left(k,p(ts)\right)}_{mq}(x). \tag{3}$$

To check ellipticity of the homogenized operator and calculate its coefficients, we describe the construction of the matrices $C^{(k,p)}$. Let us connect the interacting points of the grid $\tilde{G}$ by straight lines. We denote by K the set of points where these lines meet the faces of the parallelepiped Q with edges $d_1, d_2, \dots, d_n$ (see Section 20.1.1). If $t \in K$, then t^* stands for the point symmetric with respect to the plane passing through the center of Q and parallel to the face of Q that contains t. Owing to the periodicity of Γ, the point t^* is in K. We identify the points t and t^*. Then the lines connecting the interacting points form the graph $\hat{Y}$ with P edges and δ nodes, which are elements of G.

The graph $\hat{Y}$ has $\varkappa$ connected components. From (1) and the G-periodicity of $\tilde{A}$ and $N^{(k)}$ it follows that $\langle zC^{(k,p)}\rangle_{G^{(j)}}(x) = 0$, $j = 1,\dots,\varkappa$, $k = 1,\dots,n$. Therefore, the relations (2) are linearly dependent. We exclude $\varkappa$ linearly dependent equations and obtain $\delta - \varkappa$ equations for the matrices $C^{(k,p)}$. It is known (e.g., see ZYKOV [1], Chapter 3) that a graph with P edges, δ vertices, and $\varkappa$ connected components has $P - \delta + \varkappa$ independent cycles. Denote linear independent cycles of the graph $\hat{Y}$ by y_q, $q = 1, 2, \dots, P - \delta + \varkappa$. By (1), for any two interacting points s and t of the set G there holds the relation

$$N^{(k)}(x,t) = N^{(k)}(x,s) - l^{(k)}(ts)\, E + z(ts)\tilde{A}^{-1}(x,ts)C^{\left(k,p(ts)\right)}(x), \quad k = 1,\dots,n, \tag{4}$$

E being the identity $M \times M$ matrix. Summing up (4) over all edges of the cycle y_q, we have

$$\sum_{ts\in y_q} z(ts)\tilde{A}^{-1}(x,ts)C^{\left(k,p(ts)\right)}(x) = \sum_{ts\in y_q} l^{(k)}(ts)\, E, \quad k = 1,\dots,n. \tag{5}$$

Thus, nM^2P functions $C^{(k,p)}_{qm}$ are given by (2) and (5). If the matrices $C^{(k,p)}_0$ satisfy (2), (5), then the corresponding functions $N^{(k)}_0$ can be obtained from the equations

$$\begin{aligned} &DN^{(k)}_0(x,ts) = -l^{(k)}(ts)\, E + \tilde{A}^{-1}(x,ts)\, z(ts)\, C^{\left(k,p(ts)\right)}_0(x), \\ &s \in G, \quad t \in \gamma(s), \quad k = 1,\dots,n, \end{aligned} \tag{6}$$

while the conditions (5) are necessary and sufficient for the solvability of system (6). Using (1), one can find $C^{(k,p)}_0$. Therefore, the unique solvability of the system of equations for $C^{(k,p)}$ follows from that of system (10), 20.1. We rewrite the system (2), (5) for $N^{(k)}$ in the form

$$\hat{B}\hat{C} = \hat{W}. \tag{7}$$

Here $\hat{C}$ is a matrix consisting of $P \times n$ blocks of size $M \times M$. The q-*th* block of the k-*th* column of blocks consists of the functions $C^{(k,q)}_{mj}$. The $PM \times PM$ matrix $\hat{B}$ consists of blocks of size $M \times M$. The first $\delta - \varkappa$ rows of blocks in $\hat{B}$ correspond to the points of the set G, where are given the linear independent equations (2). If the equation (2) corresponding to the k-*th* row ($k < \delta - \varkappa$) contains the matrix $\pm C^{(j,p)}$, then the p-*th* element of the k-*th* row is $\pm E$, otherwise the p-*th* element is equal to the zero matrix. The next $P - \delta + \varkappa$ rows of blocks of the matrix $\hat{B}$ correspond to the independent cycles of $\hat{Y}$ which are subject to (6). If the interaction st with index $p(st)$ is involved in the k-*th* cycle, then the p-*th* position in the $(k - \delta + \varkappa)$-*th* row ($p = p(st)$) is taken by the matrix $z(st)\tilde{A}^{-1}(x, st)$ otherwise the matrix is replaced by the zero matrix. Finally, the matrix $\hat{W}$ contains $P \times n$ blocks of size $M \times M$. In the first $\delta - \varkappa$ rows the blocks consists of zeros. The q-*th* position of the k-*th* row of blocks with $k > \delta - \varkappa$ is taken by the matrix

$$\sum_{\tau\sigma \in y_{k-\delta+\varkappa}} l^{(q)} E.$$

The ellipticity criterion for the homogenized operator $\tilde{A}$ is related to the structure of $\hat{W}$.

Theorem 20.2.1. *The inequality*

$$\sum_{m,i=1}^{M} \hat{A}_{mi}^{(j,k)} \xi_j^{(m)} \xi_k^{(i)} > 0 \tag{8}$$

holds for any vectors $\xi_j \in \mathbb{R}^M, \xi_j \neq 0, j = 1, \dots, n$, *if and only if the rank of* $\hat{W}$ *is equal to* nM.

Proof. We change the order of summation, take account of (10), 20.1, and of the G-periodicity of the entries of the matrices $\tilde{A}$ and $N^{(k)}$ and obtain

$$\begin{aligned}\sum_{m=1}^{M} \langle l^{(k)} a_{rm} D N_{mq}^{(j)} \rangle_G &= - \sum_{m=1}^{M} \langle N_{mq}^{(j)} D^* (a_{rm} l^{(k)}) \rangle_G \\ &= \sum_{m,i=1}^{M} \langle N_{mq}^{(j)} D^* (a_{ri} D N_{im}^{(k)}) \rangle_G = - \sum_{m,i=1}^{M} \langle a_{ri} D N_{im}^{(k)} D N_{mq}^{(j)} \rangle_G.\end{aligned} \tag{9}$$

From (9) it follows that

$$\hat{A}_{rq}^{(j,k)}(x) = \sum_{m,i=1}^{M} \langle a_{ri} (\delta_{im} l^{(j)} + D N_{im}^{(j)}) (\delta_{mq} l^{(k)} + D N_{mq}^{(k)}) \rangle_G.$$

Therefore, according to (1),

$$\hat{A}_{rq}^{(j,k)}(x) = \delta^{-1} \sum_{s \in G} \sum_{t \in \gamma(s)} \sum_{m,i=1}^{M} C_{ri}^{(j,p(ts))}(x)\, h_{im}(x, ts)\, C_{mq}^{(j,p(ts))}(x),$$

h_{im} being the elements of $\tilde{A}^{-1}$. From this we derive that for any vectors $\xi_j \in \mathbb{R}^M, \xi_j \neq 0, j = 1, \dots, n$, there holds the formula

$$\begin{aligned}&\sum_{m,i=1}^{M} \hat{A}_{mi}^{(j,k)}(x) \xi_j^{(m)} \xi_k^{(i)} \\ &= \delta^{-1} \sum_{r,m,i,q=1}^{M} \sum_{s \in G} \sum_{t \in \gamma(s)} C_{ri}^{(j,p(ts))}(x)\, h_{im}(x, ts)\, C_{mq}^{(k,p(ts))} \xi_j^{(r)} \xi_k^{(q)}.\end{aligned}$$

Since the matrix $\tilde{A}^{-1}$ is positive definite, we have

$$\sum_{m,i=1}^{M} \hat{A}_{mi}^{(j,k)}(x)\xi_j^{(m)}\xi_k^{(i)} \geq \sum_{s\in G}\sum_{t\in\gamma(s)}\sum_{q,m=1}^{M}\big(C_{mq}^{(j,p(ts))}(x)\xi_j^{(m)}\big)^2. \tag{10}$$

Suppose that the rank of $\hat{W}$ equals nM. The system of equations for $C_{qk}^{(j,p)}$ is uniquely solvable and the matrix $\hat{B}$ is nonsingular. Hence, the rank of $\hat{C}$ is equal to that of $\hat{W}$ so some terms on the right in (10) differ from zero, i.e., the inequality (8) is fulfilled.

Now we assume that the rank of $\hat{W}$ and, consequently, the rank of $\hat{C}$ is less than nM. Then there exist numbers α_{mq} such that

$$\sum_{m=1}^{M} \alpha_{mq} C_{mq}^{(j,p)} = 0$$

for all $p = 1, \dots, P,\quad q = 1, \dots, M$. By virtue of (3),

$$\Xi = \sum_{m,i=1}^{M} \hat{A}_{mi}^{(j,k)}\xi_j^{(m)}\xi_k^{(i)} = \delta^{-1}\sum_{s\in G}\sum_{t\in\gamma(s)}\sum_{m,q=1}^{M} z(ts)l^{(j)}(ts)C_{mq}^{(k,p(ts))}(x)\xi_j^{(m)}\xi_k^{(q)}.$$

Choosing $\xi_k^{(q)} = \alpha_{kq}$, we obtain $\Xi = 0$ so (8) cannot be fulfilled. □

20.3 Crystalline Grid

20.3.1 Equations of the elasticity theory

In the case of $M = n = 2$ or $M = n = 3$ one can apply the relation (2), 20.1, to describe the interaction of atoms in a crystalline grid. For instance, in MOROZOV [3] and NAZAROV and PAUKSHTO [1] the interaction of atoms in a plane is modelled by means of a system of bars with longitudinal rigidity K and transverse one L. According to MOROZOV [3] and NAZAROV and PAUKSHTO [1], the elements of the matrix $\tilde{A}$ are of the form

$$a_{jk}(ts) = K(ts)\, w^{(j)}(ts)w^{(k)}(ts) + L(ts)\big(\delta_{jk} - w^{(j)}(ts)w^{(k)}(ts)\big), \tag{1}$$

$w^{(k)}$ being the projection onto the axis x_k of the unit vector directed from s to t. In view of the relation

$$\begin{aligned}\tilde{A}\xi\cdot\xi &= (\xi^{(i)}w^{(i)})^2K + \{\xi^{(i)}\xi^{(i)} - (\xi^{(k)}w^{(k)})^2\}L\\ &= (\xi^{(i)}w^{(i)})^2K + \{\xi^{(m)} - (\xi^{(j)}w^{(j)})w^{(m)}\}\{\xi^{(m)} - (\xi^{(k)}w^{(k)})w^{(m)}\}L,\end{aligned}$$

the matrix $\tilde{A}$ is positive definite. Construction of asymptotics of the solution to problem (2), (3), 20.1, implements the limit passage from the equation of the crystalline grid to the equations of solid bodies. The elasticity theory equations

$$(\partial/\partial x_k)\big(\lambda_{ikjm}(x)(\partial/\partial x_j)\, u^{(m)}(x)\big) = f^{(i)}(x) \tag{2}$$

are symmetric and elliptic. We show that the same is true for equation (15), 20.1, which provides the principal term in asymptotics of the solution to problem (2), (3), 20.1.

Theorem 20.3.1. *In the case of a crystalline grid the homogenized operator $\hat{A}$ is the operator of elasticity theory.*

Proof. Under the assumptions of Theorem 20.2.1, the operator $\hat{A}$ of equation (15), 20.1, is elliptic. Let us check that its coefficients satisfy the symmetry relations corresponding to the symmetry of λ_{ikjm}:

$$\hat{A}^{(j,k)}_{pq} = \hat{A}^{(k,j)}_{pq}, \tag{3}$$

$$\hat{A}^{(j,k)}_{pq} = \hat{A}^{(p,q)}_{jk}. \tag{4}$$

We prove (3). If a G-periodic function f satisfies $f(st) = f(ts)$, then

$$\sum_{s\in G}\sum_{t\in\gamma(s)} f(ts)h(s) = \sum_{t\in G}\sum_{s\in\gamma(t)} f(ts)h(s), \tag{5}$$

h being a G-periodic function on the set $\tilde{G}$. By virtue of (10), 20.1,

$$\sum_{t\in\gamma(s)} a_{pm}(x,ts)\, DN^{(i)}_{mq}(x,ts) = -\sum_{t\in\gamma(s)} a_{pq}(x,ts)\, l^{(i)}(ts). \tag{6}$$

We multiply (6) by $l^{(k)}(rs)$ for $i=j$ and by $l^{(j)}(rs)$ for $i=k$, sum over $s\in G$, take (5) into account, and arrive at

$$\begin{aligned}&\sum_{t\in G}\sum_{s\in\gamma(t)} \big(l^{(k)}(rs)\, a_{pm}(x,ts)\, DN^{(j)}_{mq}(x,ts)\big)\\ =&\sum_{t\in G}\sum_{s\in\gamma(t)} \big(l^{(j)}(rs)\, a_{pm}(x,ts)\, DN^{(k)}_{mq}(x,ts)\big), \quad r\in G.\end{aligned} \tag{7}$$

Setting $r=t$, we obtain (3).

Let us verify (4). Since the inner state of a crystalline grid does not depend on rotations, for any point $s\in\tilde{G}$

$$\sum_{t\in\gamma(s)} \big(l^{(j)}(ts)\, a_{pq}(x,ts)\big) = \sum_{t\in\gamma(s)} \big(l^{(q)}(rs)\, a_{pj}(x,ts)\big), \tag{8}$$

(e.g., see KOSEVITCH [1]). We multiply (8) by $l^{(k)}(rs)$ and sum over $s\in G$. By virtue of (5), there holds the identity

$$\begin{aligned}&\sum_{t\in G}\sum_{s\in\gamma(t)} \big(l^{(j)}(ts)\, l^{(k)}(rs)\, a_{pq}(x,ts)\big)\\ =&\sum_{t\in G}\sum_{s\in\gamma(t)} \big(l^{(q)}(ts)\, l^{(k)}(rs)\, a_{pj}(x,ts)\big), \quad r\in G.\end{aligned} \tag{9}$$

Setting $j=k, p=j, q=p, k=q$ in (9) and adding (9), we conclude that for $r=t$

$$\langle l^{(j)}\, l^{(k)} a_{pq}\rangle_G = \langle l^{(q)}\, l^{(p)} a_{jk}\rangle_G. \tag{10}$$

We multiply (7) by $l^{(p)}(rt)$ for $i=q, p=j, q=k$ and by $l^{(k)}(rt)$ for $i=j$. Now we sum over $s\in G$, make use of (6), and set $r=t$. Taking (10) into account, we obtain

$$\langle l^{(k)} a_{pm} DN^{(j)}_{mq}\rangle_G = \langle l^{(p)} a_{jm} DN^{(q)}_{mk}\rangle_G. \tag{11}$$

The formula (4) follows from (10), (11), and the definition of the coefficients of $\hat{A}$. Thus, (15), 20.1, is the equation of elasticity theory. □

20.3.2 Examples of homogenized operators

A grid whose cells consist of one element is called simple. In the case of a simple grid the relation $\tilde{A}(x,ts)l^{(j)}(ts) = -\tilde{A}(x,st)l^{(j)}(st)$ and equation (10), 20.1, imply that all elements of the matrix $N^{(j)}, j = 1, \dots, n$, vanish. Hence, the coefficients of the operator $\hat{A}$ satisfy

$$\hat{A}^{(j,k)}_{pq}(x) = \langle l^{(j)}\, l^{(k)} a_{pq} \rangle_G(x). \tag{12}$$

For example, let us consider a hexagonal grid on a plane (see Fig. 20.1). Suppose that elements of $\tilde{A}$ are of the form (1) while the nodes of the grid are ε apart. An elementary cell of a simple hexagonal grid is shown in Figure 20.2.

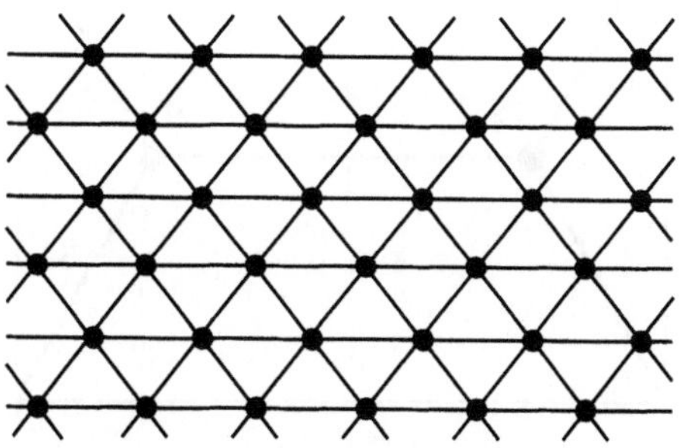

Fig. 20.1

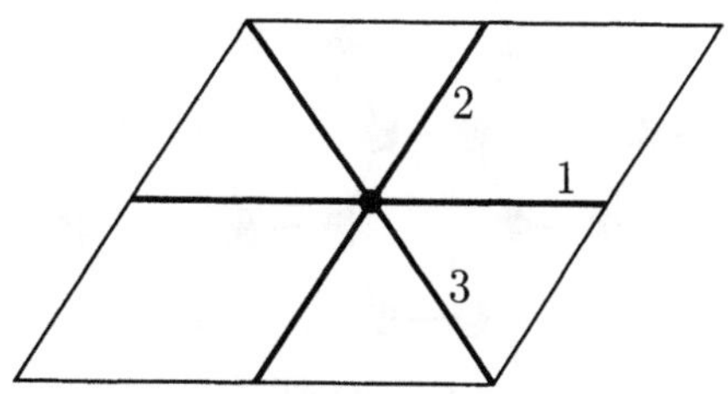

Fig. 20.2

Figures 1, 2, 3 correspond to the number $p(ts)$ of interaction ts. Since for the first interaction $w^{(1)} = 1, w^{(2)} = 0$, for the second $w^{(1)} = 1/2, w^{(2)} = 3/2$, and for the third $w^{(1)} = 1/2, w^{(2)} = -\sqrt{3/2}$, we have

$$\tilde{A}(1) = \begin{pmatrix} K & 0 \\ 0 & L \end{pmatrix}, \quad \tilde{A}(2) = \frac{1}{4}\begin{pmatrix} K+3L & \sqrt{3}(K-L) \\ \sqrt{3}(K-L) & 3K+L \end{pmatrix},$$
$$\tilde{A}(3) = \frac{1}{4}\begin{pmatrix} K+3L & \sqrt{3}(L-K) \\ \sqrt{3}(L-K) & 3K+L \end{pmatrix}, \tag{13}$$

where $\tilde{A}(p)$ stands for the value of $\tilde{A}$ at the p-*th* interaction. We substitute (13) into (12) and obtain

$$\hat{A}^{(1,1)} = \frac{3}{8}\begin{pmatrix} 3K+L & 0 \\ 0 & K+3L \end{pmatrix}, \quad \hat{A}^{(2,2)} = \frac{3}{8}\begin{pmatrix} K+3L & 0 \\ 0 & L+3K \end{pmatrix},$$
$$\hat{A}^{(1,2)} = \hat{A}^{(2,1)} = \frac{3}{8}\begin{pmatrix} 0 & K-L \\ K-L & 0 \end{pmatrix}.$$

Thus, in the case of a simple hexagonal grid the homogenized operator coincides with the operator of isotropic elasticity theory

$$\mu\nabla\cdot\nabla + (\mu+\lambda)\nabla\nabla \tag{14}$$

whose constants μ, λ are defined by

$$\mu = 3(K+3L)/8, \quad \lambda = 3(K-5L)/8.$$

As an example of a complicated grid, we consider the hexagonal grid shown in Figure 20.3. An elementary cell of the grid consists of two points s_1 and s_2, see Figure 20.4. We set -1 for the second and third interaction and 1 for the first one

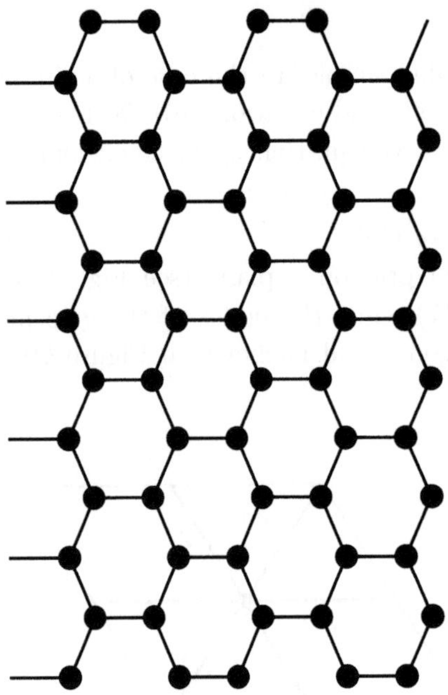

Fig. 20.3

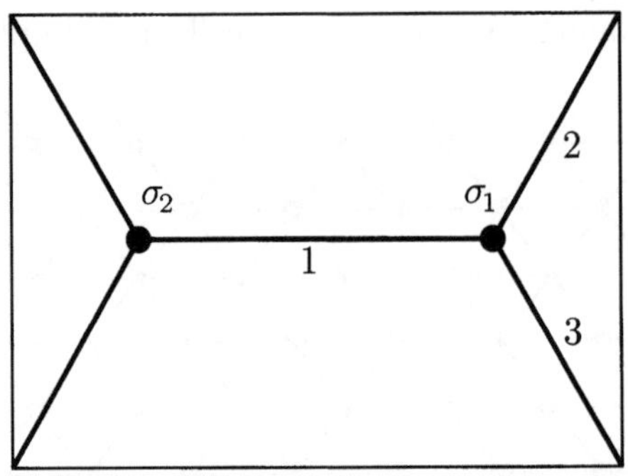

Fig. 20.4

(respectively $z(s_1 s_2) = -z(s_2 s_1)$). For the complicated hexagonal grid as well as for the simple one the entries of $\tilde{A}$ are defined by (13). However, the coefficients of $\hat{A}$ are defined by (12) while the relations (16), 20.1, do not apply because the matrices $N^{(j)}$ may differ from zero. We find the coefficients of $\hat{A}$ without calculating $N^{(j)}$. Instead we make use of (3), 20.2, and the techniques given in Section 20.2.

By virtue of (2), 20.2, the matrices $C^{(k,p)}$ satisfy the equations

$$C^{(k,p)} = C^{(k,2)} + C^{(k,3)}, \quad k = 1, 2. \tag{15}$$

For the grid shown in Figure 20.3 the relations (5), 20.2, take the form

$$\tilde{A}^{-1}(1)\, C^{(k,1)} + \tilde{A}^{-1}(2)\, C^{(k,2)} = 3^{1/k}\, E/2, \quad k = 1, 2, \tag{16}$$

$$\tilde{A}^{-1}(1)\, C^{(k,1)} + \tilde{A}^{-1}(3)\, C^{(k,3)} = (-1)^{k+1}\, 3^{1/k}\, E/2, \quad k = 1, 2. \tag{17}$$

The equations (15)–(17) form a system for determining the matrices $C^{(k,p)}, k = 1, 2, p = 1, 2, 3$. Subtracting (17) from (16) and then substituting the expression of $C^{(k,2)}$ obtained from (15), we arrive at

$$\tilde{A}^{-1}(2) C^{(k,1)} - \big(\tilde{A}^{-1}(2) + \tilde{A}^{-1}(3)\big) C^{(k,3)} = \delta_{2k} \sqrt{3} E. \tag{18}$$

Using the last equality, we find $C^{(k,3)}$ and substitute it into (17). This leads to the equation

$$\begin{aligned} &\{\tilde{A}^{-1}(1) + \tilde{A}^{-1}(3)(\tilde{A}^{-1}(2) + \tilde{A}^{-1}(3))^{-1} \tilde{A}^{-1}(2)\}\, C^{(k,1)} \\ &= (-1)^{k+1}\, 3^{k-1}\, E/2 + \sqrt{3}\, \delta_{2k}\, \{\tilde{A}^{-1}(3)\big(\tilde{A}^{-1}(2) + \tilde{A}^{-1}(3)\big)^{-1}\}. \end{aligned} \tag{19}$$

The matrices in the braces on the left and on the right in (19) are respectively equal to

$$3(L+K)\begin{pmatrix} (K(K+3L))^{-1} & 0 \\ 0 & (L(3K+L))^{-1} \end{pmatrix},$$

$$\frac{1}{2}\begin{pmatrix} 1 & (K+3L)^{-1}\sqrt{3}(k-L) \\ (3K+L)^{-1}\sqrt{3}(K-L) & 1 \end{pmatrix}.$$

Therefore, from (19) it follows that

$$\begin{aligned} C^{(1,1)} &= \big(2(K+L)\big)^{-1}\begin{pmatrix} (3L+K)K & 0 \\ 0 & (3K+L)L \end{pmatrix}, \\ C^{(2,1)} &= \big(2(K+L)\big)^{-1}(K-L)\begin{pmatrix} 0 & K \\ 0 & 0 \end{pmatrix}. \end{aligned} \tag{20}$$

Taking (18) into account we obtain

$$\begin{aligned} C^{(1,3)} &= \big(4(K+L)\big)^{-1}\begin{pmatrix} K(3L+K) & \sqrt{3}L(L-K) \\ \sqrt{3}K(L-K) & L(3K+L) \end{pmatrix}, \\ C^{(2,3)} &= \big(4(K+L)\big)^{-1}\begin{pmatrix} -\sqrt{3}L(3K+L)K & K(K-L) \\ L(L-K) & \sqrt{3}K(3L-K) \end{pmatrix}. \end{aligned} \tag{21}$$

From (15), (20), and (21) we derive that

$$\begin{aligned} C^{(1,2)} &= \big(4(K+L)\big)^{-1}\begin{pmatrix} K(3L+K) & \sqrt{3}L(K-L) \\ \sqrt{3}K(K-L) & L(3K+L) \end{pmatrix}, \\ C^{(2,2)} &= \big(4(K+L)\big)^{-1}\begin{pmatrix} \sqrt{3}L(3K+L) & K(K-L) \\ L(K-L) & \sqrt{3}K(3L+K) \end{pmatrix}. \end{aligned} \tag{22}$$

Thus, formulas (20)–(22) provide the solution to the system (15)–(17). We substitute the expressions obtained into (3), 20.2, and compute the matrices $\hat{A}^{(j,m)}$:

$$\hat{A}^{(1,1)} = 3\big(4(K+L)\big)^{-1}\begin{pmatrix} K(3L+K) & 0 \\ 0 & L(3K+L) \end{pmatrix},$$

$$\hat{A}^{(2,2)} = 3\big(4(K+L)\big)^{-1}\begin{pmatrix} L(3K+L) & 0 \\ 0 & K(3L+K) \end{pmatrix},$$

$$\hat{A}^{(1,2)} = 3\big(4(K+L)\big)^{-1}\begin{pmatrix} 0 & K(K-L) \\ L(K-L) & 0 \end{pmatrix},$$

$$\hat{A}^{(2,1)} = 3\big(4(K+L)\big)^{-1}\begin{pmatrix} 0 & L(K-L) \\ K(K-L) & 0 \end{pmatrix}.$$

In the case of the hexagonal grid shown in Figure 20.3, the homogenized operator coincides with the operator (14) of isotropic elasticity theory where

$$\mu = \big(4(K+L)\big)^{-1}3L(L+3K) \text{ and } \lambda = \big(4(K+L)\big)^{-1}3(K^2-3LK-2L^2)$$

are Lamé's coefficients.

Comments on Parts V–VII

Comments on Part V

Chapter 11

In Sections 11.1–11.3 we follow the papers by Maz'ya and Plamenevski [2] and by Nazarov [3], where more general equations were considered. Some other papers in the same direction were indicated in the survey by Kondratiev and Oleinik.

The Dirichlet problem for polyharmonic equations with data on submanifolds of various dimensions was stated and studied by Sobolev [1]. In the paper (see also Sections 14 and 15 in Sobolev [2]) the existence and uniqueness of a generalized solution were proved. Later Sternin [1] considered such problems for the general elliptic differential equations. In particular, there the Fredholm property was established and the asymptotics of solutions near the boundary submanifolds were given. In essence, the assertions collected in 11.4 were obtained in Sternin [1]. They can also be proved by methods of the works by Maz'ya and Plamenevski [2] and Nazarov [10] devoted to elliptic boundary value problems in domains with edges.

Chapter 12

At a heuristic level, the asymptotic study of solutions to some problems in the exterior of thin domains was performed long ago (see the book by Van Dyke [1]).

Rigorous approaches to problems of such a kind for $n = 2$ were developed in Geer and Keller [1], Geer [1], Ilyin [1], Homentcovshi [1] et al.In Section 12.4 the Dirichlet problem in a plane domain with thin aperture is considered as a special case of the problems studied in Part II (small nonlocal perturbations of the boundary of a domain with conical points, see Section 3.5). Works on three-dimensional boundary value problems for various equations of mathematical physics evolved at a later time that can be related to the difficulties arising from the ill-posedness of one of the limit problems (see Fedoryuk [1, 2], Maz'ya, Nazarov, and Plamenevski [6, 3]). The material of Sections 12.1 and 12.2 is a specification of the general results of Maz'ya, Nazarov, and Plamenevski [5] (see Chapter 13). Here the exposition follows Chapter 3 in Nazarov and Paukshto [1]. Sections 12.5 and 12.6 contain results from the papers by Maz'ya, Nazarov, and Plamenevski [6.3 and 2] respectively. Various questions in diffraction theory and mechanics of continua leading to the mathematical statement of the problems in domains with thin apertures and inclusions were considered in Geer [2], Myasnikov and Fedoryuk [12,3], Babich [1], Zhdanova [1–3], Morozov [1], Caillerie [1], Zorin [1, 2], Movchan, Morozov, and Nazarov [1], Movchan and Nazarov [1] et al. In particular, Section 12.4 contains the results from Zorin and Nazarov [1] on the Dirichlet problem for the Lamé system.

Chapter 13

The exposition follows Maz'ya, Nazarov, and Plamenevski [5].

Comments on Part VI

Chapter 14

The chapter originates from Maz'ya, Nazarov, and Plamenevski [12] (except for Section 13.4 that contains new material). The asymptotics of the capacity of a condenser with close plates was studied in Chew and Kong [1], Soibelman [1], et al. In Sherman [1] was found the principal term of the asymptotics of the solution to the Dirichlet problem in a half-plane with semicircular hole located near the boundary of the half-plane.

Chapter 15

In Section 15.1 the exposition follows Chapter 5 in Nazarov [16], and Sections 15.2 and 15.3 originate from Leora, Nazarov, and Proskura [1] and Nazarov and Dudnikov [1]. The material in Sections 15.4 and 15.5 was taken from Nazarov [19] and Nazarov and Chernyaev [1].

The problems solved in Sections 15.4 and 15.5 are close to the crack theory for thin beams (see Mikhailov [1], Slepyan [1], et al).

Chapter 16

To study the problems of the elasticity theory for thin domains (rods, plates, shells) there were basically used the two approaches: the asymptotic integration method (see Goldenweiser [1, 2] and the variational asymptotic method (see, for example, Berdichevski [2, 3]).

Some special problems in elasticity theory and mechanics of continua were investigated with the help of the mentioned methods and others in Friedrichs and Dressler [1], Goldenweiser and Kolos [1], Vorovich [1], Vorovich and Malkina [1], Ponyatovski [1], Berdichevski [1], Zino and Tropp [1], Kucherenko and Popov [1, 2], Bogan [1], Dudnikov and Nazarov [1], Ramm [1], Gregory and Van [1, 2], Morozov, Nazarov, and Proskura [1], Bayada and Chambat [1], Panasenko and Rescov [1], et al. The mathematical study of the boundary value problems in thin domains was initiated by Dzhavadov [1, 2]. These papers and the subsequent works by Zino and Tropp [1], Kucherenko and Popov [2], Nazarov [5, 6], Leora, Nazarov, and Proskura [1] (see also the surveys by Trenogin [1] and Markush [1]) contain various modifications of the asymptotic integration method and their justification. The variational approach in the theory of thin plates was used in Morgennstern [1] and Shoihet [1]. The algorithm in Chapter 16 (see Nazarov [5]), too, is a modification of the asymptotic integration method.

Comments on Part VII

Chapter 17

The homogenization of differential operators with rapidly oscillating coefficients was recently discussed in a number of monographs: Bensoussan, Lions, and Papanicolau [1], Bakhvalov and Panasenko [1], Sanchez-Palencia [1], in the surveys by Bakhvalov [1], Zhikov, Kozlov, Oleinik, and Kha Tjen Ngoan [1]. In Sections 17.1 - 17.4 the method of construction of the complete asymptotic expansions developed in Bakhvalov and Panasenko [1] is described for the Dirichlet and Neumann problems for a second order equation (the exposition follows Chapter 3 from Nazarov and Paukshto [1]). The study of the problem of boundary layers given in Section 17.5 is based on the papers by Nazarov [2, 7] devoted to elliptic boundary value problems with periodic coefficients (see also Kuchment [1]).

Chapter 18

The chapter is based on Maz'ya and Nazarov [1]. Some mechanical and numerical aspects of the Sapondzhyan-Babushka paradox were discussed in Panovko [1], Rajaiah and Rao [1, 2], Hanuska [1], Murrey [1], Rieder [1], and Krauthammer [1]. The asymptotics of solutions to a number of problems in domains with rapidly oscillating boundaries were described in Marchenko and Khruslov [1], Bensoussan, Lions, and Papanicolau [2], Bakhvalov and Panasenko [1], Sanchez-Palencia [1].

Chapter 19

The exposition follows Maz'ya and Slutski [1]. In comparison with Chapter 17 there is a new moment related to the asymptotic passage from a system of ordinary differential equations to the partial differential equation.

Chapter 20

The homogenization of difference equations with random coefficients on a fine rectangular lattice were studied in Kozlov [1]. Earlier at a physical level the homogenization of difference equations were used for construction of discrete models in elasticity theory (see, for example, Leibfild [1] and Kosevich [1]). Some asymptotic methods for oscillations of lattices were developed by Maslov [1].

Chapter 20 contains the results from Maz'ya and Slutskiy [2].

List of Symbols

1. Basic Symbols

$\mathbb{N}$	set of natural numbers
$\mathbb{N}_0$	$= \mathbb{N} \cup \{0\}$
$\mathbb{Z}$	set of integers
$\mathbb{Z}^*$	$= \mathbb{Z} \setminus \{0\}$
$\mathbb{R}^n$	n-dimensional Euclidian space
$\mathbb{R}$	$= \mathbb{R}^1$
$\mathbb{R}^n_+$	$= \{(x_1, \dots, x_n) \in \mathbb{R}^n : x_n > 0\}$
$\mathbb{R}_+$	$= \mathbb{R}^1_+$
$\mathbb{C}^n$	n-dimensional complex Euclidian space
$\mathbb{C}$	$= \mathbb{C}^1$
S^{n-1}	unit sphere in $\mathbb{R}^n$
$\overline{\Omega}$	closure of the set Ω
$\partial\Omega$	boundary of the set Ω
$\\|\cdot; X\\|$	norm in the space X
$\mathcal{L}(X, Y)$	space of linear bounded operators from X to Y
$\\|\cdot; X \to Y\\|$	norm in $\mathcal{L}(X, Y)$
$[A, B]$	commutator of the operators A and B
$\ker A$	kernel (null space) of the operator A
$\operatorname{coker} A$	cokernel of the operator A
$\operatorname{im} A$	range of the operator A
$\operatorname{ind} A$	index of the operator A
$\langle \cdot, \cdot \rangle_{\mathbf{L}_2(\Omega)}$	inner product (cf. 3.1.3)
$\langle \cdot, \cdot \rangle_\Omega$	inner product (cf. 7.1.1)
$(a_i)_{i=1}^n$	$= (a_1, \dots, a_n)$, n-dimensional vector
$[a_{ij}]_{i=1,j=1}^{k,l}$	$k \times l$-matrix with entries a_{ij}
$\mathbf{1}$	identity matrix, identity operator

2. Symbols for function spaces and related concepts

$\mathbf{C}^\infty(\Omega)$	set of all infinitely differentiable functions in Ω
$\mathbf{C}_0^\infty(\Omega)$	set of all functions in $\mathbf{C}^\infty(\Omega)$ with compact support
$\mathbf{W}_p^l(\Omega)$	Sobolev space (cf. 3.1.2)
$\mathbf{L}_p(\Omega)$	$= \mathbf{W}_p^0(\Omega)$
$\overset{\circ}{\mathbf{W}}{}_2^l(\Omega)$	closure of $\mathbf{C}_0^\infty(\Omega)$ in the norm of $\mathbf{W}_2^l(\Omega)$

$\mathbf{C}^{l,\alpha}(\overline{\Omega})$	3.1.2	$\mathbf{DC}^{l,\alpha}$	3.1.6
$\mathbf{C}^{k}(\overline{\Omega})$	3.1.2	$\mathbf{RC}^{l,\alpha}$	3.1.6
$\mathbf{C}^{l,\alpha}(G)$	3.2.1		
$\mathbf{C}^{l,\alpha}_\gamma(G)$	3.2.1		

$\mathbf{N}^{l,\alpha}(K)$	3.3.3

$\mathbf{V}^s_{2,\beta}(S_l)$	1.1.2	$\mathbf{DV}^l_{p,\beta}(K)$	3.2.5
$\mathbf{V}^s_{2,\gamma}(K_\alpha)$	1.2.1	$\mathbf{RV}^l_{p,\beta}(K)$	3.2.5
$\mathbf{V}^s_{2,\gamma}(\Omega)$	1.3.1, 1.6.1	$\mathbf{DV}^l_{p,\beta}(G)$	3.3.1
$\mathbf{V}^l_{p,\beta}(K)$	3.2.3	$\mathbf{RV}^l_{p,\beta}(G)$	3.3.1
$\mathbf{V}^{l-1/p}_{p,\beta}(\partial K)$	3.2.3	$\mathbf{DV}^l_{p,\beta}(\Omega(\varepsilon))$	4.1.6
$\mathbf{V}^l_{p,\beta}(G)$	3.3.1	$\mathbf{RV}^l_{p,\beta}(\Omega(\varepsilon))$	4.1.6
$\mathbf{V}^{l-1/p}_{p,\beta}(\partial G)$	3.3.1	$\mathbf{DV}_{p,\beta,\delta}$	4.2.1
$\mathbf{V}^l_{p,\beta}(\Omega(\varepsilon))$	4.1.6	$\mathbf{DV}_{p,\beta,\delta}$	4.2.1
$\mathbf{V}^{l-1/p}_{p,\beta}(\partial\Omega(\varepsilon))$	4.1.6		
$\mathbf{V}^l_{p,\beta,\gamma}(\Omega(\varepsilon))$	5.4.1		
$\mathbf{V}^{l-1/p}_{p,\beta,\gamma}(\partial\Omega(\varepsilon))$	5.4.1		
$\mathbf{V}^k_\beta(\Omega_\varepsilon)$	10.2.1		
$\mathbf{V}^{k-1/2}_\beta(\partial\Omega_\varepsilon)$	10.2.1		
$\mathbf{V}^k_\beta(S^{n-1})$	10.2.1		
$\mathbf{V}^k_\beta(\mathbb{R}^{n-1}\setminus\omega)$	10.2.1		
$\mathbf{V}^l_\beta(S^2)$	10.8.2		
$\mathbf{V}^l_\beta(\mathbb{R}^2\setminus\omega)$	10.8.2		
$\mathbf{V}^l_\beta(G_\varepsilon)$	10.8.2		
$\mathbf{V}^{l-1/2}_\beta(\partial G_\varepsilon)$	10.8.2		
$\tilde{\mathbf{V}}^l_\beta(\mathbb{R}^2\setminus\omega)$	10.8.2		

$\mathbf{W}^{l-1/p}_p(\partial\Omega)$	3.1.2	$\mathbf{DW}^l_p$	3.1.6
$\mathbf{W}^l_{p,\gamma}(G)$	3.2.1	$\mathbf{RW}^l_p$	3.1.6
$\mathbf{W}^{l-1/p}_{p,\gamma}(G)$	3.2.1	$\mathbf{RW}^l_{2,\beta}(G_\pm)$	5.6.2
$\mathbf{W}_\beta(\Omega_\varepsilon)$	10.2.1	$\mathbf{RW}^l_{2,\beta}(G(N))$	5.6.3
$\mathbf{W}_\beta(\mathbb{R}^{n-1}\setminus\omega)$	10.2.1	$\tilde{\mathbf{W}}^l_\beta(S^2)$	10.8.2
		$\tilde{\mathbf{W}}^l_\beta(G_\varepsilon)$	10.8.2

3. Symbols for functions, distributions and related concepts

$\operatorname{supp} u$	support of the function u
$\nabla u = \operatorname{grad} u$	gradient of u
$\operatorname{div} u$	divergence of u
$\tilde{u}, Fu$	Fourier transform of u (cf. 1.1.1, 3.2.1)
D_{x_j}	$= -\mathrm{i}\partial/\partial x_j$
D_x^α	$= \mathrm{D}_{x_1}^{\alpha_1} \ldots \mathrm{D}_{x_n}^{\alpha_n}$, $\alpha = (\alpha_1, \ldots, \alpha_n) \in \mathbb{N}_0^n$
$\|\alpha\|$	$= \alpha_1 + \cdots + \alpha_n$, $\alpha \in \mathbb{N}_0^n$
$\partial/\partial\nu, \partial/\partial n$	normal derivative
D_ν^k	$= (-\mathrm{i}\partial/\partial\nu)^k$
Δ	Laplace operator
δ	Dirac's measure
$f \otimes \delta^{(k)}(\partial\Omega)$	3.1.4

4. Other symbols

δ_{ij}	the Kronecker delta
mes_n	n-dimensional Lebesgue measure
$\beta + \delta$	3.2.5
r^β	3.2.5
$o(r^{a+0})$	7.1.1
$o(\rho^{a-0})$	7.1.3
$\operatorname{cap}(D)$	8.1.4

References

AKHIEZER, N.I., GLAZMAN, I.M.,

[1] *Theory of linear operators in Hilbert space*, Dover, New York 1993.

AERO, E.L., KUVSHINSKI, E.V.,

[1] Fundamental equations in theory of elasticity of media with rotating interacting particles, *Fizika tverd. tela*, 2, 7 (1960), 1399–1409. (Russian)

AGALARYAN, O.B., NAZAROV, S.A.,

[1] On the variation of the intensity factor due to soldering of a longitudinal crack in a prismatic rod, *Doklady AN Arm. SSR*, 72, 1 (1981), 18–21. (Russian)

AGMON, S., NIRENBERG, L.,

[1] Properties of solutions of ordinary differential equations in Banach spaces. *Comm. Pure Appl. Math.*, 16 (1963), 121–239.

AGMON, S., DOUGLIS, A., NIRENBERG, L.,

[1] Estimates near the boundary for solutions of elliptic partial differential equations satisfying general boundary conditions. I, *Comm. Pure Appl. Math.*, 12 (1959), 623–727, II, *Comm. Pure Appl. Math.*, 17 (1964), 35–92.

AGRANOVICH, M.S., VISHIK, M.I.,

[1] Elliptic problems with parameter and parabolic problems of general form, *Uspekhi matem. nauk*, 19, 3 (1964), 53–160. (Russian)

ALVINO, A., TROMBETTI, G.,

[1] Problemi ellittici dipendenti da un parametro in L^p, Ricerche Mat., 26, 2 (1977), 335–348.

ARUTYUNYAN, N.CH., ABRAMYAN, B.L.,

[1] *Revolution of elastic bodies*, Fizmatgiz, Moscow 1963. (Russian)

ASLANYAN, A.G., LIDSKI, V.B.,

[1] On the spectrum of elliptic equations, *Matem. zametki*, 7, 4 (1970), 495–502. (Russian)

BABICH, V.M.,

[1] Skin-effect in the case of a conductor with arbitrary cross-section, *Zapiski nauchn. seminarov Leningradskogo otdel. Matem. Instituta AN SSSR*, 128 (1983), 13–20. (Russian)

BABICH, V.M., IVANOV, M.I.,

[1] Long-wave asymptotics in scattering problems of elastic waves, *Zapiski nauchn. seminarov Leningradskogo otdel. Matem. Instituta AN SSSR*, 156 (1986), 6–19. (Russian)

BABUŠKA, I.,

[1] Stability of the domain with respect to a perturbation of the boundary of fundamental problems of the theory of partial differential equations, mainly in connection with the theory of elasticity, I. *Czech. Math. J.*, 2, 1 (1961), 76–105. II. *Czech. Math. J.*, 2, 2 (1961), 165–203. (Russian)

BAKHVALOV, N.S.,

[1] Homogenization and perturbation problems, *Comp. Meth. Appl. Sci. Eng. Proc.*, 4, Int. Symp. Versailles 1979, Amsterdam 1980, 645–658.

BAKHVALOV, N.S., PANASENKO, G.P.,

[1] *Homogenisation: averaging processes in periodic media. Mathematical problems in the mechanics of composite materials*, Kluwer, Dordrecht, 1989.

BAYADA, G., CHAMBAT, M.,

[1] The transition between the Stokes equations and the Reynolds equation, a mathematical proof, *Appl. Math. Optimiz.*, 14, 1 (1986), 73–93.

BENSOUSSAN, A., LIONS, J.L., PAPANICOLAOU, G.,

[1] *Asymptotic methods in periodic structures*, Amsterdam 1978.

BERDICHEVSKI, V.L.,

[1] High frequency long wave vibrations of plates, *Doklady AN SSSR*, 236, 6 (1977), 1319–1322. (Russian)

[2] Variational-asymptotic method of the construction of the theory of shells, *Prikladn. matem. i mekh.*, 43, 4 (1979), 664–687. (Russian)

[3] *Variational principles of the mechanics of continuous media*, Nauka, Moscow 1983.

BERS, L., JOHN, F., SCHECHTER, M.,

[1] *Partial differential equations*, Interscience, New York, London, Sydney 1964.

BERSON, G.,

[1] Comportement asymptotique des valeurs propres du Laplacien dans un domain avec un trou, *Bull. Soc. Math. France*, 113, 2 (1985), 211–230.

BIRMAN, M.SH.,

[1] Variational methods of the solution of boundary value problems which are analogous to the method of Trefftz, *Vestnik Lenigrad Univ.*, 13 (1956), 69–89. (Russian)

BOGAN, J.A.,

[1] Asymptotic behaviour of boundary value problems for an elastic ring reinforced with very stiff fibres, *Zh. prikl. mekh. i tekhn. fiz.*, 6 (1980), 118–122. (Russian)

[2] On the second boundary value problem of the theory of elasticity for an essentially anisotropic plane with an elliptic hole, *Prikl. mekh.*, 17, 9 (1981), 64–68. (Russian)

CAILLERIE, D.,

[1] The effect of a thin inclusion of high rigity in a elastic body, *Math. Meth. Appl. Sci.* 2, 3 (1980), 251–270.

CHEREPANOV, G.P.,

[1] *Mechanics of large destructions*, Nauka, Moscow 1974.

[2] *Mechanics of destruction of composite materials*, Nauka, Moscow 1983.

CHEW, W.S., KONG, J.A.,

[1] Asymptotic formulae for the capacity of two oppositely charged discs. *Math. Proc. Cambr. Phil. Soc.*, 89, 2 (1981), 373–384.

COLE, J.D.,

[1] *Perturbation methods in applied mathematics*, Blaisdell, Waltham 1968.

COSSERAT, E., COSSERAT, F.,

[1] *Theorie des corps deformaties*, Herman, Paris 1909.

COURANT, R., HILBERT, D.,

[1] *Methods of mathematical physics*, Vol. 1, Interscience, New York, 1953.

DE GIORGI, E.,

[1] Sulla differenziabilita e l'analiticita delle estremali degh integrali multiple regolari. *Mem. Acc. Sci. Torino*, 3 (1957), 1–19.

[2] Un esepio di estremali discontinue per un problema variazionable di tipo ellittico, *Bol. della Unione Mat. Italiana*, IV, 1 (1968), 135–137.

DIKANSKI, A.S.,

[1] Adjoint problems for elliptic pseudodifferential boundary value problems, *Doklady AN SSR*, 200, 5 (1971), 1020–1023. (Russian)

DUDNIKOV, V.A., NAZAROV, S.A.,

[1] Asymptotically exact equations of thin plates on the basis of Cosserat's theory, *Doklady AN SSSR* 262, 2 (1982), 306–310. (Russian) (English transl.: *Sov. Phys. Dokl.*, 27 (1982), 84–86).

[2] A relation between the intensity factors in classical and couple-stress elasticity theory, *Mekhanika tverd. tela*, 3 (1983), 185–187. (Russian)

DUNAU, J.,

[1] Fonctions d'un operatéur elliptique sur une variété compacte. *J. Math. pures et appl.* 56 (1977), 367–391.

DZHAVADOV, M.G.,

[1] Asymptotics of the solution of boundary value problems for elliptic equations in slender domains, *Doklady AN SSSR*, 163, 3 (1965), 547–550. (Russian)

[2] Asymptotics of the solution of boundary value problems for elliptic equations in slender domains, *Differenc. uravnenya*, 5, 10 (1968), 1901–1909. (Russian)

ESKIN, G.I.,

[1] *Boundary value problems for elliptic pseudodifferential equations*, AMS, Providence, RI, 1981.

FEDORYUK, M.V.,

[1] Dirichlet problem for the Laplace operator in the exterior of a slender rotating body, *Theory of cubature formulas and application of functional analysis to problems in mathematical physics, Trudy seminara S.L. Soboleva*, 1 (1980), 113–131. (Russian)

[2] Asymptotics of the solution of the Dirichlet problem for the Laplace and Helmholtz equations in the exterior of a slender cylinder, *Izv. AN SSSR*, matem., 45, 1 (1981), 167–186. (Russian)

FICHERA, G.,

[1] Existence theorems in elasticity. In: Flügge, S. (ed.) *Handbuch der Physik* V/a/2. Springer-Verlag Berlin, Heidelberg, New York 1972, 347–424.

FRIEDRICHS, K.O.,
[1] *Perturbation of spectra in Hilbert space*, American Mathem. Society, Providence 1985.

FRIEDRICHS, K., DRESSLER, R.,
[1] A boundary-layer theory for elastic plates, *Comm. Pure Appl. Math.*, 14 (1961), 1–33.

GADYLSHIN, R.R.,
[1] Asymptotics of the eigenvalue of singularly perturbed self-adjoint elliptic problems with small parameter in the boundary conditions, *Differenc. uravnenya*, 22, 4 (1986), 640–652. (Russian)

GEER, J.F.,
[1] Uniform asymptotic solutions for the two dimensional potential flow about a slender body, *SIAM J. Appl. Math.*, 26, 3 (1974), 539–553.
[2] Electromagnetic scattering by a slender body of revolution: axially incident plane wave, *SIAM J. Appl. Math.*, 38, 1 (1980), 93–102.

GEER, J.F., KELLER, J.B.,
[1] Uniform asymptotic solutions for potential flow around a thin airfoil and the electrostatic potential about a thin conductor *SIAM J. Appl. Math.*, 16, 1 (1968), 75–101.

GELFAND, I.M., SHILOV, G.E.,
[1] Verallgemeinerte Funktionen, Vol.1, VEB Deutscher Verlag der Wissenschaften, Berlin 1960.

GOHBERG I.C., KREIN, M.G.,
[1] *Introduction to the theory of linear nonselfadjoint operators*, AMS, Providence, RI, 1969.

GOHBERG, I.C., SIGAL, E.I.,
[1] Operator generalization of the logarithmic residue theorem and the theorem of Rouche, *Matem. sbornik*, 84, 3 (1971), 607–629. (Russian)

GOLDENWEISER, A.L.,
[1] Construction of an approximative plate bending theory with the help of the method of asymptotic integration of the equations in elasticity theory, *Prikl. matem. i mekh.*, 27, 6 (1962), 1057–1074. (Russian)
[2] *Theory of elastic shells.* Nauka, Moscow 1965. (Russian)

GOLDENWEISER, A.L., KOLOS, A.V.,
[1] Concerning the construction of two-dimensional equations of elastic thin plates, *Prikl. matem. i mekh.*, 29, 1 (1965), 141–161. (Russian)

GOLDSTEIN, R.V., KAPCOV, A.V., KORELSTEIN, L.B.,
[1] Asymptotic solution of three-dimensional problems of the theory of elasticity concerning a removed planar breaking crack, *Prikladn. matem. i mekh.*, 48, 5 (1984), 854–863. (Russian)

GOLDSTEIN, R.V., SHIFRIN, E.I.,
[1] Estimations and approximation formulas in the problem of the theory of elasticity concerning a planar crack with normal rupture, *Mekhanika tverd. tela*, 1 (1983), 120–127. (Russian)

GOTLIB, V.YU.,

[1] On the Raleigh asymptotics of diffraction problems, *Zapiski nauchn. seminarov Leningradskogo otdel. Matem. Instituta AN SSSR*, 62 (1976), 52–59. (Russian)

GRADSHTEYN, I.S., RYZHIK I.M.,

[1] *Table of integrals, series, and products.* Academic Press, New York, 1980. (Russian)

GREGORY, R.D., WAN, F.J.M.,

[1] On plate theories and Saint-Vanant's principle, *Int. J. Solids and Struct.*, 21, 10 (1985), 1005–1024.

[2] Edge effects in the stretching of plates, *Local Eff. Anal. Struct.* (1985), 35–54.

GRIFFITH, A.A.,

[1] The theory of rupture, *Proc. 1th Int. Congr. for Appl. Mech.*, Delft 1924.

HANUŠKA, A.,

[1] Zu den Theorien der Plattenbiegung, *Beton- und Stahlbetonbau* 9 (1969), 214–217.

HARDY, G.H., LITTLEWOOD, J.E., POLYA, G.,

[1] *Inequalities.* Cambridge University Press, New York 1934.

HOMENTCOVSCHI, D.,

[1] Conformal mapping of the domain exterior to a thin region, *SIAM J. Math. Anal.*, 10, 6 (1979), 1246–1257.

HÖRMANDER, L.,

[1] *Linear partial differential operators.* Springer-Verlag, Berlin, Göttingen, Heidelberg 1963.

[2] *The analysis of linear partial differential operators* I. Springer-Verlag 1983.

ILYIN, A.M.,

[1] A boundary value problem for an elliptic second order equations in a domain with crack, I. Two-dimensional case, *Matem. sbornik*, 99, 4 (1976), 514–537. (Russian)

[2] A boundary value problem for an elliptic second order equations in a domain with slender groove, II. Domain with a small crack, *Matem. sbornik*, 103, 2 (1977), 265–284. (Russian)

[3] Investigation of the asymptotics of the solution of an elliptic boundary value problem with a small crack, *Trudy seminara im. I.G. Petrovskogo*, 6 (1981), Moscow State University, Moscow. (Russian)

IRWIN, G.R.,

[1] *Fracture dynamics. Fracture of Metals*, ASM, Cleveland 1948.

IVANOV, M.I.,

[1] Diffraction on a solid movable inclusion, *Zapiski nauchn. seminarov Leningradskogo otdel. Matem. Instituta AN SSSR*, 156 (1986), 61–68. (Russian)

JOHN, F.,

[1] *Plane waves and spherical means applied to partial differential equations*, Interscience, New York 1955.

KAMKE, E.,

[1] *Differentialgleichungen. Lösungsmethoden und Lösungen*, Vol.I: *Gewöhnliche Differentialgleichungen*, 6th edition, Akademische Verlagsanstalt, Leipzig 1959.

KANTOROVICH, L.V., KRYLOV, V.I.,

[1] *Approximate methods of higher analysis*, Interscience, New York, 1958.

KATO, T.,

[1] *Perturbation theory for linear operators*, Springer-Verlag, Berlin, Heidelberg, New York 1966.

KISELYEV, A.N.,

[1] A small spherical radiator in a nonhomogeneous elastic medium, *Mekh. tverd. tela*, 4 (1982), 119–126. (Russian)

KOHN, J.J., NIRENBERG, L.,

[1] An algebra of pseudo-differential operators, *Comm. Pure Appl. Math.*, 18 (1965), 269–305.

KONDRATYEV, V.A.,

[1] Boundary value problems for elliptic equations in domains with conic and corner points, *Trudy Mosk. matem. obshch.* 16 (1967), 209–292. (Russian)

KONDRATYEV, V.A., OLEINIK, O.A.,

[1] Boundary value problems for partial differential equations in non-smooth domains, *Uspekhi matem. nauk* 38, 2 (1983), 3–76. (Russian)

KOSEVICH, A.M.,

[1] *Physical mechanics of real cristals.* Naukova Dumka, Kiev 1981.(Russian)

KOZLOV, S.M.,

[1] Averaging of difference schemes, *Matem. sbornik*, 129, 3 (1986), 338–357. (Russian)

KRASNOSELSKI, M.A *et al.*],

[1] *Approximate solution of operator equations*, Wolters-Noordhoff, Groningen, 1972.

KRAUTHAMMER, T.,

[1] Accuracy of the finite element method near a curved boundary, *Int. J. Computers and Structures*, 10, 12 (1979), 921–929.

KUCHERENKO, V.V., POPOV, V.A.,

[1] High frequency vibrations of plates, *Doklady AN SSSR*, 244, 4 (1979), 819–823. (Russian)

[2] Asymptotics of the solution of problems in the theory of elasticity in thin domains, *Doklady AN SSSR*, 274, 1 (1984), 58–61. (Russian)

KUCHMENT, P.A.,

[1] Flocke theory for partial differential equations, *Uspekhi matem. nauk*, 37, 4 (1982), 3–52. (Russian)

KUPRADZE, V.D.,

[1] *Potential method in the theory of elasticity.* Daniel Davey, New York, 1965.

LADYZHENSKAYA, O.A.,

[1] *The boundary value problems of mathematical physics*, Springer, New York, 1985.

LANDKOF, N.S.,

[1] *Foundations of modern potential theory*, Springer, Berlin, 1972.

LEBEDYEV, N.N., SKALSKAYA, I.P.,

[1] Application of paired integral equations to electrostatic problems for a complete conducting cylinder of finite length, *Zh. tekhn. fiziki* 43, 1 (1973), 44–51. (Russian)

LEIBFILD, G.,

[1] Gittertheorie der mechanischen und thermischen Eigenschaften der Kristalle, In: *Handbuch der Physik*, Vol. VIII, Part 2, Springer-Verlag, Berlin, Göttingen, Heidelberg 1955.

LEORA, S.N., NAZAROV, S.A., PROSKURA, A.V.,

[1] Derivation of limiting equations for elliptic problems in thin domains using computers, *Zh. vychisl. matem. i matem. fiz.*, 26, 7 (1986), 1032–1048. (Russian) (English transl.: *USSR Comput. Math. Math. Phys.*, 26, 4 (1986), 47–58).

LIEBOWITZ, H. (Ed.),

[1] *Fracture. An advanced treatise*, Vol. II: *Mathematical Fundamentals*, Academic Press, New York 1968.

LIONS, J.-L., MAGENES, E.,

[1] *Problemes aux limites non homogenenes et applications*, Vol.1, Dunod, Paris 1968.

LURIE, A.I.,

[1] *Statics of thin elastic shells*, Gostekhizdat, Moscow, Leningrad 1947. (Russian)

MARCHENKO, V.A., KHRUSLOV, E.N.,

[1] *Boundary value problems in domains with finely grained boundary*, Naukova Dumka, Kiev 1974. (Russian)

MARKUSH, I.I.,

[1] *Development of asymptotic methods in the theory of differential equations*, Vidav. UDU, Uzhgorod 1975. (Ukrainian)

MASLOV, V.P.,

[1] *Operator methods*, Nauka, Moscow 1973. (Russian)

MAZ'YA, V.G.,

[1] On the Dirichlet problem for elliptic equations of arbitrary order in unbounded domains, *Doklady AN SSSR*, 150, 6 (1963), 1221–1224. (Russian)

[2] On the behaviour near the boundary of the solution of the Dirichlet problem for second order elliptic equations in divergence form, *Matem. zametki*, 2, 3 (1967), 209–220. (Russian)

[3] Examples of non-regular solutions of quasi-linear elliptic equations with analytic coefficients, *Funcion. analiz i ego pril.*, 2, 3 (1968), 53–57. (Russian)

[4] Behaviour of solutions to the biharmonic operator at a boundary point. In: *Lecture Notes Math.* 703, Springer-Verlag, Berlin, Heidelberg, New York 1979, 250–262.

[5] *Sobolev spaces*, Leningrad University, Leningrad 1985. (Russian) (English transl.: Springer Series in Soviet Math., Berlin, New York 1985.)

[6] On the Wiener Type Regularity of a Boundary Point for the Polyharmonic Operator, *Applicable Analysis*, 71 (1999), 149–165.

MAZ'YA, V.G., MOROZOV, N.F., NAZAROV, S.A.,

[1] On the release of potential deformation energy by variation of the domain near angular stress concentrator, Preprint, *Leningrad branch of the Institute of Mechanical Engineering AN SSSR*, 4 (1989), 1–32. (Russian)

MAZ'YA, V.G., NAZAROV, S.A.,

[1] Paradoxes of limit passage in solutions of boundary value problems involving the approximation of smooth domains by polygonal domains, *Izvestya AN SSSR*, matem. 50, 6 (1986), 1156–1177. (Russian) (English transl.: *Math. USSR Izvestiya* 29 (1987), 511–533).

[2] The vertex of a cone can be nonregular in the Wiener sense for a fourth-order elliptic equation, *Matem. zametki*, 39, 1 (1986), 24–28. (Russian) (English transl.: *Math. Notes*, 39 (1986), 14–16).

[3] The asymptotic behavior of energy integrals under small perturbations of the boundary near corner points and conical points, *Trudy Moskov. matem. obshch.*, 50 (1987), 79–129. (Russian) (English transl.: *Trans. Mosc. Math. Soc.*, 50 (1988), 77–127).

[4] Singularities of solutions of the Neumann problem at a conical point, *Sibirsk. matem. zh.*, 30, 3 (1989), 52–63. (Russian) (English transl.: *Siberian Math. J.* 30, 3 (1989), 387–396).

MAZ'YA, V.G., NAZAROV, S.A., PLAMENEVSKI, B.A.,

[1] Asymptotic solutions of elliptic boundary value problems when the domain is varied close to conical points, *Doklady AN SSSR*, 249, 1 (1979), 91–94. (Russian) (English transl.: *Sov. Phys. Dokl.* 24 (1979), 904–905).

[2] Asymptotic behavior of the solutions of a quasilinear equation in nonregular perturbed domains, *Differenc. uravnenia i Primenen.*, 27, Lith. Acad. Sci., Vilnius 1980, 17–50. (Russian)

[3] On the asymptotics of solutions of the Dirichlet problem in three-dimensional domain with a slender cavity, *Doklady AN SSSR*, 256, 1 (1981), 37–39. (Russian) (English transl: *Sov. Math. Dokl.* 23 (1981), 32–35).

[4] On the asymptotic behavior of solutions of elliptic boundary value problems with irregular perturbations of the domain, *Probl. matem. anal.*, 8, Leningrad Univ., Leningrad 1981, 72–153. (Russian)

[5] Asymptotics of solutions of the Dirichlet problem in a domain with a truncated thin tube, *Matem. sbornik*, 116, 2 (1981), 187–217. (Russian) (English transl.: *Math. USSR. Sbornik*, 44 (1983), 167–194).

[6] *Asymptotics of solutions to elliptic boundary-value problems under a singular perturbation of the domain*, Tbilisi Univ., Tbilisi 1981. (Russian)

[7] Absence of De Giorgi-type theorems for strongly elliptic equations with complex coefficients, *Zapiski nauchn. seminarov Leningradskogo otdel. Matem. Instituta AN SSSR*, 115 (1982), 156–168. (Russian) (English transl.: *J. Sov. Math.*, 28 (1985), 726–734).

[8] Evaluation of the asymptotic form of the "intensity coefficients" on approaching corner or conical points, *Zh. vychisl. matem. i matem. fiz.*, 23, 2 (1983), 333–346. (Russian) (English transl.: *USSR Comput. Math. Math. Phys.*, 23, 2 (1983), 50–58).

[9] On the bending of a nearly polygonal plate with freely supported boundary, *Izvestya Vuzov*, matem. 8 (1983), 34–40. (Russian) (English transl.: *Sov. Math.* 27, 8 (1983), 40–48).

[10] On the singularities of solutions of the Dirichlet problem in the exterior of a slender cone, *Matem. sbornik*, 122, 4 (1983), 435–456. (Russian) (English transl.: *Math. USSR Sbornik*, 50 (1985), 415–437).

[11] Asymptotic expansions of the eigenvalues of boundary value problems for the Laplace operator in domains with small holes, *Izvestya AN SSSR*, matem. 48, 2 (1984), 347–371. (Russian) (English transl.: *Math. USSR Izvestiya* 24 (1985), 321–345).

[12] Dirichlet problem in domains with thin bridges, *Sibirsk. matem. zh.*, 25, 2 (1984), 161–179. (Russian) (English transl.: *Siberian Math. J.* 25 (1984), 297–313).

MAZ'YA, V.G., PLAMENEVSKI, B.A.,

[1] On the behaviour of the solutions of quasilinear elliptic boundary value problems in the neighbourhood of a conical point. *Zapiski nauchn. seminarov Leningradskogo otdel. Matem. Instituta AN SSSR*, 38 (1973), 91–97. (Russian)

[2] On elliptic boundary value problems in domains with piecewise smooth boundary, *Proceedings of Symp. of Cont. Media and Related Problems in Analysis*, 1, 171–181, Mecniereba, Tbilisi 1973. (Russian)

[3] On the coefficients in the asymptotic solutions of elliptic boundary value problems in a cone, *Zapiski nauchn. seminarov Leningradskogo otdel. Matem. Instituta AN SSSR*, 58 (1975), 110–128. (Russian)

[4] On the coefficients in the asymptotic solutions of elliptic boundary value problems in domains with conical points, *Mathem. Nachr.*, 76 (1977), 29–60. (Russian)

[5] Estimations in L_p and in Hölder classes and the Miranda-Agmon principle for the solutions of elliptic boundary value problems in domains with singularities on the boundary, *Mathem. Nachr.*, 81 (1978), 25–82. (Russian)

[6] Estimations of the Green functions and Schauder estimations of the solutions of elliptic boundary value problems in a two-edged angle, *Sibirsk. matem. zh.*, 19, 5 (1978), 1065–1082. (Russian)

[7] On the asymptotics of the fundamental solutions of elliptic boundary value problems in domains with conical points, *Problemy matem. analiza*, 7, 100–145, Leningrad Univ., Leningrad 1979. (Russian)

[8] On the maximum principle for the biharmonic equation in a domain with conical points, *Izvestiya Vuzov*, matem. 2 (1981), 52–59. (Russian)

[9] On the properties of the solutions of three-dimensional problems in the theory of elasticity and hydromachanics in domains with isolated singularities, *Izvestiya Vuzov*, matem. 2 (1981), 52–59. (Russian)

MAZ'YA, V.G., SLUTSKI, A.S.,

[1] Averaging of a differential operator on a fine periodic curved grid, *Mathem. Nachr.*, 133 (1986), 107–133.(Russian)

[2] Averaging of difference equations with fast oscillating coefficients, *Seminar Analysis* 1986/87, Akad. Wiss. DDR, Inst. Math., 63–92, Berlin 1987. (Russian)

MAZ'YA, V.G., SULIMOV, M.G.,

[1] Asymptotical solution of homogeneous difference equations with constant operator coefficients, *Matem. sbornik*, 132, 4 (1987), 451–469. (Russian)

MIKHAILOV, V.P.,

[1] Generalization of the penalty approach to problems in crack theory, *Prikl. mekh. i tekhn. fizika*, 3 (1969), 171–174. (Russian)

MIKHLIN, S.G.,

[1] *Variational methods in mathematical physics*, Macmillan, New York, 1964.

MIKHLIN, S.G., PRÖSSDORF, S.,

[1] *Singular integral operators*, Springer, Berlin, 1986.

MYASNIKOV, V.P., FEDORYUK, M.B.

[1] Raleigh approximation in the theory of elasticity, *Doklady AN SSSR*, 254, 3 (1980), 589–592. (Russian)

[2] *Radiation and scattering of elastic waves with slender axial-symmetric cavity. Earth research with new geophysical methods*, Novosibirsk 1980. (Russian)

[3] Asymptotics of the solutions of elliptic differential equations in domains with small and slender cavities, *Uspekhi matem.nauk*, 35, 5 (1980), 251–252. (Russian)

MORGENSTERN, D.,

[1] Herleitung der Plattentheorie aus der dreidimensionalen Elastizitätstheorie, *Arch. Rat. Mech. Anal.*, 4, 2 (1959), 145–152.

MOROZOV, N.F.,

[1] On the limitation of the applicability of Sneddon's formula, *Doklady AN SSR*, 259, 5 (1981), 1068–1072. (Russian)

[2] Concerning the problem about the destruction of an elastic plane, weakened by a lunaric cut, *Vestnik Leningrad Univ.*, 19 (1981), 70–74. (Russian)

[3] *Mathematical problems of crack theory*, Nauka, Moscow 1984. (Russian)

MOROZOV, N.F. NAZAROV, S.A.,

[1] On the stress-strain state in a neighbourhood of a crack setting on a grain, *Studies in Elasticity and Plasticity*, 13, Leningrad Univ., Leningrad 1980, 141–148. (Russian)

[2] About calculation of energy release in the Griffith Problem, *Studies in Elasticity and Plasticity*, 14, Leningrad Univ., Leningrad 1982, 3–9. (Russian)

MOROZOV, N.F., NAZAROV, S.A., PROSKURA, A.A.,

[1] Boundary value problems of the elasticity theory for plane domains with thin borderings, *Mekhanika deform. tel*, Nauka, Moscow 1986, 82–93. (Russian)

MORREY, C.B.,

[1] *Multiple integrals in the calculus of variations*, Springer-Verlag 1966.

MOVCHAN, A.B.,

[1] Asymptotics of eigenvalues of three-dimensional problems of the theory of elasticity for bodies with a small inclusion, *Izvestiya Vuzov*, matem. 1 (1988), 50–56. (Russian)

MOVCHAN, A.B., MOROZOV, N.F., NAZAROV, S.A.,

[1] Cracks with smoothly closing edges under plane deformation, *Prikl. matem. i mekh.*, 51, 1 (1987), 130–139. (Russian) (English transl.: J. Appl. Math. Mech. 51, 1 (1987), 99–107).

Movchan, A.B., Nazarov, S.A.,

[1] Strain-stress state of the plane domain with a thin elastic inclusion of finite sizes, *Mekhanika tverd. tela*, 1 (1987), 75–83. (Russian)

Murray, N.W.,

[1] The polygon-circle paradox and convergence in thin plate theory, *Proc. Cambr. Phil. Soc.* 73 (1973), 279–283.

Muskhelishvili, N.I.,

[1] *Some basic problems of the mathematical theory of elasticity*, Noordhoff, Groningen 1953.

Nagel, J.,

[1] On equivalent norms in functional spaces H^{μ}, *Vestnik Leningr. Univ.*, Leningrad, 7 (1974), 41–47. (Russian)

Nazarov, S.A.,

[1] On the parameter asymptotics of the solution to an elliptic boundary-value problem with periodic coefficients in a cylinder, *Differenc. uravn. i primenen.*, 30, Lith. Acad. Sci., Vilnius 1981, 27–46. (Russian)

[2] Elliptic boundary value problems with periodic coefficients in a cylinder, *Izvestya AN SSSR*, matem. 45, 1 (1981), 101–112. (Russian) (English transl.: *Math. USSR. Izvestija* 18, 1 (1982), 89–98).

[3] Vishik-Lyusternik method for elliptic boundary-value problems in regions with conical points. 1. The problem in a cone, *Sibirsk. matem. zh.*, 22, 4 (1981), 142–163. (Russian) (English transl.: *Siberian Math. J.*, 22 (1982), 594–611).

[4] Regular degeneration of elliptic systems with a small parameter multiplying the highest derivatives, *Vestnik Leningrad Univ.*, 3 (1982), 106–108. (Russian)

[5] The structure of solutions of elliptic boundary value problems in slender domains, *Vestnik Leningrad Univ.* 7 (1982), 65–68. (Russian) (English transl.: Vestn. Leningr. Univ. Math., 15 (1983), 99–104).

[6] *Introduction to asymptotic methods of the elasticity theory*, Leningrad Univ., Leningrad 1983.(Russian)

[7] On the constants in the asymptotic expansion of solutions of elliptic boundary value problems with periodic coefficients in a cylinder, *Vestnik Leningrad Univ.*, 15 (1985), 16–22. (Russian) (English transl.: *Vestn. Leningrad Univ. Math.*, 18, 3 (1985), 17–22).

[8] Justification of asymptotic expansions of the eigenvalues of nonselfadjoint singularly perturbed elliptic boundary value problems, *Matem. sbornik*, 129, 3 (1986), 307–337. (Russian) (English transl.: *Math. USSR Sbornik* 57 (1987), 317–349).

[9] Intensity factors in the problems of antiplane shear and ideal liquid flow in thin domains with a longitudinal cut, *Izvestya AN Arm. SSSR*, mekh. 40, 5 (1987), 24–34. (Russian)

[10] Estimates near the edge of the solution of the Neumann problem for an elliptic system, *Vestnik Leningrad. Univ.*, 1 (1988), 37–42. (Russian) (English transl.: *Vestn. Leningr. Univ. Math.* 21, 1 (1988), 52–59).

Nazarov, S.A., Chernyaev, P.K.,

[1] Antiplane shear of a domain with two closely located cracks, *Prikl. matem. i mekh.*, 50, 5 (1986), 815–825. (Russian) (English transl.: *J. Appl. Math. Mech.* 50 (1986), 629–637).

NAZAROV, S.A., PAUKSHTO, M.V.,

[1] *Discrete models and homogenization in problems of the elasticity theory.* Leningrad Univ., Leningrad 1984. (Russian)

NAZAROV, S.A., ROMASHEV, YU.A.,

[1] Variation of the intensity factor under rupture of the ligament between two collinear cracks, *Izvestya AN Arm. SSR*, mekh. 4 (1982), 30–40. (Russian)

NAZAROV, S.A., SEMENOV, B.N.,

[1] The connection between stress intensity factors for a plane problem of classical and couple-stress elasticity theory, *Studies in elasticity and plasticity*, 13, Leningrad Univ., Leningrad 1980, 148–154. (Russian)

NASH, J.,

[1] Continuity of solutions of parabolic and elliptic equations. *Amer. J. Math.*, 60, 4 (1958), 931–954.

NAYFEH, A.H.,

[1] *Perturbation methods*, Wiley, New York 1973.

[2] *Introduction to perturbation techniques*, Wiley, New York 1981.

NIKOLSKAYA, N.A.,

[1] On the solution of two-dimensional problems of the theory of elasticity in domains with a distant outer boundary, *Studies in Elast. and Plasticity*, 14, 150–155, Leningrad Univ., Leningrad 1982. (Russian)

NOWACKI, W.,

[1] *Theory of elasticity*, Mir, Moscow 1975. (Russian) (Polish orig.: Teoria sprezystošci, Warszawa 1970.)

OLEINIK, O.A.,

[1] On the spectra of some singularly perturbed operators, *Uspekhi matem. nauk*, 42, 3 (1987), 221–222. (Russian)

[2] On eigenvibrations of bodies with concentrated masses, *Modern problems of mathematical physics and numerical math.*, Nauka, Moscow 1988. (Russian)

OZAWA SHIN,

[1] Singular Hadamard's variation of domains and eigenvalues of the Laplacian. I, *Proc. Jap. Acad.*, 56, 6 (1980), II, *Proc. Jap. Acad.*, 57, 5 (1981), 242–246.

[2] The first eigenvalue of the Laplacian on two dimensional Riemannian manifolds, Tohoku Math. J., 34, 1 (1982), 7–14.

[3] Electrostatic capacity and eigenvalues of the Laplacian, *J. Fac. Sci. Univ. Tokyo*, Sec. 1A, 30, 1 (1983), 53–62.

PANASENKO, G.P., RESCOV, M.B.,

[1] Asymptotics of the solution of systems of equation in elasticity theory given in a nonhomogeneous plate, *Numerical methods in mathematical physics*, Nauka, Moscow 1986. (Russian)

PANOVKO, YA.,G.,

[1] *Deformation mechanics of a solid body*, Nauka, Moscow 1986. (Russian)

PARTON, W.S., PERLIN, P.I.,

[1] *Methods of the mathematical theory of elasticity*, Nauka, Moscow 1986. (Russian)

PATTISSIER, G.,

[1] Sur les fonctions d'un operateur pseudodifferentiel elliptique, *Publ. Dep. Math.*, 14, 3 (1977), 21–58.

PAZY, A.,

[1] Asymptotic expansions of solutions of ordinary differential equations in Hilbert space, *Arch. Rat. Mech. Anal.*, 24, 2 (1967), 193–218.

POLYA, G., SZEGÖ, G.,

[1] *Isoperimetric inequalities in mathematical physics*, Princeton University Press, Princeton 1951.

PONYATOVSKI, V.V.,

[1] Asymptotic bending theory of a slender rod, *Studies in Elastic. and Plasticity*, 9, 81–93, Leningrad Univ., Leningrad 1973. (Russian)

RAJAIAH, K., RAO, A.K.,

[1] On limit cases in the flexure of simply supported regular polygonal plates, *Proc. Cambr. Phil. Soc.*, 65 (1969), 831–834.

[2] On the polygon-circle paradox, *J. Appl. Mech*, 48, 3 (1969), 195–196.

RAMM,A.G.,

[1] Limit of the spectra of the interior Neumann problems when a solid domain shrinks to a plane one, *J. Math. Anal. Appl.*, 108, 1 (1985), 107–112.

REED, M., SIMON, B.,

[1] *Methods of modern physics*, IV: *Analysis of operators*, Academic Press, New York 1978.

REISSNER, E.,

[1] On the theory of bending of elastic plates, *J. Math. and Phys.*, 23 (1944).

RICE J.R., DRUCKER, D.,

[1] Energy changes in stressed bodies due to void and crack growth, *Int. J. Fracture Mech.*, 3, 1 (1967), 19–27.

RIEDER, G.,

[1] On the plate paradox of Sapondhyan and Babuška, *Mech. Res. Comm.*, 1 (1974), 51–53.

SAMARSKI, A.A.,

[1] On the influence of fastening to eigenfrequences of closed volumes, *Doklady AN SSSR* 63, 6 (1948), 631–634. (Russian)

SANCHEZ-PALENCIA, E.,

[1] *Non-homogeneous media and vibration theory*, Lecture Notes in Physics 127, Springer-Verlag, New York 1980.

[2] Perturbation of eigenvalues in thermoelasticity and vibration of systems with concentrated masses. In: *Lecture Notes Physics*, 195 (1984), 346–368.

SAPONDZHYAN, O.M.,

[1] Bending of a freely supported polygonal plate, *Izvestya AN Arm. SSR*, mekh. 5, 2 (1952), 29–46. (Russian)

[2] *Bending of thin clastic platcs*, Ajastan, Erevan 1975. (Russian)

SEDOV, L.I.,

[1] *Mechanics of continuous media*, Vol. 2, World Scientific, River Edge, NJ, 1997.

SEELEY, R.T.,
[1] Complex powers of an elliptic operator, *Proc. Symp. in Pure Math.*, 10 (1967), 228–307.

SHERMAN, D.I.,
[1] On a particular Dirichlet problem for a two-connected domain the boundary of which is completely close to a narrow zone, *Mekh. tverd. tela*, 3 (1980), 76–87. (Russian)

SHOIKHET, B.A.,
[1] An energy identity in physically nonlinear theory of elasticity and estimation of the propagation of plate equations, *Prikl. matem. i mekh.*, 40, 2 (1976), 317–326. (Russian)

SHUBIN, M.A.,
[1] *Pseudodifferential operators and spectral theory*, Springer, Berlin, 1987.

SIH, G.C., LIEBOWITZ, H.,
[1] On the Griffith energy criterion for brittle fracture, *Int. J. Solids and Structures*, 3, 1 (1967), 1–22.

SLEPYAN, L.I.,
[1] *Crack mechanics*, Sudostroenie, Leningrad 1981. (Russian)

SOBOLEV, S.L.,
[1] On a boundary value problem for polyharmic equations, *Matem. sbornik*, 2, 3 (1937), 467–500. (Russian)
[2] *Some applications of functional analysis in mathematical physics*, AMS, Providence, RI, 1991.

SOIBELMAN, YA.S.,
[1] On a asymptotic formula for the capacity of a condensator, *Doklady AN SSR*, 258, 3 (1981), 590–593. (Russian)

SOLONNIKOV, V.A.,
[1] On general boundary value problems for elliptic Douglis-Nirenberg systems I, *Izvestya AN SSSR*, matem. 28, 4 (1964), 665–706. (Russian)

STERNIN, B.YU.,
[1] Elliptic and parabolic problems on manifolds with a boundary consisting of components of different dimensions, *Trudy Mosk. matem. obshch.*, 15 (1966), 346–382. (Russian)

STRICHARTZ, R.,
[1] A functional calculus for elliptic pseudo-differential operators, *Amer. J. Math.*, 94, 3 (1972), 711–722.

SWANSON, S.A.,
[1] Asymptotic variational formulae for eigenvalues, *Cand. Math. Bull.*, 6, 1 (1963), 15–25.

TIMOSHENKO, S., WOINOWSKY-KRIEGER, S.,
[1] *Theory of plates and shells*, 2nd edition, McGraw-Hill, New York 1959.

TRENOGIN, V.A.,
[1] Development and applications of the asymptotic method of Lyusternik-Vishik, *Uspekhi matem. nauk*, 25, 4 (1970), 123–156.

VAN DYKE, M.D.,

[1] *Perturbation methods in fluid mechanics*, Academic Press, New York 1964.

VISHIK, M.I., LYUSTERNIK, L.A.,

[1] Regular degeneration and boundary layer for linear differential equations with a small parameter, *Uspekhi matem. nauk*, 12, 5 (1957), 3–122.

[2] Solution of some problems on perturbation in the case of matrices and selfadjoint and non-selfadjoint differential equations, *Uspekhi matem. nauk*, 15, 3 (1960), 3–80. (Russian)

VOROVICH, I.I.,

[1] Some mathematical questions of the theory of plates and shells. *Proc. of the 2nd All-Soviet Congress in Theor. and Appl. Mechanics*, 116–136, Nauka, Moscow 1966. (Russian)

VOROVICH, I.I., MALKINA, O.K.,

[1] Stress state of a thick plate, *Prikl. matem. i mekh.*, 31, 3 (1967), 230–241.

VAINBERG, M.M., TRENOGIN, V.A.,

[1] *Theory of branching of solutions of non-linear equations*, Noordhoff, Leyden, 1974.

WILLIAMS, M.L.,

[1] Stress singularities resulting from various boundary conditions in angular corners of plates in extensions. *J.Appl.Mech.*, 19, 4 (1951), 526–528.

ZAKHAREVICH, I.S.,

[1] On variation of the solutions of integro-differential equations of mixed problems in the theory of elasticity by variation of the domain, *Prikl. matem. i mekh.*, 49, 6 (1985), 961–968. (Russian)

ZHDANOVA, G.V.,

[1] Scattering of planar elastic waves with a slender body with free boundary, *Doklady AN SSR*, 270, 6 (1983), 1300–1305. (Russian)

[2] Scattering of planar longitutinal waves with slender cavity of revolution. The case of axial fall, *Matem. sbornik*, 134, 1 (1987), 3–27. (Russian)

ZHIKOV, V.V., KOZLOV, S.M., OLEINIK, O.A.,

[1] G-convergence of parabolic operators, *Uspekhi matem. nauk*, 36, 1 (1981), 11–58. (Russian)

ZHIKOV, V.V., KOZLOV, S.M., OLEINIK, O.A., CHA TJEN NGOAN,

[1] Averaging and G-convergence of differential operators, *Uspekhi matem. nauk*, 34, 5 (1979), 65–133. (Russian)

ZINO, I.E., TROPP, E.A.,

[1] *Asymptotic methods for problems in heat transfer and thermoelasticity*, Leningrad Univ., Leningrad 1978. (Russian)

ZORIN, I.S.,

[1] On large destruction of an elastic plane which is weakened by a slender cut, *Vestnik Leningr. Univ.*, 7 (1982), 11–16. (Russian)

[2] Concerning the problem of large destruction of an elastic plane in a two-fold stressed state, *Mekh. tverd. tela*, 1 (1985), 188–192. (Russian)

ZORIN, I.S., NAZAROV, S.A.,

[1] Strain-stress state of an elastic space with a thin toroidal inclusion, *Mekhanika tverd. tela*, 3 (1985), 79–86. (Russian)

ZYKOV, A.A.,

[1] *Theory of finite graphs*, Vol.I, Nauka, Novosibirsk 1969. (Russian)

Index

MIX
Papier aus verantwortungsvollen Quellen
Paper from responsible sources
FSC® C105338

If you have any concerns about our products,
you can contact us on
ProductSafety@springernature.com

In case Publisher is established outside the EU,
the EU authorized representative is:
Springer Nature Customer Service Center GmbH
Europaplatz 3, 69115 Heidelberg, Germany

Printed by Libri Plureos GmbH
in Hamburg, Germany